THIS BOOK PROPERTY OF AB
1515 BROAD STREET
BLOOMFIELD, N.J.

# THERMODYNAMICS
## Second Edition

# THERMODYNAMICS
## Second Edition

**N. A. Gokcen, Sc.D.**
Albany, Oregon

**R. G. Reddy, Ph.D.**
The University of Alabama
Tuscaloosa, Alabama

**Plenum Press • New York and London**

Library of Congress Cataloging-in-Publication Data

On file

If your diskette is defective in manufacture or has been damaged in transit, it will be replaced at no charge if returned within 30 days of receipt to Managing Editor, Plenum Press, 233 Spring Street, New York, NY 10013.

The publisher makes no warranty of any kind, expressed or implied, with regard to the software reproduced on the diskette or the accompanying documentation. The publisher shall not be liable in any event for incidental or consequential damages or loss in connection with, or arising out of, the furnishing, performance, or use of the software.

ISBN 0-306-45380-0

© 1996 Plenum Press, New York
A Division of Plenum Publishing Corporation
233 Spring Street, New York, N. Y. 10013

All rights reserved

10 9 8 7 6 5 4 3 2 1

No part of this book may be reproduced, stored in a retrieval system, or transmitted in any form or by any means, electronic, mechanical, photocopying, microfilming, recording, or otherwise, without written permission from the Publisher

The first edition of this book was published by Techscience, Inc., Hawthorne, California, 1975

Printed in the United States of America

To
Emel Gokcen
and
Rama Reddy
with devotion and dedication

# PREFACE

This edition of *Thermodynamics* is a thoroughly revised, streamlined, and corrected version of the book of the same title, first published in 1975. It is intended for students, practicing engineers, and specialists in materials sciences, metallurgical engineering, chemical engineering, chemistry, electrochemistry, and related fields. The present edition contains many additional numerical examples and problems. Greater emphasis is put on the application of thermodynamics to chemical, materials, and metallurgical problems. The SI system has been used throughout the textbook. In addition, a floppy disk for chemical equilibrium calculations is enclosed inside the back cover. It contains the data for the elements, oxides, halides, sulfides, and other inorganic compounds.

The subject material presented in chapters III to XIV formed the basis of a thermodynamics course offered by one of the authors (R.G. Reddy) for the last 14 years at the University of Nevada, Reno. The subject matter in this book is based on a minimum number of laws, axioms, and postulates. This procedure avoids unnecessary repetitions, often encountered in books based on historical sequence of development in thermodynamics. For example, the Clapeyron equation, the van't Hoff equation, and the Nernst distribution law all refer to the Gibbs energy changes of relevant processes, and they need not be presented as radically different relationships.

The manuscript and galley proofs were corrected not only by the authors, but also by Dr. J.A. Sommers, whose diligent efforts are gratefully acknowledged. Criticisms and suggestions during the past two decades from Professors E.F. Westrum, Jr., Y.K. Rao, D.A. Stevenson, G.R. St. Pierre, A.E. Morris, D.C. Lynch, L. Brewer, and Dr. M. Blander are also acknowledged. Help rendered by Professors S.J. Louis, F.C. Harris, Jr., and their students in developing the software package for thermodynamic calculations is greatly appreciated. They are not responsible for the content of the text, and the responsibility rests entirely with the authors who welcome further comments and criticism from readers.

Finally, we thank our families for their patience, support, and encouragement throughout the preparation of this book.

N.A. Gokcen  
Albany, Oregon

R.G. Reddy  
Tuscaloosa, Alabama

# CONTENTS

**CHAPTER I. INTRODUCTION AND DEFINITIONS** .......... 1
Introduction ............................................. 1
Terms and Symbols ....................................... 1
Thermodynamic State of a System ......................... 4
Boundaries of a System .................................. 5
Temperature ............................................. 5
Work and Energy ........................................ 8
Equilibrium ............................................. 9
Objectives of Thermodynamics ............................ 9
**Problems** ............................................. 9

**CHAPTER II. DIFFERENTIATION, INTEGRATION, AND
SPECIAL FUNCTIONS** .................................. 11
Differentials and Derivatives ............................. 11
Second Derivatives ..................................... 14
Useful Differentials .................................... 14
Maxima, Minima and Inflection Points .................... 15
L'Hôpital's Theorem .................................... 16
Homogeneous Functions ................................. 17
Euler's Theorem on Homogeneous Functions ............... 17
Homogeneous Thermodynamic Functions ................... 18
Integrals .............................................. 20
Exact Differentials ..................................... 21
Line Integrals ......................................... 23
A Graphical Example ................................... 24
Cross Differentials ..................................... 25
Lagrange's Method of Undetermined Multipliers ........... 26
Change of Independent Variables ......................... 27
Representation of Data .................................. 28
Determinants .......................................... 31

ix

x    Contents

Useful Series ............................................. 32
**Problems**............................................... 33

**CHAPTER III. THE FIRST LAW OF THERMODYNAMICS** ....... 37
Work of Compression and Expansion ......................... 38
Heat ..................................................... 39
Reversible Processes in Closed Systems .................... 40
**Application of the First Law to Ideal Gases** ........... 40
Energy of Ideal Gases ..................................... 40
Heat Capacity ............................................. 42
Processes with Ideal Gases ................................ 42
Simple Kinetic Theory of Ideal Gases ...................... 45
**Real Gases, Liquids and Solids** ........................ 47
Compressibility Factor .................................... 47
van der Waals Equation .................................... 48
Other Equations of State .................................. 51
Liquids and Solids ........................................ 53
**Enthalpy and Heat Capacity** ............................ 53
Heat Capacity of Solids ................................... 54
Empirical Representation of Heat Capacity ................. 57
Relationship between $C_p^\circ$ and $H_T^\circ - H_{298}^\circ$ .......... 57
Enthalpy Change of Phase Transformations .................. 58
**Thermochemistry** ....................................... 60
Variation of $\Delta_r H_T^\circ$ with Temperature ........ 63
Bond Energies ............................................. 65
Adiabatic Flame Temperature ............................... 66
**Problems**............................................... 67

**CHAPTER IV. THE SECOND LAW OF THERMODYNAMICS** ..... 71
The Second Law of Thermodynamics .......................... 71
Carnot Engine ............................................. 72
Carnot Theorem ............................................ 74
Kelvin Temperature Scale .................................. 75
Carnot Engine with an Ideal Gas ........................... 77
Refrigeration Engine ...................................... 79
Spontaneous Processes ..................................... 79
Reversible Cyclic Processes ............................... 79
Entropy Change in Reversible Processes .................... 83
Entropy Change in Irreversible Processes .................. 83
Method of Carathéodory .................................... 85
The Second Law of Thermodynamics .......................... 87
**Problems**............................................... 88

## CHAPTER V. ENTROPY AND RELATED FUNCTIONS ... 89
Entropy Change ... 89
Entropy of Mixing of Ideal Gases ... 91
Entropy of Phase Change and Chemical Reactions ... 92
Entropy, Randomness and Probability ... 93
Thermodynamic Equations of State ... 94
Difference Between $C_p$ and $C_V$ ... 96
Variation of $C_p$ and $C_V$ with $P$ and $V$ ... 97
Joule-Kelvin Expansion of Gases ... 97
**Problems** ... 99

## CHAPTER VI. HELMHOLTZ AND GIBBS ENERGIES ... 101
Introduction and Definitions ... 101
Partial Differential Relations ... 103
Isothermal Changes in $A$ and $G$ ... 104
Criteria for Reversibility and Irreversibility ... 105
Examples ... 107
**Problems** ... 108

## CHAPTER VII. THE THIRD LAW OF THERMODYNAMICS ... 109
Introduction ... 109
The Third Law of Thermodynamics ... 111
Entropy from Statistical Mechanics ... 113
Entropies of Supercooled Liquids ... 114
Consequences of the Third Law ... 115
Thermal Evaluation of Entropy ... 116
**Problems** ... 117

## CHAPTER VIII. PHASE EQUILIBRIA ... 119
Introduction ... 119
Two-Phase Equilibrium ... 119
Vaporization Equilibria ... 121
Variation of Vapor Pressure with Total Pressure at Constant Temperature ... 124
Representation of Phase Equilibria ... 125
Components ... 126
Variables of State and Degrees of Freedom ... 130
Partial (Molar) Gibbs Energy, or Chemical Potential ... 130
Conditions of Phase Equilibrium ... 131
Phase Rule ... 133
Other Definitions of $\overline{G}_i$ ... 135
Useful Partial (Molar) Properties ... 136
$\overline{G}_i$ and Criterion of Equilibrium ... 136
**Problems** ... 136

xii  *Contents*

**CHAPTER IX. FUGACITY AND ACTIVITY** .................. 139
Introduction ............................................. 139
Fugacity of Pure Gases ................................... 140
Alternative Equations for Fugacity ....................... 141
Variation of Fugacity with Temperature ................... 142
Definition of Activity ................................... 143
Raoult's Law ............................................. 144
Henry's Law .............................................. 145
**Problems** ............................................. 149

**CHAPTER X. SOLUTIONS** ................................. 151
**Part I  Ideal Solutions** .............................. 151
Equilibrium Between an Ideal Solution and Its Vapor ...... 152
Constant Pressure Binary Equilibrium Diagrams ............ 156
Equilibria Between Pure Immiscible Solids and Ideal Liquid Solutions ... 157
Relative Positions of Liquidus and Solidus Lines ......... 158
Depression of Freezing Point ............................. 159
Elevation of the Boiling Point ........................... 160
Determination of Molecular Weights ....................... 160
Ideal Solubilities of Gases .............................. 161
**Part II  Real Solutions** .............................. 162
Definition of Real Solutions ............................. 162
Equilibrium Between a Real Solution and Its Vapor
   at Constant Temperature ............................... 162
Equilibrium Between a Real Solution and Its Vapor
   at Constant Pressure .................................. 166
Variation of Activity and Activity Coefficients with Composition
   in Binary Solutions ................................... 170
Variation of Activities in Binary Solutions with Pressure and Temperature . 172
Dilute Solutions ......................................... 172
Molar, Partial Molar, and Excess Thermodynamic Properties of Solutions . 173
**Problems** ............................................. 177

**CHAPTER XI. PARTIAL (MOLAR) PROPERTIES** .............. 179
Introduction ............................................. 179
Partial (Molar) Properties of Binary Systems ............. 180
Excess Gibbs Energy; Binary Systems ...................... 182
Representation of $G^E$ .................................. 183
Alternative Equations .................................... 187
Regular Solutions ........................................ 189
Maximum, Minimum, and Critical Points in (11.43) ......... 190
Spinodal Points .......................................... 192
Theoretical Derivation of (11.40) ........................ 192

Effect of Temperature on $G^E$ and $g_i$ .................................. 195
Equations with Henrian Reference States ............................. 199
Wagner Interaction Parameters ........................................ 200
**Problems**.................................................................. 201

## CHAPTER XII. GIBBS ENERGY CHANGE OF REACTIONS ..... 203
Introduction .................................................................. 203
Feasibility of Chemical Reactions ..................................... 204
Equilibria in Real Gas Mixtures......................................... 209
Equilibria Involving Condensed Phases............................... 211
Determination of Standard Gibbs Energy Changes ................ 212
Method I—Determination of $\Delta G°$ from Equilibrium Constant ......... 212
Method II—Thermal Data ............................................... 214
Method III—Electromotive Force (emf) Method .................. 214
Method IV—Spectroscopic Data and Mechanics of Molecules ......... 215
Thermodynamic Equations .............................................. 215
Tabulation of Thermodynamic Data ................................... 216
Use of Tables ............................................................... 224
Other Thermodynamic Tables and Compilations ................... 228
Use of Tabular Data in Experimental Work ......................... 229
Complex Equilibria ....................................................... 238
Generalized Reactions and Their Equilibrium Constants .......... 240
**Problems**.................................................................. 244

## CHAPTER XIII. SOLUTIONS OF ELECTROLYTES ............ 247
Introduction .................................................................. 247
Activity and Activity Coefficient ...................................... 248
Debye-Hückel Theory .................................................... 250
Concentrated Electrolytes................................................ 256
Determination of Activities.............................................. 257
Weak Electrolytes ......................................................... 262
Temkin Rule................................................................. 264
**Problems**.................................................................. 265

## CHAPTER XIV. REVERSIBLE GALVANIC CELLS ............. 267
Introduction .................................................................. 267
Properties of Reversible Cells .......................................... 268
Single Electrode Reactions .............................................. 268
Convention in Notation .................................................. 269
Reaction Isotherm and emf .............................................. 271
Standard emf of Half-cells .............................................. 272
Variation of emf with Temperature and Pressure................... 276
Ionization Constant of Water ........................................... 277

xiv   *Contents*

Cells with Solid Electrolytes .................................. 280
**Problems** ..................................................... 282

## CHAPTER XV. PHASE DIAGRAMS ........................ 285
Introduction .................................................... 285
Binary Phase Diagrams ......................................... 285
Erroneous Diagrams ............................................ 288
Lever Rule ..................................................... 290
Molar Gibbs Energy of Mixing—Composition Diagrams ............ 291
$\Delta G$ Diagrams for Other Phases ........................... 294
$\Delta G$ Diagrams for Complex Systems ........................ 298
Calculation of Phase Diagrams from Thermodynamic Data .......... 299
Ternary Phase Diagrams ........................................ 302
Tielines ....................................................... 309
Thermodynamic Consideration ................................... 309
Second Order Transitions ....................................... 313
**Bibliography** ................................................ 314
**Problems** .................................................... 314
**Selected Binary Phase Diagrams** ............................. 315

## CHAPTER XVI. SPECIAL TOPICS ........................ 323
**Part I  Surface Tension** .................................... 323
Properties of Surfaces ......................................... 323
Criteria for Equilibrium ....................................... 324
Gibbs Adsorption Equation ..................................... 325
Vapor Pressure of Droplets .................................... 326
**Part II  Gravitational Electric and Magnetic Fields** ........ 327
Gravitational Field ............................................ 327
Solutions ...................................................... 328
Centrifugal Force .............................................. 329
Electric and Magnetic Fields .................................. 329
**Part III  Long-Range Order** ................................. 330
Ordering and Clustering ....................................... 331
Order-Disorder in Binary Alloys ............................... 331
Long-Range Order Parameter .................................... 333
Gorsky and Bragg-Williams (GBW) Approximation ................. 334
Heat Capacity .................................................. 336
**Problems** .................................................... 338

## APPENDIX I. GENERAL REFERENCES ...................... 339

## APPENDIX II. TABLES OF THERMODYNAMIC DATA FOR EXAMPLES AND PROBLEMS IN TEXT ........................ 341

**APPENDIX III. THERMODYNAMIC SIMULATOR (TSIM) FOR THERMODYNAMIC CALCULATIONS** ........................ 371

**APPENDIX IV. ESTIMATION OF ACTIVITIES IN MULTICOMPONENT IONIC SOLUTIONS** .................... 373

**APPENDIX V. STABILITY DIAGRAMS** ...................... 381

**APPENDIX VI. LIST OF SYMBOLS** ........................ 387

**INDEX** ................................................. 393

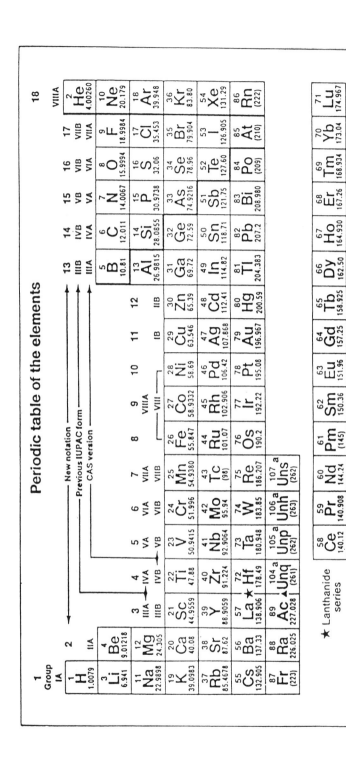

CHAPTER I

# INTRODUCTION AND DEFINITIONS

## Introduction

The subject matter of this book is the *classical* or *equilibrium thermodynamics*. The classical thermodynamics is generally known by the simpler term *thermodynamics*, a word coined from two Greek words *thermos* heat, and *dynamis* power for historical reasons. It deals broadly with the conservation and interconversion of various forms of energy, and the relationships between energy and the changes in properties of matter. The concepts of thermodynamics are based on empirical observations of the macroscopic properties of matter in physical, chemical and biological changes, and the resulting observations are expressed in relatively simple mathematical functions. These functions are subjected to the methods of ordinary and partial differential equations to derive related useful equations.

## Terms and Symbols

We shall define a *term* to represent a set of clear, concise and well devised operations or phenomena and designate it by a particular symbol. Every term is generally related to other terms by relevant mathematical formulae. The symbols used in this book are compiled and defined in Appendix VI and redefined in detail when they recur in the text.

A *thermodynamic system*, or briefly a *system*, is the macroscopic part of the universe selected for experimentation and observation. It is separated from its *surroundings*, which are the remainder of the universe, by *rigid*, or *movable boundaries*, or *walls*, through which energy may or may not pass in various forms. The boundaries do not generally constitute a part of the system since they may be made vanishingly small in mass, or their effect may be corrected in the final analysis of observation. It will usually be assumed that the boundaries do not permit the exchange of matter with the surroundings, unless specified otherwise. Systems that exchange matter with the surroundings are called the *open systems*. The pressure and the volume of a system are designated by $P$ and $V$ respectively. The pressure is expressed in terms of pascals (Pa), or bars, and the volume in $m^3$, L or $cm^3$. Various units used in this book are listed in Table 1.1.

## 2  Thermodynamics

The components of a system are defined as the minimum number of chemically pure substances with which a system can be formed, and by which its composition can be changed. For example, the components of *brine* are $H_2O$ and NaCl because they are the minimum number of pure substances constituting the system and the composition of brine may be varied by adding $H_2O$ or NaCl. In certain systems the choice of the appropriate components requires considerable knowledge about the system. For example, $CaCl_2$ is a single component system, but at high temperatures Ca and $Cl_2$ are soluble in $CaCl_2$, hence Ca and $Cl_2$ are the components, and the atomic ratio Ca/Cl = 1/2 is but one of the compositions in the two-component system Ca-Cl.

The composition is generally expressed in terms of the *mole fraction*. For $n_i$ moles of component $i$, in a given mass consisting of $c$ components, the mole fraction $x_i$ of a component $i$, is defined by

$$x_i = n_i \bigg/ \sum_1^c n_i \qquad (i = 1, 2, 3, \ldots, c) \tag{1.1}$$

When the components of a system are monatomic, or assumed to be so, the atomic fraction is sometimes used synonymously with the mole fraction.

The composition of a solution in which one of the components predominates is often expressed as *molality* or *molarity*. The molality $m_i$ is defined by the number of moles of constituent $i$ dissolved in one kilogram of the *solvent*, the predominant component. The molarity $c_i$ (not $c$) is the number of moles of $i$ dissolved in one liter of solution, and unlike molality and mole fraction, varies slightly with temperature. The less abundant components are called the *solutes*. It is a general convention to label the solvent as the component 1 and the solutes as the components 2, 3, etc. In a dilute solution, it is very easy to distinguish the solvent and the solutes, but in concentrated solutions, this is not only difficult but unnecessary. Thus when $x_1 = 0.40$, $x_2 = 0.40$ and $x_3 = 0.20$, none of the components can be identified as the solvent.

A system is homogeneous when it is completely uniform in all properties; each observable portion of such a system has the same pressure, composition and density as any other portion. Effects of natural gravity, electricity, magnetism and surface tension on homogeneity are ignored in this definition since they are considered negligible for most purposes. However, they are considered when their effects justify thermodynamic investigations. Each *homogeneous system* consists of a *single phase*; these two terms are therefore synonymous. A heterogeneous system consists of two or more homogeneous regions, i.e. phases, which are separated from each other by bounding surfaces. Sharp changes in density and composition are encountered across bounding surfaces. For example, water is homogeneous, but water and ice are heterogeneous. An emulsified mixture of fat in water, e.g. milk, is heterogeneous and it consists of two phases; although the fat is subdivided into innumerable particles, it still constitutes one phase since each particle has the same property as any other particle. Gravitational, centrifugal, electrical and

**Table 1.1** Fundamental constants, derived constants, and conversion factors.*

| Name | Symbol | Value and Units |
|---|---|---|
| *Fundamental constants:* | | |
| (Ice + water + vapor) point | | 273.1600 Kelvin |
| Molar volume of perfect gas (0°C, 1 bar) | $V^0$ | 0.02271108 m$^3$ mol$^{-1}$<br>22.71044 L mol$^{-1}$ |
| Avogadro Number | $N_A$ | 6.022137 × 10$^{23}$ mol$^{-1}$ |
| Gas constant | $R$ | 8.31451 J mol$^{-1}$ K$^{-1}$<br>1.987216 cal mol$^{-1}$ K$^{-1}$<br>0.083143 L bar K$^{-1}$ mol$^{-1}$<br>82.058 cm$^3$ atm mol$^{-1}$ K$^{-1}$ |
| Boltzmann constant | $k = R/N_A$ | 1.380658 × 10$^{-23}$ J K$^{-1}$ |
| Faraday constant | **F** | 96485.31 C mol$^{-1}$<br>23,060.54 cal mol$^{-1}$ Volt$^{-1}$ |
| Velocity of light in vacuum | $c$ | 299,792,458 m s$^{-1}$ |
| Planck constant | $h$ | 6.626076 × 10$^{-34}$ J s |
| Proton charge (-electronic charge) | $e^+$ | 1.60217733 × 10$^{-19}$ C |
| Permittivity of vacuum | $e_0 = 10^7/(4\pi c^2)$ | 8.854188 × 10$^{-12}$ F m$^{-1}$ |
| *Defined constants and conversion factors:* | | |
| Thermochemical calorie | cal | 4.1840 Joules (J) |
| Standard gravity | $g°$ | 9.80665 m s$^{-2}$ |
| Standard pressure | bar | 10$^5$ Pa |
| Atmosphere of pressure | atm | 101325 Pa (Newton m$^{-2}$) |
| Newton | N | 10$^5$ dynes |
| Joule | J | 10$^7$ ergs |
| Liter | L | 1000.028 cm$^3$ |
| Electron volt | eV | 96485.31 J mol$^{-1}$<br>23060.54 cal mol$^{-1}$ |
| Kelvin | K | 273.1500 + °C |

Fixed points (1990 scale), freezing points of pure elements in K: (In, 429.7485), (Sn, 505.078), (Zn, 692.677), (Al, 933.473), (Ag, 1234.93), (Au, 1337.33), (Cu, 1357.77).**

*I. Mills, T. Cvitas, K. Homann, N. Kallay, and K. Kuchitsu, "Quantities, Units and Symbols in Physical Chemistry", Int. Union of Pure & Appl. Chem. (IUPAC), Blackwell Scientific Publ. (1993). E. T. Cohen and P. Giacomo, Physica *146A*, 1 (1987). B. N. Taylor and E. R. Cohen, J. Res., NIST, *95*, 497 (1990).

**Bull. Alloy Phase Diag. (BAPD), *11* (2), 107 (1990). See Gen. Ref. (18) and (25) for other units and conversion factors.

magnetic forces affect the homogeneity of single phases. For example, the density of air varies continuously with the altitude and therefore necessitates a modification of our definition of *phase*. If an observable property $\mathcal{G}$ of a substance either remains constant or varies continuously within the space coordinates $x$, $y$, and $z$, then the substance consists of a single phase; where $\mathcal{G}$ is discontinuous is called the phase boundary. A *new phase* appears when $\mathcal{G}$ assumes an entirely new functional form within a given range of space coordinates.

## Thermodynamic State of a System

*The thermodynamic state of a system*, more commonly the *state of a system*, is defined in terms of a certain number of variables, called the "variables of state". A system is completely defined or described when these variables are fixed. Experimental observations on numerous closed homogeneous systems show that (a) when $P$ and $V$ are fixed, all the other observable properties such as the density, composition, viscosity, electrical conductivity etc are also fixed, (b) when $P$ and $V$ are varied continuously any observable property also varies continuously, and (c) when the system initially at $P_1$ and $V_1$ is brought into a state defined by $P_2$ and $V_2$ and then returned to $P_1$ and $V_1$, it regains the same initial observable properties. The foregoing statements show that an observable property, $\mathcal{G}$, of a single phase is a continuous, single valued function of pressure and volume; therefore,

$$\mathcal{G} = \mathcal{G}(P, V) \tag{1.2}$$

The independent variables $P$ and $V$ are called the *variables of state*, and the dependent property $\mathcal{G}$, a *thermodynamic property*. Equation (1.2) is valid for closed systems but for *the open system*, the numbers of moles of components, $n_1, n_2, \ldots, n_c$ also become variables of state; consequently,

$$\mathcal{G} = \mathcal{G}(P, V, n_1, n_2, \ldots, n_c) \tag{1.3}$$

According to this equation, the state of a system is entirely defined by its variables and *not* by the *history* of previous changes in the system, or by the changes in *time*, ordinary *velocity* and the surface area of the system, *gravitational* and *electrical fields* etc. Each of the several coexisting phases is an open phase, and each has a particular functional form of (1.3).

A change in any thermodynamic property, $\mathcal{G}$, of a system is called a *process*. It is formally defined by

$$\mathcal{G}_2 - \mathcal{G}_1 = \Delta\mathcal{G} = \mathcal{G}_2(P_2, V_2, n_1'', \ldots) - \mathcal{G}_1(P_1, V_1, n_1', \ldots) \tag{1.4}$$

The subscript "1" of $\mathcal{G}$, $P$ and $V$ refers to the initial state and "2", to the final state, and the symbol "$\Delta$" represents the *change* in $\mathcal{G}$ or $\mathcal{G}_2 - \mathcal{G}_1$. Since the subscripts to $n$ identify the components, $n_i'$ is the initial value and $n_i''$ is the final value of $n_i$. Sometimes, for clarity, the subscripts 2 and 1 are replaced by the abbreviations of final and initial respectively i.e. $\Delta\mathcal{G} = \mathcal{G}_{\text{fin}} - \mathcal{G}_{\text{in}}$. A process shown by "$\Delta$ will always refer to "the final property $\mathcal{G}_2$ minus the initial property $\mathcal{G}_1$" in this book.

*Introduction and Definitions* 5

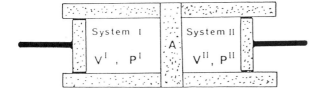

**Fig. 1.1** Two adiabatically enclosed systems separated by adiabatic rigid wall *A*.

## Boundaries of a System

The boundaries of a system may be open or closed but they must separate the system from the surroundings, or permit contact with it, depending on the nature of thermodynamic changes under investigation. There are two types of closed boundaries, or walls, which are particularly important in thermodynamics. They are called the *adiabatic walls* and the *diathermic walls*. The terms adiabatic and diathermic may evoke *heat* and *temperature*, which are to be considered as yet undefined. The adiabatic walls may be *rigid* or *movable*, i.e., deformable. The adiabatic rigid walls retain their shapes and provide a constant volume for the system. A sealed thermos bottle or a Dewar flask closely approximates an adiabatic rigid wall. A system enclosed entirely by adiabatic rigid walls is called an isolated system. The values of $G$, $P$ and $V$ in (1.2) are then fixed irrespective of any change in the surroundings. However, the adiabatic movable walls permit changes in $P$ and $V$ as a result of mechanical motions of the wall, e.g. an adiabatic piston moving in an adiabatic cylinder which contains the system, or a set of rotating paddles dissipating mechanical energy into an adiabatically enclosed system. Experimental observations show that a system inside adiabatic movable walls has two independent variables, $P$ and $V$. Let two adiabatic systems I and II touch each other along a rigid adiabatic wall (A) shown in Fig. 1.1. This assembly of two systems has four independent variables $P^I$, $V^I$, $P^{II}$ and $V^{II}$. If the adiabatic rigid wall (A) is replaced by a diathermic rigid wall it is observed experimentally that only three of the four variables are independent; hence there must exist a function $F$ among the four variables, i.e.,

$$F^{I,II}(P^I, V^I, P^{II}, V^{II}) = 0 \tag{1.5}$$

If any set of three variables in (1.5) is experimentally fixed, the fourth variable is also fixed. A non-adiabatic wall imposing the restriction expressed by (1.5) upon the two systems is called a diathermic rigid wall, and the systems I and II are said to be in diathermic equilibrium.

## Temperature

A definition of temperature based on our physiological ability to distinguish "cold" from "hot" is not only unsatisfactory but also incorrect from an operational point

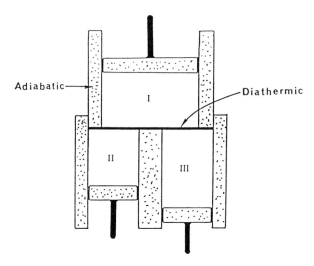

**Fig. 1.2** Three systems separated from surroundings by adiabatic walls. System I is in diathermic contact with II and III through diathermic rigid wall shown by black plate.

of view. In this section we shall give a definition of "empirical temperature" and in a subsequent chapter, a definition of "thermodynamic temperature" expressed in kelvins, K.

Consider the system I which is in diathermic equilibrium with two systems II and III as shown in Fig. 1.2. It is evident that II and III are also in equilibrium with each other since I is in equilibrium with II as well as with III. This forms the basis of the zeroth law of thermodynamics enunciated by Carathéodory* in 1909 as follows:

*"If two systems are in diathermic equilibrium with a third system, they are also in diathermic equilibrium with each other."*

The equilibria among the three systems require that

$$F^{I,II}(P^I, V^I, P^{II}, V^{II}) = 0 \tag{1.6}$$

$$F^{I,III}(P^I, V^I, P^{III}, V^{III}) = 0 \tag{1.7}$$

hence,

$$F^{II,III}(P^{II}, V^{II}, P^{III}, V^{III}) = 0 \tag{1.8}$$

According to the zeroth law, if (1.6) and (1.7) are valid, then (1.8) is also valid, for, any one of these equations is derivable from the remaining two equations.

---

*C. Carathéodory, "Gesammelte Mathematische Schriften," (Collected Mathematical Papers) II, pp. 131–177, C. H. Berk'sche Verl., München, (1955). The term "zeroth law" was properly coined by H. R. Fowler and E. A. Guggenheim, "Statistical Thermodynamics," Cambridge Univ. Press (1956).

However, it is possible to eliminate either $P^I$ or $V^I$ from (1.6) and (1.7) but not both if we attempt to derive (1.8). To eliminate both $P^I$ and $V^I$ it should be possible to write (1.6) and (1.7) as $f^I(P^I, V^I) - f^{II}(P^{II}, V^{II}) = 0$ and $f^I(P^I, V^I) - f^{III}(P^{III}, V^{III}) = 0$ respectively, but the alternative functional forms of (1.6), (1.7) are not known. We eliminate this difficulty by retaining the pressure $P^I$ of I constant and therefore (1.6) and (1.7) become

$$F^{I,II}(V^I, P^{II}, V^{II}) = 0 \tag{1.9}$$

$$F^{I,III}(V^I, P^{III}, V^{III}) = 0 \tag{1.10}$$

We solve (1.9) for $V^I$ to obtain $V^I = f^{II}(P^{II}, V^{II})$ and substitute the result in (1.10) to obtain $F^{I,III}(P^{II}, V^{II}, P^{III}, V^{III}) = F^{II,III}(P^{II}, V^{II}, P^{III}, V^{III}) = 0$, and thus derive (1.8) from (1.9) and (1.10). We now solve both (1.9) and (1.10) for $V^I$ so that

$$V^I = f^{II}(P^{II}, V^{II}) \tag{1.11}$$

$$V^I = f^{III}(P^{III}, V^{III}) \tag{1.12}$$

We note that when (1.12) is subtracted from (1.11) we obtain $f^{II} - f^{III} = 0$ and since this result also expresses diathermic equilibrium it is equivalent to (1.8), i.e. $f^{II} - f^{III} = F^{II,III} = 0$. We make the system I conveniently small in size and accurately calibrated in volume so that we can substitute for $V^{II}$ and $V^{III}$ their corresponding values from (1.11) and (1.12). As a result, the variables of state for the systems II and III may be selected as $(P^{II}, V^I)$ and $(P^{III}, V^I)$ respectively, and since $V^I$ is a variable of state of both systems, we may replace $V^I$ by a function of $V^I$ denoted by $\tau(V^I)$. We next label $\tau(V^I)$ or more simply $\tau$, as the empirical temperature, and $\tau$ is now measured by the system I which is commonly known as the constant pressure thermometer. Equations (1.11) and (1.12) may be rewritten as $\tau = f^{II}(P^{II}, V^{II})$ and $\tau = f^{III}(P^{III}, V^{III})$ respectively and as such, they are called the equations of state. We wish to emphasize here that it is not necessary to recognize that $V^I$ is related to a variable called "temperature" to develop theories and equations of thermodynamics. In fact the recognition of $V^I$ as a variable related to "temperature" is simply a labelling and scaling procedure established by tradition.

The foregoing definition of temperature was obtained from the concepts of adiabatic and diathermic walls, and the observation that the diathermic walls restrict the number of independent variables of systems in diathermic contact. The system I can be selected as an appropriate thermometric substance and a temperature scale can be devised. In the Celsius temperature scale, (°C), zero degree is assigned to the freezing point of water in air, and 100 degrees to the boiling point of water at 101325 Pa, and then this range is divided into 100 single degrees, e.g. by using gases at low pressures. The freezing point of water in air is not reproducible because of dissolved air in water; therefore the "triple point" of air-free water, the temperature at which pure solid, liquid and gaseous water coexist, is defined as +0.0100°C and 0°C is redefined to be 0.0100°C below the triple point. Practical considerations require additional fixed points, some of which are listed in Table 1.1.

It was first observed experimentally by Boyle in 1660 that the product $PV$ at sufficiently low pressures is a constant for gases at a fixed temperature. Thus, Boyle's law is simply $PV = \alpha$ where $\alpha$ is a constant at a fixed temperature, or $\alpha$ is only a function of temperature $T$, i.e. $\alpha = \alpha(T)$. Charles' law (1787) states that $P$ at a constant value of $V$ varies linearly with $T$, i.e. $P = \beta(V)T$. By multiplying this equation by $V$, we obtain $PV = V\beta(V)T$. Since $PV = \alpha(T)$ by Boyle's law, we find that $V\beta(V)T = \alpha(T)$; as a result, $V\beta(V)$ has to be a constant, independent of $V$ since $\alpha(T)$ is independent of $V$. Therefore, $\beta(V) = R/V$, where $R$ is a constant, and consequently, $RT = \alpha(T)$; hence,

$$PV = RT \tag{1.13}$$

The scale by which the values of $T$ are measured by using gases at low pressures is known as the ideal gas temperature scale and it is always represented by the symbol $T$. The constant is called the gas constant when (1.13) refers to one mole of a gas obeying this equation. For $n_i$ moles of a gas (1.13) becomes $PV = n_i RT$. The dimensions of $R$ depend on the choice of units for the remaining symbols in this equation; a number of useful values for $R$ are listed in Table 1.1. The values of $T$ are related to the Celsius scale (°C) by the following equation

$$T = 273.15 + (°C) \tag{1.14}$$

In principle, the assignment of one value for $T$ at a single point is sufficient to establish a temperature scale with (1.13).

We shall not delve any further into the subject of temperature and its measurement which requires precise methods and procedures. Our attempt here is to give a clear and concise mathematical and operational definition of temperature.*

## Work and Energy

Mechanical work is defined by the force times the displacement in the direction of force. In the Meter-Kilogram-Second System of units (MKS System), length, $l$, mass, $m$, and time, $t$, are defined quantities and acceleration, $g$, and force, N, are the derived quantities according to Newton's laws of motion. Thus, acceleration and force have the dimensions of $lt^{-2}$ and $mlt^{-2}$ respectively from $l = 0.5\,gt^2$ and $N = mg$. The unit of work in the SI System (or MKS System), Joule, is equal to the displacement of one Newton of force (N = 1) by one meter, and one Newton is the force that gives one kilogram an acceleration of $g = 1$ meter per sec$^{-2}$. *The Joule should be the universal unit of energy irrespective of the types of energy*, but other units such as calories, liter atmosphere etc are still used. For the convenience

---

*H. H. Plumb, Ed., "Temperature —Its Measurement and Control in Science and Industry" Instrument Soc. America, Pittsburgh (1972); Part 1 Edited by H. Preston-Thomas, T. P. Murray, R. L. Shepard, G. Urbain and M. Rivot; see also F. Cabannes, R. Lacroix, and G. Urbain in "Les Hautes Températures et Leurs Utilizations en Physique et en Chimie," G. Chaudron and F. Trombe, editors, Masson & Co., Paris (1973).

in reading older literature, other frequently encountered units of energy are listed in Table 1.1.

Energy is a general term representing any property that can be converted into work. The work itself is also a form of energy.

## Equilibrium

A system is in equilibrium when its state, or any of its property $G$ is dependent only on its variables of state, and independent of time. The presence of equilibrium conditions is thus implicit in the definition of "thermodynamic state of a system." It will be seen later that equilibrium is defined as the state of lowest energy, and a precise definition in this manner constitutes one of the main objectives of thermodynamics requiring a detailed treatment.

## Objectives of Thermodynamics

Thermodynamics, in the broadest sense, aims to postulate and formulate all types of relationships involving energy and matter, and correlates all measurable properties usually involving $P, V, T, n_1, n_2, \ldots, n_c$ as their variables of state.

A primary objective of thermodynamics, which is of great interest to chemists, physicists, chemical and metallurgical engineers, and life scientists, consists of the determination of conditions of equilibrium, or *the state of lowest available energy* in physical, chemical and biological processes. From a knowledge of equilibrium data, or from formal relationships between energy and chemical reactions, thermodynamics can predict whether it is feasible or entirely impossible for a given system to undergo a physical or chemical transformation. Thermodynamics can therefore treat the limiting conditions, i.e. the conditions of equilibrium. This might appear to be a serious limitation at first, but *in many biological processes at ordinary temperatures and in many chemical and physical processes at sufficiently above room temperature, equilibrium conditions are closely approached and in such cases it is possible to make valuable and labor saving calculations*. In addition, the equilibrium conditions provide a *point of reference* for the systems that are removed from equilibrium.

## PROBLEMS

1.1 (a) Find the mass of NaCl for making up 1 kg of 2 molal aqueous solution. (b) Compute the mole fraction of NaCl, and (c) write the process for making up the solution as in (b).

*Ans.:* (b) $x_{\text{NaCl}} = 0.0348$.

1.2 One mole of Ar at 298.15 K occupies 0.2479 m³ at a pressure of 10,000 Pa = 0.1 bar. (a) Calculate the value of $R$ in (1.13). Convert this value of $R$ into the units containing (b) eV, frequently used by physicists, and (c) calories.

1.3 Compute the pressure of a gas mixture containing 20 g $N_2$ and 50 g Ar in 1.00 m³ of volume at 300.00 K by using the value of $R$ in Problem 1.2.

*Ans.:* 4,902.9 Pa.

CHAPTER II

# DIFFERENTIATION, INTEGRATION, AND SPECIAL FUNCTIONS

Thermodynamics correlates energy, and physical and chemical properties of various systems by means of relatively simple equations. Mathematical review presented in this chapter is useful in understanding the derivations of these equations, and their interrelationships. Functional representation of a thermodynamic property $\mathcal{G} = \mathcal{G}(P, T, n_1, n_2, \ldots)$ and its differentiation and integration play a dominant role in thermodynamics. A thorough knowledge of the significance of each symbol and the exact mode of writing every mathematical expression greatly facilitates comprehension of thermodynamic equations. The functional form selected for $\mathcal{G}$ should be continuous, single-valued, and differentiable with respect to its variables. Discontinuities occur when there are phase transitions as will be seen in the chapters dealing with the phase equilibria.

## Differentials and Derivatives

Let $u$ be a function of the independent variables $x, y, z, \ldots$, i.e.,

$$u = f(x, y, z, \ldots) \tag{2.1}$$

Note that $x$ without subscript is a mathematical variable in contrast with $x_i$ in (1.1). The total differential of $u$ is written as $du$; it is also called the complete differential of $u$ but we shall use the first term in preference. The symbol $du$ represents the increment $f(x+\Delta x, y+\Delta y, z+\Delta z, \ldots) - f(x, y, z, \ldots)$ when $\Delta x, \Delta y, \Delta z, \ldots$ become infinitesimally small quantities. In general, $du$ is a function of the same variables, and it is expressed by

$$du = \left(\frac{\partial u}{\partial x}\right)_{y,z,\ldots} \bullet dx + \left(\frac{\partial u}{\partial y}\right)_{x,z,\ldots} \bullet dy + \left(\frac{\partial u}{\partial z}\right)_{x,y,\ldots} \bullet dz + \cdots \tag{2.2}$$

where the coefficients of $dx, dy, dz, \ldots$ are the partial derivatives of $u$ with respect to $x, y, z, \ldots$ respectively. The geometric significance of (2.2) is given in the following example.

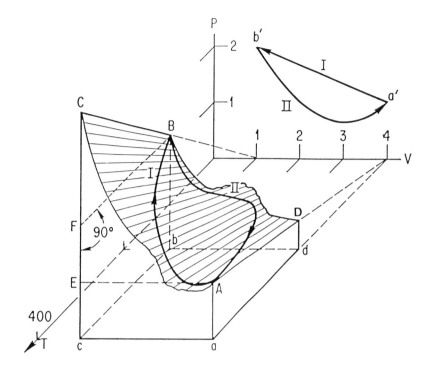

**Fig. 2.1** Surface ACBD representing P-V-T for an ideal gas.

*Example* The surface $ACBD$ in Fig. 2.1 accurately represents the single valued and continuous function, pressure $P$,

$$P = f(V, T) = T/(200V) \qquad (2.3)$$

for a gas within a limited range of volume $V$ and temperature $T$. We wish to express the total differential of $P$, $dP$, in terms of $(\partial P/\partial V)_T$, $(\partial P/\partial T)_V$, $dV$ and $dT$. The points $A, C, B$, and $D$ are such that their projection on the $T$-$V$ plane form a rectangle $acbd$ whose sides $ad$ and $cb$ are parallel to the $T$-axis. Hence, $T$ is held constant on the projection planes $AaCc$ and $DdBb$, and $V$ is held constant on $AaDd$ and $BbCc$ planes. Let a process take the system from $A$ to $B$ on the arbitrarily chosen path $ACB$. The curve $AC$ represents $P_T = f(V)_T$ at a constant temperature. The change in pressure from $A$ to $C$ is $(\Delta P)_T = P_C - P_A = EC$. In the limiting case when $A$ and $C$ are very close.

$$\lim\left(\frac{\Delta P}{\Delta V}\right)_T = \lim\left(\frac{EC}{AE}\right) = \left(\frac{\partial P}{\partial V}\right)_T \qquad (2.4)$$

Similarly, from $C$ to $B$

$$\lim\left(\frac{\Delta P}{\Delta T}\right)_V = \lim\left(\frac{CF}{FB}\right) = \left(\frac{\partial P}{\partial T}\right)_V \qquad (2.5)$$

The total pressure difference for the overall process from A to B is

$$\Delta P = P_B - P_A = bB - aA = EF$$

Since $\Delta P$ is the algebraic sum of pressure differences on the path A to C and then C to B, and observing that $CF = -FC$,

$$\Delta P = P_B - P_A = (P_C - P_A)_T + (P_B - P_C)_V$$
$$= (\Delta P)_T + (\Delta P)_V = EC + CF = EF$$

When A, C, and B are infinitesimally apart, $\Delta P$ becomes the total differential $dP$, $(\Delta P)_T$ and $(\Delta P)_V$ become $(dP)_T$ and $(dP)_V$ respectively and the preceding equation becomes

$$dP = \lim \Delta P = (dP)_T + (dP)_V$$

where lim means $\Delta P$ becomes infinitesimally small in the limit. On the triangle $AEC$,

$$\lim(\Delta P)_T = (dP)_T = \lim\left(\frac{EC}{AE}\right) \cdot AE = \left(\frac{\partial P}{\partial V}\right)_T dV$$

where $\lim(AE) = dV$ is the total change in V. Similarly, on the triangle $CFB$

$$\lim(\Delta P)_V = (dP)_V = \lim\left(\frac{CF}{FB}\right) \cdot FB = \left(\frac{\partial P}{\partial T}\right)_V dT$$

In view of the fact that $\lim(\Delta V)_T = \lim AE = dV$, and $\lim(\Delta T)_V = \lim FB = dT$, (2.5) becomes

$$dP = \left(\frac{\partial P}{\partial V}\right)_T dV + \left(\frac{\partial P}{\partial T}\right)_V dT \tag{2.6}$$

which is the desired geometrical derivation.

In expressing the total differential of $u$, $u$ was taken to be a function of independent variables $x, y, z, \ldots$. However, (2.2) is also valid even when $x, y, z, \ldots$ are not all independent variables. As a very simple example let $u = 3x + 2y + 3z$, and $z = y - x$; then from (2.2) it is seen that $du = 3dx + 2dy + 3dz$. Substituting $dz = dy - dx$ transforms $du$ into $du = 5dy$, but also the elimination of $z$ from $u$ gives $u = 5y$ from which again $du = 5dy$.

Let $u$ and $z, \ldots$ be held simultaneously constant in (2.2) so that $(du)_u$, $(dz)_z, \ldots$ become zero; the resulting equation is then

$$0 = \left(\frac{\partial u}{\partial x}\right)_{y,z,\ldots} \cdot (dx)_{u,z,\ldots} + \left(\frac{\partial u}{\partial y}\right)_{x,z,\ldots} \cdot (dy)_{u,z,\ldots}$$

from which

$$\left(\frac{dx}{dy}\right)_{u,z,\ldots} = \left(\frac{\partial x}{\partial y}\right)_{u,z,\ldots} = -\frac{(\partial u/\partial y)_{x,z,\ldots}}{(\partial u/\partial x)_{y,z,\ldots}} \tag{2.7}$$

Note that $(dx/dy)_{u,z,\ldots}$ is identical with $(\partial x/\partial y)_{u,z,\ldots}$ but the latter notation is preferred since $d$ is reserved for total differentials. Rearranging (2.7) and observing that

$$\left(\frac{\partial u}{\partial y}\right)_{x,z,\ldots} = \frac{1}{\left(\dfrac{\partial y}{\partial u}\right)_{x,z,\ldots}} \tag{2.8}$$

it is readily seen that

$$\left(\frac{\partial u}{\partial x}\right)_{y,z,\ldots} \cdot \left(\frac{\partial x}{\partial y}\right)_{u,z,\ldots} \cdot \left(\frac{\partial y}{\partial u}\right)_{x,z,\ldots} = -1 \tag{2.9}$$

For $u = f(x, y)$, it is evident that (2.9) becomes

$$\left(\frac{\partial u}{\partial x}\right)_y \left(\frac{\partial x}{\partial y}\right)_u \left(\frac{\partial y}{\partial u}\right)_x = -1 \tag{2.10}$$

The sequence of partial differentials and the restricted variables outside the parentheses in this equation should be noted.

### Second Derivatives

The first partial derivatives of $u = f_1(x, y)$ i.e., $(\partial u/\partial x)_y$ and $(\partial u/\partial y)_x$ may be differentiated again to obtain the second partial derivatives of $u$. The resulting derivatives are written as

$$\left(\frac{\partial^2 u}{\partial x^2}\right)_y, \; \left(\frac{\partial^2 u}{\partial y^2}\right)_x, \; \left[\frac{\partial(\partial u/\partial y)_x}{\partial x}\right]_y = \left[\frac{\partial(\partial u/\partial x)_y}{\partial y}\right]_x \equiv \frac{\partial^2 u}{\partial x \partial y} \tag{2.11}$$

The equality between the two derivatives before the identity sign is stated as follows: "The sequence of partial differentiation of a function with respect to its two variables is immaterial." This is possible when $u$ and its first derivatives are both continuous and differentiable. The foregoing property may also be expressed by writing $(\partial^2 u/\partial x\, \partial y) = (\partial^2 u/\partial y\, \partial x)$.

### Useful Differentials

The relationships in the preceding sections can be used to derive the differentials frequently encountered in thermodynamics. A useful theorem extending the application of these differentials, when the function consists of more than one term, is as follows: The total differential of a sum is identical with the sum of the total differentials of all terms. Accordingly, each term may be differentiated and then added to obtain the desired differential. The most frequently encountered differentials are as follows:

$$d(ax + by) = a\,dx + b\,dy \qquad (a\ \&\ b = \text{constant})$$

$$dxy = y\,dx + x\,dy$$

$$dx^n = nx^{n-1}\,dx$$

$$d(x/y) = \frac{y\,dx - x\,dy}{y^2}$$

$$de^{ax} = ae^{ax}\,dx$$

$$da^x = (a^x \ln a)\,dx$$

$$d\ln ax = \frac{dx}{x}$$

$$dx^y = yx^{y-1}\,dx + (x^y \ln x)\,dy$$

ln = Natural (or Napierian logarithm), base-$e$

log = Common logarithm, base-10

("lg" in a number of countries overseas)

$$\ln x = (\ln 10)\log x.$$

## Maxima, Minima and Inflection Points

The partial derivative of $u = x^m y^n + x^p y^q + \cdots$ with respect to $x$ represents the slope of a curve at any point when $u$ is plotted versus $x$ for any chosen fixed value of $y$, i.e., a parametric value of $y$. When $(\partial u/\partial x)_y = 0$ for a particular value of $x$, there are four possibilities: (1) The slope $(\partial u/\partial x)_y$ reaches zero from a positive value and then becomes negative as $x$ increases. This is possible only when the curve $u$ versus $x$ has a maximum. Since a plot of $(\partial u/\partial x)_y$ versus $x$ must show that $(\partial u/\partial x)_y$ decreases with increasing $x$, $(\partial^2 u/\partial x^2)$ may either be negative, or zero; if it is negative there is a maximum in the curve, therefore, a knowledge of how the slope changes sign with $x$ is not necessary. (2) The slope $(\partial u/\partial x)_y$ reaches zero from a negative value and then becomes positive as $x$ increases; hence the curve has a minimum at that value of $x$ for a particular parametric value of $y$. If $(\partial^2 u/\partial x^2)_y > 0$ with $(\partial u/\partial x)_y = 0$, the curve has a minimum and it is not necessary to know how the slope changes its sign. (3) The slope $(\partial u/\partial x)_y$ is positive before and after becoming zero. The curve $u$ versus $x$ therefore rises with increasing $x$ and has a horizontal inflection where $(\partial^2 u/\partial x^2)_y$ is also zero. (4) The slope is negative before and after becoming zero. Hence the curve $u$ versus $x$ falls with increasing $x$ and has a horizontal inflection where $(\partial^2 u/\partial x^2)_y$ is also zero.

When $(\partial^2 u/\partial x^2) = 0$, but $(\partial u/\partial x)_y \neq 0$ the curve $u$ versus $x$ has a non-horizontal inflection point, and if $(\partial u/\partial x)_y$ is positive the curve is rising, and if negative it is falling with increasing $x$. The foregoing properties are summarized

**Table 2.1** Maxima, minima and inflection at $x_o$.

| Values of $(\partial u/\partial x)_y$ | | | | | |
|---|---|---|---|---|---|
| at $x < x_o$ | at $x_o$ | at $x > x_o$ | $(\partial^2 u/\partial x^2)_y$ at $x_o$ | Remarks | Example |
| + | 0 | − | 0 (unn.) | maximum | $u = -x^4$, at $x = 0$ |
| unn. | 0 | unn. | − | maximum | $u = -x^2 + 2x$, at $x = 1$ |
| − | 0 | + | 0 (unn.) | minimum | $u = x^4$, at $x = 0$ |
| unn. | 0 | unn. | + | minimum | $u = x^2 - 2x$, at $x = 1$ |
| + | 0 | + | 0 | Horizontal Inflection | $u = x^3$, at $x = 0$ |
| − | 0 | − | 0 | Horizontal Inflection | $u = -x^3$, at $x = 0$ |
| unn. | + | unn. | 0 | Non-horiz. Inflection | $u = e^{-x^2}$, at $x = -0.707$ |
| unn. | − | unn. | 0 | Non-horiz. Inflection | $u = e^{-x^2}$, at $x = +0.707$ |

Notes: "+" means the value is positive, "−" negative. Unnecessary information is marked with "unn.". In the last four cases involving the inflection points, if $(\partial u/\partial x)_y$ is positive either at $x_o$ or in the vicinity of $x_o$, the curve rises with increasing $x$; if $(\partial u/\partial x)_y$ is negative, the curve falls with increasing $x$.

in Table 2.1 above in the succession presented in the text with $x_o$ indicating the point at which at least one of the derivatives is zero.

## L'Hôpital's Theorem

The ratio of two differentiable functions, $f_1(x)/f_2(x)$ may become indeterminate in the form of $\infty/\infty$ or $0/0$, when $x \to a$. According to L'Hôpital's theorem, the ratio is equal to $(df_1/dx) \div (df_2/dx)$ at $x = a$ provided that this is not indeterminate, but if this ratio of derivatives also becomes indeterminate then $f_1(x)/f_2(x)$ is equal to *the first* of the following expressions which is not indeterminate:

$$\frac{d^2 f_1/dx^2}{d^2 f_2/dx^2}, \frac{d^3 f_1/dx^3}{d^3 f_2/dx^3}, \text{ etc.} \tag{2.12}$$

In addition, for $x \to \infty$ or for $x \to 0$, $f_1(x)/f_2(x)$ may become $\infty/\infty$ or $0/0$ in which case the same theorem is again applicable. The proof of this theorem for $f_1/f_2 = 0/0$ is as follows. The indeterminate ratio means that when $x \to a$ both $f_1$ and $f_2$ become zero, and $\Delta f_1/\Delta x = (f_1 - 0)/(x - a)$, and as $\Delta x = x - a$ approaches zero, this equation becomes $df_1/dx = f_1/(x - a)$ and $f_1 = (x - a)(df_1/dx)$. An identical equation may also be written for $f_2$, and the results may be substituted into $f_1/f_2$ to obtain

$$\lim_{x \to a} \left(\frac{f_1}{f_2}\right) = \frac{(x-a) df_1/dx}{(x-a) df_2/dx} = \frac{df_1/dx}{df_2/dx} \tag{2.13}$$

The cancellation of $(x - a)$ requires that "$x$" may approach "$a$" as closely as one wishes with the restriction that $x$ is never set equal to $a$. This completes the proof. When $f_1/f_2 = \infty/\infty$, it is possible to write $F_1 = 1/f_1$ and $F_2 = 1/f_2$ to convert the ratio into $F_1/F_2 = 0/0$. The foregoing proof is then valid also for the indeterminate form $\infty/\infty$. For example, $\ln(x+1)/(e^x - 1)$ is $\infty/\infty$ for $x \to \infty$, and $0/0$ for $x \to 0$; by L'Hôpital's theorem the ratio is given by $1/[e^x(x+1)]$ which is zero when $x \to \infty$, and unity when $x \to 0$.

When $f_1(x) \cdot f_2(x)$ becomes $0 \cdot \infty$ or $\infty \cdot 0$ for $x \to a$, it can be converted into $f_1(x)/1/f_2(x)$ so that it becomes either $0/0$ or $\infty/\infty$. The resulting ratio can therefore be subjected to this theorem to get the limiting value of $f_1(x) \cdot f_2(x)$. A useful example in thermodynamics is $\lim(x \ln x)$ which has the indeterminate from $-0 \cdot \infty$ at $x \to 0$. The equivalent expression $\ln x/(1/x)$ becomes $-\infty/\infty$ at $x = 0$. The ratio of first derivatives is simply equal to $(-x)$ which becomes zero when $x$ approaches zero; hence

$$\lim_{x \to 0} (x \ln x) = 0$$

If $f_1(x) - f_2(x)$ assumes the indeterminate form $(\infty - \infty)$ when $x$ approaches a finite or infinite value, it may be possible to transform it into $0/0$ or $\infty/\infty$. An example is sufficient to illustrate this point: $[1/\ln(x+1)] - 1/x$ becomes $\infty - \infty$ when $x = 0$. However, it is possible to rewrite this expression as $[x - \ln(x+1)]/x \ln(x+1)$ which is $0/0$ when $x \to 0$; the ratio of first derivatives, however, is $[1 - 1/(x+1)]/[\ln(x+1) + x/(x+1)]$ which is again $0/0$. The ratio of the second derivatives gives $1/2$ as the limiting value of the original indeterminate expression.

## Homogeneous Functions

A function $u = f(x, y, z, \ldots)$ is said to be homogeneous in all its independent variables when for a parameter $k$

$$u = f(kx, ky, kz, \ldots) = k^\alpha f(x, y, z, \ldots) \tag{2.14}$$

where $\alpha$ is the degree of homogeneity. For example $u = (4x + 2y)/(x - y)$, $V = 3x^2 + y^2$, $W = 2x^3 + xy^2 - y^3$ are zeroth, first and third degree homogeneous functions respectively in both $x$ and $y$. If $u$ is homogeneous in some of its variables, as for example in $y$ and $z$; then (2.14) is valid for $y$ and $z$, i.e., $f(x, ky, kz, \ldots) = k^\alpha f(x, y, z, \ldots)$. Thus $u = (x^2 y^3 z + x^3 y z^3)/(y + z)$ is homogeneous in $y$ and $z$ to the third degree, but non-homogeneous in $x$ as can readily be verified by the application of (2.14). If a function $f(x, y, z, \ldots)$ is homogeneous in all its variables to the $\alpha$th degree, then $f(x, y, z, \ldots)/x^\alpha$ is a homogeneous function of zeroth degree.

## Euler's Theorem on Homogeneous Functions

If a function, $u = f(x, y, z)$, is homogeneous with respect to all its independent variables, it is necessary that

$$x \left( \frac{\partial u}{\partial x} \right)_{y,z} + y \left( \frac{\partial u}{\partial y} \right)_{x,z} + z \left( \frac{\partial u}{\partial z} \right)_{x,y} = \alpha u \tag{2.15}$$

It is sufficient to verify (2.15) for our purposes with an example. Let $u = u(x, y) = x^2 + xy + y^2$, which is a homogeneous function of second degree in $x$ and $y$. If we let $kx$ and $ky$ for $x$ and $y$ respectively, then it can be readily verifiable by using (2.14) that $u(kx, ky) = k^2x^2 + k^2xy + k^2y^2 = k^2u(x, y)$, confirming the degree of homogeneity. Next, we substitute $\partial u(x, y)/\partial x = 2x + y$, and $\partial u(x, y)/\partial y = x + 2y$ in (2.15) to verify that $x(2x + y) + y(x + 2y) = 2u(x, y)$.

Each successive differentiation of a homogeneous function with respect to any one of its independent variables yields a homogeneous function whose degree is one less than the preceding equation in that independent variable.

## Homogeneous Thermodynamic Functions

A thermodynamic property $\mathcal{G}$ of a single phase may be represented by

$$\mathcal{G} = \mathcal{G}(P, T, n_1, n_2, \ldots, n_c) \tag{2.16}$$

If $\mathcal{G}$ is a homogeneous function of the first degree in the number of moles $n_1, n_2, \ldots, n_c$, it is called an extensive property. For example, the volume of a system is an extensive property but the volume divided by $\sum n_i$ which is the molar volume, is an intensive property. All intensive properties are homogeneous in all $n_i$'s. For an extensive property such as $\mathcal{G}$ of (2.16), we have

$$k\mathcal{G} = f(P, T, kn_1, kn_2, \ldots) \tag{2.17}$$

According to Euler's theorem

$$n_1 \left(\frac{\partial \mathcal{G}}{\partial n_1}\right)_{P,T,n_2,} + n_2 \left(\frac{\partial \mathcal{G}}{\partial n_2}\right)_{P,T,n_1,} + \cdots = \mathcal{G} \tag{2.18}$$

For brevity, let

$$\overline{G}_1 = \left(\frac{\partial \mathcal{G}}{\partial n_1}\right)_{P,T,n_2,}, \overline{G}_2 = \left(\frac{\partial \mathcal{G}}{\partial n_2}\right)_{P,T,n_1,}, \text{ etc.} \tag{2.19}$$

where $\overline{G}_i$ is called the partial property of component "$i$". Substitution of $\overline{G}_1, \overline{G}_2, \ldots$ into (2.18) gives

$$n_1\overline{G}_1 + n_2\overline{G}_2 + \cdots = \mathcal{G} \tag{2.20}$$

Since (2.20) is equal to (2.16), each term must be homogeneous in the first degree according to (2.14), hence any one of the partial molar properties $\overline{G}_1, \overline{G}_2, \ldots$ is a homogeneous function of zeroth degree in $n_1, n_2, \ldots$ e.g., $\overline{G}_1$ must have the following functional form:

$$\overline{G}_1 = \overline{G}_1\left(P, T, \frac{n_2}{n_1}, \frac{n_3}{n_1}, \ldots, \frac{n_c}{n_1}\right) \tag{2.21}$$

Such properties that are homogeneous functions of zeroth degree in $n_i$'s are functions of $P$, $T$, and $c - 1$ independent newly selected variables $n_2/n_1, n_3/n_1, \ldots$

The mole fraction* of a component "$i$" is defined by $X_i = n_i/(n_1+n_2+\cdots+n_c)$, hence $n_2/n_1 = X_2/X_1$, and since the sum of all mole fractions is unity there are $i-1$ independent mole fractions. Therefore $c-1$ mole ratios $n_2/n_1,\ldots,$ may be expressed in terms of $c-1$ mole fractions; then (2.21) becomes

$$\overline{G}_1 = \overline{G}_1(P, T, X_1, X_2, \ldots, X_{c-1}) \tag{2.22}$$

The total differential of (2.20) is

$$d\mathcal{G} = n_1\,d\overline{G}_1 + n_2\,d\overline{G}_2 + \cdots + \overline{G}_1\,dn_1 + \overline{G}_2\,dn_2 + \cdots. \tag{2.23}$$

and that of (2.16) is

$$d\mathcal{G} = \left(\frac{\partial \mathcal{G}}{\partial P}\right)_{T,n_1,n_2,\ldots} \cdot dP + \left(\frac{\partial \mathcal{G}}{\partial T}\right)_{P,n_1,n_2,\ldots} \cdot dT$$
$$+ \left(\frac{\partial \mathcal{G}}{\partial n_1}\right)_{P,T,n_2,\ldots} \cdot dn_1 + \left(\frac{\partial \mathcal{G}}{\partial n_2}\right)_{P,T,n_1,\ldots} \cdot dn_2 + \cdots.$$

The substitution of $\overline{G}_i$ as defined by (2.19) leads to

$$d\mathcal{G} = \left(\frac{\partial \mathcal{G}}{\partial P}\right)_{T,n_1,n_2,\ldots} \cdot dP + \left(\frac{\partial \mathcal{G}}{\partial T}\right)_{P,n_1,n_2,\ldots} \cdot dT$$
$$+ \overline{G}_1\,dn_1 + \overline{G}_2\,dn_2 + \cdots. \tag{2.24}$$

Comparison of (2.23) with (2.24) gives

$$n_1\,d\overline{G}_1 + n_2\,d\overline{G}_2 + \cdots = \left(\frac{\partial \mathcal{G}}{\partial P}\right)_{T,n_1,n_2,\ldots} \cdot dP + \left(\frac{\partial \mathcal{G}}{\partial T}\right)_{P,n_1,n_2,\ldots} \cdot dT \tag{2.25}$$

At constant $P$ and $T$, the right hand side becomes zero; thus,

$$n_1\,d\overline{G}_1 + n_2\,d\overline{G}_2 + \cdots = \sum_{i=1}^{c} n_i\,d\overline{G}_i = 0, \quad \text{(constant } P \text{ and } T\text{)} \tag{2.26}$$

This is a very useful equation of general applicability, first derived by Gibbs (1875) and then independently by Duhem (1886).

Another useful form of (2.26) is obtained after dividing by $\sum_{i=1}^{c} n_i$ and then substituting $X_i$ for the resulting coefficient of each $\overline{G}_i$; i.e.,

$$X_1\,d\overline{G}_1 + X_2\,d\overline{G}_2 + \cdots = 0 \tag{2.27}$$

Similarly, (2.20) can be transformed into

$$X_1\overline{G}_1 + X_2\overline{G}_2 + \cdots = \frac{\mathcal{G}}{\sum n_i} = G \tag{2.28}$$

where each term on the left is zeroth degree in $n_i$'s and therefore $G$ is also zeroth degree in $n_i$'s. The property of $G$ is called the molar quantity and despite the fact

---

*Capital $X_i$ is used for the mole fractions only in the remaining parts of this chapter to avoid confusion with variable $x$.

that it is an intensive property, it is also called by the confusing term the "molar extensive property," presumably suggested by $\mathcal{G}/\sum n_i$ in (2.28).

The partial differential of $\mathcal{G}$ with respect to pressure is another homogeneous function of the first degree in $n_i$'s. Therefore, from Euler's Theorem,

$$\left(\frac{\partial \mathcal{G}}{\partial P}\right)_{T,n_1,n_2,} = n_1 \left[\frac{\partial (\partial \mathcal{G}/\partial P)_{T,n_1,n_2,}}{\partial n_1}\right]_{P,T,n_2,}$$
$$+ n_2 \left[\frac{\partial (\partial \mathcal{G}/\partial P)_{T,n_1,n_2,}}{\partial n_2}\right]_{P,T,n_1,} + \cdots. \qquad (2.29)$$

Since the order of differention of the right-hand side with respect to $P$ and $n_i$'s is immaterial, i.e., $\partial(\partial \mathcal{G}/\partial P)/\partial n_1 = \partial(\partial \mathcal{G}/\partial n_1)/\partial P$, with the variables other than those differentiated held constant in each step, then

$$\left(\frac{\partial \mathcal{G}}{\partial P}\right)_{T,n_1,n_2,} = n_1 \left(\frac{\partial \overline{G}_1}{\partial P}\right)_{T,n_1,n_2,} + n_2 \left(\frac{\partial \overline{G}_2}{\partial P}\right)_{T,n_1,n_2,} + \cdots. \qquad (2.30)$$

By an identical procedure, or by interchanging $P$ and $T$, it can be shown that

$$\left(\frac{\partial \mathcal{G}}{\partial T}\right)_{P,n_1,n_2,} = n_1 \left(\frac{\partial \overline{G}_1}{\partial T}\right)_{P,n_1,n_2,} + n_2 \left(\frac{\partial \overline{G}_2}{\partial T}\right)_{P,n_1,n_2,} + \cdots. \qquad (2.31)$$

**Integrals**

Integration is the reverse process of differentiation. An integral can therefore be verified by differentiation to obtain the initial equation. Integration of $du = (\partial u/\partial x)_y\, dx + (\partial u/\partial y)_x\, dy$ is written as

$$\int_{u_1}^{u_2} du = u_2 - u_1 = \left[\int_{x_1}^{x_2} \left(\frac{\partial u}{\partial x}\right)_{y_1} dx\right]_{y_1} + \left[\int_{y_1}^{y_2} \left(\frac{\partial u}{\partial y}\right)_{x_2} dy\right]_{x_2} \qquad (2.32)$$

where $u_1$ is the value at $x_1$, $y_1$ and $u_2$, at $x_2$, $y_2$; the result is a definite integral because the limits of integration are definite. The first integral in square brackets carries the function from $x_1$ to $x_2$ at a constant value of $y_1$ and the succeeding one from $y_1$ to $y_2$ at a constant value of $x_2$. Consider, for example, $du = 3x^2y^2\, dx + 2x^3y\, dy$; its integral from $x_1, y_1$ to $x_2, y_2$ is

$$u_2 - u_1 = \left[\int_{x_1}^{x_2} (3y^2)_{y_1} x^2\, dx\right]_{y_1} + \left[\int_{y_1}^{y_2} (2x^3)_{x_2} y\, dy\right]_{x_2}$$
$$= y_1^2(x_2^3 - x_1^3) + x_2^3(y_2^2 - y_1^2) = x_2^3 y_2^2 - x_1^3 y_1^2 \qquad (2.33)$$

The result is obvious since $\int du = \int d(x^3y^2) = x_2^3 y_2^2 - x_1^3 y_1^2$. The upper limits $x_2$, $y_2$ may also be variables $x$, $y$ instead of being constants, in which case the integral is indefinite. In calculus, (2.33) is written as $u = x^3y^2 + C$ where $u_1$ and $x_1^3 y_1^2$ are combined into one constant, $C$, but this is rarely practiced in thermodynamics. Equation (2.32) may also be evaluated graphically if convenience dictates it, but it is rarely recommended.

A number of frequently encountered integrals will now be presented. The general form of the integrals containing $f(y)x^m\, dx$ is

$$f(y)\int_{x_1}^{x} x^m\, dx = \frac{x^{m+1}}{m+1} f(y) - \frac{x_1^{m+1}}{m+1} f(y); \qquad (m \neq -1,\ y = \text{constant}) \tag{2.34}$$

where $m$ is any real number except $-1$. It is seen that the second term, the integration constant, is a function of $y$ only since $x_1$ is a constant.

When $m$ is $-1$ in (2.34), the integral becomes

$$f(y)\int_{x_1}^{x} \frac{1}{x}\, dx = f(y)[\ln x - \ln x_1]; \qquad (y = \text{constant}) \tag{2.35}$$

Consider an expression in the form of $u\, dv$ which may not be easy to integrate; the following identity may then be used to facilitate its integration:

$$\int_{u_1,v_1}^{u,v} u\, dv \equiv \int_{u_1,v_1}^{u,v} d(uv) - \int_{u_1,v_1}^{u,v} v\, du = uv - u_1 v_1 - \int_{u_1,v_1}^{u,v} v\, du \tag{2.36}$$

Here, it is assumed that $v\, du$ is readily integrable. The procedure shown by this equation is called *integration by parts*. For example, $xe^x\, dx$ cannot be integrated at once, but let $u = x$, and $v = e^x$ so that

$$\int_0^x xe^x\, dx = xe^x - \int_0^x e^x\, dx \tag{2.37}$$

The last integral is simply $1 - e^x$. The expressions for $u$ and $v$ must be chosen by trial so that the resulting new term can easily be integrated. For example the choice of $u = xe^x$ and $v = x$ is undesirable since it makes the resulting integral more complex by introducing the integral of $x^2 e^x\, dx$.

When an integral contains some constants as for example in $dx/(a+bx)^2$, the first step is to put $dx$ into $(1/b)d(a+bx)$ and to note that the resulting equation becomes identical in form with (2.34), after which the integral becomes $[-1/b(a+bx)] + C$, where $C$ is a constant. Likewise,

$$\int_0^x e^{ax}\, dx = \frac{e^{ax} - 1}{a} \tag{2.38}$$

$$\int_0^x (\ln ax)\, dx = x \ln ax - x \tag{2.39}$$

The forms of integrals given here are nearly all that are encountered in this book.

## Exact Differentials

Let $M$ and $N$ be continuous and differentiable functions of $x$ and $y$; the expression

$$M\, dx + N\, dy \tag{2.40}$$

is an exact differential if it is equal to the complete differential of a function $u = f(x, y)$, i.e.,

$$du = M\,dx + N\,dy \tag{2.41}$$

When this equation exists, it is evident that

$$\left(\frac{\partial u}{\partial x}\right)_y = M \quad \text{and} \quad \left(\frac{\partial u}{\partial y}\right)_x = N \tag{2.42}$$

and differentiating $M$ with respect to $y$, and $N$ with respect to $x$, an operation known as *cross differentiation*, gives

$$\left(\frac{\partial M}{\partial y}\right)_x = \frac{\partial^2 u}{\partial x\,\partial y} \quad \text{and} \quad \left(\frac{\partial N}{\partial x}\right)_y = \frac{\partial^2 u}{\partial y\,\partial x}$$

Since the sequence of differentiation of $u$ with respect to $x$ and $y$ is immaterial, the right hand sides are equal; therefore,

$$\left(\frac{\partial M}{\partial y}\right)_x = \left(\frac{\partial N}{\partial x}\right)_y \tag{2.43}$$

Equation (2.43) is known as the Euler criterion. It is the necessary condition that (2.40) be an exact differential, i.e., if the expression (2.40) is exact the criterion expressed by (2.43) is necessary.

An expression in the form of (2.40) may not be exact but it may sometimes be possible to make it exact so that it can be integrated. This may be accomplished by multiplication with a suitable factor known as the *integrating factor*, which is, in general, a function of the same variables as are $M$ and $N$. When the expression $M\,dx + N\,dy$ is equal to zero, an integrating factor always exists, though it is not always easy to find it. Therefore, the equation $dQ = M\,dx + N\,dy = 0$, whether an exact differential or not, represents a family or curves on $x$ and $y$ coordinates, with each curve separated from another by two constant parametric values of $Q$.

Consider the following simple examples in which $P$ and $V$ are pressure and volume respectively. $P\,dV + V\,dP$ is exact but $P\,dV - V\,dP$ is inexact as can be tested by (2.43). For the second example it is evident that $(\partial P/\partial P)_v = 1 \neq -(\partial V/\partial V)_p = -1$. However, $(P\,dV - V\,dP)$ becomes exact when divided by $PV$, i.e.,

$$\frac{đQ}{PV} = \frac{dV}{V} - \frac{dP}{P} = df(P, V) = d\ln\frac{V}{P} \tag{2.44}$$

where $đ$ is used whenever possible to indicate that $đQ$ is an inexact differential. The factor $1/PV$ is called an *integrating factor*, or $PV$, an *integrating denominator*. Now $df = đQ/PV$ is an exact differential; therefore, $[dF(f)/df]df$ is also an exact differential where $F(f)$ is an arbitrary function of $f$. Multiplication of both sides of (2.44) by $dF(f)/df \equiv F'(f)$, gives

$$\frac{F'(f)}{PV}đQ = F'(f)\,df, \quad \left(\text{integrating factor: } \frac{F'(f)}{PV}\right). \tag{2.45}$$

Therefore, the integrating factor $1/PV$ is not unique, and in fact once an integrating factor is found, there are an infinite number of them represented by $F'(f)/PV$. For example, if $F = e^f = V/P$ then $F' = e^f = V/P$ and $(V/P)/(PV) = 1/P^2$ is also an integrating factor. In fact, it is easily seen that

$$\frac{P\,dV}{P^2} - \frac{V\,dP}{P^2} = \left(\frac{1}{P}\right)dV + V d\left(\frac{1}{P}\right) = d\left(\frac{V}{P}\right)$$

and likewise $(V/P)^m/(PV)$, where $m$ is a constant, is also an integrating factor. Additional examples of integrating factors are given in Chapter IV.

## Line Integrals

An expression represented by

$$\int_{x_1,y_1}^{x_2,y_2} (M\,dx + N\,dy) \tag{2.46}$$

cannot always be integrated unless $x$ or $y$ is eliminated. The desired elimination is possible when a thermodynamics process can be carried out along a line or path represented by $y = f(x)$. The integral along $y = f(x)$ is then called a *line integral*. Since there are an infinite number of arbitrary functions $y = f(x)$ there are also an infinite number of integrals, each, in general, but not always different from others. Thus, for example

$$\int_{x_1=0,y_1=0}^{x_2=2,y_2=4} (y\,dx - x\,dy)$$

cannot be integrated unless a relationship between $x$ and $y$ is selected. For $y = x^2$ the integral is $-8/3$, and for $y = x^3/2$, it is $-4$. Both paths take the integral from the same initial $x_1$, $y_1$, to the same final $x_2$, $y_2$, but the integrals have different values.

Integration of (2.46) becomes quite simple when $(M\,dx + N\,dy)$ is an exact differential of a function $u = u(x, y)$; the result is

$$\int_{x_1,y_1}^{x_2,y_2} (M\,dx + N\,dy) = \int_{x_1,y_1}^{x_2,y_2} du = u_2 - u_1 \tag{2.47}$$

Equation (2.47) shows that $du$ can be integrated without the necessity of specifying a path since $u_1$ and $u_2$ are known. If a set of $u$, $x$, and $y$ coordinates are chosen, the relationship between $u$, and $x$ and $y$ is represented by a surface. On this surface $u_1$ and $u_2$ are *each* represented by *one point*, i.e., "$u$" is a *single valued function*, or *a point function*, of $x$ and $y$. The value of $u_2 - u_1$, or the integral (2.47), is therefore determined only by the initial and final values of $x$ and $y$. Hence, when the expression $M\,dx + N\,dy$ is an exact differential, its integral is *independent of the infinite number of lines or paths* that may be followed in proceeding from $x_1$, $y_1$ to $x_2$, $y_2$.

## 24  Thermodynamics

From the foregoing property of an exact differential, it follows that the integral around a closed path is zero, i.e.,

$$\oint du = \text{Path I} \int_{x_1,y_1}^{x_2,y_2} (M\,dx + N\,dy) + \text{Path II} \int_{x_2,y_2}^{x_1,y_1} (M\,dx + N\,dy) = 0 \quad (2.48)$$

The contour integral $\oint$ is defined by the sum of the two integrals at the left which take $u$ from $x_1$, $y_1$ to $x_2$, $y_2$ on one path and back to $x_1$, $y_1$ on another path.

If $M\,dx + N\,dy$ is an *inexact differential*, its integral around any closed path is in general not zero. For example $y\,dx - x\,dy$ is not an exact differential. Its integral from $(0, 0)$ to $(2,4)$ along $y = x^2$ was found to be $-8/3$. Now returning from $(2, 4)$ to $(0, 0)$ along $y = x^3/2$, the integral is $+4$; hence the integral along a *closed path* defined by $y = x^2$ and $y = x^3/2$ is not zero but equal to $+4/3$.

## A Graphical Example

A simple example for exact differentials is

$$P\,dV + V\,dP \quad (2.49)$$

which is the total differential of $u = f(P, V) = PV$. Let the function $u$ be $T$ as represented in Fig. 2.1. The integral along a closed path I–II from $A$ to $B$ and back to $A$ is

$$\oint (P\,dV + V\,dP) = 0 \quad (2.50)$$

A given path is defined by $P = f(V)$. Thus Path I is defined by its projection on the $P$-$V$ plane. Let the coordinates of $A$ be $P = 1$, $V = 4$ and those of $B$, $P = 2$, $V = 1$. Path I is then represented by

$$P = -(1/3)(V - 7) \quad (2.51)$$

and Path II by

$$P = (V^2 - 6V + 11)/3 \quad (2.52)$$

and both paths are drawn exactly to scale. The integral from $A$ to $B$ on Path I is

$$\text{Path I} \int_{V=4}^{1} (P\,dV + V\,dP) = \int_{4}^{1} \left[-\frac{1}{3}(V - 7)\,dV - \frac{1}{3} V\,dV\right] = -2$$

and from $B$ to $A$ on Path II,

$$\text{Path II} \int_{V=1}^{4} (P\,dV + V\,dP) = \frac{1}{3}\int_{1}^{4}(V^2\,dV - 6V\,dV \\ + 11\,dV + 2V^2\,dV - 6V\,dV) = 2$$

Thus the integral of (2.49) is zero along a closed path such as I–II.

The expression

$$P\,dV - V\,dP \tag{2.53}$$

was previously shown to be an inexact differential, i.e. there is no function whose total differential is $P\,dV - V\,dP$. It is therefore, *impossible to represent* this expression on the $u - P - V$ coordinates since there is no such relationship. It can be shown that the integral of (2.53) along Path I from $V = 4$ to $V = 1$ is $-7$ and along Path II back to $V = 4$ it is 4 and the integral along the closed path is $-3$

## Cross Differentials

It was shown earlier that for the exact differential

$$du = M\,dx + N\,dy \tag{2.54}$$

the Euler criterion was

$$\left(\frac{\partial M}{\partial y}\right)_x = \left(\frac{\partial N}{\partial x}\right)_y \tag{2.55}$$

Equation (2.55) is known as a *cross-differential*. Since $N = f_1(x, y)$, then

$$\left(\frac{\partial N}{\partial y}\right)_x \left(\frac{\partial y}{\partial x}\right)_N \left(\frac{\partial x}{\partial N}\right)_y = -1 \tag{2.56}$$

Multiplying (2.55) by (2.56) and eliminating $(dy)_x$ common in the denominators of both sides, and then rearranging the resulting equation gives

$$\left(\frac{\partial M}{\partial N}\right)_x = -\left(\frac{\partial y}{\partial x}\right)_N \tag{2.57}$$

Writing $M = f_2(x, y)$ and repeating the same procedure it is also seen that

$$\left(\frac{\partial N}{\partial M}\right)_y = -\left(\frac{\partial x}{\partial y}\right)_M \tag{2.58}$$

Since $M$ and $N$ are functions of $x$ and $y$, it is possible to eliminate $x$ and write $N = F(M, y)$ for which

$$\left(\frac{\partial N}{\partial M}\right)_y \left(\frac{\partial M}{\partial y}\right)_N \left(\frac{\partial y}{\partial N}\right)_M = -1$$

Solving for $(\partial M/\partial N)_y$ from this equation and substituting into (2.58) and rearranging yields

$$\left(\frac{\partial y}{\partial M}\right)_N = \left(\frac{\partial x}{\partial N}\right)_M \tag{2.59}$$

Equations (2.55), (2.57), (2.58) and (2.59) are called the cross-differentials of (2.54) and they are used in obtaining many relations in thermodynamics.

## Lagrange's Method of Undetermined Multipliers

The total differential of a function $u = f(x, y, z)$ is zero when $u$ is known to be either a maximum or a minimum for any set of infinitesimal non-zero increase in $x$, $y$ and $z$; thus,

$$du = 0 = \left(\frac{\partial u}{\partial x}\right)_{y,z} dx + \left(\frac{\partial u}{\partial y}\right)_{x,z} dy + \left(\frac{\partial u}{\partial z}\right)_{x,y} dz \qquad (2.60)$$

This equation is satisfied when

$$\left(\frac{\partial u}{\partial z}\right)_{x,y} = 0, \quad \left(\frac{\partial u}{\partial x}\right)_{y,z} = 0, \quad \left(\frac{\partial u}{\partial y}\right)_{x,z} = 0 \qquad (2.61)$$

Equations (2.61) are generally independent and they can be solved to obtain the values $x$, $y$ and $z$ at a maximum or a minimum point. When there is a restricting relationship among $x$, $y$ and $z$ in the form of $F(x, y, z) = 0$, one of the variables may be eliminated from $u$ and the same procedure may be repeated to find the maximum or minimum point specified by the restriction imposed on $u$ by $F(x, y, z) = 0$. This may not always be convenient, or it may be impossible in certain cases. An alternative method is therefore desirable. For this purpose, the total differential of $F(x, y, z) = 0$ is first written as

$$\left(\frac{\partial F}{\partial x}\right)_{y,z} dx + \left(\frac{\partial F}{\partial y}\right)_{x,z} dy + \left(\frac{\partial F}{\partial z}\right)_{x,y} dz = 0 \qquad (2.62)$$

*If it is known that u has a maximum or a minimum point* with the restriction imposed by $F$, then a combination of (2.60) and (2.62) is necessary. Multiplying (2.62) by an undetermined quantity $\lambda$ and adding the result to (2.60) gives

$$\left[\left(\frac{\partial u}{\partial x}\right)_{y,z} + \lambda \left(\frac{\partial F}{\partial x}\right)_{y,z}\right] dx + \cdots = 0 \qquad (2.63)$$

Any one of the infinitesimals $dx$, $dy$ or $dz$ may be chosen as non-zero and independent. The terms with the remaining infinitesimals may cancel one another, or become zero but they cannot cancel the terms with the independent infinitesimal for every small and different value of the infinitesimal. Hence, the coefficient of each successively chosen $dx$, $dy$ and $dz$ must be zero; i.e.

$$\left(\frac{\partial u}{\partial x}\right)_{y,z} + \lambda \left(\frac{\partial F}{\partial x}\right)_{y,z} = 0 \qquad (2.64)$$

The total differentials of $u$ and $F$ are symmetrical in their independent variables, i.e. if any two of their variables are interchanged, $u$ and $F$ are not altered; therefore, the remaining two equations similar to (2.64) may also be obtained by interchanging $x$ and $y$, and $x$ and $z$ in (2.64). This is the *principle of symmetry* which is labor saving in obtaining a family of similar derivatives. Therefore, there are two other equations similar to (2.64), which are the coefficients of $dy$ and $dz$. From $F(x, y, z) = 0$, and three equations of the type represented by (2.64), $x$, $\lambda$, $y$

and $z$ can easily be determined. If there are additional restricting relationships, their differentials are also multiplied by additional coefficients and combined with (2.60). For one more relationship $G(x, y, z) = 0$, (2.64) becomes

$$\left(\frac{\partial u}{\partial x}\right)_{y,z} + \lambda \left(\frac{\partial F}{\partial x}\right)_{y,z} dx + \xi \left(\frac{\partial G}{\partial x}\right)_{y,z} = 0 \qquad (2.65)$$

The coefficients $\lambda$ and $\xi$ are called *Lagrangian multipliers*, and they are, in general, functions of the same independent variables $x$, $y$ and $z$.

As a simple example consider $u = x^4 + y^4 + z^4$ with a restricting equation $F = 0 = x + y + z - 3$. From (2.60) and (2.64), and $\partial F/\partial x = 1$, $\partial F/\partial y = 1$, and $\partial F/\partial z = 1$, $\lambda = -4x^3$, $\lambda = -4y^3$ and $\lambda = -4z^3$, or simply $x = y = z$, and from $F = x + y + z - 3 = 0$, $x = y = z = 1$ and $\lambda = -4$. The minimum point therefore occurs at $x = y = z = 1$. The resulting solution is considerably easier than eliminating one of the variables, such as $z$, from $u$, and setting the coefficients of $dx$ and $dy$ to zero. A geometrical example is afforded by $u = x^2 + y^2$, a paraboloid, and a restricting equation $F = x + y - 1 = 0$. The latter equation is a plane parallel to the $u$-axis which cuts the $x$-axis at $x = 1$, and the $y$-axis at $y = 1$. From the preceding method, the result is $x = y = 1/2$, where the intersection of the paraboloid and the plane has a minimum. However, the function $u = x^2 + y^2$, *without restriction* has a minimum at $x = y = 0$ as may be easily seen from (2.60). In thermodynamics and statistical mechanics the nature of the particular function "$u$" is such that $du = 0$ is known to correspond to a maximum or a minimum and it is sufficient to determine the point where $du = 0$ by the method of undetermined multipliers in accordance with the conditions imposed by the restricting functions.

## Change of Independent Variables

The choice of independent variables is generally dictated by their convenience or usefulness. For example, it is more convenient to choose pressure $P$, and temperature $T$, than volume $V$ and viscosity, or $P$ and $V$ as the variables of state of a closed system. If a thermodynamic property $U$ is a function of independent variables $x$ and $y$, the total differential of $U$ is

$$dU = M\,dx + N\,dy \qquad (2.66)$$

where $dx$ and $dy$ on the right affirm that the independent variables are $x$ and $y$. It is possible to change this equation into a form in which $x$ and $N$ are the independent variables by subtracting $d(Ny) = N\,dy + y\,dN$ from both sides of (2.66). The result is

$$d(U - Ny) = dH = M\,dx - y\,dN \qquad (2.67)$$

where $H$ is a function defined by $H = U - Ny$, and the independent variables of $H$ are $x$ and $N$. For example, in thermodynamics the energy $U$, of a closed system is given by $dU = T\,dS - P\,dV$ where $S$ is a variable of state; if $d(PV)$ is

28    Thermodynamics

added on both sides, the resulting equation becomes a function of $S$ and $P$; thus

$$d(U + PV) = dH = T\,dS + V\,dP, \qquad [H \equiv U + PV = H(S, P)],$$

(2.68)

The process of introducing new functions which change the independent variables, as shown by (2.67) or (2.68), is known as the Legendre transformation.

## Representation of Data*

Thermodynamic data are usually correlated by means of appropriate mathematical equations or graphs. Various types of highly rapid computers are available for this purpose and the methods described here may also be used when the data do not require excessive labor, or a computer is not readily accessible.

Treatments of various data by statistical methods are based on two assumptions: (1) the scattering in measurements is random, i.e. the errors tend to make the data scatter equally on either side of a line, and (2) all the errors are assumed to accumulate in the dependent function, i.e. in the ordinate, and not in the abscissa. An error, $\rho_1$, is the difference between a measured value, $S_1$, and what is considered to be the most likely value, $Z$ of a property $S$; thus,

$$\rho_1 = S_1 - Z \qquad (2.69)$$

Let the most likely value $Z$ be the average value defined by

$$Z = \sum_{i=1}^{\eta} S_i/\eta \qquad (2.70)$$

where $\eta$ is the total number of measurements or determinations. An error $\rho$ is sometimes called a deviation or residual but the *error* will be preferred in this text. The term *discrepancy* should not be confused with *error* since the former means the difference between two sets of measured values of the same property either by one investigator, or more frequently, by two investigators. There are four other definitions which are frequently used in statistical interpretation of data. The *average deviation*, $\bar{a}$, is expressed by

$$\bar{a} = \sum_{i=1}^{i=\eta} |\rho_i|/\eta \qquad (2.71)$$

and the percentage of average deviation by $(\bar{a}/Z)\,100\%$. The root mean square deviation $\sigma$, is defined by

$$\sigma = \sqrt{\frac{\sum \rho_i^2}{\eta}} \qquad (2.72)$$

---

*For detailed treatments of experimental data see for example P. G. Hoel, "Introduction to Mathematical Statistics," John Wiley (1984); F. S. Acton, "Numerical Methods that Work," Math. Assoc. Am. (1990).

If this is multiplied by any number, it may also be taken as a measure of deviation. it is usually agreed that when $\sigma$ is multiplied by $\sqrt{\eta/(\eta-1)}$ the resulting expression be called the *standard deviation*, $\sigma°$; thus,

$$\sigma° = \sigma\sqrt{\frac{\eta}{\eta-1}} = \sqrt{\frac{\sum \rho_i^2}{\eta-1}} \tag{2.73}$$

When $\eta$ is large $\sigma$ and $\sigma°$ are practically the same. Even for $\eta = 10$ the difference is negligibly small, i.e. $\sigma° = 1.054\sigma$.

A range of errors from $-\rho_\varepsilon$ to $+\rho_\varepsilon$ may be chosen so that the chance of having an error in this range for a single measurement is 1.2. The error $\rho_i$ is then called the probable error; it is defined by

$$\rho_\varepsilon = 0.6745\sigma \tag{2.74}$$

This definition is based on the probability integral which will not be discussed here.

The method of least squares is generally used for the correlation of experimental data. It is seen from (2.69) and (2.70) that the average value Z, is the value which makes $\sum \rho_i$ zero for the entire set of measurements. However, the sum of the squares of errors $\sum \rho_i^2$ is not zero but it is a minimum for a particular value of Z in (2.69). Thus

$$\sum_{i=1}^{\eta} \rho_i^2 = (S_1 - Z')^2 + (S_2 - Z')^2 + \cdots + (S_\eta - Z')^2$$

$$= \text{a minimum} \tag{2.74a}$$

where $Z'$ is called the "most probable value" defined by this equation. Differentiating with respect to $Z'$, it is seen that

$$\frac{d \sum \rho_i^2}{dZ'} = -2(S_1 - Z') - 2(S_2 - Z') - \cdots - 2(S_\eta - Z')$$

$$= -2\left(\sum_1^\eta S_i - \eta Z'\right) = 0$$

This equation is satisfied when $Z' = \sum S_i/\eta = Z$. Hence the average value is also the most probable value which satisfies the requirement of the method of least squares.

The foregoing presentation is a simple application of the method of least squares to the measurement of a single unknown $x$, such as the length of a bar or the temperature of a set thermostat. When there are two unknown quantities $x$ and $y$, and both $x$ and $y$ are measured at various intervals, then the dependence of $y$ on $x$ may be expressed as a function of $x$ by the same method. Assume first that the functional relationship between $x$ and $y$ is linear, i.e.

$$y = a + bx \tag{2.75}$$

It is now necessary to find the values of $a$ and $b$ so that (2.75) represents a given set of data according to the method of least squares. The error for each measurement of $y$ is

$$\rho_i = y_i - (a + bx_i)$$

where $y_i$ and $x_i$ are the experimentally measured values of $x$ and $y$, and $(a + bx_i)$ is the value of $y_i$ from (2.75) in terms of $a$ and $b$ regarded as unknown at this stage. The method requires that

$$\sum_{i=1}^{\eta} \rho_i^2 = [y_i - (a + bx_1)]^2 + [y_2 - (a + bx_2)]^2 + \cdots + [y_\eta - (a + bx_\eta)]^2$$
$$= \text{a minimum} \qquad (2.76)$$

or, stated otherwise, $a$ and $b$ must be chosen so that this equation assumes the smallest possible value. This is satisfied when the partial differentials of $\sum \rho_i^2$ with respect to $a$ and to $b$ are zero, i.e.

$$\left(\frac{\partial \sum \rho_i^2}{\partial a}\right)_{b, y_i, x_i} = 0 \quad \text{and} \quad \left(\frac{\partial \sum \rho_i^2}{\partial b}\right)_{a, y_i, x_i} = 0 \qquad (2.77)$$

From (2.76) and (2.77) it can be shown that

$$\eta a + b \sum x_i = \sum y_i$$
$$a \sum x_i + b \sum x_i^2 = \sum x_i y_i \qquad (2.78)$$

These equations are then solved to obtain

$$a = \frac{\left(\sum x_i^2\right)\left(\sum y_i\right) - \left(\sum x_i\right)\left(\sum x_i y_i\right)}{\left(\eta \sum x_i^2\right) - \left(\sum x_i\right)^2} \qquad (2.79)$$

$$b = \frac{\left(\eta \sum x_i y_i\right) - \left(\sum x_i\right)\left(\sum y_i\right)}{\left(\eta \sum x_i^2\right) - \left(\sum x_i\right)^2} \qquad (2.80)$$

If $y$ is given by $y = a + bx + cx^2 + dx^3 + \cdots$ then there are as many derivatives represented by (2.77) as the number of constants $a, b, c, d, \ldots$ yielding independent equations of the type (2.78) from which the constants can be determined.

**Example** The property "$y$" is measured at various values of "$x$" and tabulated as follows:

y: 0.12  0.20  0.28  0.36
x: 0.10  0.20  0.30  0.40

It is known from theoretical considerations that the relationship between $x$ and $y$ is linear, i.e., $y = a + bx$. Substitution of these values in (2.79) and (2.80) gives

$$a = \frac{0.3 \times 0.96 - 1 \times 0.28}{4 \times 0.3 - 1} = 0.04$$

$$b = \frac{4 \times 0.28 - 1 \times 0.96}{4 \times 0.3 - 1} = 0.8$$

The result is therefore

$$y = 0.04 + 0.8x$$

## Determinants

The process of fitting a set of data with an analytical equation requires solving for the unknown coefficients of $x, y, z, \ldots$ of a set of *simultaneous linear equations*. For brevity we consider two such equations as follows:

$$\begin{Bmatrix} a_1 x + b_1 y = k_1 \\ a_2 x + b_2 y = k_2 \end{Bmatrix} \tag{2.81}$$

An "array" shown by

$$D = \begin{vmatrix} a_1 & b_1 \\ a_2 & b_2 \end{vmatrix} \tag{2.82}$$

is called the "determinant of the coefficients" of (2.81). The determinant (2.82) is of the second order, and contains two rows, two columns, and four elements. We denote $D_1$ and $D_2$ as the determinants obtained by replacing the first and second columns of $D$ by $k_1$ and $k_2$, i.e.

$$D_1 = \begin{vmatrix} k_1 & b_1 \\ k_2 & b_2 \end{vmatrix} \quad \text{and} \quad D_2 = \begin{vmatrix} a_1 & k_1 \\ a_2 & k_2 \end{vmatrix} \tag{2.83}$$

The solution of simultaneous equations (2.81) is given by

$$x = D_1/D \quad \text{and} \quad y = D_2/D \tag{2.84}$$

The rule for expansion of $D$ and $D_1$ is

$$D = a_1 b_2 - a_2 b_1, \quad D_1 = k_1 b_2 - k_2 b_1, \text{ etc.} \tag{2.85}$$

A determinant of the third order is

$$D = \begin{vmatrix} a_1 & b_1 & c_1 \\ a_2 & b_2 & c_2 \\ a_3 & b_3 & c_3 \end{vmatrix} = a_1 \begin{vmatrix} b_2 & c_2 \\ b_3 & c_3 \end{vmatrix} - b_1 \begin{vmatrix} a_2 & c_2 \\ a_3 & c_3 \end{vmatrix} + c_1 \begin{vmatrix} a_2 & b_2 \\ a_3 & b_3 \end{vmatrix} \tag{2.86}$$

where the right side shows the rule for expansion in terms of the elements of the first row. Equations (2.85) and (2.86) give the rule for the numerical evaluation of the determinants. The determinant which is the coefficient of $a_1$ in (2.86) is obtained by striking out the row and the column on which $a_1$ is located, and similarly for the remaining determinants on the right. The determinants $D_1$, $D_2$ and $D_3$ are obtained by replacing the first, second and third columns with $k_1$, $k_2$ and $k_3$ respectively as in (2.83). The values of the three unknowns are then $D_1/D$, $D_2/D$ and $D_3/D$ since a third order determinant is for three simultaneous equations. Simple computer programs are available for solving $n$ equations with $n$ unknowns. For hand-held calculators, see for example Hewlett-Packard "Solve Equation Library" for 48G and 48GX).

## Useful Series

The power series for one independent variable is

$$u = a_0 + a_1 x + a_2 x^2 + \cdots + a_n x^n \tag{2.87}$$

The values of $a_i$ may be determined by successive differentiations of $u$ with respect to $x$ and then setting $x$ to zero; thus,

$$u(\text{at } x = 0) = u(0) = a_0, \quad \frac{du}{dx}(\text{at } x = 0) = u'(0) = a_1,$$

$$\frac{d^2 u}{dx^2}(\text{at } x = 0) = u''(0) = 2 a_2, \text{ etc.} \tag{2.88}$$

As a result, we may rewrite (2.87) as follows:

$$u = u(0) + u'(0) x + \frac{u''(0)}{2!} x^2 + \frac{u'''(0)}{3!} x^3 + \cdots. \tag{2.89}$$

The power series in this form is known as the Maclaurin series. The general form of (2.89) is called the Taylor series, which is the expansion of a function in powers of $x - c$, i.e.,

$$u = a + a_1 (x - c) + a_2 (x - c)^2 + \cdots. \tag{2.90}$$

The coefficients $a_i$ are determined similarly so that

$$u = u(c) + u'(c)(x - c) + \frac{u''(c)}{2!}(x - c)^2 + \frac{u'''(c)}{3!}(x - c)^3 + \cdots. \tag{2.91}$$

where the successive derivatives are evaluated at $x = c$. When $c = 0$ the Taylor series become identical with the Maclaurin series. Additional useful series are

$$e^x = 1 + x + \frac{x^2}{2!} + \frac{x^3}{3!} + \cdots. \tag{2.92}$$

$$(x + a)^n = x^n + \frac{n x^{n-1}}{1!} a + \frac{n(n-1)}{2!} x^{n-2} a^2$$

$$+ \frac{n(n-1)(n-2)}{3!} x^{n-3} a^3 + \cdots + a^n \tag{2.93}$$

$$\ln(1 + x) = x - \frac{x^2}{2} + \frac{x^3}{3} + \frac{x^4}{4} + \cdots + (-1)^{n-1} \frac{x^n}{n}; [-1 < x \leq 1] \tag{2.94}$$

The limits $-1 < x < 1$, for which (2.94) is useful, are called the "limits of convergence" of the series. The most convenient and easy test for convergence is the ratio test which requires that the ratio of $(n + 1)$th term to the $n$th term must be smaller than $|1|$ as $n$ approaches infinity.

The succeeding chapters contain numerous applications of the foregoing equations.

## PROBLEMS

2.1 Give the total differentials of the following equations

(a) $u = \dfrac{xyz}{x+y+z}$

(b) $u = \dfrac{x}{y} + \dfrac{y}{z} + \dfrac{z}{x}$

(c) $u = e^{x+y^2}$

(d) $u = (xy + yz + x^2z)^{1/3}$

(e) $u = x \ln y + y \ln x$

(f) $z = \dfrac{x+y}{\ln xy}$

2.2 For $u = (x/y) + (y/x)$, express $(\partial y/\partial x)_u$ in terms of $x$ and $y$ by means of (2.7).

2.3 From $u = x^2 + xy + y^2$, (a) express and $\partial u/\partial x)_y$ and $(\partial u/\partial y)_x$ and (b) show that (2.10) is satisfied.

2.4 Find $(\partial x/\partial y)_u$ and $(\partial u/\partial x)_y$ for $u^2 + x^2 + uxy = 0$ at $x = -y = 2$.

2.5 If $G = f_1(P, T)$ and also $G = f_2(V, T)$ show that

$$\left(\frac{\partial V}{\partial P}\right)_T \left(\frac{\partial T}{\partial G}\right)_P = \left(\frac{\partial V}{\partial P}\right)_G \left(\frac{\partial T}{\partial G}\right)_V$$

2.6 Find the limits of $(e^{-x} - 1) \ln x$, $x/(a^x - 1)$, $e^{-x}x$, $x^{-1} \ln(x + 1)$ when $x \to 0$ and $x \to \infty$.

2.7 Which of the following expressions are homogeneous in $x$ and $y$, and which are homogeneous in $x$, $y$ and $z$; what are the degrees of homogeneity?

(a) $x^2 + y^2 + z^2$,

(b) $2xyz + 3yz^2 + z^3$,

(c) $3x^3y^2z + 6x^2y^3 - x^4yz^2$

(d) $e^{(x/y^3)+(x^2/y^3)}$,

(e) $3yz + xyz^2 - yz^4$,

(f) $x \ln x + (1 - x) \ln(1 - x)$

2.8 For $u = (x^3/y) + (y^3/x) + xy$, show that (a) $(\partial u/\partial x)_y$ and $(\partial^2 u/\partial x^2)_y$ are homogeneous functions of first and zeroth degree in $x$ and $y$, and (b) verify that

$$x\left(\frac{\partial^2 u}{\partial x^2}\right)_y + y\left(\frac{\partial^2 u}{\partial x \partial y}\right) = \left(\frac{\partial u}{\partial x}\right)_y.$$

2.9 Show that for $u = x^2y^2/(x^2 + y^2)$ the derived functions $\partial^2 u/\partial x^2$, $\partial^2 u/\partial y^2$ and $\partial^2 u/\partial x \partial y$ are homogeneous functions of zeroth degrees in $x$ and $y$ and that they may be expressed as functions of one independent parameter $t = x/y$. Observe that $u$ is symmetrical in $x$ and $y$, therefore $(\partial u/\partial y)_x$ may be obtained from $(\partial u/\partial x)_y$, vice versa, by interchanging $x$ and $y$.

2.10 Repeat problem 2.9 for $u = (x^3/y) + (y^3/x) + xy$.

2.11 For a two-component system certain partial properties are

$$\overline{G}_1 = a(1 - X_1)^2 \quad \text{and} \quad \overline{G}_2 = bX_1^2$$

where $X_1$ is the independent composition variable in mole fraction, and $a$ and $b$ are constants at constant pressure and temperature.

(a) Show that the Gibbs-Duhem relation is satisfied when $a = b$.

(b) Express the molar property $G$ in terms of $a$ and $X_1$.

2.12 The partial properties of a two-component phase are represented by $\overline{G}_1 = a \ln X_1 + bX_2$ and $\overline{G}_2 = a \ln X_2 + cX_1$ where $a, b$ and $c$ are constants. By means of the

## 34  Thermodynamics

Gibbs-Duhem relation and by eliminating either $X_1$ or $X_2$, find the numerical values of $b$ and $c$. Hint: The resulting equation must be satisfied for all possible values of $X_1$ or $X_2 = 1 - X_1$.

2.13 For a phase consisting of three components the partial properties are expressed by $\overline{G}_1 = a \ln X_1$, $\overline{G}_2 = b \ln X_2$ and $\overline{G}_3 = c \ln X_3$ where $a$, $b$ and $c$ are constants. Show, by Lagrange's method of undetermined multipliers, the relationships among $a$, $b$ and $c$ which satisfy the Gibbs-Duhem relation. Note that the restricting function is $X_1 + X_2 + X_3 = 1$.

2.14 Integrate $dy = (dx/x^2) + (dx/x) + 2x \ln x \, dx + xe^x dx$ from $x = 1$ to $x = 2$. The last two terms can be integrated by parts, i.e., use $2x \ln x \, dx = d(x^2 \ln x) - x \, dx$, for integrating $2x \ln x \, dx$.

Ans.: $y_2 - y_1 = 9.85$

2.15 Integrate $(\partial u/\partial x)_y = x^2 y + x^{0.5} y^2$ from $x = 0$ to $x$, and then (a) find the numerical value of $u(x, y) = u(4, 2)$ assuming that $u(0, 2) = 0$, and (b) compute $(\partial u/\partial y)_x$ at $x = 4$, $y = 2$ if $[\partial u(0, y)/\partial y)]_x = y$.

2.16 (a) Which of the following expressions are exact differentials:

(i)  $xy \, dx + xy \, dy$

(ii) $xy^2 \, dx - x^2 y \, dy$

(iii) $\dfrac{x}{y^2} dx - \dfrac{x}{y^3} dy$

(iv) $\dfrac{dx}{y} - \dfrac{x}{y^2} dy$

(b) Integrate the exact differentials among the preceding expressions from $(x = y = 0)$ to $(x, y)$.

(c) Find the integrating factors for the inexact differentials by trial.

2.17 Evaluate $\oint$ for the expression (i) in Problem 2.16 from $x = 0$, $y = 0$ to $x = 3$, $y = 9$ along $y = x + 6$ and back to $x = 0$, $y = 0$ along $y = x^2$.

2.18 Give the derivations of the following equations for the total differential $dU = T \, dS - P \, dV$:

$$\left(\frac{\partial U}{\partial T}\right)_V = T\left(\frac{\partial S}{\partial T}\right)_V, \quad \left(\frac{\partial U}{\partial T}\right)_P = T\left(\frac{\partial S}{\partial T}\right)_P - P\left(\frac{\partial V}{\partial T}\right)_P$$

$$\left(\frac{\partial T}{\partial V}\right)_S = -\left(\frac{\partial P}{\partial S}\right)_V, \quad \left(\frac{\partial T}{\partial P}\right)_S = \left(\frac{\partial V}{\partial S}\right)_P$$

$$\left(\frac{\partial P}{\partial T}\right)_V = \left(\frac{\partial S}{\partial V}\right)_T, \quad \left(\frac{\partial V}{\partial T}\right)_P = -\left(\frac{\partial S}{\partial P}\right)_T$$

2.19 (a) Show graphically and analytically that $z^2 = x^2 + y^2$ represents two cones on the $x - y - z$ coordinates, symmetrically located on both sides of the $x$-$y$ plane, with their apexes at $x = y = z = 0$, by projecting them on $x$-$y$, $x$-$z$ and $y$-$z$ planes.

(b) Show analytically that the intersection of one of these cones with the surface $x + y = 1$, parallel, to the $z$-axis, is a hyperbola. Find the minimum point in the first octant, i.e., where $x > 0$, $y > 0$ and $z > 0$, for $z^2 = x^2 + y^2$ with the restricting equation $x + y = 1$, by (c) Lagrange's method of undetermined multipliers and (d) analytically.

2.20 For $u = v^2 + x^2 + y^2 + z^2$, when $du = 0$ there is a minimum. Find the values of $v$, $x$, $y$ and $z$ at the minimum for the restricting condition $v + 2x + 3y + 4z = 30$, by the following methods: (a) Lagrange's method of undetermined multipliers, and

(b) by eliminating one of the independent variables in $u$ by means of the restricting equation and then setting the coefficients of the total differentials $dv, dx \ldots$ to zero.

Ans.: $v = 1, x = 2, y = 3, z = 4$.

2.21 The following data represent simultaneous measurements of $x$ and $y$:

$x$: 0.00  1.04  1.52  2.05  2.60  3.11
$y$: 2.02  5.10  6.60  8.21  9.80  11.33

(a) Correlate $x$ and $y$ by a linear equation, using the method of least squares.
(b) Compute $\bar{a}, \sigma, \sigma°$ and $\rho_\varepsilon$. [cf. 2.71–2.74].

2.22 Derive the relationships similar to (2.78) for the representation of a set of data by means of $y = a + bx + cx^2$.

2.23 Solve the following equations by using their determinants:

$$2x - y + z = 8; \quad x + y - 3z = -8; \quad 5x + 7y - z = 0$$

Ans.: $x = 2, y = -1, z = 3$.

CHAPTER **III**

# THE FIRST LAW OF THERMODYNAMICS

The first law of thermodynamics is a special consequence of the law of conservation of energy in physics. We state the law of conservation of energy in the following simple form: *A given amount of a particular type of energy may be converted into another type, and the process may be reversed to obtain the initial type of energy without destroying or augmenting any portion of the initial amount.* The historical and outmoded enunciation of the first law as "the perpetual motion of the first kind is impossible, i.e., it is impossible to construct an engine that does work without using any form of energy" is not *incorrect*, but it is unsatisfactory to the thermodynamicist. We enunciate the first law of thermodynamics in the following simple mathematical form:*

> *There is a function $U$, called the energy of a system, which is a function of its variables of state, and the change in the value of $U$, $\Delta U$, from a given initial state to a final state is always equal to the same amount of energy irrespective of the path followed by the system.*

The function $U$ of a closed system, in which the variables of state may be chosen as the pressure $P$, and volume $V$, is written as

$$U = U(P, V) \tag{3.1}$$

A change in $U$ corresponding to the changes in $P$ and $V$ is

$$\Delta U = U_2 - U_1 = U(P_2, V_2) - U(P_1, V_1) \tag{3.2}$$

The first law requires that $\Delta U$ be always the same for the same values of $P_1$, $V_1$ and $P_2$, $V_2$, irrespective of the path followed by the system in going from $P_1$, $V_1$ to $P_2$, $V_2$; hence $U$ is a point function and $dU$ is an exact differential. An important consequence of the first law is expressed by

$$\Delta U = Q + W + W' + W'' + \cdots \tag{3.3}$$

---

*See Gen. Refs. (4, 22, 27) and R. A. Alberty and R. J. Silbey, "Physical Chemistry," John Wiley and Sons, Chapters 1–3 (1992); J. M. Smith and H. C. Van Ness, "Introduction to Chemical Engineering Thermodynamics," McGraw-Hill (1987).

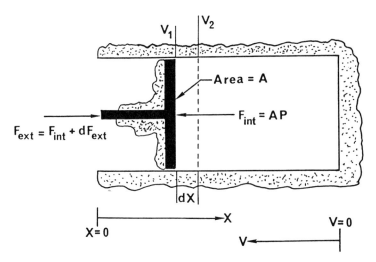

**Fig. 3.1** Work of compression on an adiabatically enclosed system. $F_{ext}$ is $F$ in (3.5).

or for an infinitesimal change in $U$,

$$dU = dQ + dW + dW' + dW'' + \cdots \tag{3.4}$$

where $dW$ is the mechanical work done on the system, and the remaining terms are other forms of energy, all positive in their numerical values when received by the system and negative when lost from the system. The differential $d$ is used before $Q$ and $W$ to emphasize that $dQ$ and $dW$ are inexact differentials.

### Work of Compression and Expansion

An exchange of work with the surroundings occurs when a system is deformed by the application of external forces as shown in Fig. 3.1. The work of compression results in an increase of $U$, but the work of expansion causes a decrease in $U$. Hence, $\Delta U$ and $W$ are both positive for the work of compression, and negative for the work of expansion.

We consider a system such as a gas or any readily compressible substance contained in a cylinder with a piston as shown in Fig. 3.1. If (a) the piston is idealized to move without friction, and (b) the external force, $F_{ext}$, applied on the piston is infinitesimally greater than the counter-acting force of the system, i.e., $F_{ext} = AP + dF_{ext}$, where the force exerted by the system is the pressure $P$ times the cross-sectional area $A$, $AP$, and (c) the piston moves very slowly, then the resulting work is called "reversible." The work $dW$, when the force $F_{ext}$ moves the cylinder by an infinitesimal distance $dx = x_2 - x_1$, is given by

$$dW = F_{ext}(x_2 - x_1) = F_{ext}\,dx = AP\,dx + dF_{ext}\,dx = AP\,dx \tag{3.5}$$

where $dF_{ext}\,dx$ is negligible since it is an infinitesimal of the second order. We note that $dx$ is positive since $x_2 > x_1$ on the selected scale for $x$, but positive for $V$ from the right to the left so that $Ax_2 - Ax_1 = -(V_2 - V_1)$ because $V_2 < V_1$ while $Ax_2 > Ax_1$. We denote $(V_2 - V_1)$ by $dV$ when the difference is infinitesimal, and substitute the result into (3.5) to obtain

$$dW = -P\,dV \tag{3.6}$$

The negative sign in this equation is the result of the fact that when $x$ increases, $V$ decreases. It may similarly be shown that this equation is also valid for the work of expansion. The change in $U$ for an adiabatically enclosed system, subjected to the work of expansion or compression, is then

$$dU = -P\,dV \tag{3.7}$$

The system is always uniform in pressure, temperature and composition during a reversible compression or expansion.

## Heat

We wish to examine a set of three adiabatic systems joined by rigid diathermic walls as in Fig. 1.2 in which one of the systems (e.g. III) may be a constant pressure thermometer so that $T = f(V)$. The mass of the thermometer may be made very small without affecting its accuracy. It is an experimental fact that when work is done on one of the systems by a piston, i.e., when $dU = dW > 0$, the temperature of the system rises, and when the system does work by expanding, i.e., $dU = dW < 0$, the temperature decreases. If System I receives work not only its temperature increases but the temperature of the second system also increases, hence System II receives energy from System I through the diathermic wall. The energy thus received by System II from System I is called "heat." The set of two systems is adiabatic; hence, the sum of the heat leaving System I, $dQ^I$, and the heat acquired by System II, $dQ^{II}$, is zero, i.e.,

$$dQ = 0 = dQ^I + dQ^{II}; \qquad [dQ^{III} \approx 0 \text{ for small III}]$$

where $dQ$ refers to the set of Systems I and II, (III is negligibly small) and since the set is adiabatically enclosed, $dQ$ is zero. The first law for System I may be written as

$$dU^I = dQ^I + dW = dQ^I - P\,dV \tag{3.8}$$

and since the energy of System II is $dU^{II} = dQ^{II}$ the overall energy change is $dU = dU^I + dU^{II} = -P\,dV$ as expected. When the change in $W$ is reversible, $Q^I$ and $Q^{II}$ are also reversible. Both $dQ$ and $dW$ are inexact differentials because they are not functions of the variables of state as will be seen later.

Various forms of energy, $Q$, $W$, $W'$, ... in (3.3) are the *experimentally measurable quantities* that cross the boundary of a system from which $\Delta U$ may be calculated. A number of processes, generally used for the evaluation of $dQ$ and $dW$ are described in the succeeding sections.

### Reversible Processes in Closed Systems

The change in energy, $dU$, of a system confined by rigid walls in diathermic equilibrium with the surroundings is given by (3.8) when $dV$ is zero, i.e.,

$$dU_V = dQ_V, \quad \text{and} \quad \Delta U_V = (U_2 - U_1) = Q_V, \tag{3.9}$$

where the subscript $V$ indicates that the process takes place at a constant volume. Likewise, at a fixed pressure, or under an isobaric condition, $dU$ is given by

$$\boxed{dU_P = dQ_P - P\,dV} \tag{3.10}$$

The isobaric work is then

$$W_P = -\int_{V_1}^{V_2} P\,dV = -P(V_2 - V_1) \tag{3.11}$$

and $\Delta U_P$ is simply

$$\Delta U_P = (U_2 - U_1)_P = Q_P - P(V_2 - V_1) \tag{3.12}$$

When the temperature of the system is held constant the process is called isothermal; $dU$ is then written as

$$dU_T = dQ_T - P_T\,dV_T, \quad \Delta U_T = Q_T - \int_{V_1}^{V_2} P_T\,dV_T \tag{3.13}$$

It is always possible to express $P$ as a function of $V$ at a given temperature and evaluate the integration shown by the second equation. An adiabatic process was defined earlier by the condition that $dQ$ be zero, and the change in $U$ was given by (3.7). Integration of (3.7) may be carried out by using the functional dependence of $P$, $V$ and $T$. Simple examples are provided by gases obeying simple equations of state.

## APPLICATION OF THE FIRST LAW TO IDEAL GASES

### Energy of Ideal Gases

It was shown in Chapter I that the equation of state for gases having sufficiently low boiling points was $PV = RT$ at values of $P$ not much larger than one bar. In general, the lower the boiling point and the lower the pressure, the more closely this equation is obeyed. The gases obeying this equation are called ideal gases. The energy of any closed system may be represented by a function of volume and temperature, i.e., $U = U(V, T)$. Let two adiabatically isolated containers be in diathermic contact with each other as in Fig. 3.2 so that System I contains an ideal gas at a few bars of pressure and System II is completely evacuated. A valve, initially in closed position, connects Systems I and II. By means of an apparatus similar to that shown in Fig. 3.2, Gay-Lussac (1807) and later Joule (1844) found

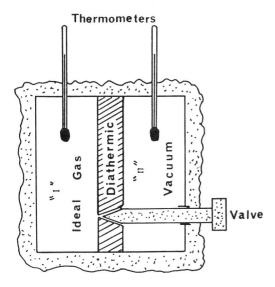

**Fig. 3.2** Irreversible adiabatic expansion of ideal gas into vacuum.

that when the valve was opened and thus the gas was expanded into a vacuum, there was virtually no change in temperature. Although their experiments were not sufficiently precise to detect a small change in temperature, it was found later that the closer the gas obeys $PV = RT$, the smaller is the temperature change. The expansion of such a gas into a vacuum involves no work; hence $W$ is zero, and the gas is adiabatically isolated; therefore $Q$ is also zero. The first law then requires that

$$\Delta U = U(V_2, T) - U(V_1, T) = Q + W = 0 \tag{3.14}$$

where $V_1$ is the initial volume prior to opening the valve and $V_2$ is the volume after expansion, or $V_2$ is the combined volumes of both systems. Equation (3.14) states that the energy of an ideal gas is independent of volume; hence

$$U = U(T); \quad \text{(ideal gas)} \tag{3.15}$$

The equation of state $PV = RT$, and $U = U(T)$ are also applicable to mixtures of ideal gases in which there are no chemical reactions. The partial pressure $P_i$ of a constituent gas in the mixture is defined as the pressure when the gas occupies the entire volume $V$ of the mixture in the absence of other gases. The ideality therefore requires that each constituent gas behave independently of others. Thus, we may write $P_i V = x_i RT$ for each gas and for all the "$c$" components in one mole of mixture with $x_i$ as the mole fraction of $i$

$$\sum_{i=1}^{c} P_i V = V \sum_{i=1}^{c} P_i = PV = RT \sum x_i = RT \tag{3.16}$$

where $P$ is the sum of all $P_i$; this is the classical statement of Dalton's law. Further,

$$x_i = \frac{P_i}{P} \tag{3.17}$$

The energy $U$ of the gas mixture is also the sum of all $U_i$, so that $U = U_1(T) + U_2(T) + \cdots$ where $T$ is the temperature of the mixture.

## Heat Capacity

The heat capacity $C_V$ of any substance at constant volume is defined by

$$C_V = \left(\frac{\partial U}{\partial T}\right)_V = \left(\frac{\partial Q}{\partial T}\right)_V \tag{3.18}$$

where $(dU)_V = (dQ)_V$, because $dW$ at constant volume is always zero. For an ideal gas, $(dU)_V = dU$ because $U$ is a function of $T$ only and it is not necessary to impose the constant volume restriction on $dU$; therefore,

$$C_V = \frac{dU}{dT}; \quad \text{(ideal gas)} \tag{3.19}$$

In experimental measurements of $C_V$ for any substance, including an ideal gas, $Q$ is the measured energy input and it is essential that this input be carried out at a fixed volume as required by (3.18). According to the kinetic theory of gases, which will be discussed later, the value of $C_V$ is independent of volume and temperature for ideal gases, and for a monatomic ideal gas, $C_V = 3R/2$ and for a diatomic ideal gas, $C_V = 5R/2$.

The heat capacity of any substance at constant pressure is designated by $C_p$ and defined by

$$C_p = \left(\frac{\partial Q}{\partial T}\right)_P = \left(\frac{dU + P\,dV}{dT}\right)_P = \left(\frac{\partial U}{\partial T}\right)_P + P\left(\frac{\partial V}{\partial T}\right)_P \tag{3.20}$$

For an ideal gas, $(\partial U/\partial T)_P = dU/dT = C_V$ and $P(\partial V/\partial T)_P = R$; it follows, therefore, that

$$C_p = C_V + R; \quad \text{(ideal gas)} \tag{3.21}$$

Hence, $C_p = 5R/2$ for an ideal monatomic gas, and $C_p = 7R/2$ for an ideal diatomic gas. In all the preceding equations in this section $U$ is the molar energy, i.e., energy per mole, and $C_V$ and $C_p$ are the molar heat capacities, although for brevity they are always called the "heat capacities."

## Processes with Ideal Gases

An infinitesimal change in $U$ of one mole of ideal gas is given by

$$\boxed{dU = C_V\,dT = dQ - P\,dV} \tag{3.22}$$

where we have only the work of compression or expansion. Integration of this equation gives

$$\Delta U = C_V(T_2 - T_1) = Q - \int_{P_1,V_1}^{P_2,V_2} P\, dV \tag{3.23}$$

The integral is zero if $Q$ is received by the system at a constant volume. If a process is carried out at constant pressure the resulting work is $-P(V_2 - V_1)$ according to (3.11) and since $PV_2 = RT_2$ and $PV_1 = RT_1$, then

$$W_p = -R(T_2 - T_1) \tag{3.24}$$

The change in the energy at constant pressure is

$$dU = C_V\, dT = dQ_p - R\, dT_p \tag{3.25}$$

From this equation and from the definition of the heat capacity at constant pressure, $C_p$, where $Q_p = C_p(T_2 - T_1)$, it is evident that $dU = C_p\, dT - R\, dT_p$, and $\Delta U = C_p(T_2 - T_1) - R(T_2 - T_1) = C_V(T_2 - T_1)$. For an isothermal process $dU = 0$, or $\Delta U = 0$ and

$$dW_T = -dQ_T = -P\, dV_T \tag{3.26}$$

Substituting for pressure, $P = RT/V$, (3.26) becomes

$$dW_T = -RT(dV/V)$$

and then integration gives

$$W_T = -RT \int_{V_1}^{V_2} \frac{dV}{V} = -RT \ln \frac{V_2}{V_1} \tag{3.27}$$

or, considering that $P_1 V_1 = P_2 V_2$ for an isothermal process,

$$W_T = -RT \ln \frac{V_2}{V_1} = RT \ln \frac{P_2}{P_1} \tag{3.28}$$

For a reversible adiabatic process $dQ = 0$, and the equation for $dU$ is

$$dU = C_V\, dT = -P\, dV = -RT(dV/V) \tag{3.29}$$

Separation of the variables gives

$$C_V(dT/T) = -R(dV/V)$$

and integration from $V_1, T_1$ to $V_2$ and $T_2$ yields

$$C_V \ln \frac{T_2}{T_1} = -R \ln \frac{V_2}{V_1}, \quad \text{or} \quad \frac{T_2}{T_1} = \left(\frac{V_1}{V_2}\right)^{R/C_V}, \quad \text{(adiabatic)} \tag{3.30}$$

Since $P_1 V_1 / P_2 V_2 = T_1/T_2$ and $V_1/V_2 = T_1 P_2 / T_2 P_1$, (3.30) becomes

$$\frac{T_2}{T_1} = \left(\frac{P_2}{P_1}\right)^{R/C_p} \tag{3.31}$$

## 44    Thermodynamics

and similarly the elimination of $T$ from (3.30) or (3.31) gives

$$\frac{P_2}{P_1} = \left(\frac{V_1}{V_2}\right)^{C_p/C_V} \tag{3.32}$$

$$P_1(V_1)^{C_p/C_V} = P_2(V_2)^{C_p/C_V} = \text{Constant}, \quad \text{(adiabatic)} \tag{3.33}$$

The change in energy, is equal to work as shown by (3.29); integration of this equation yields

$$\Delta U = C_V(T_2 - T_1) = W \tag{3.34}$$

or in terms of $T_1$, $P_1$ and $P_2$

$$\Delta U = C_V T_1 \left[\left(\frac{P_2}{P_1}\right)^{R/C_p} - 1\right] \tag{3.35}$$

***Example***   The following reversible processes are carried out with one mole of argon: (I) Adiabatic compression from 1 bar and 298 K to 10 bars. (II) Isothermal compression from 1 bar (100,000 Pa) and 298 K to 10 bars and then isobaric expansion to the final state as in the preceding adiabatic compression. Compute the values of $\Delta U$, $Q$ and $W$ for each process, assuming that the gas is ideal, $C_V = 1.5R$ and $C_p = 2.5R$.

(I) For the adiabatic compression, $Q$ is zero and the final temperature from (3.31) is

$$T_2 = 298(10)^{2/5} = 748.5 \text{ K}$$

where $R/C_p = 2/5$, since $C_p = 2.5R$. The change in $U$ and $W$ is

$$\Delta U = W = 1.5R(748.5 - 298) = 675.8R = 5619 \text{ Joules}$$

(II) The isothermal compression to 10 bars requires

$$W = -Q = -RT \ln \frac{P_1}{P_2} = 5705 \text{ J}$$

$\Delta U$ for this process is obviously zero. The subsequent isobaric expansion is accomplished by heating reversibly from 298 to 748.5 K. The accompanying change in the energy is

$$\Delta U = \int_{298}^{748.5} C_V \, dT = 1.5R(748.5 - 298) = 5619 \text{ J}$$

The work $W_p$ and the heat absorbed, $Q_p$ are

$$W_p = -R(748.5 - 298) = -3746 \text{ J}$$

$$Q_p = \Delta U - W_p = 5619 + 3746 = 9365 \text{ J}$$

Total $Q$ and $W$ for the isothermal compression plus the isobaric expansion are

$$Q = -5705 + 9365 = 3660 \text{ J}$$

$$W = 5705 - 3746 = 1959 \text{ J}$$

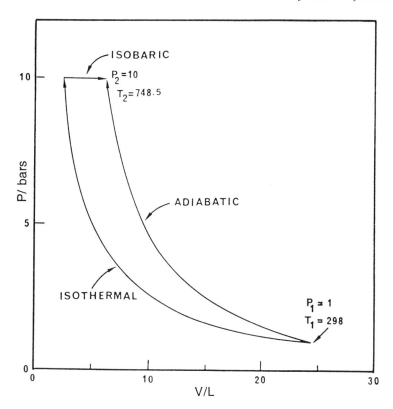

**Fig. 3.3** Reversible isothermal compression and isobaric expansion to $P_2$ and $T_2$. Change in energy of system is identical when reversible adiabatic compression also brings enclosed gas to $P_2$ and $T_2$. System is one mole of argon.

The overall $\Delta U$ in Part II is the same as in Part I because the initial and the final states are the same. The paths followed by all the processes are shown in Fig. 3.3, on a $P$-$V$ diagram.

## Simple Kinetic Theory of Ideal Gases

Thermodynamics of ideal gases can be developed by the classical method of experimental observations without theories or molecular hypotheses. However, an elementary knowledge of the kinetic theory of gases greatly enhances the understanding of ideal gases, which play an important role in the development of thermodynamics.

According to the kinetic theory of gases, the molecules of an ideal gas are point particles randomly moving in space without exerting attractive or repulsive forces on one another. Consider one mole of an ideal monatomic gas consisting of $N_A$ molecules and occupying $V$ cm$^3$. An isolated cube of 1-cm shown in Fig. 3.4

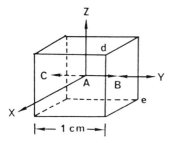

**Fig. 3.4** A one-centimeter cube containing an ideal gas.

contains $N_A/V$ molecules of this gas. Let $m$ be the mass of each molecule and $u$ its average velocity and assume that one third of all molecules move in each of the $x$, $y$, and $z$ directions. A molecule at "A" may move toward "B" and collide with the surface at "B" and then rebound and travel to "C" and return to "A" again to repeat its journey. In this process, each molecule travels a distance of 2 cm for each collision on the selected surface $dBe$.

When a molecule rebounds from a surface its average velocity changes from $+u$ to $-u$ and the change of its momentum at the surface becomes $mu - (-mu) = 2mu$. According to Newton's laws of motion, the force exerted on a surface by a moving body is equal to the rate of change of momentum and since this force acts on $dBe$ which has an area of 1 cm², it also represents the pressure. The time necessary for each collision is $(2 \text{ cm})/(u \text{ cm} \cdot \text{sec}^{-1})$, therefore the rate of change of momentum is $2mu \div (2/u) = mu^2$, which is the pressure due to each molecule. For all the molecules moving in the direction of $y$, i.e., $N_A/(3V)$ the pressure is $P = [V_A/(3V)]mu^2$. Multiplying both sides by $V$ and observing that $PV = RT$, we obtain

$$PV = (N_A/3)mu^2 = (2N_A/3)(mu^2/2) = RT \tag{3.36}$$

This equation shows that the temperature is directly proportional to the kinetic energy of molecules.

The kinetic theory states that the internal energy of a monatomic gas consists entirely of the kinetic energy of its molecules $E_{\text{kin}}$, and an increase in temperature simply causes a corresponding increase in the velocity, hence, an increase in the kinetic energy of molecules; therefore,

$$E_{\text{kin}} = N_A(mu^2/2) = 1.5RT \tag{3.37}$$

It is evident from this equation that $C_V = 3R/2$. According to the principle of equipartition of energy in statistical mechanics, the energy of a monatomic gas in (3.37) may be regarded as consisting of $0.5RT$ of contribution from each of the three cartesian directions of molecular motion. Equation (3.37) and $C_V = 3R/2$ appears to be obeyed very closely for large temperature ranges. For example,

$C_V$ of helium is $3R/2$ from about 50 K to 1000 K, and $C_V$ of Ne, Ar, Kr, and monatomic gases of metals is also $3R/2$ for wide ranges of temperature.

A diatomic molecule may, in addition to the translational motion, rotate with respect to the two axes perpendicular to the line joining the atoms or vibrate along the line joining the atoms. The principle of equipartition of energy states that rotation about each axis contributes $0.5RT$ to the energy. On the other hand, the energy of vibration consists of kinetic and potential terms, each of which contributes $0.5RT$. At ordinary temperatures only rotation is permissible, therefore, the energy of a diatomic gas is

$$U = E_{\text{kin}} + E_{\text{rot}} = (3/2)RT + (2/2)RT = (5/2)RT; \quad \text{(diatomic)} \quad (3.38)$$

from which $C_V = 5R/2$. At high temperatures the rotation and vibration are both permissible so that $U = E_{\text{kin}} + E_{\text{rot}} + E_{\text{vib}} = 7RT/2$ and $C_V = 7R/2$. According to the kinetic theory, the change in $C_V$ from $5R/2$ to $7R/2$ may occur at a particular temperature but this actually occurs over a range of temperature. At very low temperatures, only translational motion is permissible; for example, the energy of hydrogen is the same as that of a monatomic gas, and $C_V = 3R/2$ in the neighborhood of 50 K. For hydrogen at ordinary temperatures $C_V = 5R/2$; at 1000 K, $C_V \approx 2.67R$; and at 2500 K, $C_V \approx 3.27R$, increasing almost linearly with increasing temperature above 1000 K.

The description given here is rather simple and there are notable exceptions such as diatomic iodine gas. Appropriate motions that can be assigned to a given molecule require precise interpretations of molecular spectra, and other relevant data beyond the scope of this text.

## REAL GASES, LIQUIDS AND SOLIDS

Real gases deviate considerably from ideality with increasing pressure and decreasing temperature. Over one hundred equations of state have been proposed to correlate the relationships among $P$, $V$ and $T$ for real gases, but we shall be concerned with a limited number of such equations most frequently encountered in thermodynamics. The procedure for obtaining the equation of state for liquids and solids is largely empirical in nature and requires specific data for each substance.

### Compressibility Factor

The compressibility factor, $Z$, is a dimensionless empirical factor defined by

$$Z = PV/RT \quad (3.39)$$

It varies as a function of pressure and temperature and represents a measure of deviation from ideality. With decreasing pressure, $Z$ approaches unity and the gas becomes ideal. At low temperatures, $Z$ first decreases and then increases with increasing pressure after having reached a minimum; however, with increasing temperature the minimum disappears as shown in Fig. 3.5 where $Z$ for hydrogen

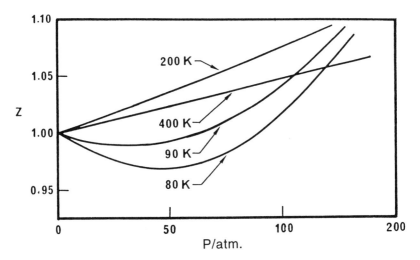

**Fig. 3.5** Compressibility factor $Z$ for hydrogen, based on Gen. Ref. (6).

is plotted versus pressure. The temperature at which $Z = 1$ from $P = 0$ to a few bars is called the Boyle temperature, and in this limited region $\Delta Z/\Delta P = 0$. For $H_2$, the Boyle temperature is 107 K.

The values of $Z$ for a number of important gases have been tabulated at various pressures and temperatures [Gen. Ref. (6)]. For example, $Z = 1 - 0.00486P - 26 \times 10^{-6} P^2$ for $CO_2$ at 300 K and up to 40 bars. The isothermal work of compression from $P_1$ to $P_2$ can be calculated by eliminating $V$ with $V = ZRT/P$ in $dW = -P\,dV$ and integrating it; the result is

$$W = -\int_{P_1}^{P_2} P\,dV = 300R\left[\ln\frac{P_2}{P_1} + 13 \times 10^{-6}\left(P_2^2 - P_1^2\right)\right] \tag{3.40}$$

The isobaric compression can be obtained by first expressing $Z$ as a function of temperature as $Z = 1 - P(\alpha/T) - P^2(\beta/T^2) - \ldots$, and then integrating $dW = -P\,dV$. It will be seen later in this chapter that the coefficients of $P$ in this equation are closely related to the virial coefficients.

### van der Waals Equation

van der Waals (1873), using theoretical considerations, modified the ideal gas equation to make it applicable to real gases at moderately high pressures. His equation is relatively simple and widely used to express the behavior of real gases in the absence of extensive accurate data. For one mole of a gas, the van der Waals equation is

$$\left(P + \frac{a}{V^2}\right)(V - b) = RT \tag{3.41}$$

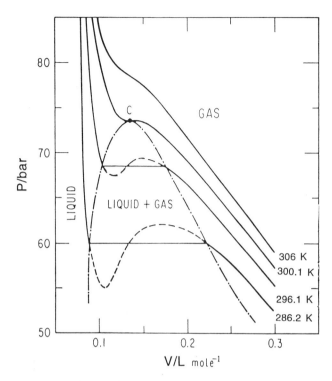

**Fig. 3.6** van der Waals isotherms for carbon dioxide near critical point C. Broken lines correspond to physically nonexisting states. Gas and liquid phases coexist in area inside curve shown by dot-dash line. Horizontal straight lines connect coexisting liquid and gas, called a tie lines.

where "$a$" and "$b$" are constants for a particular gas. The term $a/V^2$ accounts for the intermolecular attractive forces, which increases with decreasing temperatures and increasing pressures, and accounts for the liquefaction of gases. The correction represented by "$b$" arises from the fact that the molecules of a gas have a finite volume which is negligible at low pressures but not at high pressures. The inherent weakness of (3.41) is that both "$a$" and "$b$" are not truly constants but vary with pressure and temperature, therefore (3.41) is approximate in nature but the results obtained from it are often fairly close, and further, (3.41) is simple and convenient in mathematical handling of related thermodynamic equations.

The van der Waals equation is cubic in $V$, and the isotherms of this equation, represented in Fig. 3.6 for carbon dioxide, may have one to three real values of $V$ for each chosen set of $P$ and $T$. For the second curve from the top, the temperature is such that at the pressure $P_c$, the three values of $V$ are identical and the curve has a horizontal inflection point. The point is called the critical point and $P_c$, $V_c$, and

50   Thermodynamics

**Table 3.1**   Critical point data and van der Waals constants.*

| Formula | $T_c$/K | $P_c$/bar | $V_c$/cc mol$^{-1}$ | $10^{-6} a$/cc$^2$ bar mol$^{-1}$ | $b$/cc mol$^{-1}$ |
|---|---|---|---|---|---|
| He    | 5.3   | 2.29   | 57.6 | 0.0346 | 23.7 |
| H$_2$ | 33.3  | 12.97  | 65.0 | 0.247  | 26.6 |
| Ne    | 44.5  | 26.24  | 41.7 | 0.218  | 17.6 |
| N$_2$ | 126.1 | 33.94  | 90.0 | 1.41   | 39.1 |
| CO    | 134.0 | 35.46  | 90.0 | 1.51   | 39.9 |
| A     | 151   | 48.6   | 75.2 | 1.36   | 32.2 |
| O$_2$ | 154.4 | 50.4   | 74.4 | 1.38   | 31.8 |
| CH$_4$| 190.7 | 46.4   | 99.0 | 2.59   | 42.7 |
| CO$_2$| 304.2 | 74.0   | 95.7 | 3.64   | 42.7 |
| NH$_3$| 405.6 | 113.0  | 72.4 | 4.23   | 37.1 |
| H$_2$O| 647.2 | 220.6  | 45.0 | 5.53   | 30.5 |

*Adapted from: J. O. Hirschfelder, C. F. Curtiss, and R. B. Bird, "Molecular Theory of Gases and Liquids," John Wiley & Sons, Inc., New York 1964.

$T_c$ are called the critical pressure, molar volume and temperature respectively. To obtain the critical point, (3.41) is first written as

$$P = \frac{RT}{V-b} - \frac{a}{V^2} \qquad (3.42)$$

and then differentiated in succession to obtain

$$\left(\frac{\partial P}{\partial V}\right)_T = -\frac{RT}{(V-b)^2} + \frac{2a}{V^3} \qquad (3.43)$$

$$\left(\frac{\partial^2 P}{\partial V^2}\right)_T = \frac{2RT}{(V-b)^3} - \frac{6a}{V^4} \qquad (3.44)$$

At the critical point, (3.43) and (3.44) become zero and they may be solved with (3.42) to obtain

$$P_c = \frac{a}{27b^2}, \qquad V_c = 3b \quad \text{and} \quad T_c = \frac{8a}{27Rb} \qquad (3.45)$$

For the isotherms above the one for $T_c$, there is only one value of $V$ for each value of $P$ but for those below $T_c$, there are three distinct values of $V$ for the range of pressure determined by the maximum and minimum points on each curve. At temperatures below $T_c$ the gas may begin to liquefy when the pressure is larger than $P_c$, and the gas and the liquid follow the horizontal lines instead of the hypothetical broken curves, which do not represent the actual behavior of the system. The critical temperature may be estimated roughly, in the absence of any data, from the Guldberg-Waage rule expressed by $T_c \approx 1.6 T_b$ where $T_b$ is the boiling point of the substance at one bar. The critical points of some gases are listed in Table 3.1 with constants $a$ and $b$.

The pressure, molar volume and temperature of a gas may be expressed by

$$P = \pi P_c, \qquad V = \varphi V_c \quad \text{and} \quad T = \tau T_c \qquad (3.46)$$

where the dimensionless quantities $\pi, \varphi, \tau$ are called the reduced pressure, volume and temperature respectively. Substitution of these quantities into (3.45) and elimination of $P_c$, $V_c$ and $T_c$, transforms (3.41) into

$$\left(\pi + \frac{3}{\varphi^2}\right)(3\varphi - 1) = 8\tau \qquad (3.47)$$

This equation is called the "reduced equation of state." Any two real gases at the same $\pi, \varphi$ and $\tau$ are in "corresponding states," and (3.47) is a formal expression of the "law of corresponding states" [Gen. Ref. (19)].

The reversible isothermal work for a van der Waals gas is given by

$$W = RT \ln\left(\frac{V_1 - b}{V_2 - b}\right) + a\left(\frac{1}{V_1} - \frac{1}{V_2}\right) \qquad (3.48)$$

The internal energy of this gas is a function of two variables, e.g., $P$, and $T$, therefore the determination of $\Delta U$ for any process requires a knowledge of variation of $U$ with any set of two variables.

## Other Equations of State

Another equation of state with thermodynamic applications is the Berthelot equation, which can be derived from (3.41) by expanding and rearranging it so that

$$PV = RT + Pb - \frac{a}{V} + \frac{ab}{V^2} \qquad (3.49)$$

It is known experimentally that if "$a$" is taken as inversely proportional to temperature, i.e., $a = A/T$, a better representation of $P$-$V$-$T$ data is possible. The term $ab/V^2$ may be neglected since it has the product of two small quantities "$a$" and "$b$" provided that $V$ is not too small. Replacing "$a$" by $A/T$, and $1/V$ by its approximate equivalent $P/RT$ gives the Berthelot equation, i.e.,

$$PV = RT\left(1 + \frac{Pb}{RT} - \frac{AP}{R^2T^3}\right) \qquad (3.50)$$

The values of "$A$" and "$b$" may be determined in terms of $P_c$, $V_c$, and $T_c$ as described for the constants of the van der Waals equation by using $RT = (P + A/TV)(V - b)$. The resulting expressions are then modified slightly to provide a better fit so that $A = 16 P_c V_c^2 T_c/3$ and $b = V_c/4$. Substitution of these quantities with $R = 32 P_c V_c/9 T_c$ into (3.50) gives a much more useful form of the Berthelot equation:

$$PV = RT\left[1 + \frac{9}{128} \cdot \frac{PT_c}{P_c T}\left(1 - 6\frac{T_c^2}{T^2}\right)\right] \qquad (3.51)$$

**Table 3.2** Selected values of second and third virial coefficients* at 25°C.

| Formula | $B_2$/cc mol$^{-1}$ | $B_3$/cc$^2$ mol$^{-2}$ | Formula | $B_2$/cc mol$^{-1}$ | $B_3$/cc$^2$ mol$^{-2}$ |
|---|---|---|---|---|---|
| $H_2$ | +14.1 | 350 | Ar | −15.8 | 1,160 |
| He | +11.8 | 121 | CO | −8.6 | 1,550 |
| $N_2$ | −4.5 | 1,100 | $CO_2$ | −123.6 | 4,930 |
| $O_2$ | −16.1 | 1,200 | $NH_3$ | −261 | — |

*Adapted from J. H. Dymond and E. B. Smith, "The Virial Coefficients of Pure Gases and Mixtures," Oxford University Press, England (1980). See also H. V. Kehiaian in Gen. Ref. 25.

The virial equation of state (virial = force) proposed by Kammerlingh–Onnes (1901) is a power series in $1/V$, i.e.,

$$PV/B_1 = 1 + (B_2/V) + (B_3/V^2) + \cdots = Z \tag{3.52}$$

where $B_1 = RT$, $B_2$, $B_3$, ... are the first, second, third, ... virial coefficients respectively. Selected values of $B_2$ and $B_3$ are in Table 3.2. An equation of state as a power series in $P$ is

$$PV/RT = 1 + b_2 P + b_3 P^2 + \cdots = Z \tag{3.52a}$$

where $b_2 \approx B_2/RT$ and $b_3 \approx (B_3 - B_2^2)/R^2T^2$ for the quadratic forms of (3.52) and (3.52a) as will be seen in Chapter IX. The coefficients $B_2$, $B_3$, ... are usually expressed as power series in $T^{-1}$, e.g., $B_i = B_\circ + \alpha T^{-1} + \beta T^{-2} + \cdots$, as suggested by J. A. Beattie and O. C. Bridgeman (1927). These equations will be discussed in greater detail in Chapter IX* in conjunction with fugacities.

*Example* We use the data in Table 3.2 for $CO_2$ to obtain an equation for $Z$ as a function of $P$ in bars. Thus, $b_2 = B_2/RT = -123.6 \times 10^{-6}/(298.15 \times 8.3145) = -4.99 \times 10^{-8}$. (Note that cc has been converted into m$^3$). The value of $B_3$ is $4930 \times 10^{-12}$, and $b_3 = (B_3 - B_2^2)/(RT)^2 = -1.648 \times 10^{-15}$. Considering that 1 bar = $10^5$ Pa, the foregoing coefficients need to be multiplied by $10^5$ and $10^{10}$ respectively in order to express $Z$ in terms of $P$ in bars; thus,

$$Z = 1 - 0.00499 P - 1.684 \times 10^{-5} P^2; \quad (P \text{ in bars})$$

The value of $Z$ at 10.1325 bars is 0.9477, in nearly perfect agreement with Gen. Ref. 6. The values of $B_3$ from different sources, quoted by Dymond and Smith, disagree considerably with one another due to the sensitivity of $B_3$ to experimental errors.

It is appropriate to consider briefly mixtures of real gases at moderate pressures. Dalton's law of additivity of pressures when each gas occupies a given volume by itself cannot be used for mixtures of real gases. Instead, a simple rule

---

*For other equations of state, see J. O. Hirschfelder, et al. loc. cit., General Ref. (19), and G. W. Toop, Met. Mat. Trans., 26B, 577 (1995).

called the *additive pressure law* may be used to compute the partial pressure of each gas. According to this rule,

$$P_i = P x_i \tag{3.53}$$

where $P_i$ and $x_i$ are the partial pressure and the mole fraction of $i$ respectively, and $P$ is the pressure of gas (not the mixture) when one mole of the gas $i$ alone occupies the volume of one mole of mixture.

The volume of an ideal gas mixture is equal to the sum of the volumes of component gases when each gas is at the pressure of the mixture. This rule, sometimes called the additive volume law, or Amagat's law (1893), is also applicable with a fair degree of approximation to real gases at moderate pressures.*

## Liquids and Solids

For liquids and isotropic solids the equation of state may be written in power series as

$$V = V_\circ(1 + \alpha t + \alpha' t^2 + \cdots) + \beta P + \beta' P^2 + \cdots \tag{3.54}$$

where $\alpha, \alpha', \ldots, \beta, \beta', \ldots$ are empirical constants; $t$ is in °C and $V_\circ$ is the volume when $t$ and $P$ are zero. Often an equation with one term for temperature and one for pressure is adequate for most purposes. At a fixed temperature, the equation $V = V_\circ + \beta P$ is valid at several thousand bars. Substitution of this equation into $dU = -P\, dV$ and integration from $P_1 \to 0$ to $P_2$ gives

$$U_2 - U_1 = -0.5\beta P_2^2 = -0.5 P_2 (V_2 - V_\circ) \tag{3.55}$$

which is called the Hugoniot equation. In general, the changes in $U$ for one or two bars of change in pressure are negligible for most condensed substances and are usually neglected. The corresponding equation for anisotropic solids is too complicated since the thermal expansion, and the mechanical compression are different for different directions.

## ENTHALPY AND HEAT CAPACITY

An overwhelming number of processes occur at constant pressure of one bar, and the accompanying exchange of heat, $Q_P$, is related to $\Delta U$ and $W$ by $Q_p = \Delta U + P(V_2 - V_1)$, or

$$Q_p = [(U_2 + PV_2) - (U_1 + PV_1)]_P \tag{3.56}$$

The expression $(U + PV)$ occurs frequently; therefore, for convenience and brevity, $H$ is defined by

$$H = U + PV \tag{3.57}$$

---

*See General Refs. (19, 22) for more accurate calculations, based on various methods.

The function $H$ is called the molar *enthalpy*. Introduction of $H$ into (3.56) gives

$$Q_P = (H_2 - H_1)_P = \Delta H_P \tag{3.58}$$

and for an infinitesimal quantity of heat exchange at constant pressure, the corresponding equation is $dQ_P = dH_P$. It must be emphasized that the change in enthalpy is equal to the absorbed heat, *only in a process that occurs at a constant pressure*. In any other process the heat absorbed by the system *may not necessarily be equal* to the change in enthalpy. The heat capacity at constant pressure for one mole of a closed system was previously defined by $C_p = (dQ/dT)_P$; hence, from (3.58), $C_p$ is redefined by

$$C_p = \left(\frac{\partial H}{\partial T}\right)_P \tag{3.59}$$

For an ideal gas, $U = U(T)$ and $PV = RT$, and $H = U(T) + RT = H(T)$, and $H$ is only dependent on temperature; therefore,

$$C_p = \left(\frac{\partial Q}{\partial T}\right)_P = \frac{dH}{dT} = \frac{dU}{dT} + R = C_V + R \tag{3.60}$$

The total differential of $U = U(V, T)$ for one mole of *any closed* system (ideal or nonideal) is

$$dU = \left(\frac{\partial U}{\partial V}\right)_T dV + \left(\frac{\partial U}{\partial T}\right)_V dT \tag{3.61}$$

Under constant pressure condition and dividing by $(dT)_P$, leads to

$$\left(\frac{\partial U}{\partial T}\right)_P = \left(\frac{\partial U}{\partial V}\right)_T \left(\frac{\partial V}{\partial T}\right)_P + \left(\frac{\partial U}{\partial T}\right)_V \tag{3.62}$$

From the definitions of $C_p$ and $H$, we have

$$C_p = \left(\frac{\partial H}{\partial T}\right)_P = \left(\frac{\partial U}{\partial T}\right)_P + P\left(\frac{\partial V}{\partial T}\right)_P \tag{3.63}$$

Combining the preceding two equations with $(\partial U/\partial T)_V = C_V$, yields

$$C_p - C_V = \left(\frac{\partial V}{\partial T}\right)_P \left[P + \left(\frac{\partial U}{\partial V}\right)_T\right] \tag{3.64}$$

which is a general equation applicable to any closed system. For an ideal gas $(\partial V/\partial T)_P = R/P$ and $(\partial U/\partial V)_T = 0$, and it follows that $C_p - C_V = R$.

## Heat Capacity of Solids

It was found empirically by Dulong and Petit (1819) that the heat capacities of solids at ordinary temperatures were $3R$, or about 24.94 J per degree per gram atom for solid elements. Theoretical justification for the Dulong and Petit rule was made by Boltzmann on the basis of the kinetic theory of matter (1871). However, the heat capacities of solids increase from zero, near 0 K, toward $3R$ at sufficiently high temperatures. Einstein (1907) obtained the first successful theoretical equation expressing the heat capacity of solids as a function of temperature. He assumed that each atom in a crystal vibrates in three directions as a harmonic oscillator so that one mole of a solid consists of $3N$ independent oscillators with a single frequency of $v$. With this assumption, and by the application of the quantum theory, he derived

$$C_V = 3R\left(\frac{\theta_E}{T}\right)^2 \frac{e^{\theta_E/T}}{(e^{\theta_E/T} - 1)^2} \qquad (3.65)$$

where $\theta_E$ is the Einstein characteristic temperature defined by $\theta_E = hv/k$, with $h$ as the Planck constant, and $k = R/N_A$, the Boltzmann constant. The values of $\theta_E$ for most solids range from 100 to 300 K, and the appropriate value for each solid may be regarded as an empirical constant based on at least one experimental value if we expect to have success in representing $C_V$ by (3.65) for a significant range of temperature. At high temperatures $\theta_E/T$ approaches zero and the application of L'Hôpital's theorem to $(\theta_E/T) \div (e^{\theta/T} - 1)$ in (3.65) yields $C_V = 3R$ as required by the rule of Dulong and Petit. At sufficiently low temperatures $-1$ in the denominator of (3.65) is negligible and $C_V = 3R(\theta_E/T)^2 \exp(-\theta_E/T)$, and the exponential factor becomes dominant. However, experimental data do not support this equation at low temperatures, since $C_V$ varies with $T^3$ as will be seen later. Debye (1912) modified the Einstein equation by assuming that a crystal as a whole is an elastic vibrating medium with a broad spectrum of frequencies. His equation for low temperatures, i.e., $T < \theta/10$, is

$$C_V = 2.4\pi^4 R\left(\frac{T}{\theta}\right)^3 = 233.782 R\left(\frac{T}{\theta}\right)^3 \qquad (3.66)$$

At sufficiently high temperatures,

$$C_V = 3R\left[1 - \frac{1}{20}\left(\frac{\theta}{T}\right)^2 + \frac{1}{560}\left(\frac{\theta}{T}\right)^4 + \cdots\right] \qquad (3.67)$$

When $\theta/T$ is close to zero, this equation again yields $C_V = 3R$.

The Debye equation is not perfect is representing $C_V$ as a function of temperature because $\theta$ is not a constant but varies somewhat with temperature. In addition, $\theta$ may have different values along different crystal axes of solids. Therefore, empirical modifications of (3.66) and (3.67) have been made.*

The values of $\theta$ refer to the crystalline elements or the compounds, not to the glassy or amorphous substances. Some representative values of $\theta$, Debye temperatures, are Al, 418; C (diamond), 2000; Cu, 339; Si, 658; $H_2O$, 315; MgO, 820.

The Debye equation ignores an additional heat capacity term, $C_{el}$, due to the electronic contribution. This is *relatively* important in metals at very low

---

*For low temperature $C_V$, see E. F. Westrum, G. T. Furukawa and J. P. McCollough in "Experimental Thermodynamics," Edited by J. P. McCollough and D. W. Scott, Butterworths (1968).

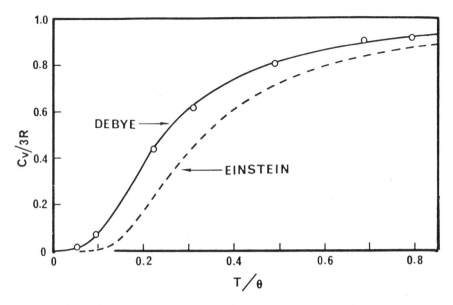

**Fig. 3.7** Variation of heat capacity with temperature according to Einstein and Debye equations. Points represent experimental data for silver with $\theta = 210$ K. $C_V/3R$ is 0.952 at $T = \theta$, and 0.998 at $T = 5\theta$ according to Debye equation.

and high temperatures. According to the electron theory of metals, atoms are imbedded in an atmosphere of relatively free electrons referred to as the "electron gas." Sommerfeld, on the basis of the application of Fermi-Dirac statistics to the electron gas, found that

$$C_{el} = \gamma_{el} T \qquad (3.68)$$

Equation (3.68) is called the Sommerfeld equation. The total value of $C_V$ is therefore equal to $C_V$ (Debye) $+ C_{el}$. For example $C_V$ (total) $= 1.45 \times 10^{-3} T + 2.99 \times 10^{-5} T^3$ J K$^{-1}$ mol$^{-1}$ at temperatures below $\theta/10$ for aluminum. The electronic contribution at 5 K is 0.0073 and the Debye contribution, 0.00374 but at 40 K, the electronic contribution is 0.058 and it is much smaller than the Debye contribution of 1.914. The values of $C_V$ at high temperatures should be $3R + \gamma_{el} T$ where $3R$ is the limiting value of $C_V$ from the Debye equation. However, there are complications which cause disparity between the computed and the experimental values, necessitating empirical modifications.

The heat capacity of a compound may be approximated as the sum of the heat capacities of its constituent atoms when experimental data for the compound are not available. This approximation is called the Kopp or Kopp-Neumann rule (1865). For a reasonable degree of approximation at ordinary temperatures, Kopp suggested the following empirical values of the atomic heat capacities: hydrogen

9.6, boron 11.3, carbon 7.5, oxygen 16.7, fluorine 21.0, silicon 15.9, phosphorus 22.6, sulphur 22.6, and for the elements of higher atomic weights, 25.1, all in J mol$^{-1}$ K$^{-1}$. For example, according to this rule, the molar heat capacities of $Fe_2O_3$, $CaCO_3$ and $CaSO_4$ are 100.3, 82.7, and 114.5 respectively, whereas the actual values are 104.6, 82.0 and 100.8, all in J K$^{-1}$ mol$^{-1}$. In the absence of data, a similar approximation may also be used for solutions. An approximation from the known neighboring compounds or solutions is always preferable.

A quantitative theoretical formulation of the heat capacities of liquids has not been made thus far because of the complexity of molecular motions in liquids. Heat capacities of liquids are on the average greater than those of solids and gases.

## Empirical Representation of Heat Capacity

It is frequently convenient to express the heat capacity as a function of temperature by an empirical equation such as

$$C_p^\circ = \alpha + 2\beta T + 6\lambda T^2 + \cdots \tag{3.69}$$

where the number of terms is limited by the range of temperature and the accuracy of data for $C_p^\circ$. A simpler equation with three terms for temperatures above 298 K has been proposed by Maier and Kelley (Gen. Ref. 2):

$$C_p^\circ = \alpha + 2\beta T - 2\epsilon/T^2 \tag{3.70}$$

and a compromise between these equations, e.g.,

$$C_p^\circ = \alpha + 2\beta T + 6\lambda T^2 + \cdots - (2\epsilon/T^2) \tag{3.71}$$

has been used for computer fitting of experimental data [cf. Gen. Ref. (1)]. The equation used in Gen. Ref. (2) is usually (3.70).

## Relationship between $C_p^\circ$ and $H_T^\circ - H_{298}^\circ$

The change in enthalpy at a constant pressure $(H_T^\circ - H_{298}^\circ)$ is the experimentally measured quantity, and the heat capacity is the derived quantity by $[\partial (H_T^\circ - H_{298}^\circ)/\partial T]_P = C_P^\circ$. The enthalpy change is related to $C_p^\circ$ by the following equation:

$$H_T^\circ - H_{298}^\circ = \int_{298}^{T} C_p^\circ \, dT \tag{3.72}$$

where the superscript "$\circ$" refers to the pure substance under one bar of pressure and the subscript 298 refers to 298.15. When (3.71) with four terms is substituted in (3.72) we obtain

$$H_T^\circ - H_{298}^\circ = \alpha T + \beta T^2 + 2\lambda T^3 + (2\epsilon/T) - I \tag{3.73}$$

where $-I$ is the integration constant obtained by substituting 298.15 for $T$ in the preceding terms. For example $C_p^\circ$ and $H_T^\circ - H_{298}^\circ$ for chlorine are given by

$$C_p^\circ = 36.93 + 0.000736T - (285{,}767/T^2), \text{ J mol}^{-1}\text{ K}^{-1} \tag{3.74}$$

$$H_T^\circ - H_{298}^\circ = 36.93T + 0.000368T^2 + (285{,}767/T) - 12{,}002 \tag{3.75}$$

58    Thermodynamics

The same equations are in terms of calories in Gen. Ref. (2). Equation (3.75) expresses the more accurate tabular values of $H_T^\circ - H_{298}^\circ$ within approximately 0.5% in the range of 298.15 to 3000 K.

## Enthalpy Change of Phase Transformations

In the preceding section, the change in enthalpy of *a single phase* was expressed as a function of temperature when there was no phase transformation in the temperature range of interest. Under equilibrium conditions, the phase transformation occurs at constant temperature and pressure and the macroscopic properties of the substance change abruptly. Let $H_T^\circ(I)$ be the enthalpy of Phase-I, and $H_T^\circ(II)$ that of Phase-II and let Phase-II be stable above the transition temperature, $T'$ under one bar (1 atm. in Gen. Ref. 2). The reaction

$$\text{Phase-I} \rightleftharpoons \text{Phase-II} \tag{3.76}$$

represents the reversible or equilibrium phase transformation, i.e., an infinitesimal increase in $T$ would transform all of I into II, but an infinitesimal decrease in $T$ would reverse the transformation. The forward transformation is accompanied by the corresponding change in enthalpy expressed by

$$H_{T'}^\circ(II) - H_{T'}^\circ(I) = \Delta H_{T'}^\circ(I \to II) = \Delta_{tr} H_{T'}^\circ = Q_{P,T'} \tag{3.77}$$

where the superscript "$\circ$" refers to the pure phases under one bar of pressure and $T'$ is the transformation temperature. Differentiation of (3.77) with respect to $T$, after replacing $T'$ by the variable temperature $T$ gives

$$\left( \frac{\partial \Delta H_T^\circ}{\partial T} \right)_P = C_p^\circ(II) - C_p^\circ(I) \tag{3.78}$$

integration of this equation gives a universal relationship for any isothermal phase transition (3.76), i.e.,

$$\Delta H_T^\circ - \Delta H_{T'}^\circ = \int_{T'}^{T} \left[ C_p^\circ(II) - C_p^\circ(I) \right] dT \tag{3.79}$$

**It will be seen later that this equation is also applicable to chemical reactions.**

The enthalpy of a pure stable substance is usually tabulated at one bar of standard pressure as $H_T^\circ - H_{298}^\circ$ where $H_{298}^\circ$ refers to 298.15 K, and $H_T^\circ$ **refers to** $T$, [Gen. Refs. (1) and (2)]. The values of $H_T^\circ - H_{298}^\circ$ correspond to the following equation for $298.15 < T' < T'' < T$ and for $H_T^\circ = H_T^\circ(III)$ which refers to Phase III as the stable phase at $T$, and $H_{298}^\circ = H_{298}^\circ(I)$ for Phase I:

$$H_T^\circ(III) - H_{298}^\circ(I) \equiv \left[ H_{T'}^\circ(I) - H_{298}^\circ(I) \right] + \Delta H_{T'}^\circ(I \to II)$$
$$+ \left[ H_{T''}^\circ(II) - H_{T'}^\circ(II) \right] + \Delta H_{T''}^\circ(II \to III)$$
$$+ \left[ H_T^\circ(III) - H_{T''}^\circ(III) \right] \tag{3.80}$$

**Table 3.3** Calcium, Ca (Reference State).* $C_p^\circ$ in J mol$^{-1}$ K$^{-1}$, and $H_T^\circ - H_{298}^\circ$ in J mol$^{-1}$.

| $T, K$ | $C_p^\circ$ | $H_T^\circ - H_{298}^\circ$ | $T, K$ | $C_p^\circ$ | $H_T^\circ - H_{298}^\circ$ |
|---|---|---|---|---|---|
| 0 | 0 | −5,736 | 1300 | 35.000 | 42,852 |
| 100 | 19.504 | −4,756 | 1400 | 35.000 | 46,352 |
| 200 | 24.542 | −2,487 | 1500 | 35.000 | 49,852 |
| 298.15 | 25.929 | 0 | 1600 | 35.000 | 53,352 |
| 300 | 25.946 | 0,048 | 1700 | 35.000 | 56,852 |
| 400 | 26.868 | 2,683 | 1773.658 | 35.000 | 59,430 |
| 500 | 28.487 | 5,448 | 1773.658 | 20.837 | 208,477 |
| 600 | 30.382 | 8,391 | 1800 | 20.845 | 209,026 |
| 700 | 32.406 | 11,529 | 1900 | 20.888 | 211,112 |
| 716.000 | 32.737 | 12,051 | 2000 | 20.953 | 213,204 |
| 716.000 | 31.503 | 12,981 | 2100 | 21.046 | 215,304 |
| 800 | 33.818 | 15,723 | 2200 | 21.173 | 217,415 |
| 900 | 36.713 | 19,249 | 2300 | 21.343 | 219,540 |
| 1000 | 39.706 | 23,069 | 2400 | 21.561 | 221,685 |
| 1100 | 42.763 | 27,192 | 2500 | 21.836 | 223,854 |
| 1115.000 | 43.226 | 27,837 | 2600 | 22.173 | 226,054 |
| 1115.000 | 35.000 | 36,377 | 2700 | 22.578 | 228,291 |
| 1200 | 35.000 | 39,352 | 2800 | 23.057 | 230,572 |
| | | | 2900 | 23.614 | 232,905 |
| | | | 3000 | 24.251 | 235,298 |

*Adapted from Gen. Ref. (1). Crystal I → Crystal II occurs at 716.00 K; melting point is 1115.0 K, and the boiling point at 1 bar is 1773.658 K. The subscript 298 is for 298.15 K. Reference State tables are only for the <u>elements</u> in their stable standard states; they contain the discontinuities in $C_p^\circ$ and $H_T^\circ - H_{298}^\circ$ at the first order phase transition temperatures.

The identity can be verified by substituting $\Delta H_T^\circ(\text{I} \to \text{II}) \equiv H_{T'}^\circ(\text{II}) - H_{T'}^\circ(\text{I})$, etc. for the phase transitions. This equation corresponds to

$$H_T^\circ(\text{III}) - H_{298}^\circ(\text{I}) \equiv \int_{298}^{T'} C_p^\circ(\text{I})\, dT + \Delta H_{T'}^\circ(\text{I} \to \text{II}) + \int_{T'}^{T''} C_p^\circ(\text{II})\, dT$$

$$+ \Delta H_{T''}^\circ(\text{II} \to \text{III}) + \int_{T''}^{T} C_p^\circ(\text{III})\, dT \qquad (3.81)$$

Each integral refers to the region where a particular phase is stable; thus the first integral in (3.81) is equal to $H_{T'}^\circ(\text{I}) - H_{298}^\circ(\text{I})$.

The usefulness of (3.80) is illustrated by Table 3.3 for calcium taken from General Ref. (1). (A detailed description of such tables will be presented in Chapter XII). This reference state table will now be used to show the values of each set of terms in (3.80) and (3.81). The listed values of $H_T^\circ - H_{298}^\circ$ from 298.15 to $T$

for $\alpha$ or Crystal I are given in J mol$^{-1}$ by

$$H_T^\circ - H_{298}^\circ = \int_{298}^T C_p^\circ(\text{I})\, dT$$
$$= 16.428T + 0.01101T^2 - \frac{261{,}600}{T} - 4{,}999 \quad (3.82)$$

(For temperatures below 298.15, i.e., $T < 298.15$, the values of $H_T^\circ - H_{298}^\circ$ are negative). The value at $T' = 716$ K from this equation is 12,042 J mol$^{-1}$, in close agreement with Table 3.3. The next term $\Delta H_{716}^\circ(\text{I} \to \text{II})$ is $930 = 12{,}981 - 12{,}051$ J mol$^{-1}$. The next term is

$$H_T^\circ(\text{II}) - H_{716}^\circ(\text{II}) = 9.965T + 0.0149T^2 - 14{,}774; \text{ J mol}^{-1} \quad (3.83)$$

where the maximum value of the upper limit is 1115 K which is the melting point. In Crystal II, or $\beta$-state, the enthalpy with reference to Crystal I, is obtained by adding $H_{716}^\circ(\text{II}) - H_{298}^\circ(\text{I}) = 12{,}981$ to (3.83) to obtain

$$H_T^\circ(\text{II}) - H_{298}^\circ(\text{I}) = 9.965T + 0.0149T^2 - 1{,}793; \text{ J mol}^{-1} \quad (3.84)$$

The procedure for the liquid at 1115.00 to 1773.658 K and the gas above 1773.658 K is identical.

## THERMOCHEMISTRY*

Thus far, the thermodynamic processes affecting two independent variables in $G = f(P, T)$ have been considered, i.e., compression, expansion, heating and cooling. In all such processes, the moles of each species in the system were assumed to remain unchanged. A chemical reaction is a thermodynamic process which changes the number of species constituting the system. Thermochemistry deals with the energy and the enthalpy changes accompanying chemical reactions and phase transformations, and in fact both processes are essentially the same, and all thermodynamic generalizations that are applicable to one process are also applicable to the other.

Consider a chemical reaction that converts the *reactants* into the *products* under either constant volume and temperature or constant pressure and temperature. If the volume and temperature are kept constant, as in a bomb calorimeter shown in Fig. 3.8, then the heat of reaction, $Q_V$, is expressed by the following relation:

$$Q_V = \Delta U_V = \left[\sum U_{\text{products}} - \sum U_{\text{reactants}}\right]_{V,T} \quad (3.85)$$

---

*For methods of measurements see F. D. Rossini, Ed. "Experimental Thermochemistry," Vol. I, Interscience Publi. (1956); Vol. II, H. A. Skinner, Ed. (1962); E. F. Westrum, Jr., in "Proceedings NATO Advanced study Institute on Thermochemistry at Viana do Castello, Portugal, edited by M. A. V. Ribeiro de Silva, Reidel Publ. Co. (1994); see also Gen. Ref. (27).

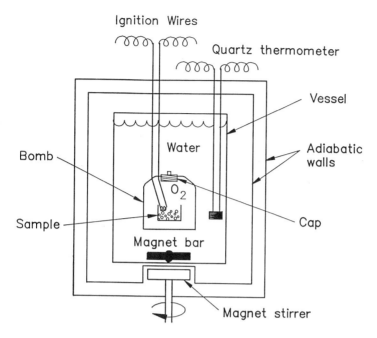

**Fig. 3.8** Oxygen Bomb Calorimeter. The bomb contains oxygen at a high pressure, and a sample touching the ignition wires inside a nonreacting crucible. The rise in water temperature after ignition is a direct measure of $Q_V$ at constant volume, which is converted into $Q_p = \Delta_r H°$. The reaction products are analyzed to correct for the effects of side reactions. The calorimeter is calibrated with a known amount of electrical energy dissipation (not shown). A fluorine bomb calorimeter is similar in design and operating procedure. (see footnotes on the preceding page).

For a general chemical reaction

$$\mathbf{a}A + \mathbf{b}B + \cdots = \mathbf{m}M + \mathbf{n}N + \cdots \qquad (3.86)$$

$\Sigma U_{\text{products}}$ is $(\mathbf{m}U_M + \mathbf{n}U_N + \cdots)$ where $U_i$ is the molar energy of any element or compound "$i$," and similarly, $\Sigma U_{\text{reactants}}$ is $(\mathbf{a}U_A + \mathbf{b}U_B + \cdots)$. If instead, the pressure and temperature are kept constant, the heat of reaction is

$$Q_P = \Delta H_P = \left(\sum H_{\text{products}} - \sum H_{\text{reactants}}\right)_{P,T} \qquad (3.87)$$

The quantities $Q_V$ and $Q_P$ refer to the complete reactions as written or the completely converted portion of the reactants into the products, and further, $Q_P = Q_V + P\Delta_r V$, where $\Delta_r V$ is the change in volume due to the reaction.

Enthalpies of reactions are expressed according to the following convention: (1) The enthalpy effects refer to the reactions under the constant pressure of one bar unless specified otherwise. If the experimentally measured quantity refers to the constant volume condition, the result is converted into the constant pressure of one bar. (2) The states of aggregation of the reactants and products are specified as pure substances at one bar for any given temperature. The state of a substance defined in this manner is called the *standard state*. For solids, the pure stable crystalline form, and for liquids the pure substance at one bar are the respective standard states. The standard state for gases is the pure ideal gas at one bar. If the gas is not ideal, it is first expanded to a low pressure where it becomes ideal, and then compressed to one bar as a hypothetical ideal gas. (3) The *reference temperature* is arbitrarily chosen as 25°C for the purpose of tabulation. Certain substances, e.g., the ions of electrolytes, do not exist in the pure state; hence their standard states refer to the infinitely dilute solution, as will be seen later.

The proper notation used in writing a reaction and the corresponding change in standard enthalpy are illustrated as follows:

$$C(\text{graphite}) + O_2(g) = CO_2(g); \qquad \Delta_r H^\circ_{298} = -393{,}522 \text{ J(mol CO}_2)^{-1} \tag{3.88}$$

The standard state for carbon is graphite, and for oxygen and carbon dioxide, ideal gases at one bar. The quantity $\Delta_r H^\circ_{298}$ is the standard enthalpy of reaction, indicated with "°", at 298.15 K for the formation of one mole of $CO_2$. The reaction is called exothermic if $\Delta_r H^\circ < 0$, or energy is lost to the surroundings, and endothermic, if $\Delta_r H^\circ > 0$, or energy must be supplied to the reaction. Reaction (3.88) is sometimes written as $C(\text{graphite}) + O_2(g) = CO_2(g) - \Delta_r H^\circ$ so that the equality of both sides represents the "material balance" as well as the "enthalpy balance," but this notation will not be used in this text. When all the reactants and products in (3.86) are ideal gases, then,

$$Q_V = \Delta U, \quad \text{and} \quad Q_P = \Delta_r H = \Delta_r U + P \Delta V = Q_V + P \Delta V \tag{3.89}$$

Since the energy of ideal gases is independent of volume and pressure, the restrictions on $\Delta_r U$ and $\Delta_r H$ are unnecessary. For reactions involving only liquids and solids, $P \Delta V$ is negligibly small and $\Delta_r H \approx \Delta_r U = Q_V$; hence $P \Delta V$ usually refers to the gaseous reactants and products when reactions involve gases and condensed phases. In (3.89) $\Delta V$ represents the sum of the volumes of products minus the sum of the volumes of reactants at the same temperature and pressure, and therefore $P \Delta V$ becomes identical with $\Delta n RT$. The effect of deviations from ideality on $P \Delta V$ is very small and can be obtained from the equations of state.

The standard enthalpy of formation of a compound is the enthalpy *of reaction* when the *reactants are the elements* in their standard states, and the *product* is the compound in its standard state. Thus, $\Delta_r H^\circ_{298}$ in (3.88) is also the standard enthalpy of formation, $\Delta_f H^\circ_{298}$, of carbon dioxide at 298.15 K.

$\Delta_r H^\circ_T$ for a reaction is independent of any intermediate change in temperature and pressure that may occur while the reaction proceeds, provided that the initial and final pressure and temperature are the same before and after the overall reaction. The nature and the number of the intervening reactions are immaterial so long as the same ultimate products are obtained from the same reactants.* For example, $\Delta_f H^\circ_{298}$ of CO can be determined from the combination of the readily measurable $\Delta_f H^\circ_{298}$ of $CO_2$ and the enthalpy of combustion of CO, i.e.,

$$\begin{array}{ll} C(\text{graphite}) + O_2(g) = CO_2(g), & \Delta_f H^\circ_{298} = -393{,}522 \text{ J} \\ -[CO(g) + 0.5\,O_2(g) = CO_2(g)], & -[\Delta_r H^\circ_{298} = -282{,}995 \text{ J}] \\ \hline C(\text{graphite}) + 0.5 O_2(g) = CO(g), & \Delta_f H^\circ_{298} = -110{,}527 \text{ J} \end{array} \quad (3.90)$$

This is particularly important in indirect determination of the standard enthalpies of formation for which direct experimental observations are difficult or impossible as for CO. Likewise, it is experimentally impossible to form $C_2H_2$ by directly combining graphite and hydrogen in a calorimeter in order to determine its $\Delta_f H^\circ_{298}$, but, from $\Delta_f H^\circ_{298}$ for water and for carbon dioxide, and from the enthalpy of combustion of $C_2H_2$, $\Delta_f H^\circ_{298}$ for $C_2H_2$ can be obtained.

$\Delta_r H^\circ$ for a reaction can be obtained from the tabulated standard enthalpies of formation by the following simple convention:

$$CO(g) + 0.5\,O_2(g) = CO_2(g), \quad \Delta_r H^\circ_{298} = -393{,}522 - (-110{,}527)$$

$$(-110{,}527), \quad (0), \quad (-393{,}522) \qquad\qquad = -282{,}995 \qquad (3.91)$$

where the quantities under the compounds represent the standard enthalpies of formation at the reaction temperature. This procedure is derived from the algebraic summation of reactions and their enthalpy effects [cf. (3.90)]. *The standard enthalpy of formation of a stable element* at any temperature is zero by convention. The standard enthalpy of reaction is then the sum of the standard enthalpies of formation of the products, minus the sum of the standard enthalpies of formation of the reactants.

## Variation of $\Delta_r H^\circ_T$ with Temperature†

The changes in the values of $\Delta_r H^\circ$ with temperature can be obtained by a simple application of the first law. As an example we take a hypothetical reaction at 298.15 K:

$$a\mathbf{A}(s) + b\mathbf{B}(l) = m\mathbf{M}(s) + n\mathbf{N}(l), \qquad (\text{at } 298.15 \text{ K}) \qquad (3.92)$$

and we assume that $\Delta_r H^\circ_{298}$ is known. We wish to compute $\Delta_r H^\circ_T$ at $T$ where the foregoing reaction becomes

$$a\mathbf{A}(l) + b\mathbf{B}(g) = m\mathbf{M}(g) + n\mathbf{N}(g), \qquad (\text{at } T) \qquad (3.93)$$

where **A** is now liquid and **B**, **M** and **N** are gaseous. We first follow a path in which we start with the reactants in (3.92) at 298.15 K and permit the reaction to

---
*For numerous energy and material balances see, e.g. Y. K. Rao, "Stoichiometry and Thermodynamics of Metallurgical Processes," Cambridge U.P. (1985).
†See Gen. Refs. (4 and 7) and D. R. Gaskell, "Introduction to Metallurgical Thermodynamics," McGraw-Hill (1981); O. F. Devereux, "Topics in Metallurgical Thermodynamics," Krieger (1989).

proceed to obtain **m** moles of **M**(s) and **n** moles of **N**(l) at 298.15 K, and then heat the products to their gaseous states at $T$ and thus obtain the right side of (3.93) at $T$ from the left side of (3.92) at 298.15 K. The resulting enthalpy effect is

$$\Delta_r H°_{298} + \mathbf{m}\left[H°_T(g) - H°_{298}(s)\right]_\mathbf{M} + \mathbf{n}\left[H°_T(g) - H°_{298}(l)\right]_\mathbf{N} \tag{3.94}$$

Next, we follow a second path with the reactants in (3.92) at 298.15 K, and heat them to $T$ to obtain **a** moles of liquid **A** and **b** moles of gaseous **B** and then permit the reaction (3.93) to proceed, and thus obtain the right side of (3.93) at $T$ from the left side of (3.92) at 298.15 K. The enthalpy effect accompanying the second path is

$$\mathbf{a}\left[H°_T(l) - H°_{298}(s)\right]_\mathbf{A} + \mathbf{b}\left[H°_T(g) - H°_{298}(l)\right]_\mathbf{B} + \Delta_r H°_T \tag{3.95}$$

Both paths take the system from the same initial state to the same final state; hence (3.94) and (3.95) are equal and from this equality, we obtain,

$$\Delta_r H°_T = \Delta_r H°_{298} + \sum_{\text{products}} \nu_i \left(H°_T - H°_{298}\right)_i$$

$$- \sum_{\text{reactants}} \lambda_j \left(H°_T - H°_{298}\right)_j \tag{3.96}$$

where $\nu_i$ and $\lambda_j$ are the stoichiometric coefficients of products and reactants respectively. The summation for the products represents the terms after $\Delta_r H°_{298}$ in (3.94), and that for the reactants, the terms preceding $\Delta_r H°_T$ in (3.95), and $H°_T - H°_{298}$ refer to one mole of any species in its pure stable state. Equation (3.96) is a general relationship in thermodynamics of reactions and it is applicable to any property $G$ for any reaction.

It is possible to differentiate (3.96) with respect to $T$ at constant $P$, with $P = 1$ bar, in the *absence* of *phase transitions* to obtain

$$\frac{\partial \Delta_r H°_T}{\partial T} = \sum_{\text{products}} \nu_i C°_p - \sum_{\text{reactants}} \lambda_j C°_p \equiv \Delta_r C°_p \tag{3.97}$$

We now integrate (3.97) to obtain a special form of (3.96):

$$\Delta_r H°_T = \Delta_r H°_{298} + \int_{298}^T \Delta_r C_p \, dT \tag{3.98}$$

It is interesting to note that (3.98) is identical in form with (3.79). When there are phase transitions, the general relation shown by (3.96) should be used instead of (3.98). See (3.81) for a complete form of (3.98) when phase transitions occur.

**Example** Generally, it is preferable and more convenient to use (3.96) to calculate $\Delta_r H°_T$ at $T$ instead of using (3.98), even when there is no phase transition. Here, only as a simple exercise, we calculate $\Delta_r H°_T$ at 2000 K by using (3.98) for the following reaction:

$$CO(g) + 0.5\, O_2(g) = CO_2(g); \qquad \Delta_r H°_{298} = -282{,}995 \text{ J mol}^{-1}$$

where $\Delta_r H^\circ_{298}$ was given in (3.90). The standard molar heat capacities adapted from Gen. Ref. (2) are as follows:

CO: $C_p^\circ = 28.066 + 0.004628T - 25{,}941/T^2$; J mol$^{-1}$ K$^{-1}$

$O_2$: $C_p^\circ = 30.250 + 0.004209T - 189{,}117/T^2$; J mol$^{-1}$ K$^{-1}$

$CO_2$: $C_p^\circ = 45.367 + 0.008686T - 961{,}902/T^2$; J mol$^{-1}$ K$^{-1}$

*Solution* From the foregoing heat capacities, $\Delta C_p^\circ$ for the reaction is

$$\Delta C_p^\circ = 2.176 + 0.0019535T - 841{,}403/T^2; \text{ J mol}^{-1} \text{ K}^{-1}$$

The substitution in (3.98) and integration yields

$$\Delta_r H_T^\circ = \Delta_r H^\circ_{298} + 2.176T + 9.7675 \times 10^{-4} T^2$$
$$+ (841{,}403/T) - 3{,}558; \text{ J mol}^{-1}$$

where $-3{,}558$ is the lower integration limit in (3.98). The substitution of $T = 2000$ K, and $\Delta_r H^\circ_{298} = -282{,}995$ in this equation gives $\Delta_r H^\circ_{2000} = -277{,}873$ J mol$^{-1}$. The tables in Gen. Ref. (1) give $\Delta_f H^\circ_{2000}(CO_2) - \Delta_f H^\circ_{2000}(CO) = -396{,}784 + 118{,}896 = -277{,}888$ J mol$^{-1}$, within 15 J of the preceding computed value.

## Bond Energies

Pauling* has shown by a systematic analysis that a bond such as H—C or C—C has nearly the same bond energy irrespective of the gaseous compound in which it may be present. It is therefore possible to predict the enthalpy of formation of a new compound from the monatomic gaseous elements within a few kJ mol$^{-1}$ by adding all the bond energies.

The bond energy represents the average enthalpy of breaking a bond in a gaseous compound, into its gaseous atoms. Thus the bond energy of H—H is the enthalpy of the reaction $H_2(g) = 2$ H(g), i.e., 436 kJ mol$^{-1}$ at 25°C. Likewise the bond energy of C—H is one quarter the enthalpy of reaction for $CH_4(g) = C(g) + 4$ H(g), since there are four C—H bonds in $CH_4$. The energy for removing each of the four hydrogen atoms in $CH_4$ is not the same but it is convenient to define an average value from the enthalpy of dissociation into atoms, and name it the bond energy for C—H.

When the enthalpy of formation of a compound from monatomic gaseous elements is not known, it can be estimated by adding the bond energies and changing the sign. For example, the bond energy for C—H is 413 kJ, for C—C, 348 kJ, and for H—H, 436 kJ, and the enthalpy of formation of gaseous $C_3H_8$ from C(g) and H(g) is equal to the negative of the sum of 8(C—H) and 2(C—C) bond energies.

---

*L. Pauling, "The Nature of the Chemical Bond," Third Edition, Cornell University Press, Ithaca, N.Y. (1961).

The result is $-8 \times 413 - 2 \times 348 = -4{,}000$ kJ mol$^{-1}$. The standard enthalpy of formation of $C_3H_8$ from the stable elements at 25°C, $\Delta_f H^\circ_{298}$, can be obtained by adding the enthalpy of sublimation of carbon, $3 \times 718$ and the bond energy of 4 (H—H) bonds, $4 \times 436$ so that $\Delta_f H^\circ_{298} = -4{,}000 + 3 \times 718 + 4 \times 436 = -102$ kJ. The experimentally determined value of $\Delta_f H^\circ_{298}$ is $-103.8$ kJ mol$^{-1}$, and thus the agreement in this case is excellent. In general, the enthalpy of formation computed in this manner is not expected to yield accurate results, particularly when the compound is complex and contains multiple bonds. Further, the results estimated in this manner, are obtained by the subtraction of *very large values* and therefore the errors in bond energies and in the assumption regarding their equality in all compounds may become cumulative. Several methods of estimation, very useful in the absence of experimental data, are available as summarized by Janz and others.*

## Adiabatic Flame Temperature

The combustion of a substance at a certain temperature, e.g., 25°C, gives the products at the same temperature, and a certain quantity of negative heat (enthalpy) which was called the enthalpy of combustion. When *this heat is returned to the system*, the temperature of the *products* must increase and the overall process must then be adiabatic because the net exchange of heat with the surroundings is zero; hence,

$$-\Delta H^\circ_{298} = \sum_{\text{products}}^{i} \nu_i \left( H^\circ_T - H^\circ_{298} \right)_i = \int_{298}^{T} \nu_i C^\circ_{pi} (\text{products}) \, dT \qquad (3.99)$$

Note that unlike [(3.96) and (3.98)], (3.99) does not contain the reactants. The adiabaticity of the process is based on the fact that $\Delta H^\circ_{298} + \sum \nu_i (H^\circ_T - H^\circ_{298})_i = Q_P = 0$. The resulting final temperature from (3.99) is called the adiabatic flame temperature which can be determined from the tabular values of $H^\circ_T - H^\circ_{298}$, or by integrating $\nu_i C_{pi} \, dT$ for all the combustion products if there is no phase change. For example, for complete combustion of one mole of CO with 0.5 mole of $O_2$, $\Delta_r H^\circ_{298} = -282{,}995$ J; the reaction product is simply $CO_2$; hence (3.99) is $+282{,}995 = (H^\circ_T - H^\circ_{298})_{CO_2}$. Tables in Appendix show that $+282{,}995$ J mol$^{-1}$ for $H^\circ_T - H^\circ_{298}$ lies between 5000 and 5100 K. A linear interpolation between these temperatures gives 5058 K for the adiabatic flame temperature. If we burn one mole of CO with air instead, 1.881 moles of nitrogen must accompany 0.5 mole of oxygen since air contains 0.21 mole fraction of oxygen and 0.79 mole fraction of nitrogen. We combine $H^\circ_T - H^\circ_{298}$ for one mole of $CO_2$ and 1.881 moles of $N_2$ and make a special table of enthalpy contents of both gases to find that $\sum \nu_i (H^\circ_T - H^\circ_{298})_i$ is 274,721 at 2600 K, and 287,809 at 2700 K, and therefore, the adiabatic flame temperature is 2663 K, which is much lower than 5058 K when pure oxygen is used for combustion. This analysis provides the reason for using oxygen in oxyacetylene,

---

*See for example G. J. Janz, "Thermodynamic Properties of Organic Compounds. Estimation Methods, Principles and Practice," Academic Press (1967); P. I. Gold and G. J. Ogle, Chem. Eng., 76, 122 (1969); Gen. Ref. (7).

oxyhydrogen and oxygas torches and for oxygen enriching the air used in high temperature processes. The adiabatic flame temperature is further decreased by excess air and by incomplete combustion. In practice a slight excess of air is beneficial in assuring combustion without unduly decreasing the flame temperature. In practical computations of this type, it is frequently convenient to use $v_i C_{pi}^\circ$ as functions of temperature instead of the tables as illustrated by Problems 3.11 and 3.12.

The flame temperature increases if the reactants are preheated. The procedure for obtaining the flame temperature in this case is similar. For this purpose, the following steps are necessary: (1) Cool the reactants to 298.15 K and change the sign of extracted heat and add it to $-\Delta H_{298}^\circ$ in (3.99), and thus increase the positive quantity $-\Delta H_{298}^\circ$ by the amount corresponding to the preheating of the reactants, and (2) use (3.99) as before with $H_T^\circ - H_{298}^\circ$ for all the gases in the combustion products. For example, if in the preceding example the air were preheated to 1000 K, it would contribute $0.5(22,703) + 1.881(21,463) = 51,723$ J to $-\Delta H_{298}^\circ = 282,995$ and then the resulting flame temperature would be 3058 K or about 400 K higher than without the preheating.

*Example* Calculate the adiabatic flame temperature, $T_{flm}$, for CO burned completely with 0.5 mole of $O_2$, given $\Delta_r H_{298}^\circ = -282,995$ J mol$^{-1}$ in (3.90), $H_{3000}^\circ - H_{298}^\circ = 152,882$ J mol$^{-1}$ from Gen. Ref. (1), and an approximate equation for $C_p(CO_2)$, valid in the range 3000–5000 K, i.e.,

$$C_P \approx 59.50 + 9.1 \times 10^{-4} T; \text{ J mol}^{-1}\text{K}^{-1}$$

The integration of this equation yields

$$H_T^\circ - H_{3000}^\circ = 59.50\, T + 4.55 \times 10^{-4}\, T^2 - 182,595; \text{ J mol}^{-1}$$

We add $H_{3000}^\circ - H_{298}^\circ = 152,852$ to both sides of this equation to obtain

$$H_T^\circ - H_{298}^\circ = 59.50\, T + 4.55 \times 10^{-4}\, T^2 - 29,743$$

[Examine a similar equation in Gen. Ref. (2) for $CO_2$ for the range 2000–3000 K]. The left side of the preceding equation has to be equal to $-\Delta_r H_{298}^\circ = 282,995$; hence, after rearrangement, we obtain

$$4.55 \times 10^{-4}\, T^2 + 59.50\, T - 312,738 = 0$$

The solution is $T_{flm} = 5060.3$ K, in good agreement with 5058 K in the text.

## PROBLEMS

3.1 Calculate $\Delta U$, $\Delta H$, $Q$ and $W$ when one mole of helium is heated from 25°C and 1 bar to 100°C (a) at constant volume, (b) at constant pressure, and (c) by adiabatic compression.

3.2 An ideal diatomic gas at 600 K and 2 bars is cooled to 300 K at constant volume. The same final conditions are also attained by first cooling the gas at constant pressure of 2 bars to 300 K and expanding it isothermally to 1 bar. Calculate $\Delta U$, $\Delta H$, $Q$ and $W$ for each step.

**68**   *Thermodynamics*

3.3 One mole of an ideal monatomic gas at 300 K and 4 bars is expanded adiabatically to 1 bar. What is the work of expansion? Is this work larger or smaller than the isothermal work of expansion (a) to the same pressure, and (b) to the same volume? Indicate the results on a $P$-$V$ diagram.

3.4 An ideal gas is compressed adiabatically from 1 bar and 300 K to 5 bar and 475 K. Calculate the molar heat capacity of the gas at constant volume.

*Ans.*: $C_V = 5R/2$

3.5 Complete the following tabular summary for various processes with an ideal gas in terms of $C_V$, $C_p$, and $R$.

| Process | Initial State | Final State | $\Delta U$ | $\Delta H$ | $Q$ | $W$ |
|---|---|---|---|---|---|---|
| Constant $V$ | $V_1, T_1$ | $V_1, T_2$ | $C_V(T_2 - T_1)$ | $C_p(T_2 - T_1)$ | — | — |
| Constant $P$ | $P_1, T_1$ | $P_1, T_2$ | — | — | — | — |
| Constant $T$ | $T_1, P_1$ | $T_1, P_2$ | — | — | $RT_1 \ln \dfrac{P_1}{P_2}$ | — |
| Adiabatic (Reversible) | $T_1, P_1$ | $P_2$ | — | — | — | — |

3.6 Verify the following equation for an ideal gas:

$$C_p - C_V = \left[V - \left(\frac{\partial H}{\partial P}\right)_T\right]\left(\frac{\partial P}{\partial T}\right)_V$$

3.7 If $U$ is written as a function of $PV/R$ for an ideal gas, show that

$$\left(\frac{\partial U}{\partial P}\right)_V - \frac{VC_V}{R} \quad \text{and} \quad \left(\frac{\partial H}{\partial P}\right)_V = \frac{VC_p}{R}$$

(*Hint*: $U = C_V T = C_V PV/R$).

3.8 At the critical pressure, volume, and temperature, $P_c$, $V_c$, and $T_c$

$$\left(\frac{\partial P}{\partial V}\right)_T = 0 \quad \text{and} \quad \left(\frac{\partial^2 P}{\partial V^2}\right)_T = 0$$

for the van der Waals equation. Express these derivatives in terms of $P_c$, $V_c$ and $T_c$, and show in detail that

$$P_c = a/27b^2, \quad V_c = 3b \quad \text{and} \quad T_c = 8a/27Rb.$$

3.9 Calculate the compressibility factor, $Z$, for $H_2$ at 100 bars and 200 K with the Berthelot equation and compare the result with that scaled off from Fig. 3.5.

*Ans.*: $Z = 1.0752$

3.10 Express $C_p - C_V$ for a liquid by using the following empirical equations in (3.64):

$$V = V_\circ[1 + a(T - 273)] + bP$$
$$U - U_{273} = AT + BV$$

3.11 Express the enthalpies of liquid and gaseous zinc with reference to solid zinc from the following data:

$$C_p^\circ(s) = 22.36 + 0.0104\, T; \qquad \Delta_{fus} H^\circ = 7{,}322\ \text{J mol}^{-1} \text{ at } 692.73\ \text{K}$$

$$C_p^\circ(l) = 31.38; \qquad \Delta_{vap} H^\circ = 115{,}311\ \text{J mol}^{-1} \text{ at } 1180.17\ \text{K}$$

3.12 Calculate the adiabatic flame temperature for CO burned completely with the theoretically sufficient amount of air by using the following data: $\Delta H_{298}^\circ$ (combustion) $= -282{,}995$ J mol$^{-1}$ of CO, $C_p^\circ(CO_2) = 44.225 + 0.00878T$ and $C_p^\circ(N_2) = 28.58 + 0.00376T$. Assume that the air contains 20% $O_2$ and 80% $N_2$.

*Ans.: 2567 K*

3.13 What is the amount of zinc at 25°C to be added into one mole of liquid zinc at 1180.17 K to obtain equal amounts of liquid and solid. (See Problem 3.11 for required data).

*Ans.: 0.255 mole of Zn at 298.15 is needed per mole of liquid Zn.*

3.14 A mixture of steam and oxygen at 1000 K and 1 bar react with a column of graphite in a furnace also at 1000 K to yield a mixture of $H_2$ and CO at the same temperature. Calculate the composition of the steam-oxygen mixture assuming that the furnace is adiabatic, i.e., temperature is not affected by the reaction. (See Appendix for required $\Delta H^\circ$).

*Approximate ans.: 69% CO and 31% $H_2$.*

3.15 Compute the available heat for a process temperature of 2000 K with one mole of gas containing 35% CO and 65% $N_2$ by using air with 20% $O_2$ and 80% $N_2$, and the air enriched to contain 40 and 60% $O_2$. Available heat is defined by the heat of combustion at 25°C minus the heat contents of gases leaving the processing furnace. Use the tables in Appendix II.

3.16 (a) Calculate the change in $U$ in Problem 3.11 accompanying the evaporation of one mole of Zn at 1180.2 K. Assume that Zn(g) is ideal and the density of Zn(l) is 7g cm$^{-3}$.

*(Ans.: $\Delta U = 105{,}500$ J mol$^{-1}$)*

(b) Express $H_T^\circ - H_{298}^\circ$ in Problem 3.11 for Zn(g) by using $C_p^\circ(g) = 20.79$ J mol$^{-1}$K$^{-1}$.

CHAPTER IV

# THE SECOND LAW OF THERMODYNAMICS

The first law of thermodynamics expresses the existence of the energy function $U$ as a consequence of the conservation of energy; therefore, the change in energy, $\Delta U$, for any process must take place without violating the conservation of energy whether a process is possible or impossible. For example, the first law does not rule out the spontaneous heat transfer from a lower temperature to a higher temperature so that a pencil at a uniform temperature may become hotter at one end and colder at the other without external interference. Such processes do not violate the first law so long as the conservation of energy prevails, i.e., the decrease in energy of one end of the pencil is compensated by the increase in energy of the other end. It is the second law of thermodynamics that sets the criterion for the impossibility of such processes.

## The Second Law of Thermodynamics

The second law of thermodynamics has been stated in numerous equivalent forms since the time it was enunciated implicitly by Carnot (1824) and explicitly first by Clausius (1850) and later independently by Kelvin (1851). In accordance with Clausius, Kelvin, and as reformulated by Planck, the following statement is chosen to enunciate the second law of thermodynamics:

> *It is impossible to construct a cyclic engine that can convert heat from a reservoir at a uniform temperature into mechanical energy without leaving any effect elsewhere.*

The engine must be cyclic, i.e., it must return to its initial position after each stroke or revolution so that by any number of revolutions or cycles any amount of work can be generated. Were it possible to construct such an engine, the vast amount of heat stored in the atmosphere, the oceans and the earth could be converted into mechanical energy and thus perpetual generation of mechanical work could be accomplished. Such a hypothetical engine is said to be capable of performing "the perpetual motion of the second kind" in contrast with "the perpetual motion of the first kind," which violates the law of conservation of energy by creating work without consuming any form of energy. Many attempts have been made, particularly in the 18th and 19th centuries, to devise such mechanisms without success.

An engine utilizing heat to generate mechanical work in a cyclic manner is called a *heat engine*, and the substance contained in it, the *working substance*. A heat engine takes up a certain amount of heat, $Q_2$, from a high temperature reservoir at $T_2$, and converts a part of it into mechanical work and discharges the remaining part, $Q_1$, to the surroundings at $T_1$ and then returns to its original position to repeat the cycle. According to the second law of thermodynamics, $Q_1$ cannot be zero; otherwise, it would be possible to convert all the heat from one reservoir into mechanical work without leaving any effect elsewhere. It will be convenient to assume that the mass of the engine is infinitesimally small so that the system under consideration is the working substance, and the engine provides the boundaries. The value of $Q_2$ is positive and $Q_1$ negative because $Q_2$ is received and $Q_1$ is discharged by the system. The heat reservoirs are very large, and constant and uniform in temperature. Therefore, after each cycle the working substance returns to its initial state, and by the first law,

$$\Delta U = 0 = Q_2 + Q_1 + W \tag{4.1}$$

The amount of heat converted into mechanical work is therefore $Q_2 + Q_1$ which is equal to $-W$, or simply the work leaving the engine. The fraction of heat converted into mechanical work is called the efficiency of a heat engine; thus

$$\text{efficiency} = \frac{Q_2 + Q_1}{Q_2} = -\frac{W}{Q_2}; \quad (W < 0) \tag{4.2}$$

All the foregoing processes are assumed to be reversible, i.e., the engine operates very slowly and it is idealized to have no mass and no friction during its operation.

## Carnot Engine

A type of idealized heat engine having far-reaching consequences in the development of thermodynamics was discovered by S. Carnot (1824). The engine bearing his name is a reversible engine that operates between two large reservoirs at constant and uniform temperatures $T_2$ and $T_1$, and follows a cycle called the Carnot cycle. The working substance is any substance capable of undergoing cyclic changes.

Figure 4.1 shows the Carnot cycle in schematic detail. It consists of four steps or reversible processes performed usually in the following sequence: (A) The engine has diathermic walls and its working substance is at $P_1$, $V_1$, $T_2$, represented by (1). At this point, the engine is immersed in the hot reservoir which is also at $T_2$. The engine does work by absorbing $Q_2$ units of heat and expanding isothermally from (1) to (2). (B) At (2), the engine is removed and covered with adiabatic materials to perform reversible adiabatic work until it reaches (3) where the temperature drops to $T_1$. (C) The adiabatic materials are removed and the engine is immersed in a reservoir also at $T_1$ for isothermal compression from (3) to

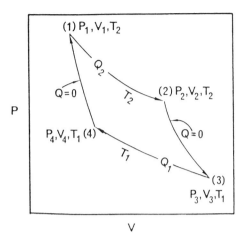

**Fig. 4.1** Carnot cycle.

(4). During this step, $Q_1$ units of heat are rejected by the engine. (D) The engine is removed from the cold reservoir, and adiabatically isolated from the surroundings and then compressed adiabatically until it reaches the initial state (1). The work necessary for compressing the system in steps (C) and (D) is supplied from steps (A) and (B). Point (4) must be chosen in such a way that the adiabatic compression restores the system to (1). The first law for steps (A) and (B) gives

$$\int_{(1)}^{(3)} dU \equiv U_3 - U_1 = Q_2 + \int_{(1)}^{(2)} dW + \int_{(2)}^{(3)} dW$$

$$\equiv Q_2 - \int_{(1)}^{(2)} P\, dV - \int_{(2)}^{(3)} P\, dV$$

where, for brevity, the limits of integration are shown by (1), (2), ... instead of $(P_1, V_1), (P_2, V_2), \ldots$ respectively. Likewise, for steps (C) and (D), the corresponding equation is

$$\int_{(3)}^{(1)} dU \equiv U_1 - U_3 = Q_1 + \int_{(3)}^{(4)} dW + \int_{(4)}^{(1)} dW$$

$$\equiv Q_1 - \int_{(3)}^{(4)} P\, dV - \int_{(4)}^{(1)} P\, dV$$

The sum of these two equations gives $\Delta U = 0$ since the system returns to its initial state and the result may be written as

$$Q_2 + Q_1 = -W \equiv \int_{(1)}^{(2)} P\, dV + \int_{(2)}^{(3)} P\, dV + \int_{(3)}^{(4)} P\, dV + \int_{(4)}^{(1)} P\, dV$$

The first two integrals on the right are positive in numerical value and their sum is the area under the curves (1)–(2) and (2)–(3). The last two integrals are negative

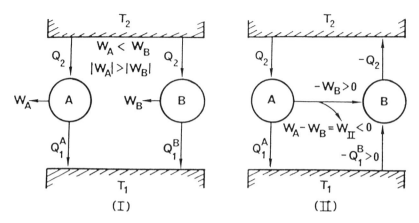

**Fig. 4.2** Proof of Carnot theorem. (I) Two reversible engines A and B operate to yield work. (II) Work $W_A$ from A is used to operate B in reverse and to pump $-Q_1^B$ from reservoir at $T_1$ into reservoir at $T_2$; $T_2 > T_1$.

in value and their sum is the area under the curves (3)–(4) and (4)–(1). The sum of the four integrals is therefore the area inside the cycle which represents $Q_2 + Q_1$, the amount of heat converted into work. In summary, the Carnot cycle consists of four reversible processes, two of which are the isothermal steps 1–2 and 3–4, and the two others are the adiabatic steps 2–3 and 4–1. There are other types of engines using different cycles, such as the Rankine cycle but they are not useful for our purposes because they have no particular advantage over the Carnot engine.

## Carnot Theorem

The Carnot theorem places an upper limit on the efficiencies of all heat engines operating between two temperatures. The statement chosen to enunciate the Carnot theorem is as follows:

*All reversible cyclic heat engines operating between two fixed temperatures have the same efficiency.*

The proof is obtained in the following manner. Consider any two reversible engines A and B operating between two reservoirs at $T_2$ and $T_1$ as shown in Fig. 4.2, (I). The engines may run several cycles in such a way that they return to their initial states after having absorbed the same amount of heat, $Q_2$ per cycle. The quantities of heat discharged by Engines A and B are designated by $Q_1^A$ and $Q_1^B$ respectively. If the engines A and B do not have the same efficiency, assume first that Engine A is more efficient than Engine B, or simply

$$\frac{Q_2 + Q_1^A}{Q_2} > \frac{Q_2 + Q_1^B}{Q_2} \tag{4.3}$$

Since $Q_2 > 0$, it follows from this equation that

$$Q_1^A > Q_1^B \tag{4.4}$$

The quantities of heat rejected by the engines are negative in value; therefore, $|Q_1^A| < |Q_1^B|$, or the more efficient engine rejects less heat than the less efficient engine. Let Engine B operate in reverse as shown in Fig. 4.2, (II) by coupling it with Engine A by a suitable gear mechanism. Engine A is capable of reversing Engine B since it produces more than enough work by virtue of its higher efficiency. Engine B will take up $-Q_1^B$ quantity of heat at $T_1$ and $-W_B$ amount of work from Engine A and discharge $-Q_2$ quantity of heat at $T_2$. The work $W_{II}$ is what remains from $W_A$ after $-W_B$ goes into Engine B. The heat $Q_2$ absorbed by engine A is completely returned to the hot reservoir by Engine B and therefore there is no loss of heat from the hot reservoir. Hence, the cyclic operation of the coupled engines causes no change in the energy of hot reservoir. The application of the first law to the coupled engines gives

$$Q_2 + Q_1^A + \left(-Q_2 - Q_1^B\right) + W_{II} = 0$$

from which

$$-W_{II} = Q_1^A - Q_1^B \tag{4.5}$$

Since the right hand side of this equation is positive according to (4.4), $Q_1^A - Q_1^B$ units of heat are absorbed by the coupled engines and entirely converted into mechanical work. The coupled engines operating in cycles therefore take up heat only from the cold reservoir and convert it entirely into work without causing any change in the hot reservoir. This is in contradiction with the second law of thermodynamics; therefore, Engine A cannot be more efficient than Engine B. Next, assume that Engine A is less efficient than Engine B. By proceeding in the same manner except reversing the Engine A instead of B, it can be shown that this assumption also leads to the violation of the second law. Since the efficiency of one engine cannot be greater or smaller than the efficiency of the other, it must be concluded that all reversible engines operating between the same temperatures have the same efficiency irrespective to the working substance and the mode of operation.

The Carnot engine is also a reversible engine; therefore, the efficiency of any reversible engine operating between two temperatures is the same as the efficiency of a Carnot engine operating between the same temperatures. These arguments are applicable only to the reversible engines; they do not preclude the possibility of one engine being less efficient than the other because of various degrees of irreversibility such as friction, or heat leak to the surroundings. However, lower efficiencies due to irreversible processes are not in disagreement with the second law.

## Kelvin Temperature Scale

The Carnot theorem proves that the efficiencies of all reversible engines operating between two temperatures are the same, i.e., the efficiency is independent of the

76   Thermodynamics

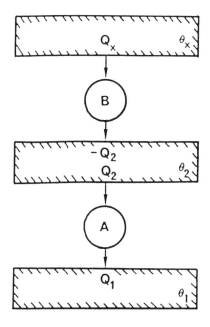

**Fig. 4.3** Carnot engines for establishing Kelvin temperature scale.

working substance and the modes of operation of the engines provided that the operation is reversible. The efficiency is therefore a function of the temperatures of reservoirs; hence

$$\frac{Q_2 + Q_1}{Q_2} = \varphi(\theta_1, \theta_2) \tag{4.6}$$

where $\varphi$ is a function of $\theta_1$ and $\theta_2$ representing the temperatures of the cold and hot reservoirs respectively on some arbitrary scale. Simplification of (4.6) gives

$$Q_1/Q_2 = \varphi(\theta_1, \theta_2) - 1 = f(\theta_1, \theta_2) \tag{4.7}$$

Figure 4.3 shows a Carnot engine A for which (4.6) and (4.7) are applicable. Let there be another such engine operating between any arbitrary temperature $\theta_x$ and $\theta_2$ and discharging as much heat as that absorbed by engine A. The combined engines A and B have no effect on the reservoir at $\theta_2$ and therefore they are equivalent to a single Carnot engine operating between $\theta_x$ and $\theta_1$; therefore,

$$Q_1/Q_x = f(\theta_1, \theta_x) \tag{4.8}$$

The function $f$ here has the same form as in (4.7) since it is related to the efficiency function. Similarly for engine B,

$$-Q_2/Q_x = f(\theta_2, \theta_x) \tag{4.9}$$

Dividing (4.9) by (4.8) and comparing with (4.7) gives

$$\frac{Q_2}{Q_1} = -\frac{f(\theta_2, \theta_x)}{f(\theta_1, \theta_x)} = f(\theta_1, \theta_2) \tag{4.10}$$

where the last equality is from (4.7). Equation (4.10) is independent of the arbitrarily chosen temperature $\theta_x$; therefore $\theta_x$ must cancel out of the ratio in the center. This is possible if and only if $f(\theta_2, \theta_x) = \phi(\theta_2)/g(\theta_x)$ and $f(\theta_1, \theta_x) = \phi(\theta_1)/g(\theta_x)$ so that

$$Q_2/Q_1 = -\phi(\theta_2)/\phi(\theta_1) \tag{4.11}$$

Now designate $\phi(\theta)$ by $\tau$ which is a new arbitrary temperature scale and label it as the thermodynamic temperature scale. Equation (4.11) now assumes the simple form

$$Q_2/Q_1 = -\tau_2/\tau_1 \tag{4.12}$$

The size of a degree on this scale is completely defined if $\tau_1$ or $\tau_2$ is assigned an arbitrary number corresponding to a single fixed point. If $\tau_1$ is assigned a fixed value, then the value of $\tau_2$ is obtained from

$$\tau_2 = -\tau_1(Q_2/Q_1) \tag{4.13}$$

by substituting the measured values of $Q_2$ and $Q_1$ from the isothermal processes of a Carnot engine. This entails the actual construction of a Carnot engine and its operation between a fixed point $\tau_1$ and an unknown temperature, $\tau_2$. It is not possible to operate such an engine accurately enough for highly reliable temperature measurements, but it is possible to operate such an engine hypothetically if an ideal gas is used as the working substance as in the following section.

## Carnot Engine with an Ideal Gas

A Carnot engine using one mole of an ideal gas as its working substance follows a cycle shown in Fig. 4.4 where the temperatures are expressed on the ideal gas scale. The stages 1–2 and 3–4 are isothermal processes, and 2–3 and 4–1 are

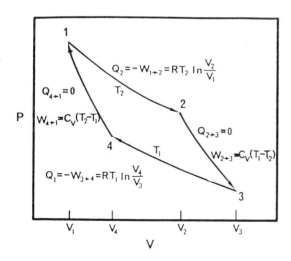

**Fig. 4.4** Carnot engine using one mole of an ideal gas as working substance.

adiabatic processes. The work and the heat in each stage are shown in Fig. 4.4 and summarized as follows:

| Stage | Heat | Work |
|---|---|---|
| 1–2 | $Q_2 = RT_2 \ln \dfrac{V_2}{V_1}$ | $-RT_2 \ln \dfrac{V_2}{V_1}$ |
| 2–3 | 0 | $C_V(T_1 - T_2)$ |
| 3–4 | $Q_1 = RT_1 \ln \dfrac{V_4}{V_3}$ | $-RT_1 \ln \dfrac{V_4}{V_3}$ |
| 4–1 | 0 | $C_V(T_2 - T_1)$ |

The resulting conversion of heat into work is represented graphically by the area inside the lines 1–2–3–4 and expressed by

$$Q_2 + Q_1 = -W = RT_2 \ln(V_2/V_1) + RT_1 \ln(V_4/V_3) \tag{4.14}$$

It is seen that the work in the adiabatic stage 2–3 cancels that in stage 4–1. For stage 2–3 and stage 4–1, the adiabatic relationship gives

$$\left(\frac{T_1}{T_2}\right)^{C_V/R} = \frac{V_2}{V_3}, \quad \text{and} \quad \left(\frac{T_2}{T_1}\right)^{C_V/R} = \frac{V_4}{V_1}$$

from which

$$V_2/V_1 = V_3/V_4 \tag{4.15}$$

Substitution of (4.15) into (4.14) gives

$$Q_2 + Q_1 = RT_2 \ln(V_2/V_1) - RT_1 \ln(V_2/V_1)$$

From this equation, and from $Q_2 = RT_2 \ln(V_2/V_1)$, the efficiency becomes

$$\boxed{\text{efficiency} = \frac{Q_2 + Q_1}{Q_2} = \frac{T_2 - T_1}{T_2}} \tag{4.16}$$

The efficiency approaches unity when $T_2$ is very large, or when $T_1$ is very small. Thus, for $T_2 = 800$ K and $T_1 = 400$, the Carnot efficiency is 0.50, whereas for $T_2 = 600$ and $T_1 = 400$, it drops to 0.33.

Equation (4.16) is valid within the assumptions concerning ideal gas behavior, and it may be simplified into

$$Q_1/Q_2 = -T_1/T_2 \tag{4.17}$$

This equation is in a form identical with (4.12), therefore

$$\tau_1/\tau_2 = T_1/T_2 \tag{4.17a}$$

The ideal gas and the thermodynamic scales of temperature become identical if the triple point of water (0.0100°C) is taken to be 273.16 on both scales. It is quite evident that a single fixed point is sufficient for the establishment of a thermodynamic temperature scale and its measurement by (4.17) as first suggested by Kelvin (1854); therefore, $T$ on this scale is called the Kelvin temperature scale, and designated by K, (formerly °K).

## Refrigeration Engine

When the Carnot engine is operated in reverse, it becomes a refrigeration engine which receives work, and absorbs $Q_{1r}$ at a low temperature and discharges $Q_{2r}$ at a high temperature. Thus in effect the engine pumps up heat from a cold reservoir by receiving work, and therefore, refrigerates or maintains a low temperature. Since both engines follow the same steps reversibly but in opposite directions, (4.16) is again valid. For each complete cycle, again $Q_{2r} + Q_{1r} + W_r = 0$ from (4.14), and the solution for $W_r/Q_{1r}$ gives

$$\frac{W_r}{Q_{1r}} = \frac{T_2 - T_1}{T_1} \tag{4.18}$$

where $W_r$ is the amount of work received by the engine to pump up $Q_{1r}$ joules from a lower temperature reservoir to a higher temperature reservoir (see Problem 4.5). The amount of work necessary to remove a fixed amount of heat $Q_{1r}$ to a fixed temperature $T_2$ increases with decreasing $T_1$ and approaches infinity with $T_1 \to 0$ according to this equation. This conclusion is called the principle of unattainability of 0 K because of the infinite amount of work required to accomplish the necessary refrigeration.

## Spontaneous Processes

All processes occurring in nature without the interference of externally directed forces are called *spontaneous*. For example, the flow of heat from a hot to a cold body, the flow of water from a high to a low level, the expansion of a compressed gas into the atmosphere, etc., are spontaneous processes. All systems have tendencies to reach a certain stable level after a spontaneous process. It is a universal experience that these processes do not reverse themselves without external interference. It will now be shown that this is a necessary consequence of the second law.

Assume first that a spontaneous transfer of heat from a cold body to a hot body is possible. In that event, $Q_1$ rejected by engine A in Fig. 4.2 can go entirely into the hot reservoir at $T_2$ through a hypothetical medium. As a result, it would be possible to convert heat from the reservoir at $T_2$ into mechanical work without leaving any effect in the reservoir at $T_1$. This obviously contradicts the second law. Likewise, if it were possible for water in a reservoir to move spontaneously upward against gravity, into a tank, the conservation of energy would require that its temperature must decrease to compensate for the gain in potential energy. When the water is returned to the reservoir, mechanical energy could be generated without leaving any effect elsewhere in contradiction with the second law. The criterion for the possibility of occurrence of a spontaneous process will now be formulated.

## Reversible Cyclic Processes

For any reversible engine that takes the system from the initial state 1 to the final state 3 and back to the initial state, as shown in Fig. 4.1, (4.17) may be rewritten as

$$\frac{Q_2}{T_2} + \frac{Q_1}{T_1} = 0 \tag{4.19}$$

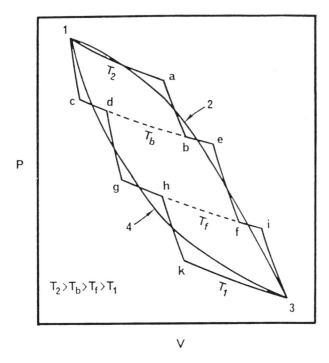

**Fig. 4.5** Three Carnot engines in succession for proving existence of a function $dS = dQ_{\text{rev}}/T$.

where $Q_2$ is the heat absorbed when the system is taken along 1–2–3, and $Q_1$ along 3–4–1. It should be emphasized that $Q_2$ and $Q_1$ are the quantities of heat exchanged between the working substance and the reservoirs at $T_2$ and $T_1$ respectively, and the temperatures of the reservoirs are assumed to be constant and uniform. Instead of having two heat reservoirs and one engine that takes the system from 1 to 3 as in Fig. 4.1, it is possible to have four reservoirs and three engines for the same purpose as shown in Fig. 4.5. The temperatures $T_2$ and $T_1$ are the same in both figures, and $T_b$ and $T_f$ are for the intermediate cycle $cegf$. The application of (4.18) to all three cycles gives

$$\left(\frac{Q_{1a}}{T_2}\right)_{1ab} + \left(\frac{Q_{bd} + Q_{dc}}{T_b}\right)_{bdc1} = 0, \quad \text{(Cycle: } 1abc1\text{)}$$

$$\left(\frac{Q_{db} + Q_{be}}{T_b}\right)_{dbef} + \left(\frac{Q_{fh} + Q_{hg}}{T_f}\right)_{fhgd} = 0, \quad \text{(Cycle: } defgd\text{)}$$

$$\left(\frac{Q_{hf} + Q_{fi}}{T_f}\right)_{hfi3} + \left(\frac{Q_{3k}}{T_1}\right)_{3kh} = 0, \quad \text{(Cycle: } hi3kh\text{)}$$

These equations may be added together to obtain an important relationship. Since $Q_{bd} = -Q_{db}$, the term $(Q_{bd} + Q_{db})/T_b$ is zero, and likewise, $(Q_{fh} + Q_{hf})/T_f$ is also zero; the dotted lines $db$ and $hf$ are therefore cancelled in the final sum, and the result becomes

$$\left(\frac{Q_{1a}}{T_2}\right)_{1ab} + \left(\frac{Q_{be}}{T_b}\right)_{bef} + \left(\frac{Q_{fi}}{T_f}\right)_{fi3} + \left(\frac{Q_{3k}}{T_1}\right)_{3kh}$$
$$+ \left(\frac{Q_{hg}}{T_f}\right)_{hgd} + \left(\frac{Q_{dc}}{T_b}\right)_{dc1} = 0$$

The first three terms take the system from 1 to 3 along the zig-zag path $1abefi3$ as shown by the succession of subscripts, or for brevity, $1e3$. Likewise the last three terms return the system from 3 to 1 along $3khgdc1$, or for brevity, $3g1$; hence the foregoing equation can be regarded as follows:

$$\left[\frac{Q_{1a}}{T_2} + \frac{Q_{be}}{T_b} + \frac{Q_{fi}}{T_f}\right]_{1e3} + \left[\frac{Q_{3k}}{T_1} + \frac{Q_{hg}}{T_f} + \frac{Q_{dc}}{T_b}\right]_{3g1} = 0 \quad (4.20)$$

If, instead of three cycles, there were an infinite number of cycles, it would have been hypothetically possible to approximate the curve 1–2–3 by $1e3$, and the curve 3–4–1 by $3g1$. The resulting equations for the curves would then be

$$\text{Path 1–2–3} \sum_{i=1}^{\infty} \frac{\Delta Q_i}{T_i} + \text{Path 3–4–1} \sum_{j=1}^{\infty} \frac{\Delta Q_j}{T_j} = 0 \quad (4.21)$$

where $\Delta Q_i$ and $\Delta Q_j$ are small increments in $Q$ at each temperature. The sums may be replaced by the integrals since $\sum \Delta Q/T = \int dQ/T$; hence,

$$\text{Path 1–2–3} \int_{(1)}^{(3)} \frac{dQ}{T} + \text{Path 3–4–1} \int_{(3)}^{(1)} \frac{dQ}{T} = \oint \frac{dQ}{T} = 0 \quad (4.22)$$

The circular integral takes the system from 1 to 3 along 1–2–3 and returns it to 1 along 3–4–1, and since this integral is zero $dQ/T$ is the differential of an exact function $S$; i.e.,

$$\boxed{dS = dQ_{\text{rev}}/T} \quad (4.23)$$

where the subscript "rev" is to emphasize that all values of $Q$ refer to reversible processes. The factor $1/T$ is therefore an integrating factor of inexact differential $dQ$, or $T$ is an integrating denominator. It was shown in Chapter III that $dQ$ is not an exact differential because it depends on the path followed by the system. For example in Fig. 3.3 when the system was taken from the initial state $P_1, T_1$ to the final state $P_2, T_2$ along the isothermal and isobaric curves, $Q$ was 3660 J whereas along the adiabatic path to the same final state, $Q$ was zero, and along the closed path, the isothermal and isobaric processes shown by the arrows and returning from $P_2, T_2$ to $P_1, T_1$ along the reversed adiabatic path, $Q$ is not equal to zero. Therefore, $dQ = dU + P\,dV = C_V\,dT + RT\,d\ln V$ for an ideal gas is not exact. This is also evident from the fact that the cross differentials are not equal, i.e., $\partial C_V/\partial \ln V = 0 \neq R\partial T/\partial T = R$, but $T$ is an integrating denominator in view of the fact that

$$dQ/T = C_V d\ln T + R d\ln V = d\ln(V^R T^{C_V})$$

where the right side is a function of $V^R T^{C_V}$ which has a unique value when integrated from $V_1, T_1$ to $V_2, T_2$ irrespective of the path followed.

It should be emphasized that the first law requires that if a system gains energy, it must go from an initial state $P_1, V_1$ to a final state $P_2, V_2$ along a path set by the method of furnishing the energy; hence the energy acquired by the system is functionally related to the variables of state, or the function $U = U(P, V)$ exists and when the system returns to the initial state along another path, the same amount of energy must be lost by the system so that the circular integral $\oint dU(P, V)$ always becomes zero and therefore $dU$ is an exact differential. Likewise the circular integral of $dQ_{\text{rev}}/T$ is zero as required by the second law and shown by (4.22); hence $dQ_{\text{rev}}/T$ is an exact differential of some function $S = S(P, V)$. Both laws therefore establish functions useful in thermodynamics.

Integration of (4.23) along any path that joins $P_1, V_1$ to $P_2, V_2$ gives the same value of $S(P_2, V_2) - S(P_1, V_1)$ since $dS$ is exact. This is also evident from (4.22) when the sign of the second integral is changed by $-\int_{(3)}^{(1)} = +\int_{(1)}^{(3)}$ and (4.22) is rearranged so that

$$S(P_2, V_2) - S(P_1, V_1) = \text{Path 1-2-3} \int_{(1)}^{(3)} \frac{dQ}{T}$$

$$= \text{Path 1-4-3} \int_{(1)}^{(3)} \frac{dQ}{T} \qquad (4.24)$$

Thus the integral of $dQ/T$ for reversible exchanges of heat is independent of the path and dependent on the initial and final values of the independent variables of state for the system. In the simplest case of a single Carnot cycle $S_3 - S_1 = Q_2/T_2 = R\ln(V_2/V_1)$ along 1-2-3 in Fig. 4.4 and again $S_3 - S_1 = -Q_1/T_1 = R\ln(V_3/V_4)$ along 1-4-3 and since $V_2/V_1 = V_3/V_4$ by (4.15), the equality is evident. [The terms $R\ln(V_i/V_j)$ must be multiplied by $n$ if $n$ moles of an ideal gas were used instead of one mole in the Carnot cycle.] The function $S$, like $U$ and $H$, is an intensive property, and it may be represented as a function of the same set of independent variables of state.

The property $S$ was named *entropy* by Clausius (1850), a word signifying *change* in Greek. Its existence is clearly a consequence of the second law of thermodynamics. The substitution of (4.23) into the equation for the first law, i.e., $dU = dQ - PdV$, gives

$$\boxed{dU = TdS - PdV} \qquad (4.25)$$

which is frequently considered to be a formal summary of both laws. It is one of the most important equations in thermodynamics.

## Entropy Change in Reversible Processes

The entropy change of the system in reversible processes is the integral of $dQ_{rev}/T$ and the entropy of the surroundings is the integral of $-dQ_{rev}/T$ because $dQ_{rev}$ for the system is equal to $-dQ$ for the surroundings. Hence, for the reversible processes

$$(S_2 - S_1)_{sys} + (S_{fin} - S_{in})_{surr} = 0 \tag{4.26}$$

The entropy change of the system is zero for a complete reversible cycle, and since $\Delta S_{sys} = -\Delta S_{surr}$ according to (4.26), the entropy change of the surroundings is also zero.

The entropy change of the surroundings in an *irreversible* process is expressed by an equation similar to (4.24), i.e.,

$$(S_{fin} - S_{in})_{surr} = -\int_{in}^{fin} \frac{dQ_{irr}}{T} \tag{4.27}$$

because the surroundings can be restored to their initial state by reversing the direction of heat flow irrespective of the reversibility.

## Entropy Change in Irreversible Processes

When there is any irreversibility in the Carnot cycle, such as friction in stage 1–2, the efficiency would be less because a certain quantity of work must be used to overcome the friction. Therefore, $Q_2(rev) > Q_2(irr)$ since $Q_2(irr)$ is a smaller quantity because friction would convert some of the mechanical work back into heat. In the limiting case when friction converts all the work into heat, the net heat exchange $Q_2(irr)$ would obviously be zero. Dividing $Q_2(rev) > Q_2(irr)$ by $T_2$ and subtracting $[Q_2(rev)/T_2] + [Q_1(rev)/T_1] = 0$ from the inequality gives

$$-\frac{Q_2(irr)}{T_2} - \frac{Q_1(rev)}{T_1} > 0 \tag{4.28}$$

The first term in this inequality is the entropy change of the surroundings, $(S_3 - S_1)_{surr}$, and the second term is the entropy change of the system $(S_3 - S_1)_{sys}$ so that

$$(S_3 - S_1)_{surr} + (S_3 - S_1)_{sys} > 0 \tag{4.29}$$

Thus, when the system goes to (3) along 1–2–3 in Fig. 4.1 with any degree of irreversibility, the entropy change of the system plus the surroundings is greater than zero. Now let the system return reversibly to (1) along 3–4–1. Since $Q_1/T_1$ is $(S_1 - S_3)_{sys}$ and $-Q_1/T_1 = (S_1 - S_3)_{surr}$, it follows that

$$(Q_1/T_1) - (Q_1/T_1) = 0 = (S_1 - S_3)_{sys} + (S_1 - S_3)_{surr} \tag{4.30}$$

Addition to this equation to (4.29) makes the entropy change for the system zero since the system returns to its initial state but the surroundings gain a net amount of entropy greater than zero; consequently for a complete cycle containing irreversibility,

$$(\Delta S_{sys} + \Delta S_{surr})_{cycle} = \Delta S_{surr} > 0 \tag{4.31}$$

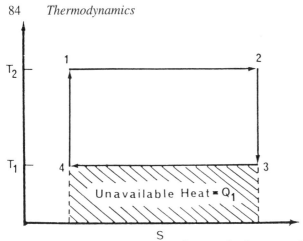

**Fig. 4.6** Temperature-entropy diagram for Carnot cycle and unavailable heat.

where Δ refers *in this case* to a complete cycle. This generalization is also applicable to Fig. 4.5 where more than one Carnot cycle takes the system through a complete cycle. From (4.29) and (4.30), it is evident that for any process one may write

$$(S_{\text{fin}} - S_{\text{in}})_{\text{sys}} + (S_{\text{fin}} - S_{\text{in}})_{\text{surr}} \geq 0 \tag{4.32}$$

where the equal sign refers to a reversible process and the greater sign, to an irreversible process. In view of the fact that all natural processes are irreversible, Clausius (1865) stated that the entropy of the universe is continually increasing. Although this is true within the portion of the universe which can be observed with confidence, it cannot be applied with absolute certainty to the entire universe which is too large to be subjected to a thermodynamic investigation.

The second law of thermodynamics may be restated by means of (4.32) in the following form: *There is no process by which the entropy of a system and its surroundings can be decreased.* In terms of the cyclic engines, it may also be stated that "it is impossible to construct a cyclic engine which can decrease the sum of entropies of the hot and cold reservoirs."

The Carnot cycle may also be represented on the temperature-entropy coordinates as suggested by Gibbs, and shown in Fig. 4.6. The paths 1–2 and 3–4 are isothermal, hence they are perpendicular to the $T$-axis, but 2–3 and 4–1 are adiabatic, during which there is no change in entropy. For this reason, a reversible adiabatic process is also called an isentropic process. The quantity $Q_1$ may be expressed from Fig. 4.6, by

$$T_1(S_4 - S_3) \equiv T_1 \Delta S_1 = Q_1$$

where $\Delta S_1$ is identical with $(S_4 - S_3)$, and the numerical value of $Q_1$, hence $\Delta S_1$, is negative as is shown by the crosshatched area. If $\Delta S_1$ were zero, all the heat absorbed at $T_2$ would be converted into work. But this is against the second law;

therefore the heat engines are said to "discharge entropy" at the sink where the quantity of heat $Q_1$ is no longer *available* for work. The expression $T_1 \Delta S_1$, i.e., the product of temperature and entropy change, is therefore called the "unavailable heat." The part of $Q_2$ converted into work is then

$$Q_2 + T_1 \Delta S_1 = -W$$

or since $\Delta S_1 = -\Delta S_2$,

$$Q_2 - T_1 \Delta S_2 = -W$$

If, for example, there is friction in the stage 3-4, the value of $|Q_1|$ would increase, i.e., the amount of unavailable heat and the discharged entropy would also increase.

## Method of Carathéodory*

An alternative statement of the second law of thermodynamics, formulated by Carathéodory (1909), is based on the properties of Pfaff equations, i.e.,

$$L\,dx + M\,dy + N\,dz + \cdots = đQ = 0; \quad \text{(adiabatic)} \tag{4.33†}$$

where $L$, $M$ and $N$ are some functions of the variables $x, y, z, \ldots$. The differential $đQ$ may or may not be exact, and if it is not exact, it may or may not have an integrating denominator $\lambda$. If there is an integrating denominator, then $đQ/\lambda = dF(x, y, z \ldots)$ is exact and $F$ is a single-valued function of $x, y, z, \ldots$.

A Pfaff equation with two variables is

$$L\,dx + M\,dy = đQ = 0 \tag{4.34}$$

This equation may be exact or inexact, but it is always integrable. Therefore, when it is inexact, it is always possible to find an integrating denominator that would make it exact so that (2.43) is satisfied. Thus, $xy\,dx + x^2\,dy = đQ = 0$ is not exact because (2.43) becomes $\partial(xy)/\partial y \equiv x \neq \partial x^2/\partial x \equiv 2x$. However $x$ is an integrating denominator, or $1/x$ is an integrating factor, which transforms the equation into $y\,dx + x\,dy = d(xy) = 0$. Alternatively the variables are separable; hence $dy/y = -dx/x$, or $d\ln(xy) = 0 = d(xy)$, and $xy = $ constant. The solution for this Pfaff differential equation is then a family of hyperbolas that fill the $x$-$y$ coordinates and satisfy both $đQ$ and $đQ/x$.

*Pfaff Equations with Three Variables:* A Pfaff equation with three independent variables is

$$L\,dx + M\,dy + N\,dz = đQ = 0 \tag{4.35}$$

---

*This method is for advanced readers.
†It is important to note that the differential before the thermodynamic quantities $Q$ and $W$ should be $đ$, not $d$, since $đQ$ and $đW$ are inexact differentials in thermodynamics. In the mathematical examples of Chapter II occasional uses of $d$ for $đ$ should not unduly disturb the reader.

If $dQ$ is an exact differential the condition of exactness consists of three equations, i.e.,

$$\left(\frac{\partial L}{\partial y}\right)_{x,z} = \left(\frac{\partial M}{\partial x}\right)_{y,z}, \left(\frac{\partial M}{\partial z}\right)_{x,y} = \left(\frac{\partial N}{\partial y}\right)_{x,z}, \left(\frac{\partial N}{\partial x}\right)_{y,z} = \left(\frac{\partial L}{\partial z}\right)_{x,y} \tag{4.36}$$

It is far more frequent that equations in the form of (4.35) are inexact differentials, $dQ$, and therefore (4.36) is not obeyed. On rare occasions an inexact equation of the form given by (4.35) has an integrating denominator $\lambda$ so that

$$\frac{L}{\lambda}dx + \frac{M}{\lambda}dy + \frac{N}{\lambda}dz = \frac{dQ}{\lambda} = dF = 0 \tag{4.37}$$

and this equation satisfies the condition of exactness so that

$$N\frac{\partial(L/\lambda)}{\partial y} = N\frac{\partial(M/\lambda)}{\partial x}, L\frac{\partial(M/\lambda)}{\partial z} = L\frac{\partial(N/\lambda)}{\partial y}, M\frac{\partial(N/\lambda)}{\partial x} = M\frac{\partial(L/\lambda)}{\partial z}$$

where both sides of each equation is multiplied by the same symbol. When these equations are summed, the terms containing $\lambda$ cancel out, and the following equation is derived:

$$L\left(\frac{\partial M}{\partial z} - \frac{\partial N}{\partial y}\right) + M\left(\frac{\partial N}{\partial x} - \frac{\partial L}{\partial z}\right) + N\left(\frac{\partial L}{\partial y} - \frac{\partial M}{\partial x}\right) = 0 \tag{4.38}$$

This equation is called the *condition of integrability*; it assures that an integrating denominator exists, even when the condition of exactness of (4.36) is not obeyed.

**Examples** (a) The condition of exactness is not obeyed by $dx + x\,dy - x\,dz = 0$, because (4.36) gives $0 \neq 1$, $0 = 0$, and $-1 \neq 0$, since $L = 1$, but the condition of integrability (4.38) is obeyed. However, in this simple case $1/x$ is an integrating factor because it separates the variables to give $(dx/x) + dy - dz = 0$, or $F = y - z + \ln x$. (b) The following equations

$$dx + dy + y\,dz = 0 \tag{4.39}$$

$$y\,dx - z\,dy + x\,dz = 0 \tag{4.40}$$

$$y\,dx + dy - x\,dz = 0 \tag{4.41}$$

are not integrable as can readily be verified. The overwhelming majority of equations with three or more independent variables are inexact as well as non-integrable. In fact, the existence of an integrating factor is quite rare.

If the Pfaff equation $L\,dx + M\,dy + N\,dz = dQ = 0$ is inexact but possesses an integrating factor $1/\lambda$, then $dQ/\lambda = dF$ is exact, and both $dQ$ and $dF$ are satisfied by a one parameter family of surfaces represented by $F(x, y, z) =$ constant in the $x$-$y$-$z$ space. If we start from a point $(a, b, c)$ along any path so that $dQ = 0$, we must move on a surface satisfying $dQ$ and $dF$, and therefore it is not possible to reach any arbitrary point in space. If, however, $dQ$ is non-integrable, there are *no surfaces capable of representing đQ = 0, and therefore it should be possible to reach any arbitrary point from any selected point* $(a, b, c)$.

## The Second Law of Thermodynamics

The statement chosen by Carathéodory for the second law of thermodynamics is, in its essential form, as follows: "In the close neighborhood of an arbitrarily selected state of a given system, there are states that cannot be reached by any adiabatic process whatever." The states that cannot be reached from the selected state may be chosen as close to the selected point as possible. This statement is Carathéodory's enunciation of the second law, in contrast with that by Clausius, Kelvin, or Planck. This statement assures that (4.35) possesses an integrating denominator. The significance of this statement may be illustrated by one mol of an ideal gas for which the adiabatic equation is

$$đQ = dU + P\,dV = C_V\,dT + P\,dV \tag{4.42}$$

The conditions $đQ = 0$, and $P(\text{external}) = P(\text{internal})$ and very slow changes in $V$ define a reversible adiabatic process. If we dissipate mechanical energy irreversibly by friction as for example by rotating a paddle wheel while keeping $V$ constant, it is evident that $dU$ as well as $đQ$ would increase.

It is evident that this equation is not exact, but a Pfaff equation with two independent variables is always integrable, and in this trivial case, $T$ (or $PV/R$) is an integrating denominator, and $đQ/T$ is now exact. Also, in this case, there are points that cannot be reached from a selected initial point by any adiabatic process whatever, and this statement redundantly assures us that the foregoing equation has an integrating denominator. In order to have a non-trivial case, it is necessary to have a composite system with three independent variables. Such a composite system could be that in Fig. 1.1, wherein the common partition A must now be made *diathermic*. The independent variables are then $T$, $V^{\mathrm{I}}$, and $V^{\mathrm{II}}$. The Pfaff equation is then

$$đQ = đQ^{\mathrm{I}} + đQ^{\mathrm{II}} = \left(C_V^{\mathrm{I}} + C_V^{\mathrm{II}}\right)dT + P^{\mathrm{I}}\,dV^{\mathrm{I}} + P^{\mathrm{II}}\,dV^{\mathrm{II}} \tag{4.43}$$

Carathéodory's enunciation of the second law assures that, in the vicinity of a selected point, there are points that cannot be reached by any adiabatic process whatever; hence, $đQ$ must have an integrating denominator. Since $đQ^{\mathrm{I}}/T$ and $đQ^{\mathrm{II}}/T$ are both exact, and $T$ is common to both systems, we conclude that $T$ is also the integrating denominator of $đQ$, since the sum of two exact differentials, $đQ^{\mathrm{I}}/T$ and $đQ^{\mathrm{II}}/T$ is also exact; hence,

$$\frac{đQ}{T} = \frac{(đQ^{\mathrm{I}} + đQ^{\mathrm{II}})}{T} = dS \tag{4.44}$$

The property $đQ/T$ may then the denoted by $dS$, the differential of "entropy". A more rigorous treatment with geometrical examples is in the first edition of this book, but the resulting conclusions are the same.

## PROBLEMS

4.1 A Carnot cycle using helium as the working substance is described by the following data: $T_2 = 500$ K, $T_1 = 300$ K, $P_1 = 4$ bar, $V_1 = 25$ L, before expansion at 500 K, and $P_3 = 1$ bar and $V_3 = 60$ L after the adiabatic expansion to 300 K. Compute the values of $Q$, $W$ and the changes of entropy for each stage, and the efficiency of the engine. Draw the Carnot cycle to the exact scale on $(V, P)$ coordinates.

4.2 (a) Draw the Carnot cycle for one mole of an ideal diatomic gas to the exact scales on $(V, S)$ and $(T, P)$ coordinates for $T_2 = 700$ K, $P$ (initial) $= 10$ bars, $P$ final $= 3$ bars for the upper isothermal expansion, and $T_1 = 315$ K for the lower isothermal compression. (b) What fraction of heat absorbed at $T_2$ is converted into mechanical energy? (c) What is the change in the entropy of surroundings per cycle when 20% of the work for the reversible cycle is lost by friction in the third stage?

4.3 Prove the Carnot theorem by assuming that the engines in Fig. 4.2 (I) discharge the same amount of energy at $T_1$, i.e., $Q_1^A = Q_1^B$, but take up different amounts of heat at $T_2$.

4.4 A house is maintained at 25°C by air conditioning while outside temperature is 38°C. Assume that the air conditioner operates reversibly between 15 and 50°C. Calculate the amount of work necessary to remove 7 kJ of heat from the house.

Ans.: 850 J.

4.5 Show that $x\,dy - (y + x^2)\,dx = 0$ is an inexact differential and verify that $x^2$ is an integrating denominator. Give the functional form of other possible integrating denominators.

4.6 Show that $zy\,dx - xz\,dy + 2xy\,dz = đQ = 0$ is inexact but integrable and $z/y^2$ is the integrating factor, and integrate $z\,đQ/y^2$.

Ans.: $F = xz^2/y$.

4.7 Show whether or not the following Pfaff equations are, (i) exact, (ii) inexact, and (iii) integrable:

(a) $(x^2 + y)\,dx + (y^2 + x)\,dy = 0$,  (b) $y^2\,dx - x^2\,dy = 0$

(c) $x\,dy - y\,dx - zxy\,dz = 0$,  (d) $2yz\,dx + 3xz\,dy + 3xy\,dz = 0$

4.8 One mole of argon at $T_1$ and two moles of nitrogen at $T_2$ are isolated by adiabatic walls from the surroundings and from each other. Compute $\Delta S$, $\Delta U$ and $T$ (final) if initially $T_2 = 2T_1$, and when both systems are brought in contact with each other through a diathermic rigid wall. Assume that Ar and $N_2$ are ideal and $C_V$ is $3R/2$ for argon and $5R/2$ for nitrogen.

CHAPTER V

# ENTROPY AND RELATED FUNCTIONS

## Entropy Change

The entropy is an extensive property dependent on the variables of state $P, T, n_1, n_2, \ldots$, where $n_i$ is the number of moles of $i$, and the molar entropy is an intensive property dependent on $P, T, x_1, x_2, \ldots$ where $x_i$ is the mole fraction of $i$. For a simple closed system the variables of state are any set of two variables out of $P, V$ and $T$, and the change of entropy is

$$dS = \frac{dU + P\, dV}{T} \tag{5.1}$$

This equation is often rewritten as $dU = T\, dS - P\, dV$ and it is sometimes called an explicit statement of the first law of thermodynamics and an implicit statement of the second law because it recognizes the existence of entropy.

The change in $Q$ at a constant pressure is $đQ_P = d(U + PV)_P = dH_P = C_p\, dT$; hence,

$$dS_P = \frac{dH_P}{T} = \frac{C_p}{T} dT = C_p\, d\ln T \tag{5.2}$$

Integration of this equation gives

$$S(P, T_2) - S(P, T_1) = (S_2 - S_1)_P = \int_{T_1}^{T_2} \frac{C_p}{T} dT = \int_{T_1}^{T_2} C_p\, d\ln T \tag{5.3}$$

where the left side may also be represented by $\Delta S_P$. This is a very useful equation because most processes take place at a constant pressure. The corresponding equations for a constant volume process may be obtained by starting with $đQ_V = dU_V = C_V\, dT$. Numerical evaluation of the integral in (5.3) can be made analytically by means of the appropriate heat capacity equations. The use of empirical equation $C_p = \alpha + 2\beta T + 6\lambda T^2 - 2\epsilon/T^2$ [cf. (3.71)] in (5.3) for solids, liquids or gases gives

$$\Delta S_P = \alpha \ln T + 2\beta T + 3\lambda T^2 + \epsilon/T^2 + S_\theta \tag{5.4}$$

where $S_\theta$ is the integration constant for the right side.

Cross differentiation of $dU = T\,dS - P\,dV$ as given by (2.55) yields, after rearrangement, the following equations:

$$\left\{\begin{array}{ll} \left(\dfrac{\partial S}{\partial P}\right)_V = -\left(\dfrac{\partial V}{\partial T}\right)_S, & \left(\dfrac{\partial S}{\partial V}\right)_P = \left(\dfrac{\partial P}{\partial T}\right)_S \\[2mm] \left(\dfrac{\partial S}{\partial V}\right)_T = \left(\dfrac{\partial P}{\partial T}\right)_V, & \left(\dfrac{\partial S}{\partial P}\right)_T = -\left(\dfrac{\partial V}{\partial T}\right)_P \end{array}\right\} \quad (5.5)$$

These equations are called the Maxwell relations. The third of these relations may be integrated to express variation of $S$ with $V$, and the fourth, variation of $S$ with $P$; the latter may be integrated to obtain

$$S(P_2, T) - S(P_1, T) = (S_2 - S_1)_T = -\int_{P_1}^{P_2} (\partial V/\partial T)_P \, dP \quad (5.6)$$

For brevity, the left side of this equation may be represented by $\Delta S_T$. If the compressibility data are available, the derivative $(\partial V/\partial T)_P$ can be evaluated and (5.6) can be integrated. An equation of state such as $PV = RT$ for ideal gases and the Berthelot equation for nonideal gases permit a convenient analytical integration of (5.6).

The usefulness of the combined results of (5.3) and (5.6) is easily illustrated by writing the left sides of these equations and adding the results as follows:

$$\begin{array}{ll} S(P_1, T_2) - S(P_1, T_1) & \text{(from 5.3)} \\ \underline{S(P_2, T_2) - S(P_1, T_2)} & \text{(from 5.6)} \\ S(P_2, T_2) - S(P_1, T_1) & \end{array}$$

The combined results from (5.3) and (5.6) therefore permit computation of $\Delta S$ from an initial state $P_1, T_1$ to any final state $P_2, T_2$.

It is frequently desirable to correct the observed value of entropy for a real gas at one bar to obtain the entropy of the hypothetical ideal gas at one bar, $S°$, because all the tabular values for gases refer to $S°$. For this purpose (5.6) may be written as

$$S_2(\text{ideal } P_2 \to 0) - S_1(\text{obs. 1 bar}) = \int_{P_2}^{1} \left(\frac{\partial V}{\partial T}\right)_P dP \quad (5.7)$$

where $S_1$ is the observed entropy at 1 bar and $S_2$ is the entropy at a very low pressure, $P_2$, where the gas is ideal. Compression of the gas as a hypothetical ideal gas from $P_2$ to 1 bar gives

$$S°(\text{ideal, 1 bar}) - S_2(\text{ideal}) = -\int_{P_2}^{1} (R/P) \, dP \quad (5.8)$$

The desired correction is obtained by adding (5.7) and (5.8); i.e.,

$$S°(\text{ideal, 1 bar}) - S_1(\text{obs. 1 bar}) = \int_{P_2}^{1} \left[\left(\frac{\partial V}{\partial T}\right)_P - \frac{R}{P}\right] dP \quad (5.9)$$

The expression for $(\partial V/\partial T)_P$ can be obtained from the compressibility data or an appropriate equation of state. The most convenient equation for this purpose is the Berthelot equation; which may be substituted in (5.9) to obtain

$$S°(\text{ideal, 1 bar}) - S_1(\text{obs. 1 bar}) = \frac{27R}{32P_c}\left(\frac{T_c}{T}\right)^3. \tag{5.10}$$

Thus, for oxygen $P_c = 50.4$ bars, $T_c = 154.3$ K and the difference in entropies from this equation is about 0.700 J mol$^{-1}$ K$^{-1}$ at the boiling point of oxygen, i.e., 90.2 K, and at 200 K, 0.064 J mol$^{-1}$ K$^{-1}$.

The preceding relationships apply to any *closed* system real or ideal, homogeneous or heterogeneous. For ideal gases the entropy relations can be expressed in very simple forms. From $dQ = dU + P\,dV = C_V\,dT + P\,dV$, and from $P = RT/V$, the entropy change for one mole of ideal gas is

$$dS = \frac{C_V\,dT}{T} + R\frac{dV}{V}$$

from which

$$S(V_2, T_2) - S(V_1, T_1) = C_V \ln(T_2/T_1) + R\ln(V_2/V_1) \tag{5.11}$$

The term $V_2/V_1$ in this equation may be replaced by $T_2 P_1/T_1 P_2$ to obtain

$$\boxed{S(P_2, T_2) - S(P_1, T_1) = C_p \ln(T_2/T_1) - R\ln(P_2/P_1)} \tag{5.12}$$

From (5.11) and (5.12), it is readily seen that $\Delta S_V = C_V \ln(T_2/T_1)$, $\Delta S_T = R\ln(V_2/V_1) = -R\ln(P_2/P_1)$ and $\Delta S_p = C_p \ln(T_2/T_1)$. For $n$ moles of an ideal gas the foregoing equations must be multiplied by $n$. For example, (5.12) may be used to compute the entropy change when one mole of argon at 273 K and 1 bar is brought to 373 K and 2.182 bars; the result is $S_2 - S_1 = (5R/2)\ln(373/273) - R\ln 2.182 = 0$, where $C_p = 5R/2$, and in this simple case the increase in the entropy with temperature is cancelled by the decrease in entropy with compression to 2.182 bars. If the final pressure were 0.5 bar instead, $S_2 - S_1$ would be $1.4734R$.

## Entropy of Mixing of Ideal Gases

The entropy of mixing of ideal gases refers to the process in which *each of the component gases is initially at* $(P, T)$ *and after mixing, the mixture is also at* $(P, T)$. The process of mixing may also be described as follows: (1) Each species of $n_1, n_2, \ldots, n_c$ moles of ideal gas occupies the corresponding volumes $V_1, V_2, \ldots, V_c$; (2) the gases are at the same temperature and the same pressure; (3) the vessel is rigid and completely isolated from the surroundings; and (4) after the interconnection or the removal of partitions, there is no chemical reaction, and the pressure and temperature remain the same, but each gas occupies the entire volume of the mixture. The gases are therefore completely *mixed* under *adiabatic* and *isothermal* conditions. The entropy change for any gas $i$ is the same as that accompanying the isothermal expansion from $V_i$ to the total volume of the mixture $\sum_{i=1}^{c} V_i = V$ so that from (5.11)

$$\Delta S_i = -n_i R \ln(V_i/V) \qquad (5.13)$$

The ratio of the volume of each gas "$i$" to the total volume can be expressed as

$$V_i/V = n_i \Big/ \sum_{i=1}^{c} n_i = x_i$$

where $x_i$ is the mole fraction of the $i$th gas. Thus (5.13) becomes

$$\Delta S_i = -n_i R \ln x_i \qquad (5.14)$$

The total entropy change for the entire process is the sum of the contributions $S_i$ from all the species of gases, so that

$$\sum_{i=1}^{c} \Delta S_i = -\sum_{i=1}^{c} (n_i R \ln x_i) \qquad (5.15)$$

Dividing both sides by the total number of moles of gases transforms this equation into

$$\boxed{\Delta S = -R \sum_{i=1}^{c} x_i \ln x_i} \qquad (5.16)$$

where $\Delta S$ is the molar entropy of mixing, i.e., entropy of mixing for one mole of mixture. The value of $x_i$ is smaller than unity so that $\ln x_i$ is negative and the entropy of mixing is always positive. The adiabatic, isothermal mixing is spontaneous and it is always accompanied by an increase in the entropy of the system in accordance with the second law of thermodynamics.

## Entropy of Phase Change and Chemical Reactions

Under equilibrium conditions, the phase change occurs at a constant temperature and pressure. For a pure substance $A$ the phase change from the $\alpha$-state to the $\beta$-state, and the accompanying heat effects are

$$A(\alpha) \rightleftharpoons A(\beta), \quad \text{and} \quad (Q_{\text{rev}})_p = (H_\beta - H_\alpha)_p$$

the entropy of phase change is

$$(S_\beta - S_\alpha)_p = \left(\frac{Q_{\text{rev}}}{T_{\alpha\beta}}\right)_p = \left(\frac{H_\beta - H_\alpha}{T_{\alpha\beta}}\right)_p \qquad (5.17)$$

where $T_{\alpha\beta}$ is the transformation temperature. It has been pointed out by H. Crompton (1897) and J. W. Richards (1897) that the entropy of fusion of metals is a constant and on the order of $1.1 R \approx 9.1$ J mol$^{-1}$ K$^{-1}$. This statement, often called the *Richards rule*, is applicable to true metals which have closely packed crystal structures with large coordination numbers (number of nearest neighbors) of 8 to 12. In metals, the atoms are imbedded in an *electron gas*, and the bonding is called *metallic*. In non-metals, such as C, Si, Ge, As, Sb, Bi, Se, Te, the bonding is covalent, i.e., electrons are shared by the adjoining atoms, and their entropy of fusion is greater than $1.1R$. The molar entropies of fusion of substances with very low melting points, such as, $H_2$, He, $N_2$, $O_2$, A, are 1.01, 0.75, 1.37, 0.98, and $1.70R$ respectively, hence they vary by a factor of more than two from one element to another. The entropies of fusion of the same type of compounds, e.g., NaCl, KCl, NaF, are the same. There is no other approximate rule for compounds, and in fact for organic substances the Richards rule does not apply at all.

The molar entropy of vaporization at 1 bar is approximately $11R = 91.5$ J mol$^{-1}$ K$^{-1}$ for metals and many compounds with molecular weights not too far from 100. This is known as *Trouton's rule* (F. Trouton, 1884). It is applicable to vaporizing substances which do not associate or dissociate in the gaseous state.

If the experimental data on the enthalpies of fusion $\Delta_{fus}H$ and vaporization $\Delta_{vap}H$ are not available, they can be estimated by these rules if the melting point and the boiling point are known. For example, the melting and the boiling points of zinc are 692.7 and 1180.2 K, and from the Richards rule

$$\Delta_{fus}H = 9.1 \times 692.7 = 6{,}304 \text{ J mol}^{-1}$$

and the experimental value is 7322 J mol$^{-1}$; similarly, from Trouton's rule

$$\Delta_{vap}H = 91.5 \times 1180.2 = 107{,}988 \text{ J mol}^{-1}$$

whereas the actual value is 115,311 J mol$^{-1}$. A better agreement may sometimes be obtained by taking the entropies of fusion and vaporization of the neighboring metals in the periodic table, if experimental data for those metals are available.

The entropy change in chemical reactions can be written as

$$\boxed{\Delta_r S = \sum S_{products} - \sum S_{reactants}} \tag{5.18}$$

The methods used to evaluate the entropy change of chemical reactions will be the subject of the succeeding chapters. This relationship is used extensively in chemical thermodynamics.*

## Entropy, Randomness and Probability

An increase in the entropy of a system is always accompanied by a corresponding increase in the randomness of the system. Thus, a solid crystalline substance, in which the atoms are arranged in some sort of geometric pattern, becomes more random upon melting since the molecules can move more freely in the liquid state. When vaporization occurs, the molecules of the gaseous substance move more

---

*For experimental measurements see F. D. Rossini and H. A. Skinner, "Experimental Thermochemistry," *loc. cit.*; for real gases, General Ref. (19).

randomly than in the liquid state. Likewise, expansion of a gas into a vacuum, formation of a mixture of gases, or diffusion are each accompanied by an increase in entropy and in randomness. Consequently, it is useful to associate entropy with randomness or disorder. The correlation of entropy and probability is a subject of statistical mechanics; however, a simple relationship between the two concepts would elucidate the significance of entropy and its relation to randomness. For this purpose it is first necessary to establish the mathematical significance of probability. Let there be 2 boxes, one of which contains a marble. A blind-folded person reaching for any one box has on the average one chance out of two to find the marble. (If there were 3 boxes instead of 2, the chance of finding the marble in one attempt would be one out of three, or stated differently, only one out of three attempts would be successful). Let $V_1$ represent the volume of the particular box, $V_2$ that of all the boxes, and the probability of finding the marble in $V_1$ is defined by

$$\psi_1 = V_1/V_2 \tag{5.19}$$

Assume that there are two marbles, one black, the other white, and that the two boxes are interconnected to permit a free passage of the marbles when they are moved by shaking. The chance of finding both marbles in one box after each shake is one out of four, or $(1/2)^2$; the chance of finding only the white marble in the particular box is of course the same. In other words any configuration has equal probability of happening. If there are three marbles instead of two, the probability of any configuration in one box is one out of eight, or $(1/2)^3$. From the foregoing considerations, the probability, $\psi_1$, of finding $N_A$ marbles in a box having a volume of $V_1$ out of any number of interconnected boxes representing a volume of $V_2$ is

$$\psi_1 = (V_1/V_2)^{N_A} \tag{5.20}$$

This generalization may now be extended to an ideal gas to establish a relationship between entropy and probability when the marbles represent the molecules which are assumed to be distinguishable and $N_A$ is the Avogadro number. The probability of finding all the molecules in the entire volume, $V_2$, is of course unity, i.e.,

$$\psi_2 = (V_2/V_2)^{N_A} = 1 \tag{5.21}$$

Let $S_2$, represent the entropy corresponding to $\psi_2$, and $S_1$, that corresponding to $\psi_1$; and then let it be postulated that

$$S_2 - S_1 = -k \ln(\psi_2/\psi_1)^{N_A} = kN_A \ln(V_2/V_1) \tag{5.22}$$

where $kN_A$ is the same as the gas constant $R$. It is noteworthy that this equation is identical with $\Delta S_T$ from (5.11); therefore, the relationship (5.22) between entropy and probability is justified. Thus an increase in probability from $\psi_1$ to $\psi_2$ is the same as an increase in randomness or disorder.

### Thermodynamic Equations of State

The first law and the existence of entropy as given by $dU = T\,dS - P\,dV$ may be converted into the following relationship by imposing the constant-temperature condition and dividing by $(dV)_T$:

$$\left(\frac{\partial U}{\partial V}\right)_T = T\left(\frac{\partial S}{\partial V}\right)_T - P$$

Rearranging this equation and substituting $(\partial S/\partial V)_T = (\partial P/\partial T)_V$, from (5.5) yields

$$P = T\left(\frac{\partial P}{\partial T}\right)_V - \left(\frac{\partial U}{\partial V}\right)_T \tag{5.23}$$

Likewise, starting with $dH = dU + d(PV) = T\,dS + V\,dP$ and writing

$$\left(\frac{\partial H}{\partial P}\right)_T = T\left(\frac{\partial S}{\partial P}\right)_T + V$$

and then using (5.5) gives

$$V = T\left(\frac{\partial V}{\partial T}\right)_P + \left(\frac{\partial H}{\partial P}\right)_T \tag{5.24}$$

Equations (5.23) and (5.24) are called the thermodynamic equations of state.

Variation of $U$ and $H$ at constant temperature for any system may be determined by these equations. For example, from (5.24),

$$(H_2 - H_1)_T = \int_{P_1}^{P_2}\left[V - T\left(\frac{\partial V}{\partial T}\right)_P\right]dP; \quad \text{(constant } T\text{)} \tag{5.25}$$

This equation may be integrated from the compressibility data or by using the Berthelot equation. If the compressibility data are available, it is evident from $PV = ZRT$ that

$$(H_2 - H_1)_T = -RT^2\int_{P_1}^{P_2}\left(\frac{\partial Z}{\partial T}\right)_P\frac{dP}{P}, \quad \text{(constant } T\text{)} \tag{5.26}$$

For an ideal gas $Z = 1$ and this equation is zero as expected. Otherwise, from the Berthelot equation, (5.25) can be transformed into

$$(H_2 - H_1)_T = \frac{9}{128}\frac{RT_c}{P_c}\left(1 - 18\frac{T_c^2}{T^2}\right)(P_2 - P_1) \tag{5.27}$$

Variation of $U$ and $H$ with pressure and volume is small. Thus for $P_2 = 1$ and $P_1 \to 0$, (5.27) yields, with $T_c = 154.3$ K and $P_c = 50.4$ bar, $H_2 - H_1 = -0.823R = -6.84$ J mol$^{-1}$ at 25°C for oxygen. For solids and liquids at ordinary temperatures and a few bars of pressure, $T(\partial V/\partial T)$ is about 1% of $V$; hence the right side of (5.25) becomes $V(P_2 - P_1)$ because $V$ is also virtually independent of pressure. For example, $V = 0.0071$ L mol$^{-1}$ for iron at 25°C and for $P_2 - P_1 = 10$ bars, $H_2 - H_1 = 0.07$ liter bar $= 0.84R = 7.0$ J mol$^{-1}$. For higher pressures, an equation of state for liquids and solids is necessary for an accurate evaluation of the integral in (5.25).

## Difference Between $C_p$ and $C_V$

It was shown earlier that the theoretical heat capacity equations yield $C_V$, whereas $C_p$ is the more useful thermodynamic property. The value of $C_p$ from $C_V$ and other measurable quantities can be obtained from an alternative expression of (3.64), viz.:

$$C_p - C_V = \left(\frac{\partial V}{\partial T}\right)_P \left[P + \left(\frac{\partial U}{\partial V}\right)_T\right]$$

which assumes the following useful form when $P$ is replaced by the right side of (5.23), i.e.,

$$C_p - C_V = T\left(\frac{\partial V}{\partial T}\right)_P \left(\frac{\partial P}{\partial T}\right)_V \tag{5.28}$$

Utilizing the Berthelot equation, the foregoing relationship becomes

$$C_p - C_V = R\left[1 + \frac{27}{16}\left(\frac{T_c}{T}\right)^3 \frac{P}{P_c}\right] \tag{5.29}$$

It may be shown that for oxygen, $C_p - C_V = 1.236R$ at $25°C$ and $50.7$ bars, and $1.0047R$ for $1.01325$ bar when $T_c$ and $P_c$ from Table 3.1 are substituted in (5.29). Solving for $(\partial P/\partial T)_V$ from

$$\left(\frac{\partial P}{\partial T}\right)_V \left(\frac{\partial T}{\partial V}\right)_P \left(\frac{\partial V}{\partial P}\right)_T = -1$$

and substituting the result into (5.28) yields the following alternative equation:

$$C_P - C_V = -T\left(\frac{\partial V}{\partial T}\right)_P^2 \bigg/ \left(\frac{\partial V}{\partial P}\right)_T \tag{5.30*}$$

For solids and liquids the cubical coefficient of thermal expansion, $\dot{\alpha}$, and the compressibility coefficient $\dot{\beta}$ are defined by

$$\dot{\alpha} = \frac{1}{V}\left(\frac{\partial V}{\partial T}\right)_P \tag{5.31}$$

$$\dot{\beta} = -\frac{1}{V}\left(\frac{\partial V}{\partial P}\right)_T \tag{5.32}$$

Substitution of (5.31) and (5.32) into (5.30) yields

$$C_p - C_V = \dot{\alpha}^2 TV/\dot{\beta} \tag{5.33}$$

For example the value of $\dot{\alpha}$ for iron in the vicinity of room temperature is $35.1 \times 10^{-6}$ per K and that of $\dot{\beta}$, $0.52 \times 10^{-6}$ per bar, the molar volume is $0.0071$ L mol$^{-1}$ so

---

*For measurements of partial derivatives in this equation, see for example, J. C. Petit and L. T. Minassian, J. Chem. Thermodyn., 6, 1139, (1974); see also General Ref. (19), and J. O. Hirschfelder, et al., *loc. cit.*, for calculations.

that
$$C_p - C_V = \frac{(35.1 \times 10^{-6})^2 \times 298 \times 0.0071}{0.52 \times 10^{-6}} = 0.00503 \text{ L bar mol}^{-1} \text{ K}^{-1}$$
$$= 0.503 \text{ J mol}^{-1} \text{ K}^{-1}$$

## Variation of $C_p$ and $C_V$ with P and V

Variation of $C_p$ with pressure can be obtained from (5.25), which may be written as

$$\left(\frac{\partial H}{\partial P}\right)_T = V - T\left(\frac{\partial V}{\partial T}\right)_P \tag{5.34}$$

Differentiation of the left side of this equation with respect to $T$ at constant pressure gives $\partial(\partial H/\partial P)/\partial T = \partial(\partial H/\partial T)/\partial P = (\partial C_p/\partial P)_T$ since the sequence of differentiation is immaterial, and likewise, differentiation of the right side with respect to $T$ gives $-T(\partial^2 V/\partial T^2)_P$; hence

$$\left(\frac{\partial C_p}{\partial P}\right)_T = -T\left(\frac{\partial^2 V}{\partial T^2}\right)_P \tag{5.35}$$

Integration of this equation by using the Berthelot equation gives

$$C_p - C_p(P \to 0) = -T\int_{P \to 0}^{P} \left(\frac{\partial^2 V}{\partial T^2}\right)_P dP = R\frac{81}{32}\left(\frac{T_c}{T}\right)^3 \cdot \frac{P}{P_c} \tag{5.36}$$

The heat capacities of gases vary considerably with pressure at very low temperatures, but negligibly at high temperatures.

Similarly it can be shown that

$$\left(\frac{\partial C_V}{\partial V}\right)_T = T\left(\frac{\partial^2 P}{\partial T^2}\right)_V \tag{5.37}$$

and

$$C_V - C_V(V \to \infty) = T\int_{V \to \infty}^{V} \left(\frac{\partial^2 P}{\partial T^2}\right)_V dV \tag{5.38}$$

Again the compressibility data or an equation of state can be used to evaluate this integral.

## Joule-Kelvin Expansion of Gases

The Joule-Kelvin, (or Joule-Thomson) expansion of a gas is of special interest in the liquefaction of cryogenic gases. Such an expansion of a gas at $P_1$, $T_1$, in a chamber is carried out *adiabatically and irreversibly* through an adiabatic porous plug into another chamber where the gas is at $P_2$ and $T_2$ as shown in Fig. 5.1. The initial pressure $P_1$ is obviously greater than $P_2$ and there is no exchange of heat between the two chambers. The pressures are kept constant by adiabatic frictionless pistons. The pistons are at $A$ and $B$ and the plug is closed by a suitable mechanism

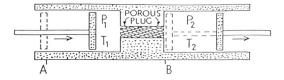

**Fig. 5.1** Joule-Kelvin expansion.

before the expansion is started. The experiment may be terminated when one mole of gas has passed to the right-hand side chamber. This type of expansion is essentially a throttled adiabatic expansion. The net amount of work received by the gas is $P_1V_1 - P_2V_2$, since at the left side, the amount of isobaric, isothermal work received by the system is $P_1V_1$ and at the right $-P_2V_2$. Application of the first law of thermodynamics to this process gives

$$U_2 - U_1 = P_1V_1 - P_2V_2, \quad (Q=0)$$

from which

$$U_1 + P_1V_1 = U_2 + P_2V_2, \quad \text{or} \quad H_1 = H_2 \tag{5.39}$$

This equation merely states that the enthalpy of a real gas before and after a Joule-Kelvin expansion is the same, i.e., the process is isenthalpic. The same process can be accomplished without the pistons if the throttled expansion is carried out between two reservoirs large enough so that the pressures $P_1$ and $P_2$ remain virtually constant when one mole of a gas is throttled through the plug.

The total differential of $H = f(P, T)$ is

$$dH = \left(\frac{\partial H}{\partial P}\right)_T dP + \left(\frac{\partial H}{\partial T}\right)_P dT$$

During a Joule-Kelvin expansion $dH$ is zero so that the foregoing equation becomes

$$\left(\frac{\partial H}{\partial P}\right)_T (dP)_H + \left(\frac{\partial H}{\partial T}\right)_P (dT)_H = 0$$

from which

$$\left(\frac{\partial T}{\partial P}\right)_H = -\left(\frac{\partial H}{\partial P}\right)_T \bigg/ \left(\frac{\partial H}{\partial T}\right)_P \tag{5.40}$$

where $(\partial T/\partial P)_H$ is called the Joule-Kelvin coefficient and denoted by $\mu_{JK}$. It represents the rate of variation of temperature with pressure in a throttled expansion. Since $C_p = (\partial H/\partial T)_p$, (5.40) becomes

$$\mu_{JK} = -\frac{1}{C_p}\left(\frac{\partial H}{\partial P}\right)_T \tag{5.41}$$

This equation is applicable to any gas, real or ideal. The enthalpy of an ideal gas is a function of temperature only, therefore $(\partial H/\partial P)_T$ is zero, and since $1/C_p$ is finite, $\mu_{JK}$ is also zero.

Elimination of $(\partial H/\partial P)_T$ from (5.24) and (5.41) gives

$$\mu_{JK} = \frac{1}{C_p}\left[T\left(\frac{\partial V}{\partial T}\right)_P - V\right] \tag{5.42}$$

The Joule-Kelvin coefficient may be determined by inserting in (5.42) the value of $(\partial V/\partial T)_P$ from the $P$-$V$-$T$ data or from the appropriate equation of state. For the van der Waals equation, it may readily be shown that

$$\mu_{JK} \approx \frac{1}{C_p}\left(\frac{2a}{RT} - b - \frac{3abP}{R^2T^2}\right) \tag{5.43}$$

The right-hand side of this equation may also be expressed in terms of the reduced pressure temperature and volume. The approximate equality $\approx$ is due to the experimental fact that $a$ and $b$ are not independent of temperature. Equation (5.43) is quadratic in $T$; hence for each pressure there are two temperatures at which $\mu_{JK} = (\partial T/\partial P)_H$ becomes zero. These temperatures are called the upper and lower inversion temperatures. Between these temperatures $\mu_{JK}$ is positive, or simply a throttled expansion is accompanied by a decrease in temperature. At low temperatures the magnitude of Joule-Kelvin coefficient is large enough to be useful for liquefaction of gases.*

## PROBLEMS

5.1 Show that for any system

$$\left(\frac{\partial S}{\partial U}\right)_V = \frac{1}{T}, \quad \left(\frac{\partial S}{\partial H}\right)_P = \frac{1}{T}, \quad \left(\frac{\partial S}{\partial P}\right)_H = \frac{-V}{T},$$

and $\quad \left(\frac{\partial (P/T)}{\partial U}\right)_V = \left(\frac{\partial (1/T)}{\partial V}\right)_U$

5.2 Integrate (5.6) by using the Berthelot and van der Waals equations.

5.3 Two moles of Zn(s) at 25°C is added adiabatically into one mole of Zn(l) at 700 K. Calculate the entropy change from the following data: $C_p^\circ(\text{solid}) = 22.38 + 0.0100\,T$ J mol$^{-1}$ K$^{-1}$, $\Delta_{fus}H^\circ = 7322$ J mol$^{-1}$ at 692.7 K and $C_p^\circ(\text{liquid}) = 31.38$ J mol$^{-1}$ K$^{-1}$.

5.4 A mixture of 2 moles of neon and 3 moles of nitrogen at 0.5 bar and 25°C is brought to 100°C and 1.5 bar. Compute the total change in entropy by taking $C_p^\circ = 2.5R$ for neon and $C_p = 3.5R$ for nitrogen.

---

*For measurements of $\mu_{JK}$ and $H = H(P, T)$ see for example, R. A. Dawe and P. N. Snowdon, J. Chem. Thermodyn., 6, 65 (1974).

5.5 The values of $C_p/R$ for hydrogen at one bar and at various temperatures are as follows:

| $T/K$: | 40 | 60 | 80 | 100 | 120 | 150 |
|---|---|---|---|---|---|---|
| $C_p^\circ/R$: | 2.56 | 2.56 | 2.61 | 2.72 | 2.86 | 3.06 |

Express $C_p^\circ$ as a quadratic function of $T$ and then determine the change of entropy of one mole of hydrogen from 40 to 150 K.

5.6 Calculate the molar entropy of mixing $\Delta S$ of two ideal gases $A$ and $B$ at 0.2, 0.4, 0.6, and 0.8 mole fraction of $B$ and represent the results graphically on a plot of $\Delta S$ versus $x_B$. Differentiate $\Delta S$ with respect to $x_B$ and find the value of $x_B$ for which $\Delta S$ is a maximum.

5.7 One mole of a mixture of 21% oxygen and 79% nitrogen by volume at 1 bar is separated into 0.21 mole of oxygen at 0.21 bar and 0.79 mole of nitrogen at 0.79 bar. (a) Assuming that the mixture is ideal, what is the entropy change? (b) Compare the result with the entropy of *unmixing*, i.e., forming $N_2$ and $O_2$ each at 1 bar from air at the same pressure.

*Ans.:* (a) zero; (b) entropy of unmixing $= -0.514R$

5.8 One mole of argon at 1 bar and $T_1$ and 3 moles of helium at 3 bars and at the same temperature are first mixed and then compressed isothermally to 4 bars then heated to $T_2 = 2T_1$ at constant pressure. What is the entropy change for each step and for the entire process?

5.9 Use the adiabatic (isentropic) equations (3.32) and (3.33) in which $P_2$ and $V_2$ are variables, and $PV = RT$, to verify the following relationships:

$$C_p = T\left(\frac{\partial V}{\partial T}\right)_P \left(\frac{\partial P}{\partial T}\right)_S; \qquad \frac{C_p}{C_V} = \left(\frac{\partial P}{\partial V}\right)_S \left(\frac{\partial V}{\partial P}\right)_T$$

CHAPTER VI

# HELMHOLTZ AND GIBBS ENERGIES

## Introduction and Definitions

The criterion for equilibrium as expressed by $(dS)_{U,V} \geqq 0$ in Chapter IV states that the entropy change of an isolated system is zero or positive. An isolated system is not a very convenient and useful system in thermodynamics, hence it is desirable to investigate the state of a system as affected by the surroundings, usually under constant temperature and pressure. The first and second laws are formulated by $dW = dU - T\,dS$, which may be integrated at constant temperature for a reversible process to obtain

$$\sum W_T = \Delta U_T - T\Delta S_T, \qquad (dT = 0) \tag{6.1}$$

where $W_T$ is the reversible isothermal work of any type, e.g., work of expansion or compression. An extensive thermodynamic property $A$ may be defined in terms of the properties $U$, $T$ and $S$ as follows:

$$A = U - TS \tag{6.2}$$

The main purpose of introducing $A$ is primarily for brevity and the usefulness of the resulting derived equations. Comparison with the foregoing equations show that

$$\sum W_T = \Delta A_T \tag{6.3}$$

For a reversible process $\sum W_T = \Delta A_T$ and for this reason the function $A$ is known as the *work function* or the *Helmholtz energy*. (From Arbeit = Work in German.) It is seen that the definition of $A$ as related to a measurable quantity $\sum W_T$ is similar to the definition of $\Delta H_P$ as related to $Q_P$.

The first law of thermodynamics with the work of expansion $dW = -P\,dV$ and any other type of work, $\underline{W'}$, may be rewritten as

$$dU = T\,dS - P\,dV + d\underline{W'} \tag{6.4}$$

At constant pressure and temperature, integration and rearrangement gives

$$\underline{W'}_{P,T} = \Delta U_{P,T} + P\Delta V_{P,T} - T\Delta S_{P,T}, \qquad (dT = dP = 0) \tag{6.5}$$

101

## 102   Thermodynamics

Again for brevity, it is permissible to define

$$G = U + PV - TS = H - TS; \qquad (H = U + PV) \tag{6.6}$$

The change in $G$ at constant pressure and temperature is

$$\Delta G_{P,T} = \Delta U_{P,T} + P\Delta V_{P,T} - T\Delta S_{P,T} \tag{6.7}$$

Since the right-hand sides of (6.5) and (6.7) are equal, it is evident that

$$\underline{W}'_{P,T} = \Delta G_{P,T} \tag{6.8}$$

The function $G$ is called the Gibbs energy, and it is equal to the work exchange with the surroundings other than the work of expansion at constant $P$ and $T$. It is clear that $G$, like $A$ and $H$, is a purely brief and convenient definitional function. The total amount of work from (6.5) is $(\underline{W}' - P\Delta V)_{P,T} = \sum W_{P,T}$ and the work of expansion is $(-P\Delta V)_{P,T}$ hence $\underline{W}'_{P,T} = $ total work − work of expansion, i.e.,

$$\underline{W}'_{P,T} = \Delta G_{P,T} = \left(\sum W_{P,T}\right) + (P\Delta V)_{P,T} \tag{6.9}$$

Thus the change in Gibbs energy in a process at constant temperature and pressure is equal to the work other than that expressed by $-P\,dV$ namely, the net work or useful work. Note the fact that $+P\Delta V_{P,T}$ in any process, ordinarily under atmospheric pressure, is used up in exchanging work with the atmosphere and this quantity of work is not useful. For example, a reversible galvanic cell operating at constant temperature and atmospheric pressure, for which the chemical reaction is

$$Zn(c) + 2H^+ = Zn^{++} + H_2(g)$$

produces hydrogen and yields a certain quantity of electricity capable of doing useful work which may be designated by $\underline{W}'$. The total amount of work is the sum of the electrical work and the work of expansion of one mole of hydrogen against atmospheric pressure. Thus $\sum W = \underline{W}' - (P\Delta V)_{P,T}$ and

$$\sum W = \underline{W}' + W(\text{expansion}) = \underline{W}' - P\Delta V_{P,T}$$

Therefore, the Gibbs energy change is

$$\boxed{\Delta G_{P,T} = \left(\sum W\right) + P\Delta V_{P,T} = \text{net work}} \tag{6.9a}$$

The electrical work, $\underline{W}'$ is the net work because it may entirely be converted into useful work. The work of expansion against the atmospheric pressure, on the other hand, is not available for doing useful work. Another example is the isothermal isobaric vaporization of water at 100°C, confined in a vertical cylinder fitted with a weightless and frictionless piston. The Gibbs energy change for this process is given by (6.9) where $\sum W$ is $-P\Delta V_{P,T}$; hence there is no net work in this process.

The available work in (6.1) and (6.5) contains the unavailable energy term given by $-T\Delta S$ which was historically called the bound energy, and for this reason $A$ was formerly called the Helmholtz free energy and $G$, the Gibbs free energy. The term *free* is now obsolete.

## Helmholtz and Gibbs Energies

The procedure outlined by (6.1) to (6.9) for justifying the definitions of the Helmholtz and the Gibbs energies is historical in nature. A mathematically elegant procedure is based on the change of independent variables in $dU = T\,dS - P\,dV$. The function $U$ in $dU = T\,dS - P\,dV$ is a dependent function of $S$ and $V$ because the differentials of $S$ and $V$ appear on the right side; hence, $U = U(S, V)$. However, $S$ and $V$ are not very convenient independent variables, and when the differentials of $PV$ and $-TS$ are added to both sides of $dU = T\,dS - P\,dV$ in appropriate succession, the independent variables are changed as follows:

$$dU = T\,dS - P\,dV \tag{6.10}$$

$$dU + d(PV) = d(U + PV) \equiv dH = T\,dS + V\,dP \tag{6.11}$$

$$d(U - TS) \equiv dA = -S\,dT - P\,dV \tag{6.12}$$

$$\boxed{d(U + PV - TS) = d(H - TS) \equiv dG = -S\,dT + V\,dP} \tag{6.13}$$

The identity signs define the functions $H \equiv U + PV$, $A \equiv U - TS$ and $G \equiv H - TS \equiv U + PV - TS$, and the variables preceded by the derivative $d$ on the right sides of these equations are the independent variables so that $U = U(S, V)$, $H = H(S, P)$, $A = A(T, V)$ and $G = G(T, P)$. The independent variables $T$ and $P$ for $G$ are the most convenient variables from a practical point of view. It will be seen later that for this reason $G$ is the most useful function. Other functions such as, the Massieu function, $J = -A/T$, the Planck function, $Y = -G/T$ will not be used in this book. The processes for obtaining (6.11), (6.12) and (6.13) from (6.10) are elementary examples of Legendre transformations of (6.10).

### Partial Differential Relations

Combination of $A = U - TS$ and $(\partial A/\partial T)_V = -S$ gives

$$A = U + T\left(\frac{\partial A}{\partial T}\right)_V \tag{6.14}$$

and a process in which $A_2 - A_1 = \Delta A$, it is evident that

$$\Delta A = \Delta U + T\left(\frac{\partial \Delta A}{\partial T}\right)_V \tag{6.15}$$

Likewise from $G = H - TS$ and $(\partial G/\partial T)_P = -S$,

$$G = H + T\left(\frac{\partial G}{\partial T}\right)_P \tag{6.16}$$

$$\Delta G = \Delta H + T\left(\frac{\partial \Delta G}{\partial T}\right)_P \tag{6.17}$$

Equations (6.14) to (6.17) are called the Gibbs-Helmholtz equations, first derived by J. W. Gibbs (1875) and later independently by H. von Helmholtz (1882).

104  Thermodynamics

Dividing $G = H - TS$ by $T$ and differentiating with respect to $T$ at constant pressure gives

$$\left(\frac{\partial (G/T)}{\partial T}\right)_P = -\frac{H}{T^2} + \frac{1}{T}\left(\frac{\partial H}{\partial T}\right) - \left(\frac{\partial S}{\partial T}\right)_P \quad (6.18)$$

The last two terms on the right cancel each other since $(\partial S/\partial T)_P = C_p/T$ and $(\partial H/\partial T)_P = C_p$; hence,

$$\left(\frac{\partial (G/T)}{\partial T}\right)_P = -\frac{H}{T^2} \quad (6.19)$$

This is a very useful equation in thermodynamics, and it is sometimes converted into

$$\left(\frac{\partial (G/T)}{\partial (1/T)}\right)_P = H \quad (6.20)$$

upon multiplying (6.19) by $\partial T/\partial (1/T) = -T^2$. In (6.19) and (6.20), it is also possible to replace $G$ and $H$ by $\Delta G$ and $\Delta H$ where $\Delta$ may refer to such processes as phase changes, and chemical reactions. The corresponding equations for $A$ are obtained by differentiating $A/T$ with respect to $T$ or $1/T$ at constant volume; thus

$$\left(\frac{\partial (A/T)}{\partial T}\right)_V = -\frac{U}{T^2}, \quad \text{and} \quad \left(\frac{\partial (A/T)}{\partial (1/T)}\right)_V = U \quad (6.20a)$$

## Isothermal Changes in *A* and *G*

Integration of $dA = -S\,dT - P\,dV$ and $dG = -S\,dT + V\,dP$ at constant temperature gives

$$(A_2 - A_1)_T = -\int_{V_1}^{V_2} P\,dV \quad (6.21)$$

and

$$(G_2 - G_1)_T = \int_{P_1}^{P_2} V\,dP \quad (6.22)$$

For an ideal gas these equations become

$$(A_2 - A_1)_T = -\int_{V_1}^{V_2} \frac{RT}{V}\,dV = RT \ln \frac{V_1}{V_2}, \quad \text{(ideal gas)} \quad (6.23)$$

$$(G_2 - G_1)_T = \int_{P_1}^{P_2} \frac{RT}{P}\,dP = RT \ln \frac{P_2}{P_1}, \quad \text{(ideal gas)} \quad (6.24)$$

Since $P_2/P_1 = V_1/V_2$ for an ideal gas at constant temperature, (6.23) and (6.24) give

$$(A_2 - A_1)_T = (G_2 - G_1)_T, \quad \text{(ideal gas)} \quad (6.25)$$

If the lower and upper integration limits in (6.24) for an ideal gas are taken as $P_1 = 1$, and $P_2 = P$, where $P$ is any pressure, then $(G_2 - G_1)_T = RT \ln P$, and $G°(T)$ may be substituted for $G_1$ at $P_1 = 1$ bar and $G(P, T)$ for $G_2$ to obtain

$$G(P, T) = G°(T) + RT \ln P \qquad (6.26)$$

where $G°(T)$ is a function of $T$ only and the pressure dependence of $G(P, T)$ appears in the second term on the right. It will be seen later that (6.26) is one of the most useful equations in thermodynamics.

## Criteria for Reversibility and Irreversibility

The forces affecting any change in the system differ infinitesimally from the counteracting forces within the system when a process occurs in a reversible manner. The system is therefore always under equilibrium conditions during a reversible process. When an ideal gas is expanded reversibly and isothermally, $dS = dQ_{rev}/T$ where $dQ_{rev}$ is positive in value so that $dQ$ absorbed by the system is converted into work since $dU = 0$, and $dQ_{rev} = -P\,dV$. If the expansion is slow but there is friction during expansion, the amount of $dQ$ absorbed irreversibly is smaller than $dQ_{rev}$. For an adiabatic reversible process $dQ_{rev}$ (adiab) is zero and since a combination of an isothermal reversible or irreversible process and an adiabatic reversible process can take a system from an initial state $P_1, T_1$ to any arbitrary final state $P_2, T_2$, it is evident that for any differential change in state

$$dS_{system} = \frac{dQ_{rev}}{T}, \quad \text{(reversible, i.e., under equilibrium)}$$

$$dS_{system} > \frac{dQ_{irr}}{T}, \quad \text{(irreversible, i.e., spontaneous)}$$

These relationships are also valid for compression and for any real substance instead of an ideal gas. For any process in which $Q$ is reversible or irreversible, these two equations can be summarized by

$$dS_{system} \geqq \frac{dQ}{T} \qquad (6.27)$$

The first law of thermodynamics may be written as $dU = dQ - P\,dV + d\underline{W}'$, or

$$dQ = dU + P\,dV - d\underline{W}' \qquad (6.28)$$

where $dQ$ is the quantity of reversible or irreversible heat, and $d\underline{W}'$ the work other than that expressed by $-P\,dV$, e.g., electrical work. It must be stressed here that $-P\,dV$ is meaningful and can be evaluated only when the work of expansion or compression is reversible. Combination of (6.27) and (6.28) yields

$$dU + P\,dV - T\,dS - d\underline{W}' \leqq 0 \qquad (6.29)$$

where $P\,dV$ must be replaced by the appropriate quantity when the work of expansion involves irreversibility. This is the general criterion of equilibrium or reversibility, and of spontaneity or irreversibility; the equal sign refers to equilibrium and the smaller than zero, to spontaneity. There are criteria for equilibrium derivable from (6.29), e.g., in the absence of any type of work $dV$ and $d\underline{W}'$ are both zero, and (6.29) becomes

$$(dU - T\,dS)_{V,\underline{W}'} \leqq 0 \tag{6.30}$$

In addition, at constant energy this equation reduces further to

$$dS_{U,V,\underline{W}'} \geqq 0 \tag{6.31}$$

which signifies that in a completely isolated system, for any reversible process, the entropy change is zero, but for an irreversible, or spontaneous process, the entropy change is positive. *In other words, after a spontaneous process the system becomes more stable by acquiring an increase in entropy or randomness, or simply the entropy is a maximum for an isolated system.* Likewise, at constant pressure, the first two terms of (6.29) represent $dH_p$ and if $d\underline{W}' = 0$, it is evident that

$$(dH - T\,dS)_{P,\underline{W}'} \leqq 0 \tag{6.32}$$

Combination of $dA = dU - T\,dS - S\,dT$ with (6.29) gives $dA \leqq -P\,dV - S\,dT + d\underline{W}'$, from which

$$dA_{T,\underline{W}'} \leqq -P\,dV, \quad \text{and} \quad dA_{V,T,\underline{W}'} \leqq 0 \tag{6.33}$$

Likewise, $dG = dU + P\,dV + V\,dP - S\,dT - T\,dS$ and (6.29) give $dG \leqq V\,dP - S\,dT + d\underline{W}'$ from which

$$dG_{T,\underline{W}'} \leqq V\,dP \tag{6.34}$$

$$\boxed{dG_{P,T,\underline{W}'} \leqq 0} \tag{6.35}$$

This relationship is more useful than the others because an overwhelming majority of physical and chemical changes occur at constant temperature and pressure. According to (6.34), at constant $P$, $T$ and $\underline{W}'$ a process is possible if $dG$ is zero or negative, and impossible when $dG$ is positive. For a finite process, (6.35) may be integrated to obtain

$$\boxed{\Delta G_{P,T,\underline{W}'} \leqq 0} \tag{6.36}$$

or, since $\Delta G = G_{\text{final}} - G_{\text{initial}}$,

$$(G_{\text{final}})_{P,T,\underline{W}'} \leqq (G_{\text{initial}})_{P,T,\underline{W}'} \tag{6.37}$$

Equations (6.33) to (6.34) are the most convenient criteria for equilibrium since $A$ and $G$ are functions of very convenient variables of state. According to (6.36) a process is reversible, or proceeds under equilibrium, when the Gibbs energies in the final and initial states are equal; irreversible or spontaneously possible, if the final state has a lower Gibbs energy than the initial state, and impossible if the Gibbs energy is greater in the final state than in the initial state. Thus, (6.34) and (6.35) state that the Gibbs energy is a minimum at equilibrium. In general $\underline{W'}$ is zero and unless its existence is specified, it will be ignored when the foregoing equations are used in the succeeding chapters.

A great number of thermodynamic formulas, such as those in this chapter, exist among the first and second derivatives of the eight common thermodynamic properties $P, T, V, U, H, S, A,$ and $G$. Many of these relationships are rarely used in thermodynamics. Several methods have been devised by Bridgman (1926), Koenig,* and others to obtain all such formulas which may be of interest to unusually inquiring readers.

## Examples

Simple applications of the foregoing equations will now be illustrated by the following numerical examples.

*Example (1)* One mole of an ideal gas, at 300 K and initially confined in a 1-liter container, is expanded into a connected 3-liter evacuated container. Calculate $\Delta A$ and $\Delta G$ and show that the criteria of equilibrium given by (6.33) and (6.34) are obeyed.

*Solution* From $dA = -S\,dT - P\,dV$, [cf. (6.12)] and from the fact that the process is isothermal and $P = RT/V$, $dA_T = -P\,dV = -RT\,d\ln V$; hence, $\Delta A_T = -RT\ln V_2/V_1 = -RT\ln 4$, since $V_2$ is 4. Whether the process is reversible or not $\Delta A$ is a function of variables of state and a reversible process may be used in evaluating $\Delta A_T$. However, $-P\,dV$ in (6.33), which is the work of expansion, is zero because the gas was expanded into a vacuum. Therefore the integral of (6.33) gives $\Delta A_T = -RT\ln 4 < 0$. For an ideal gas $dA_T = -P\,dV = dG_T = V\,dP$ because $V\,dP + P\,dV = 0$ at a constant temperature; hence, again $\Delta A_T = \Delta G_T < 0$. It should be noted that $V\,dP$ in (6.34) obtained by adding $d(PV)$ on both sides of (6.33) is zero since the work of expansion in (6.33) is zero and $d(PV)$ for isothermal expansion is also zero.

*Example (2)* One mole of steam at 99.63°C and 0.5 bar is compressed isothermally with some friction into water at 1 bar. The energy of condensation of water is $\Delta U_{\text{cond}} = -4517R$. Calculate (a) $\Delta G_T$ from 0.5 to 1 bar and (b) $\Delta H_{T,P}$ and $\Delta G_{T,P}$ for isothermal isobaric compression into water assuming that steam is an ideal gas.

*Solution* (a) The value of $\Delta G_T$ is obtained by integrating $dG_T = V\,dP = RT\,d\ln P$ from $P_1 = 0.5$ to $P_2 = 1$; the result is $\Delta G_T = RT\ln 2 > 0$, therefore

---

*F. O. Koenig, J. Chem. Phys., 56, 4556 (1972).

the process requires external aid or forced compression to proceed. The work required, $dW$ is greater than $-P\,dV$, i.e., $dW_{irr} > -P\,dV$, because of friction and if $d(PV)_T = 0$ is added to both sides, then $dW_{irr} > V\,dP$; therefore $\Delta G_T < W_{irr}$ and the process is possible. (b) $\Delta H_{P,T} = \Delta U_{cond,T} + P\Delta V$, and assuming that the volume of water is negligible then $\Delta V = -V_{steam}$ and $P\Delta V = -RT$, and $\Delta H_{P,T} = -4517R - 372.78R = -4890R$. The entropy of condensation is $\Delta S_{P,T} = \Delta H_{P,T}/372.78$ and $\Delta G_{P,T} = \Delta H_{P,T} - T\Delta S_{P,T} = 0$, but because of irreversibility $dW_{irr} > (V\,dP)_T = 0$, or $\Delta G_{T,P} < W_{irr}$.

## PROBLEMS

6.1 Differentiate $A/T$ at constant volume, and $G/T$ and $S/T$ at constant pressure with respect to $T$ and $1/T$.

6.2 It will be seen later that $\Delta H$ and $\Delta S$ for chemical reactions are insensitive to moderate changes of temperature. Derive $\partial(\Delta A/T)/\partial(T^{-1})$ and $\partial(\Delta G/T)/\partial(T^{-1})$ by assuming that $\Delta H$ and $\Delta S$ are independent of temperature and compare the details of derivations with those in Problem 6.1.

6.3 Calculate $\Delta S_T$ and $\Delta G_T$ when 0.21 mole of oxygen and 0.79 mole of nitrogen, each at 1 bar and 25°C are mixed to form a gas mixture at 1 bar. Ideal gas behavior may be assumed for the gases and the mixture.

*Ans.:* $\Delta S = 0.514R$, and $\Delta G = -153R$

6.4 One mole of liquid water at 25°C is injected into an evacuated copper chamber immersed in a thermostat at 25°C, and vaporized completely to form steam at 2,112 Pa of pressure. Equilibrium vapor pressure of water is 3,168 Pa at 25°C temperature. Assuming that steam is an ideal gas, what is the change in Gibbs energy of water for this process? (Note that a reversible path may be devised to carry out this process in order to evaluate $\Delta G_T$ as follows: Vaporize the water at 3,168 Pa, and then expand it isothermally to 2,112 Pa.

*Ans.:* $\Delta G_{P,T} = -1,005$ J mol$^{-1}$

6.5 One mole of an ideal gas under 2 bars of pressure is expanded adiabatically and spontaneously into vacuum to decrease its pressure to 1 bar, and then isothermally and reversibly compressed to 4 bars. Calculate $\Delta G$ and $\Delta S$ for each step.

*Ans.:* For adiabatic expansion into vacuum $\Delta S = 0.693R$, and $\Delta G = -0.693RT$. For isothermal compression 1 to 4 bars, $\Delta S = -1.386R$, and $\Delta G = +1.386RT$.

6.6 (a) Differentiate $G$ and $A$ with respect to $T$ to obtain two important relationships for $S$ and (b) differentiate $G/T$ and $A/T$ with respect to $T$ to obtain the equations for $C_p$ and $C_V$ in terms of $G$, $A$, $T$ and their derivatives.

CHAPTER VII

# THE THIRD LAW OF THERMODYNAMICS

## Introduction

The entropy of a pure substance at any temperature is expressed by

$$S_T = S_0 + \int_0^T \frac{dU + P\,dV}{T} \tag{7.1}$$

where $S_0$ is the entropy at 0 K. The value of $S_0$ for a pure substance in internal stable equilibrium, i.e., in a stable molecular configuration, is the subject of the third law of thermodynamics.

The discovery of the third law and its full significance required accurate data on chemical equilibria and on heat capacities at very low temperatures early in the twentieth century. Continued research at low temperatures has confirmed the validity of this law and provided interesting interpretations of atomic and molecular behavior of substances in gaseous and condensed states. Statistical and quantum mechanics greatly enhanced the validity of the third law and accounted for the discrepancies between statistically calculated and experimentally measured values of entropy.

The third law and its consequences are based on a wealth of data from two experimentally distinct types of investigations down to very low temperatures: (i) chemical equilibrium, and (ii) calorimetry for the enthalpies of transformation and the heat capacities of various allotropes of pure substances. It is appropriate to discuss each set of these results briefly before enunciating the third law of thermodynamics.

(i) *Chemical Equilibrium*: It was shown by T. W. Richards (1902), W. Nernst (1906), and later by numerous other investigators, that the change in entropy tends to zero at 0 K for chemical reactions, involving pure reactants and pure products in their stable states.* The entropy change for a reaction is usually determined by measuring the equilibrium constant but alternative methods, e.g., electrochemical methods, are also available as will be discussed in the forthcoming

---

*Controversies in initial development of the third law were resolved largely by W. F. Giauque, G. N. Lewis, and F. E. Simon; see F. E. Simon, "Year Book of the Physical Society," p. 1, Phys. Soc., London (1956). See also J. Wilks, "The Third Law of Thermodynamics," Oxford Univ. Press (1961).

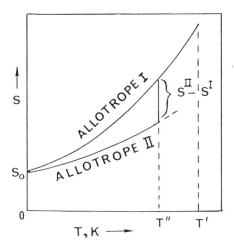

Fig. 7.1 Entropies of two allotropes as functions of temperature.

chapters. It is sufficient to note here that the equilibrium constant $K_p$ is related to the standard Gibbs energy change $\Delta_r G°$ of reaction corresponding to the reactants and products in their pure stable states under one bar of pressure by $-RT \ln K_p = \Delta_r G° = \Delta_r H° - T \Delta_r S°$ as will be seen in Chapters IX and XII. If the standard enthalpy of reaction, $\Delta_r H°$, is known from calorimetric measurements, a single and accurate value of $K_p$ is sufficient to determine $\Delta_r S°$. However, when $\Delta_r H°$ is not known an adequate number of values of $K_p$ in a small range of temperature is necessary to determine both $\Delta_r H°$ and $\Delta_r S°$. When the values of $K_p$ are sufficiently precise and the temperature range is relatively small, a plot of $RT \bullet \ln K_p$ versus $T$ gives a straight line whose slope is $\Delta_r S°$. The resulting value of $\Delta_r S°$ may be extrapolated to $T = 0$ provided that the necessary heat capacity equations and all the data on phase transitions of all the reactants and the products are known. All the reliable values of $\Delta_r S°$ extrapolated to 0 K show that $\Delta_r S° \to 0$ as $T \to 0$. The experimental observation that $\Delta_r S°$ for a chemical reaction is zero as the temperature approaches zero is known historically as the Nernst heat theorem.

(ii) Calorimetric data on the allotropic forms of pure elements or pure compounds follow the pattern shown in Fig. 7.1 where the entropy of each allotrope has been obtained from the calorimetric data on the heat capacity under one bar of pressure. If we start with a sample at $T'$, where Allotrope I is stable, and cool the sample to $T''$ very slowly, Allotrope I transforms into Allotrope II, and the enthalpy of transformation divided by $T''$ gives the entropy of transformation $S^{II} - S^{I} = \Delta_{tr} S_{T''}$ for the reaction I $\to$ II. The measurement of the heat capacity of II down to sufficiently low temperatures provides the entropy curve for II which may confidently be extrapolated to 0 K. On the other hand if I is cooled sufficiently rapidly, it might not transform into II, and remain in internal thermodynamic equilibrium or stable molecular configuration. Although I below $T''$ is thermodynamically unstable with respect to II, it is stable with respect to small motions of molecules and as such all its thermodynamic properties are functions of its variables of state. The measurements of the heat capacity of I, as in the case of II, provide the curve for the entropy of I down to 0 K. Accurate measurements on numerous allotropic pairs of the same element or compound show that all such pairs follow the pattern shown in Fig. 7.1, where it is important to note that both allotropes have the same entropy at 0 K. Therefore, the entropy of transformation $S^{II} - S^{I}$ tends to zero at 0 K for I $\rightarrow$ II in precisely the same manner as in a chemical reaction. A phase transition of the type I $\rightarrow$ II may also be regarded as a chemical reaction which transforms one allotrope into another allotrope. The experiments in (i) and (ii), while similar from the thermodynamic point of view, are experimentally distinct because in (i) $\Delta_r S^\circ$ is determined from the equilibrium constant while in (ii) $\Delta_{tr} S = S^{II} - S^{I}$ is obtained by calorimetry.

Experimental results also show that $S^{II} - S^{I}$ at 0 K is zero if the measurements are made at any other pressure, and all such curves as the pair in Fig. 7.1 converge into the same intercept. Therefore, $S_0^I(T = 0, P_1) = S_0^I(T = 0, P_2)$ and $S_0^{II}(T = 0, P_1) = S_0^{II}(T = 0, P_2)$, and thus $S_0$ for any stable phase of a pure component is independent of pressure, i.e.,

$$S_0^I(T \rightarrow 0, P_1) = S_0^I(T \rightarrow 0, P_2) = S_0^{II}(P_1 \text{ or } P_2, T \rightarrow 0) \tag{7.2}$$

A vast number of experiments shows that, whatever the variables of state, $S_0(T \rightarrow 0)$ is independent of them at 0 K for a pure single component; i.e.,

$$S_0(T \rightarrow 0, P_1, Y_1, Z_1, \ldots) = S_0(T \rightarrow 0, P_2, Y_2, Z_2, \ldots)$$
$$= S_0(T \rightarrow 0) \tag{7.3}$$

## The Third Law of Thermodynamics

On the basis of the foregoing types of experimental observations, we select the following statement to enunciate the third law of thermodynamics:

*The change of entropy for a chemical reaction or a phase transformation involving pure elements and pure compounds, all in their internal equilibrium states, tends to zero at zero Kelvin.*

It was pointed out in Fig. 7.1 that the entropy change for the phase transformation I $\rightarrow$ II was zero at 0 K, i.e., $S_0^{II} - S_0^I = 0$; hence, $S_0^{II} = S_0^I$ and if we set $S_0^I$ equal to an arbitrary value $\alpha$, then $S^{II}$ must also be set equal to $\alpha$. The fact that the entropies of various allotropes of the same pure substance are identical plays a very important role in stating the third law. Likewise, for a chemical reaction such as C(graphite) + $O_2$(g) = $CO_2$(g), the entropy change is zero if the equilibrium data are extrapolated to 0 K with all the appropriate phase changes so that the reactants and the products are pure solids at 0 K, but again this is not the case for the enthalpy of reaction which is a very large negative quantity, on the order of 418,000 J mole$^{-1}$, at 0 K. If we assign an arbitrary numerical value $\alpha$ for the molar entropy of C(graphite) and likewise $\beta$ for $O_2$(solid), both at 0 K, then the entropy of

$CO_2$(solid) would be $\alpha+\beta$. We could thus assign arbitrary values to the entropies of elements at 0 K; the entropy of a compound would then be the sum of the entropies of the component elements. If we assign the same arbitrary values including zero to the entropies of all the stable elements at 0 K, the result that $\Delta_r S_0^\circ$ be zero is not affected, and this is also true for any allotropic transformation. For convenience, it is therefore permissible to assign zero to $S_0$ for all the pure elements and compounds in their stable configurations at 0 K as first suggested by Planck.* This is a very useful *convention*, which is based on the third law of thermodynamics.

The assignment of zero entropy at 0 K is also justified by statistical mechanics, and in anticipating this, we may state a useful convention, called the third law convention, as follows:

*The entropy of each pure element or pure compound in its stable molecular configurations may be taken to be zero at zero Kelvin.*

Certain types of entropy, such as the nuclear spin entropy are never zero but chemical reactions and phase transformations do not alter these entropies.

Pure elements ordinarily contain various concentrations of their isotopes. For naturally occurring concentrations of isotopes, chemical reactions always distribute the isotopes in the same manner; hence it is an established convention to ignore the isotopes in the measurements of entropy. We now give two examples justifying the third law of thermodynamics.

*Example (1)* Sulfur exists in the rhombic form up to 368.54 K and the monoclinic form, above this temperature. The monoclinic form can readily be supercooled to avoid transition and its heat capacity can be measured down to 15 K. The heat capacity of rhombic sulfur, was also measured above room temperature [Gen. Ref. (1–4)]. $\Delta_{tr}H^\circ$ for Rhombic $\rightarrow$ Monoclinic at 368.54 K is 401.7 J g atom$^{-1}$. From these measurements, the entropy of rhombic sulfur at 368.54, obtained using the heat capacity of rhombic sulfur, is

$$S_{368.54}(\text{rhombic}) = 1.49 \times 10^{-4} \times 15^3 + \int_{15}^{298.15} \frac{C_P^\circ}{T} dT$$

$$+ \int_{298.15}^{368.54} \frac{C_P^\circ}{T} dT = 36.86 \text{ J g atom}^{-1} \text{ K}^{-1} \qquad (7.4)$$

The value of the first term on the right is 0.50 from the Debye heat capacity extrapolation for 0 to 15 K, the second term is 31.40 from a graphical integration. The third term is 4.96, which is obtained by analytical integration using a heat capacity equation for rhombic sulfur and the total value is 36.86 J g atom$^{-1}$ K$^{-1}$. The entropy change for Rhombic $\rightarrow$ Monoclinic is $401.7/368.54 = 1.09$; hence the entropy of monoclinic sulfur at 368.54 K is $36.86 + 1.09 = 37.95$ J g atom$^{-1}$ K$^{-1}$. On the other hand, the entropy computed from the heat capacity data on supercooled monoclinic sulfur is 37.82, which is in very good agreement with the preceding

---

*M. Planck, "Treatise on Thermodynamics," translated by A. Ogg, Dover Publications (1945).

value of 37.95, well within the estimated experimental errors of 0.40, (all entropies are in J atom-mol$^{-1}$ K$^{-1}$).

Other typical examples similar to sulfur are, He, C, P, Mn, Sn, PH$_3$ and cyclohexanol. The entropy calculated in the manner illustrated by the preceding example is called the *third law entropy*.

*Example (2)* The entropy change calculated from the third law convention (Gen. Ref. 1) for the Haber process, 0.5N$_2$(g) + 1.5 H$_2$(g) = NH$_3$(g) is $\Delta_r S° = -13.65R$ at 673.15 K. The equilibrium constant, $K_p$, determined experimentally by various investigators agree well (Gen. Ref. 1). The average values of $K_p$ between 10 to 50 bars are 0.0260, 0.0176, 0.0124, 0.0089, and 0.0065 at 350, 375, 400, 425, and 450°C respectively. A plot of $RT \ln K_p$ (or $-\Delta_r G°$) versus $T$ yields a straight line having a slope of $\Delta_r S° = -13.62R$ at the average temperature of 673.15 K, in excellent agreement with $-13.65R$ by the third law convention. The value of $\Delta_r S°$ determined by using $K_p$ is called the *second law value*.

## Entropy from Statistical Mechanics

The entropy of gases may also be obtained by entirely different methods, i.e., by the methods of statistical mechanics, which are based on the mechanics of molecules, largely deduced from interpretations of various spectra of molecules. Table 7.1 shows a number of examples (Gen. Ref. 1 and 2). It is a remarkable fact that the agreement is excellent in many cases. Comparisons of the calorimetric and the statistical values have contributed to accurate interpretations of motions in complex molecules. In general, statistical entropy for gases is preferred to the calorimetric entropy when there is no uncertainty in the interpretations of molecular vibrations and rotations.

The statistical equation for the entropy of noble gases is

$$S°/R = 1.5 \ln M + 2.5 \ln T - 1.1518, \quad \text{(at 1 bar)} \tag{7.5}$$

where $S°$ refers to the entropy per mole at 1 bar. This equation is known as the *Sackur-Tetrode equation*, first obtained empirically by O. Sackur (1911) and later theoretically by H. Tetrode (1912). The values of $S°/R$ from this equation are listed in Table 7.1. For other monatomic gases such as F, Cl, etc., an additional term is required in (7.5). A few comments on Table 7.1 are in order. The disagreements for CO, NO and N$_2$O between the calorimetric and the statistical values in Table 7.1 have not yet been explained satisfactorily. It has previously been assumed that in the solid state, the crystals of these compounds are not arranged perfectly; (perfect arrangement is CO–CO–CO ..., i.e., each carbon atom has only oxygen as the nearest neighbors and each oxygen atom only carbon). However, this argument is greatly in doubt because carbon and oxygen atoms tend to arrange themselves properly in solid CO even at fairly low temperatures.* It appears that new calorimetric measurements are necessary to resolve these discrepancies, and

---

* E. K. Gill and J. A. Morrison, J. Chem. Phys. *45*, 1585 (1966).

**Table 7.1** Standard entropies of gases at 298.15 K.

| Gas | $S°/R$* Statistical | $S°/R$ Calorimetric | |
|---|---|---|---|
| Ar | 18.62 | 18.60 | ± 0.10 |
| Kr | 19.73 | 19.72 | 0.05 |
| Ne | 17.60 | 17.63 | 0.05 |
| Xe | 20.41 | 20.49 | 0.15 |
| $O_2$ | 24.67 | 24.72 | 0.05 |
| $N_2$ | 23.04 | 23.12 | 0.08 |
| HCl | 22.47 | 22.40 | 0.08 |
| $H_2O$ | 22.71 | 22.30 | |
| CO | 23.82 | 23.27 | |
| $CO_2$ | 25.71 | 25.73 | 0.05 |
| NO | 25.34 | 25.02 | |
| $N_2O$ | 26.45 | 25.89 | 0.05 |
| HBr | 47.49 | 47.59 | 0.15 |

*Accuracy on this column is generally better than ±0.01.

the statistical values are the accepted values of entropies for these substances. Other examples of discrepancies are given in Gen. Ref. 1 and 2 and an interesting explanation for methane, ethane, and diborane are given by Ko and Steele.* For molecules having atoms of highly dissimilar diameters, such as HCl and HBr the arrangement of molecules in the solid is perfect, i.e., there is no randomness, and the calorimetric and the statistical values are in good agreement. The discrepancy for $H_2O$ has been satisfactorily explained by structural analysis of ice.

Numerous substances, particularly paramagnetic materials transform from a more random state to a more ordered state upon cooling at a temperature called the Curie point or the lambda point as shown in Fig. 7.2 for $FeNH_4(SO_4)_2 \cdot 12H_2O$, ($FeNH_4$-alum), and for Tutton salt. The lambda point for liquid $^4He$ is about 2 K,[†] for MnS, 152 K and for Fe, 1000 K [Gen. Ref. (1 and 2)]. When the heat capacity measurements are not low enough to include a low temperature transition such as the lambda transition, the calorimetric entropy is lower than the true entropy from statistical calculations.

### Entropies of Supercooled Liquids

Certain substances having fairly low enthalpies of fusion can be supercooled below the equilibrium fusion temperature to retain various degrees of randomness that exist in the liquid state. Such substances are said to be in a glassy state. If the cooling is rapid, residual entropy remains down to 0 K. When a glassy substance

*H. W. Ko and W. A. Steele, J. Chem. Phys. *51*, 4595 (1969).
†K. R. Atkins and M. J. Buckingham and W. M. Fairbank; presented in "Introduction to Liquid Helium" by J. Wilks, Clarendon Press, Oxford (1970).

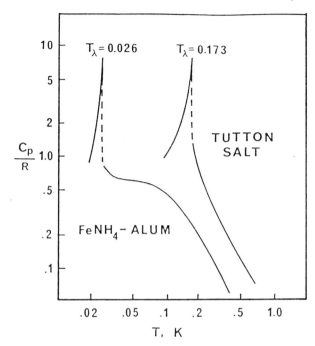

**Fig. 7.2** Heat capacities of $FeNH_4(SO_4)_2 \cdot 12H_2O$, (FeNH$_4$-alum) and $Mn(NH_4)_2(SO_4)_2 \cdot 12H_2O$ (Tutton salt). Note that coordinates are logarithmic to show details of each curve. [Adapted from O. E. Vilches and J. C. Wheatley, Phys. Rev. *148*, 509 (1966)].

is cooled sufficiently slowly for its molecules to assume their "annealed" configuration, then the entropy should be zero at 0 K. However, the mobility of molecules at low temperatures is too slow for the glassy state to assume a stable, or annealed configuration. The only supercooled liquid substance that remains in an equilibrium configuration down to 0 K is helium, and the measurements on the liquid, as well as the solid, show that the entropies of both phases approach each other at 0 K.

### Consequences of the Third Law

The entropy of a substance is related to its heat capacity by

$$T\left(\frac{\partial S}{\partial T}\right)_p = C_p \tag{7.6}$$

as required by (7.1). Since $S$ is a finite and continuous function of temperature (except when there is a phase change at a known value of $T$) then $(\partial S/\partial T)_p$ is finite and the left side of (7.6) is therefore zero for $T \to 0$; hence,

$$C_p \to 0 \quad \text{when} \quad T \to 0 \tag{7.7}$$

A similar relationship can also be obtained for $C_V$.

116    Thermodynamics

Some of the most important consequences of the third law are obtained by rewriting (7.3) as follows:

$$S_0(T \to 0, P', Y', Z', \ldots) - S_0(T \to 0, P'', Y'', Z'', \ldots) = 0 \qquad (7.7a)$$

If the difference between two values of a variable is infinitesimal, e.g., $Y' - Y'' = dx$, then the left side of this equation is also an infinitesimal difference $dS$, and therefore, in general,

$$\lim \frac{\Delta S_0}{Y' - Y''} = \frac{\partial S_0}{\partial Y} = 0; \qquad (T \to 0) \qquad (7.8)$$

The first and second laws may be written as

$$dU = T\,dS - P\,dV + \mathcal{H}\,d\mathcal{M} + \sigma\,d\underline{A} \qquad (7.9)$$

where $\mathcal{M}$ is the magnetic moment per mole; $\mathcal{H}$, the magnetic field; $\sigma$ the surface tension; and $\underline{A}$, the area. Additional work terms may be added to this equation if they are present. Cross differentials from (7.9) yield

$$\left(\frac{\partial S}{\partial P}\right)_{T \to 0} = -\left(\frac{\partial V}{\partial T}\right)_P = 0; \quad \left(\frac{\partial S}{\partial V}\right)_{T \to 0} = \left(\frac{\partial P}{\partial T}\right)_V = 0$$

$$\left(\frac{\partial S}{\partial \mathcal{H}}\right)_{T \to 0} = \left(\frac{\partial \mathcal{M}}{\partial T}\right)_\mathcal{H} = 0; \quad \left(\frac{\partial S}{\partial \underline{A}}\right)_{T \to 0} = -\left(\frac{\partial \sigma}{\partial T}\right)_{\underline{A}} = 0 \qquad (7.10)$$

The left hand sides of these equations are zero by (7.8) and the right-hand sides are directly measurable. The first of these, i.e., $(\partial V/\partial T)_P$, is the volume coefficient of expansion which is found to be zero for crystals and for liquid helium in the vicinity of 0 K. This is also true for glasses and other unstable phases with stable or unstable configurations. Likewise $(\partial \mathcal{M}/\partial T)_\mathcal{H}$ is known to vanish for ferromagnetic substances and it is expected to vanish for other substances as well. The surface tension becomes independent of temperature by the last equation, and measurements of the surface tension of liquid helium obeys this relationship.

## Thermal Evaluation of Entropy

Thermal evaluation of entropy is made by using (7.1) and the details of procedure are as follows: (a) the entropy change of a solid, denoted for example, as Crystal II, is calculated from 0 K to the lowest temperature $T$ for which experimental data on heat capacity are available so that the heat capacity can be expressed by the Debye-Sommerfeld equation and $\int_0^{T'} (C_P/T)\,dT$ can be evaluated analytically with $T'$ usually in the neighborhood of 5 to 20 K. (b) The heat capacity measurements from $T'$, to the phase transformation temperature $T_{tr}$(Cryst. II $\to$ I) permits analytical evaluation of the entropy change. (c) The entropy of phase transformation from Crystal II to I is obtained from $(H_I - H_{II})/T_{tr}$. Further increase in entropy by heating, melting and vaporization is obtained in the same manner. (d) When the final state of aggregation is gaseous, correction for departure from ideal behavior is obtained by (5.9). The resulting entropy of the ideal gas at 1 bar and at $T$, can be obtained from

$S_T^\circ$(ideal gas, 1 bar)

$$= \int_0^{T'} \frac{C_p(\text{Cryst. II, Debye-Som.})}{T} dT + \int_{T'}^{T_{tr}} \frac{C_p(\text{II})}{T} dT + \frac{\Delta H_{\text{II}\to\text{I}}}{T_{tr}}$$

$$+ \int_{T_{tr}}^{T_m} \frac{C_p(\text{I})}{T} dT + \frac{\Delta_{\text{melt}} H}{T_{\text{melt}}} + \int_{T_m}^{T_{\text{vap}}} \frac{C_p(\text{liq.})}{T} dT + \frac{\Delta_{\text{vap}} H}{T_{\text{vap}}}$$

$$+ \int_{P_2}^{1} \left(\frac{\partial V}{\partial T} - \frac{R}{P}\right) dP + \int_{T_{\text{vap}}}^{T} C_p^\circ(\text{ideal gas}) \, d\ln T \qquad (7.11)$$

The entropy calculated in this manner is called "the entropy from the third law." The following example illustrates the application of (7.11) to oxygen for which stepwise calculations, based on the data of Giauque and Johnston,* are as follows, with $S$ expressed in terms of J mol$^{-1}$ K$^{-1}$:

$S_{11.75}^\circ$(crystals III) = 1.34; extrapolation, Debye $C_p$.
$S_{23.66}^\circ$(III) $- S_{11.75}^\circ$, (III) = 7.11; experimental $C_p$.
$\Delta S_{23.66}^\circ$(III, II) = 93.81/23.66 = 3.96; phase transformation, experimental.
$S_{43.75}^\circ$(II) $- S_{23.66}^\circ$, (II) = 19.50; experimental $C_p$.
$\Delta S_{43.75}^\circ$(II, I) = 743.1/43.75 = 16.99; phase transformation, experimental.
$S_{54.39}^\circ$(I) $- S_{43.75}^\circ$(I) = 10.04; experimental $C_p$.
$\Delta S_{54.39}^\circ$(melting) = 444.8/54.39 = 8.18; experimental.
$S_{90.13}^\circ(l) - S_{54.39}^\circ(l)$ = 27.03; experimental $C_p$.
$S$(non-ideal gas at 1 bar) $-S^\circ(l)_{90.13}$ = 6,814.9/90.13 = 75.61; vaporization at 1 bar, experimental.
$S^\circ$(ideal, 1 bar)$-S$(non-ideal, 1 bar)]$_{90.13}$ = 0.71; correction to ideal state from non-ideal state obtained from $P$-$V$ data at 90.13 K and from (5.9) in Chapter V.
$S_{298}^\circ - S_{90.13}^\circ$ = 34.94; $C_p^\circ$ of ideal diatomic gas.
The sum gives $S_{298}^\circ$ = 205.41 $\pm$ 0.42 J mol$^{-1}$ K$^{-1}$.
The spectroscopic data gives $S_{298}^\circ$ = 205.12 J mol$^{-1}$ K$^{-1}$, listed in Table 7.1.
The necessary data up to 90 K are given in Problem 7.7 for detailed computation of entropy at 90 K.

## PROBLEMS

7.1  Evaluate $\Delta S_{298}^\circ$ for the following reactions from the tables in Appendix II:

$$H_2(g) + \tfrac{1}{2}O_2(g) = H_2O(g)$$
$$C(\text{graph}) + O_2(g) = CO_2(g)$$
$$2CO(g) = CO_2 + C(\text{graph})$$
$$N_2(g) + 3H_2(g) = 2NH_3(g)$$

*From a summary by K. K. Kelley and E. G. King, Bull. 592, U.S. Bureau of Mines (1961).

7.2 Plot $\Delta_f G° = -RT \ln K_p$ by using the values of $K_p$ in the text for the formation of $NH_3$, and evaluate $\Delta_f S°$ at 673.15 K.

Ans.: $\Delta_f S° = -113.26$ J mol$^{-1}$ K$^{-1}$

7.3 A pure substance transforms from Cryst. I to Cryst. II at 24 K on cooling. The equations for $C_p°$ are

$$C_p°(\text{Cryst. I}) = 7.5 \times 10^{-4} T^3 \text{ and } C_p°(\text{Cryst. II}) = 4.6 \times 10^{-4} T^3 \text{ J mol}^{-1} \text{ K}^{-1}$$

Calculate the enthalply and entropy of transformation at 24 K.

Ans.: $\Delta S° = \Delta H°/24 = -1.336$ J mol$^{-1}$ K$^{-1}$

7.4 Compute the entropy of mixing of one mole of liquid $^3$He with 10 moles of $^4$He at 5 K assuming that the solution is ideal, i.e., similar to an ideal gas mixture.

Ans.: $\Delta S = 3.351 R$

7.5 Compute $S°$(ideal 1 bar) $-$ $S$(real, 40 bars) for carbon dioxide from the following data for compressibility:

| $P$, bar: | 1 | 4 | 7 | 10 | 40 |
|---|---|---|---|---|---|
| $Z$: | 0.99517 | 0.9804 | 0.9651 | 0.9493 | 0.7648 |

Hint: Express $Z$ as a quadratic function of pressure first.

Ans.: $3.471 R$

7.6 Correct or amend the following statements if necessary:

(a) The entropy of a pure crystalline compound varies slightly with small changes in pressure at 0 K.

(b) For a pure crystalline metal $C_p° = \alpha T + 3\beta T^3$ and $S° = \alpha T + 3\beta T^3$ up to 20 K.

(c) The value of $\partial C_p°/\partial T = 0.01$ and is constant at 300–600 K, and $S°_{600} - S°_{300} = 31.00$; therefore, $C_p°$ (at 450 K) $= 41.90$. Hint: $C_p° = a + 0.01$ T.

7.7 Calculate $S°$ for liquid oxygen at 90.0 K from the following data (see also the text):

| $T$, K: | 11.75 | 15.12 | 15.57 | 16.94 | 18.32 | | |
|---|---|---|---|---|---|---|---|
| $C_p°/R$: | 0.483 | 0.805 | 0.901 | 1.13 | 1.36 | | |
| $T$, K: | 20.26 | 21.84 | 22.24 | 25.02 | 30.63 | | |
| $C_p°/R$: | 1.76 | 2.11 | 2.19 | 2.73 | 3.49 | | |
| $T$, K: | 37.59 | 42.21 | 46.00 | 52.00 | 60.00 | 70.00 | 90.00 |
| $C_p°/R$: | 4.57 | 5.40 | 5.55 | 5.56 | 6.40 | 6.43 | 6.53 |

Phase transition at 23.66 K: $\Delta_{tr} H°/R = 11.28$ K
Phase transition at 43.75 K: $\Delta_{tr} H°/R = 89.4$ K
Fusion at 54.39 K: $\Delta_{fus} H°/R = 53.5$ K

7.8 Derive the equation for $S°(\text{Ar}) - S°(\text{Ne})$ at 300 K. Is this difference temperature dependent?

CHAPTER VIII

# PHASE EQUILIBRIA

## Introduction

The phase equilibria are a vast area of interest in chemistry, physics, metallurgy, geology and related fields such as chemical engineering, materials science, biology, and oceanography. We shall first consider the phase equilibria for single component systems and then proceed to the multicomponent systems. It is well known from observations that in a single component system two phases may coexist under equilibrium over a range of temperature and the corresponding range of pressure. Let I and II represent the coexisting phases of a substance $A$ for which the phase transformation is

$$A(\text{I}) = A(\text{II}) \tag{8.1}$$

Under equilibrium conditions, an infinitesimal change in the molar Gibbs energy is zero, i.e., $dG_{P,T} = 0$; hence, integration of this equation for the phase transformation (8.1) is

$$\Delta G_{P,T} = (G^{\text{II}} - G^{\text{I}})_{P,T} = 0 \tag{8.2}$$

or, simply, at equilibrium,

$$G^{\text{I}} = G^{\text{II}} (P \text{ and } T \text{ of both phases are equal}) \tag{8.3}$$

where the superscripts refer to the coexisting phases. When there are three coexisting phases in equilibrium,

$$\begin{aligned} G^{\text{II}} - G^{\text{I}} &= 0 \\ G^{\text{III}} - G^{\text{II}} &= 0 \end{aligned} \tag{8.4}$$

hence, $G^{\text{I}} = G^{\text{II}} = G^{\text{III}}$, or the molar Gibbs energies of three phases are the same at the same pressure and temperature. The usual notation of phases with $G$, is $G(\text{I})$, $G(\text{II})$ and $G(\text{III})$, but the use of superscripts in this chapter facilitates the derivation of important relationships.

## Two-Phase Equilibrium

The total differential of (8.3) for two phases in equilibrium is

$$dG^{\text{I}} = dG^{\text{II}} \tag{8.5}$$

120   Thermodynamics

**Table 8.1**   Effect of pressure on transformation temperature.*

| Compound | Transformation | T/K | $\Delta H$/ J mol$^{-1}$ | $V^I$/ cm$^3$ mol$^{-1}$ | $V^{II}$/ cm$^3$ mol$^{-1}$ | $dT/dP$/ K bar$^{-1}$ |
|---|---|---|---|---|---|---|
| Bi | Melting | 544.59 | 11,297 | 21.5(s) | 20.8(l) | −0.0034 |
| BF$_3$ | Melting | 281.95 | 12,029 | 42.39(s) | 48.16(l) | +0.0135 |
| H$_2$O | Melting | 273.15 | 6,009 | 19.649(s) | 18.018(l) | −0.0074 |
| H$_2$O | Vaporization | 372.78 | 40,671 | 18.76(l) | 30,140(g) | +27.61 |

*Data are mostly from General Refs. (1), (8), (25), and (26).

For each phase, $dG = V\,dp - S\,dT$, where $G$, $V$, and $S$ are the molar properties, and from (8.5), it follows that

$$V^I\,dP - S^I\,dT = V^{II}\,dP - S^{II}\,dT$$

and therefore,

$$\frac{dP}{dT} = \frac{S^{II} - S^I}{V^{II} - V^I} = \frac{\Delta S}{\Delta V} \tag{8.6}$$

This relationship is called the Clapeyron equation first derived by B. P. E. Clapeyron (1832) and later rigorously by R. J. E. Clausius (1850). Table 8.1 gives the values of $dT/dP$ computed from (8.6) for fusion and vaporization of a number of interesting substances. The values of $dT/dP$ were computed by noting that $\Delta S = \Delta H/T$ and converting (8.6) into

$$\frac{dT}{dP} = \frac{(V^{II} - V^I)T}{\Delta H} \tag{8.7}$$

For dimensional consistency, $\Delta H$ should be expressed in J · mol$^{-1}$, $V$ in m$^3$ mol$^{-1}$ when $T$ is in K and $P$ in Pa. For example, for BF$_3$ in Table 8.1, it is evident that

$$\frac{dT}{dP} = \frac{(48.16 - 42.39)10^{-6} \times 281.95}{12,029} = 1.35 \times 10^{-7}\ \text{K} \cdot \text{Pa}^{-1}$$

$$= 0.0135\ \text{K} \cdot \text{bar}^{-1}$$

If it is assumed that $dT/dP \approx \Delta T/\Delta P$, an increase of 100 bars would increase the melting point of BF$_3$ by 1.35°C. For Bi and H$_2$O, $\Delta T/\Delta P$ is negative because the volume decreases upon melting. Graphite to diamond transformation at high temperatures is a special case, not listed as an example in Table 8.1, and for this phase transformation, i.e.,

Graphite = Diamond;   ($\Delta$ = Diamond − Graphite)

the equilibrium pressure $P$ as a function of temperature is simply*

$$P(\text{bar}) = 7{,}100 + 27.00\,T; \qquad (\text{high } T) \tag{8.7a}$$

From 600 to 2000 K, $dP/dT$ is 27.00 bar K$^{-1}$. The change of volume is $\Delta V = V(\text{diamond}) - V(\text{graphite}) = -1.77$ cm$^3$ mol$^{-1}$; hence from (8.6) and (8.7),

$$\frac{dP(\text{Pa})}{dT} = 27.00 \times 10^5 = \frac{\Delta S}{-1.77 \times 10^{-6}} = \frac{\Delta H \times 10^6}{-1.77\,T}$$

from which $\Delta S = -4.78$ J g atom$^{-1}$ K$^{-1}$, and $\Delta H = -4.78\,T$ in J·mol$^{-1}$ in agreement with the indirect calorimetric data obtained from the enthalpies of combustion of graphite and diamond. The entropy versus $T$ for graphite and for diamond must converge to each other as required by the third law and illustrated by Fig. 7.1 in the preceding chapter, i.e., $\Delta S/\Delta V$ and $\Delta S/\Delta P$ become zero at 0 K according to (7.7a), $S_0(\text{diamond}) - S_0(\text{graphite}) = 0$, and consequently,

$$dP/dT = (\Delta S/\Delta V)_{T \to 0} = 0$$

and the pressure versus temperature curve for the graphite-diamond transformation approaches zero slope at $T \to 0$. Another interesting experimental verification of this equation is provided by the pressure versus temperature curve for freezing of liquid $^4$He$^{II}$. Below about 25 bars, $^4$He$^{II}$ is liquid down to 0 K but above 25 bars, it becomes solid and $\Delta P/\Delta T \approx dP/dT$ is virtually zero below 1 K, confirming that $\Delta S/\Delta V$ also becomes zero.

Unlike the condensed phase transitions, vaporization is accompanied by a large volume change, i.e., $\Delta V = V(\text{vapor}) - V(\text{liquid})$ is very large, and therefore a small change in pressure causes a large change in the boiling point, e.g., for water, a change of 0.1 bar in pressure causes an increase of approximately 2.8°C in the boiling temperature.

## Vaporization Equilibria

Application of the Clapeyron equation to the vaporization of liquids gives

$$\frac{dP}{dT} = \frac{\Delta_{\text{vap}} H}{T(V^g - V^l)} \tag{8.8}$$

where $\Delta_{\text{vap}} H$ is the enthalpy of vaporization of Chapters VI and VII. Equation (8.8) is exact and its integration requires functional representation of $\Delta_{\text{vap}} H$ and $V^g - V^l$ in terms of the independent variables $P$ and $T$. Under certain conditions (8.8) can be integrated. Thus at ordinary pressures the volume of vapor is much larger than that of liquid, i.e., $V^g \gg V^l$ so that $V^g - V^l \approx V^g$. For example, the volume of

---

*R. Bearman and F. E. Simon, Z. Elektrochemie, 59, 333 (1955); See also pioneering work of P. W. Bridgman, "Collected Experimental Papers," Harvard Univ. Press (1964); F. P. Bundy, J. Chem. Phys., 38, 631 (1963); F. P. Bundy, H. P. Bovenkerk, H. M. Strong, and R. H. Wentorf, Jr., ibid., 35, 383 (1961); See also I. A. Bulgak, A. S. Skoropanov, and A. A. Vecher, Russian J. Phys. Chem., 62, 125 (1988); K. E. Spear and J. P. Dismukes, editors, "Synthetic Diamond," John Wiley and Sons (1994); J. E. Field, Ed.,"Properties of Natural and Synthetic Diamond," Academic Press (1992).

steam is roughly 1700 times larger than that of water at 100°C and 1 bar; therefore the error in neglecting $V^l$ is very small. Assuming that the vapor behaves as an ideal gas, i.e., $V^g = RT/P$, $V^g$ may be eliminated from (8.8) after ignoring $V^l$, and then the Clapeyron equation becomes

$$\frac{dP}{P} = \frac{\Delta_{\text{vap}} H \, dT}{RT^2} \tag{8.9}$$

This equation is also applicable to sublimation. Equation (8.9) is usually called Clausius-Clapeyron equation. It can be integrated to obtain $P$ as a function of $T$ after expressing $\Delta_{\text{vap}} H$ as a function of temperature. The procedure of integrating (8.9) for gases is followed in texts adhering to the historical development of thermodynamics. A simple way of obtaining the equivalent integral is by writing $\Delta G = G^g - G^l$, where $G^g = G^{\circ,g}(T) + RT \ln P$ as given by (6.26). For pure liquids, $G^l$ is written as $G^l = G^{\circ,l}(P, T)$; hence, $\Delta G$ is given by

$$\Delta G = 0 = G^{\circ,g}(T) + RT \ln P - G^{\circ,l}(P, T) \tag{8.10}$$

or upon rearrangment,

$$\Delta_{\text{vap}} G^\circ \equiv G^{\circ,g}(T) - G^{\circ,l}(P, T) = -RT \ln P \tag{8.11}$$

where $\Delta_{\text{vap}} G^\circ$ refers to vaporization. From the definitional equation $\Delta_{\text{vap}} G^\circ \equiv \Delta_{\text{vap}} H^\circ - T \Delta_{\text{vap}} S^\circ$ and from (8.11), it follows that

$$\ln P = -\frac{\Delta_{\text{vap}} H^\circ}{RT} + \frac{\Delta_{\text{vap}} S^\circ}{R} \tag{8.12}$$

Historically, this is called the integral of Clausius-Clapeyron equation but it is also called the van't Hoff equation when it refers to other equilibria which differ only in detail but not in basic thermodynamic processes as will be seen later. When $P$ is 1 bar, $T$ becomes the baric boiling point, $T_b$, and $\Delta_{\text{vap}} S^\circ = \Delta_{\text{vap}} H^\circ/T_b$ and if the Trouton rule is valid, i.e., $\Delta_{\text{vap}} S^\circ/R \approx 11$, then (8.12) becomes

$$\ln P \approx 11(T - T_b)/T \tag{8.13}$$

This equation permits estimation of the vapor pressure from a knowledge of the boiling point only, or the boiling point from any single value of the vapor pressure at any temperature. For the sublimation process, $\Delta_{\text{vap}} H^\circ$ and $\Delta_{\text{vap}} S^\circ$ are replaced by $\Delta_{\text{sub}} H^\circ$ and $\Delta_{\text{sub}} S^\circ$ respectively in (8.12) and 11 is replaced by 12.1 which is the value of $\Delta_{\text{sub}} S^\circ/R$ from the combined rules of Richards and Trouton.

It is frequently possible, particularly in measurements of vapor pressure at high temperatures, to assume that $\Delta_{\text{vap}} H^\circ$, hence $\Delta_{\text{vap}} S^\circ$, is independent of temperature. In general, they are functions of temperature as follows:

$$\Delta_{\text{vap}} H^\circ_T = \Delta_{\text{vap}} H^\circ_{T_o} + \int_{T_o}^{T} \Delta_{\text{vap}} C^\circ_p \, dT \tag{8.14}$$

$$\Delta_{\text{vap}} S^\circ_T = \Delta_{\text{vap}} S^\circ_{T_o} + \int_{T_o}^{T} \frac{\Delta_{\text{vap}} C^\circ_p}{T} \, dT \tag{8.15}$$

where the constant terms preceding the integrals are experimentally determined at $T_\circ$. It is possible to carry out the integration in these equations if $C_p^\circ$ of the liquid and of the vapor are known as functions of temperature. Substitution of the following empirical equation

$$\Delta_{vap} C_p^\circ = \Delta\alpha + 2\Delta\beta T + 6\Delta\lambda T^2 - 2\Delta\epsilon T^{-2} \tag{8.15a}$$

into (8.14) and (8.15) gives

$$\Delta_{vap} H^\circ = \Delta_{vap} H_\theta^\circ + \Delta\alpha T + \Delta\beta T^2 + 2\Delta\lambda T^3 + 2\Delta\epsilon T^{-1} \tag{8.16}$$

$$\Delta_{vap} S^\circ = \Delta_{vap} S_\theta^\circ + \Delta\alpha \ln T + 2\Delta\beta T + 3\Delta\lambda T^2 + \Delta\epsilon T^{-2} \tag{8.17}$$

where $\Delta_{vap} H_\theta^\circ$ and $\Delta_{vap} S_\theta^\circ$ are the constant terms plus the integration constants on the right sides in (8.14) and (8.15) respectively. Substitution of (8.16) and (8.17) into (8.12) gives the following useful equation:

$$R \ln P = -\frac{\Delta_{vap} H_\theta^\circ}{T} + \Delta_{vap} S_\theta^\circ - \Delta\alpha + \Delta\alpha \ln T$$
$$+ \Delta\beta T + \Delta\lambda T^2 - \Delta\epsilon T^{-2} \tag{8.18}$$

The terms $R \ln P - \Delta\alpha \ln T - \Delta\beta T - \Delta\lambda T^2 + \Delta \epsilon T^{-2}$ are set equal to $\Sigma$ for brevity in notation so that

$$\Sigma = -\left(\Delta_{vap} H_\theta^\circ / T\right) + \left(\Delta_{vap} S_\theta^\circ - \Delta\alpha\right) \tag{8.18a}$$

When $\Sigma$ is plotted versus $1/T$, a linear correlation is obtained wherein $-\Delta_{vap} H_\theta^\circ$ is the slope, and $(\Delta_{vap} S_\theta^\circ - \Delta\alpha)$, the intercept. Such a correlation, known as the *sigma plot*, permits a linear fitting of a set of data by a simple method of least squares. Alternative computations with (8.18) and various interpretations of the vapor pressure data are available as will be seen in Chapter XII in conjunction with the chemical equilibria, since $P$ in (8.18) is the equilibrium constant for the vaporization reaction of a pure component, i.e.,

(Condensed phase) = (Vapor phase)

As a numerical example, consider mercury with the following data: baric boiling point, $T_\circ = 629.84$ K, $\Delta_{vap} H_\circ^\circ$ (at $T_\circ$) = 59,205 J mol$^{-1}$, $\Delta_{vap} S_\theta^\circ = 59,205/629.84 = 94.000$, and

$$C_p^\circ(l) = 26.280 + 8.46 \times 10^{-4} T + 131,400/T^2$$

$$C_p^\circ(g) = 20.786 \text{ (both in J mol}^{-1} \text{ K}^{-1})$$

$$\Delta\alpha = -5.494; \quad \Delta\beta = -4.23 \times 10^{-4}; \quad \Delta\epsilon = +65,700$$

Therefore, (8.16) and (8.17) at $T_\circ = 629.84$ become

$$59,205 = \Delta_{vap} H_\theta^\circ - 5.494 \, T_\circ - 4.23 \times 10^{-4} \, T_\circ^2 + 131,400/T_\circ$$

$$94.000 = \Delta_{vap} S_\theta^\circ - 5.494 \ln T_\circ - 8.46 \times 10^{-4} \, T_\circ + 65,700/T_\circ^2$$

124    Thermodynamics

Substitution of $T_o = 629.84$ in these equations yields $\Delta_{vap} H_\theta^\circ = 62{,}625$ and $\Delta_{vap} S_\theta^\circ = 129.779$. Thus, (8.18) becomes

$$R \ln P(\text{bar}) = -\frac{62{,}625}{T} + 129.779 + 5.494 - 5.494 \ln T$$
$$- 4.23 \times 10^{-4} T - 65{,}700/T^2$$

The value of $P$ at 25°C from this equation is 0.260 Pa ($\approx 1.95 \times 10^{-3}$ mm Hg).

A frequently used empirical equation with three constants $A$, $B$, and $C$, which are adjusted to represent the vapor pressure data, is

$$\ln P = -\frac{A}{T+B} + C \qquad (8.19)$$

which is known as the Antoine equation (1888). When $B$ is zero, $A$ and $C$ are thermodynamically meaningful as required by (8.12), otherwise $A$, $B$, and $C$ must be regarded as empirically adjustable parameters. Extensive tabulations of vapor pressures of hydrocarbons and related compounds in terms of the Antoine equation are presented in Gen. Ref. (14).

## Variation of Vapor Pressure with Total Pressure at Constant Temperature

Under equilibrium conditions, the pressure over each phase of a single component system is obviously the same. However, when an inert gas is introduced, the property of the vapor itself is not altered if the vapor and the gas form an ideal mixture. The pressure over the condensed phase, i.e., liquid or solid phase, is the sum of the partial pressures of the vapor and the inert gas. At a constant temperature, therefore, the change in the Gibbs energy of liquid is equal to that of vapor, i.e., $dG^l = dG^v$; hence $V^l(dP^l)_T = V^v(dP^v)_T$. Rearrangement of this equation gives

$$\left(\frac{\partial P^v}{\partial P^l}\right)_T = \frac{V^l}{V^v} \qquad (8.20)$$

which is known as the Poynting equation (1881). If the vapor is assumed to be ideal, i.e., $V^v = RT/P^v$, then (8.20) becomes

$$\left(\frac{\partial P^v}{\partial P^l}\right)_T = \frac{V^l}{RT} P^v$$

For moderate variations of the external pressure on the liquid, the left hand side is adequately expressed by $\Delta P^v / \Delta P^l$; therefore

$$\Delta P^v \approx (V^l/RT) P^v \Delta P^l \qquad (8.21)$$

For example, water boils under its own vapor pressure of 1 bar ($10^5$ Pa) at 372.78 K, i.e., $P^v = 1$, the molar volume of liquid is $1.88 \times 10^{-5}$ m$^3$ mol$^{-1}$, and when it

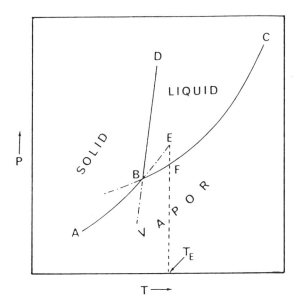

**Fig. 8.1** Phase equilibria for one-component system.

vaporizes in a closed vessel under 10 bars of nitrogen, the total pressure over the liquid is 11 bars and the increase in the vapor pressure, $\Delta P^v$ is given by

$$\Delta P^v \approx 1.88 \times 10^{-5} \times 10^5 \times (11 - 1) \times 10^5/(8.31451 \times 372.78)$$

$$\approx 607 \text{ Pa} \approx 0.006 \text{ bar}.$$

Therefore, $\Delta P^v$ represents an increase of 0.6% in the vapor pressure. The calculation for mercury gives an increase of 0.3% in the vapor pressure at 630 K. Thus the change in equilibrium vapor pressure with pressure is very small in most cases. The actual cases are complicated because the gas mixture is not ideal, and the inert gas is usually dissolved to a limited extent in the liquid.

### Representation of Phase Equilibria*

Equilibria among various phases of a single component are shown in Fig. 8.1, which is a phase diagram for a single component system. Solid and vapor phases coexist on $AB$; liquid and vapor on $BC$; and, solid and liquid on $BD$. The areas represent the regions of stability for the single phases, and the lines, the coexisting two phases. At $B$, all three phases are in equilibrium; therefore $B$ is called a triple point. The relative slopes of the lines at $B$ are significant in thermodynamics. The

---

*R. J. Borg and G. J. Dienes, "The Physical Chemistry of Solids," Academic Press (1992); I. M. Klotz and R. M. Rosenberg, "Chemical Thermodynamics," John Wiley & Sons (1994).

slope of $BD$, from (8.7) is given by

$$\frac{dP}{dT} = \left(\frac{\Delta_{\text{fus}} H}{T_B}\right)\left(\frac{1}{V^l - V^s}\right) \tag{8.22}$$

where $\Delta_{\text{fus}} H$ is the molar enthalpy of fusion, and $T_B$ is the triple point temperature. If $V^l > V^s$, the slope is positive, which is the usual case; but if $V^l < V^s$ as for As, Sb, Bi, $H_2O$, in which the volume decreases upon melting, then the slope is negative. From (8.9), the slope of the sublimation curve $AB$ is

$$\frac{dP}{dT} = \left(\frac{\Delta_{\text{sub}} H}{RT_B^2}\right) P \tag{8.23}$$

where $\Delta_{\text{sub}} H$ is the molar enthalpy of sublimation, and similarly for $BC$,

$$\frac{dP}{dT} = \left(\frac{\Delta_{\text{vap}} H}{RT_B^2}\right) P \tag{8.24}$$

According to the first law of thermodynamics, the enthalpy of sublimation at $T_B$ is equal to the sum of the enthalpies of fusion and vaporization, i.e., $\Delta_{\text{sub}} H = \Delta_{\text{fus}} H + \Delta_{\text{vap}} H$; hence $\Delta_{\text{sub}} H > \Delta_{\text{vap}} H$ and thus from (8.23) and (8.24), it is evident that at the triple point $B$,

slope $AB$ > slope $BC$

Consequently the lines $AB$ and $BC$ must intersect each other so that the supercooled liquid and superheated solid must have higher vapor pressures than the corresponding equilibrium phases. For example, at $T_E$, $P_E$ is greater than $P_F$, and since $G^g = G^{\circ,g} + RT \ln P$ then $G^g$ (over solid at $E$) > $G^g$ (over liquid at $F$) and for the coexisting phases $G^g$ (at $E$) = $G^s$ and $G^g$ (at $F$) = $G^l$; therefore the preceding inequality is identical with $G^s > G^l$ at $T_E$. The solid at $E$ must therefore transform into liquid at $F$ because the phase with the lower Gibbs energy is the liquid phase. Thus, the extended portions of all the curves terminate in regions where they represent nonequilibrium conditions and this requires that the angle between any pair of curves at $B$ must be smaller than $180°$.

Numerous pure substances have fairly complicated phase diagrams, particularly under very high pressures as illustrated by Fig. 8.2 for $H_2O$. Ordinary ice, or Ice I, is less dense than liquid water, but all other ices are denser than liquid water as indicated by the slopes of liquid-solid equilibrium curves. An empirical correlation of each curve as $P = f(T)$ yields $dP/dT$ at any point from which $\Delta H$ and $\Delta S$ may be computed from (8.6) and (8.7) provided that the volume change $\Delta V$ is known. The Ice III = Ice II transformation indicated by the straight line $CE$ is quite similar to the Graphite = Diamond transition at 600–2000 K, graphite corresponding to Ice III, and Diamond, to Ice II, [cf. (8.7a)].

## Components

Thus far, phase equilibria in single component systems have been considered. To examine multicomponent systems, it is first necessary to define the term

Phase Equilibria    127

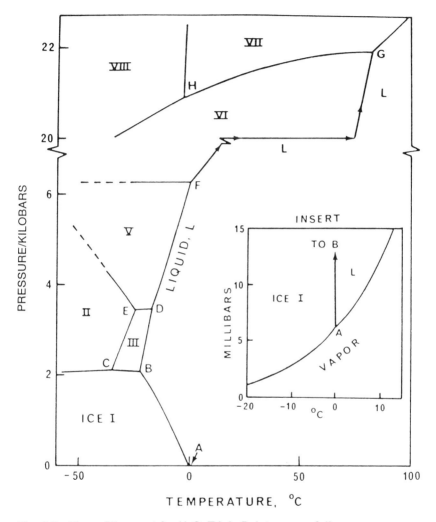

**Fig. 8.2** Phase Diagram* for $H_2O$. Triple Points are as follows:

| Points: | A | B | C | D | E | F | G | H |
|---|---|---|---|---|---|---|---|---|
| °C: | 0.0 | −22.0 | −34.7 | −17.0 | −24.3 | +0.16 | +81.6 | −3 |
| kilobars: | $6.1 \times 10^{-6}$ | 2.074 | 2.13 | 3.46 | 3.44 | 6.26 | 21.9 | 20.9 |

*Points $G$ and $H$ are from C. W. F. T. Pistorius, E. Rapoport, and J. B. Clark, J. Chem. Phys., 48, 5509 (1968). Ice IV was never confirmed. Insert shows Ice I/L (Liquid)/Vapor equilibria at much lower pressures. See also, D. Eisenberg and W. Kauzmann, "Structure and Properties of Water," Oxford Univ. Press (1969). Ice XI and XII exist below −140°C (without IX and X) according to N. N. Sirota and T. B. Bizhigitov, Russ. J. Phys. Chem., 62, 134 (1988).

*components.* The components may be added to or taken away from an *open system* to change the composition of the system. While thermodynamic properties of single component systems and certain single phase but multicomponent systems may be investigated when the system is a *closed system*, it is necessary to have an *open system* if the composition is to be a variable of state.

*The number of components of a system is the minimum number of simple elements or compounds from which all phases can be prepared and the compositions of phases can be expressed.* This definition might appear simple at first, but in many cases the choice of an appropriate number of components requires a complete knowledge of the system in order to avoid confusion. For example $CaCl_2$ may be considered as a single component system if the ratio of calcium to chlorine is constant in the temperature range of interest. At high temperatures, the composition of $CaCl_2$ varies with the pressure of gaseous chlorine and gaseous calcium, therefore, $CaCl_2$ may no longer be considered as a component since the composition of chloride itself must now be expressed in terms of two simpler substances which are calcium and chlorine. The composition of the phase corresponding to $CaCl_2$ actually exists under a certain chlorine and calcium pressure at high temperatures but it merely represents a particular composition of the phase with a measurably wide composition range. When a limited variation in the stoichiometric ratio is not of interest, such compounds may be chosen as a component for the sake of simplicity provided that the consequences of such a choice are fully understood.

An element or a compound may occur in different molecular, atomic and ionic species but only a certain number of species can be taken as the components of a system. As an example, consider hydrogen at very high temperatures where partial dissociation into monatomic hydrogen occurs. The mixture of monatomic and diatomic hydrogen is still a one-component system in view of the fact that when the pressure and temperature are fixed, the system is completely defined and the ratio of $H/H_2$, hence the composition is entirely fixed. Likewise, liquid water may contain $H_2O$, $H^+$, $OH^-$, $H_3O^+$, etc., but at a given temperature and pressure all the physical and chemical properties and the composition of pure water are completely fixed. It is impossible to change the composition of various species in pure water independently, and there exist no such free ionic species from which water can be formed. Furthermore, the state of water is entirely fixed when its pressure and temperature are fixed, and consequently water consists of a single component.

The presence of a chemical reaction involving one or more components affects the choice of the number of components. For example in a system containing $j$ moles of $CaCO_3$, $k$ moles of $CaO$ and $l$ moles of $CO_2$ at sufficiently high temperatures, equilibrium exists among the three species according to the following reaction:

$$CaCO_3(s) = CaO(s) + CO_2(g) \qquad (8.25)$$
$$[j] \qquad\quad [k] \qquad\;\; [l]$$

where the moles of constituents are shown in brackets. The system consists of two components because it can be prepared from $j + k$ moles of $CaO$ and $j + l$ moles of $CO_2$, because $j$ moles of CaO and $j$ moles of $CO_2$ would react completely to yield $j$ moles of $CaCO_3$ and leave $k$ moles of CaO and $l$ moles of $CO_2$. Likewise, the system can be obtained from $j + l$ moles of $CaCO_3$ and $k - l$ moles of CaO; or from $j + k$ moles of $CaCO_3$ and $l - k$ moles of $CO_2$. In the last case, $k$ moles of $CaCO_3$ out of $j + k$ moles would dissociate completely to yield $k$ moles of CaO and $k$ moles of $CO_2$, and $l - k$ plus $k$ moles of $CO_2$ yield $l$ moles of $CO_2$ to make up the system consisting of $j, k,$ and $l$ moles of $CaCO_3$, CaO, and $CO_2$ respectively. In this case, it may be pointed out, that when $k$ is larger than $l$, the number of moles of $CO_2$ becomes negative, but this is of no concern since the system is open and the components may be added or removed. However, it is always possible to select positive numbers of moles as the added components, and this is the usual convention. When it is possible to suppress the reaction $CaCO_3 =$ CaO + $CO_2$ at sufficiently low temperatures, the system can be prepared by using three components, i.e., all three species.

As an additional example, consider a gas phase consisting of $(j)$ moles of $H_2O$, $(k)$ moles of $H_2$ and $(l)$ moles of $O_2$ at sufficiently high temperatures, e.g., 1500 K. These species are under equilibrium according to $H_2O = H_2 + 0.5O_2$. Any of the two species may be chosen as the components from which the system may be prepared, i.e., $(j+k)$ moles of $H_2$ and $[(j/2)+l]$ moles of $O_2$; or from $(j+2l)$ moles of $H_2O$ and $(k-2l)$ moles of $H_2$; or $(j+k)$ moles of $H_2O$ and $[l-(k/2)]$ moles of $O_2$. For instance, in the last case, out of $(j+k)$ moles of $H_2O$, $k$ moles dissociate to yield $(k)$ moles of $H_2$ and $(k/2)$ moles of $O_2$, and the required number of moles of $O_2$ should therefore be $[l - (k/2)]$. The components of the system are then a properly selected set of two species out of three. The two components, as expressed in terms of the number of moles, are the extensive variables. The number of independent composition variables is one, and the state of the system is completely defined when the mole fraction of one of the components and the pressure and temperature are specified and the system is closed. The reason for this will become clear when the equilibrium constant, $K_p$, is considered in Chapter XII. For $H_2O = H_2 + 0.5\,O_2$, $K_p = P^{0.5} \cdot x_1 \cdot x_2^{0.5}/x_3$ where $P$ is the total pressure, and $x_1, x_2,$ and $x_3$ are the mole fractions of $H_2$, $O_2$, and $H_2O$ respectively. The numerical value of $K_p$ is fixed at a *given temperature* and when the total pressure $P$ is fixed, there is only one independent variable in view of the fact that $x_1+x_2+x_3 = 1$, i.e., when $T$ is fixed $K_p$ is fixed, when $P$ is fixed, $x_1 \cdot x_2^{0.5}/x_3$ is fixed, and with $x_1 + x_2 + x_3 = 1$, there is only one independent variable which may be $x_1$, or $x_2$, or $x_3$.

A gaseous mixture of $H_2O$, $H_2$, and $O_2$, however, does not react at ordinary temperatures in the absence of a catalyst and the system behaves like a mixture of three nonreacting gases. Consequently, when the reaction $H_2 + 0.5O_2 = H_2O$ is suppressed, the system consists of three components.

As a more complicated example, consider a system at moderately high temperatures consisting of graphite, gaseous carbon monoxide, carbon dioxide, hydrogen, and steam which make up 5 species. There are three possible dependent chemical reactions among these species, i.e., $C + CO_2 = 2CO$; $C + H_2O = CO + H_2$, and $H_2 + CO_2 = CO + H_2O$. Any two reactions are evidently independent since the remaining one can be obtained from the others. The number of components in this case is 3 since the system can be made up from any three species, for, each independent reaction eliminates one species as a component. The foregoing examples make it amply clear that the number of components $c$ in a given system is

$$c = \text{all species} - \text{all independent reactions} \tag{8.25a}$$

If there are insignificant concurrent reactions, ignored in these examples for simplicity, they may be included if their consideration is necessary for special reasons. The basic arguments, however, in selecting the number of components are the same.

## Variables of State and Degrees of Freedom

An extensive thermodynamic property $\mathcal{G}$ of an open system consisting of one phase may be expressed as a function of pressure, temperature, and the numbers of moles $n_1, n_2, \ldots, n_c$, i.e.,

$$\mathcal{G} = f(P, T, n_1, n_2, \ldots, n_c) \tag{8.26}$$

An intensive property $\overline{G}_i$ of component $i$ in a phase, is a function of the intensive variables $P$, $T$, and $c - 1$ independent composition variables $x_i$; thus,

$$\overline{G}_i = \overline{G}_i(P, T, x_1, x_2, \ldots, x_{c-1}). \tag{8.27}$$

It follows from (8.27) that the number of independent intensive variables of state, or briefly, the variables of state, for a single phase consisting of $P$, $T$, and $c-1$ mole fractions, is equal to $1 + c$. Under specified conditions, such as fixing $P$, $T$, and any number of composition variables, and the coexistence of several phases, the number of independent variables of state is reduced. *The number of independent variables of state of any system under equilibrium is called the degrees of freedom, or variance.*

If magnetic, electric, and gravitational fields were also variables of state in (8.26) and (8.27), the number of independent variables would increase by 3 and the degrees of freedom would be $4 + c$ instead of $1 + c$. Additional variables of state do not present particular difficulties, and for simplicity, it is sufficient to consider the variables of state given in (8.27).

## Partial (Molar)* Gibbs Energy, or Chemical Potential

It was seen in Chapter II that any extensive thermodynamic property $\mathcal{U}$, $\mathcal{H}$, $\mathcal{S}$, $\mathcal{A}$, $\mathcal{G}$, etc., of a phase is a homogeneous function of the first degree in $n_1, n_2, \ldots, n_c$ of the $c$ components. The partial (molar) property $\overline{G}_1$ of the component 1 was defined by

$$\overline{G}_1 = \left(\frac{\partial \mathcal{G}}{\partial n_1}\right)_{P, T, n_2, n_3, \ldots} \tag{8.28}$$

---

*The term (molar) in "partial molar" is used for emphasis; it should be omitted.

where $\overline{G}_1$ is a homogeneous function of zeroth degree in the number of moles, therefore, it is an intensive property with a functional form as in (8.27).

The most useful partial (molar) property in expressing equilibria is the partial Gibbs energy $\overline{G}_i$ of component $i$ defined by (8.28) and historically denoted by $\mu_i$ and called the chemical potential by J. W. Gibbs (1875). We shall use $\overline{G}_i$ instead of $\mu_i$ in honor of J. W. Gibbs, and thus retain the symmetry in notation* in such equations as $\overline{G}_i = \overline{H}_i - T\overline{S}_i$. The total differential of the Gibbs energy $\mathbf{G} = \mathbf{G}(P, T, n_1, n_2, \ldots, n_c)$ is

$$d\mathbf{G} = \left(\frac{\partial \mathbf{G}}{\partial P}\right)_{T,n_1,\ldots} \bullet dP + \left(\frac{\partial \mathbf{G}}{\partial T}\right)_{P,n_1,\ldots} \bullet dT + \overline{G}_1\, dn_i + \overline{G}_2\, dn_2 + \cdots \quad (8.29)$$

The first two terms represent the contributions to the Gibbs energy when the number of moles of all components is constant, or when the system is a closed system; hence from $d\mathbf{G} = \mathbf{V}\, dP - \mathbf{S}\, dT$, $\partial \mathbf{G}/\partial P = \mathbf{V}$, and $\partial \mathbf{G}/\partial T = -\mathbf{S}$, it follows that

$$d\mathbf{G} = \mathbf{V}\, dP - \mathbf{S}\, dT + \overline{G}_1\, dn_1 + \overline{G}_2\, dn_2 + \cdots \quad (8.30)$$

where boldface symbols refer to the extensive properties for $n$ moles of a system. It was shown in Chapter II by means of Euler's theorem that the Gibbs energy of a system is related to the partial (molar) Gibbs energies by

$$\mathbf{G} = n_1\, \overline{G}_1 + n_2\, \overline{G}_2 + \cdots \quad (8.31)$$

## Conditions of Phase Equilibrium

It is now possible, with the foregoing equations, to consider equilibrium in a system consisting of $c$ components and $\phi$ phases, each phase being considered as an open system with respect to the others. The conditions for equilibrium for the system require that

$$d\mathbf{G}_{P,T,n} = 0 \quad (8.32)$$

where $\mathbf{G}$ is the Gibbs energy of the entire system or simply the sum of the Gibbs energies of all phases, i.e.,

$$\mathbf{G} = \mathbf{G}^{I} + \mathbf{G}^{II} + \cdots + \mathbf{G}^{\phi} \quad (8.33)$$

where the superscripts refer to the coexisting phases. The system is equilibrium under *four specific conditions* according to (8.32): (i) the change in Gibbs energy is zero, (ii) pressure and (iii) temperature are fixed and uniform from one phase to another, and (iv) the system must be closed, and therefore the amount of each

---

*Most texts use both $\overline{G}_i$ and $\mu_i$ for the partial (molar) Gibbs energy, but the use of two symbols for the same property is undesirable.

component $n_1, n_2, \ldots, n_c$ is fixed. The last condition requires that

$$
\left.\begin{aligned}
n_1 &= n_1^\text{I} + n_1^\text{II} + \cdots + n_1^\phi \\
n_2 &= n_2^\text{I} + n_2^\text{II} + \cdots + n_2^\phi \\
&\ldots\ldots\ldots\ldots\ldots\ldots\ldots\ldots \\
n_c &= n_c^\text{I} + n_c^\text{II} + \cdots + n_c^\phi
\end{aligned}\right\} \tag{8.34}
$$

or

$$
\left.\begin{aligned}
dn_1 &= 0 = dn_1^\text{I} + dn_1^\text{II} + \cdots + dn_1^\phi \\
dn_2 &= 0 = dn_2^\text{I} + dn_2^\text{II} + \cdots + dn_2^\phi \\
&\ldots\ldots\ldots\ldots\ldots\ldots\ldots\ldots\ldots \\
dn_c &= 0 = dn_c^\text{I} + dn_c^\text{II} + \cdots + dn_c^\phi
\end{aligned}\right\} \tag{8.35}
$$

The Gibbs energy change for each phase is

$$d\mathcal{G}^\text{I} = \mathcal{V}^\text{I}\, dP - \mathcal{S}^\text{I}\, dT + \overline{G}_1^\text{I}\, dn_1^\text{I} + \overline{G}_2^\text{I}\, dn_2^\text{I} + \cdots + \overline{G}_c^\text{I}\, dn_c^\text{I} \tag{8.36}$$

where the pressure and temperature refer to all phases, since under equilibrium conditions, the temperature and pressure of all phases must be the same. The total differential of (8.33) is

$$d\mathcal{G} = d\mathcal{G}^\text{I} + d\mathcal{G}^\text{II} + \cdots + d\mathcal{G}^\phi \tag{8.37}$$

Substitution of (8.36) for all phases into (8.37) gives

$$
\begin{aligned}
d\mathcal{G} = {}&(\mathcal{V}^\text{I} + \mathcal{V}^\text{II} + \cdots + \mathcal{V}^\phi)\, dP - (\mathcal{S}^\text{I} + \mathcal{S}^\text{II} + \cdots + \mathcal{S}^\phi)\, dT \\
&+ \overline{G}_1^\text{I}\, dn_1^\text{I} + \overline{G}_1^\text{II}\, dn_1^\text{II} + \cdots + \overline{G}_1^\phi\, dn_1^\phi \\
&+ \overline{G}_2^\text{I}\, dn_1^\text{I} + \overline{G}_2^\text{II}\, dn_1^\text{II} + \cdots + \overline{G}_2^\phi\, dn_1^\phi \\
&\ldots\ldots\ldots\ldots\ldots\ldots\ldots\ldots\ldots\ldots\ldots \\
&+ \overline{G}_c^\text{I}\, dn_c^\text{I} + \overline{G}_c^\text{II}\, dn_c^\text{II} + \cdots + \overline{G}_c^\phi\, dn_c^\phi
\end{aligned} \tag{8.38}
$$

According to (8.32), $d\mathcal{G}$ in (8.38) is zero at equilibrium when the system is closed and both $P$ and $T$ are fixed, i.e., $dn$, $dP$, and $dT$ are zero. It can be shown by Lagrange's method of indeterminate multipliers that this is possible if $\overline{G}_i$ of each component is the same in all phases. The proof is as follows (cf. Chapter II). Multiply the first equation in (8.35) by $\overline{G}_1^\text{I}$, the second by $\overline{G}_2^\text{I}$, the third by $\overline{G}_3^\text{I}$, etc., and subtract them from (8.38) at $dP = 0$ and $dT = 0$ for a closed system at equilibrium so that $d\mathcal{G}_{P,T,n} = 0$; the result is

$$0 = \left(\overline{G}_1^{II} - \overline{G}_1^{I}\right) dn_1^{II} + \left(\overline{G}_1^{III} - \overline{G}_1^{I}\right) dn_1^{III} + \cdots + \left(\overline{G}_1^{\phi} - \overline{G}_1^{I}\right) dn_1^{\phi}$$
$$+ \left(\overline{G}_2^{II} - \overline{G}_2^{I}\right) dn_2^{II} + \left(\overline{G}_2^{III} - \overline{G}_2^{I}\right) dn_2^{III} + \cdots + \left(\overline{G}_2^{\phi} - \overline{G}_2^{I}\right) dn_2^{\phi}$$
$$\cdots\cdots\cdots\cdots\cdots\cdots\cdots\cdots\cdots\cdots\cdots\cdots\cdots\cdots\cdots\cdots\cdots$$
$$+ \left(\overline{G}_c^{II} - \overline{G}_c^{I}\right) dn_c^{II} + \left(\overline{G}_c^{III} - \overline{G}_c^{I}\right) dn_c^{III} + \cdots + \left(\overline{G}_c^{\phi} - \overline{G}_c^{I}\right) dn_c^{\phi}$$
$$\tag{8.39}$$

Equations (8.35) are the required restrictive equations for the Lagrange method. From (8.35), it is seen that for each line there are $\phi - 1$ independent $n$'s because the total number of moles of each component is fixed. For every possible value of every $dn_i^j$, the foregoing equation must be identically zero. This is possible if, and only if, the coefficients of every $dn_i^j$ is zero; hence,

$$\left.\begin{array}{l} \overline{G}_1^{I} = \overline{G}_1^{II} = \cdots = \overline{G}_1^{\phi}, (\phi \text{ terms}) \\ \overline{G}_2^{I} = \overline{G}_2^{II} = \cdots = \overline{G}_2^{\phi} \\ \cdots\cdots\cdots\cdots\cdots\cdots \\ \cdots\cdots\cdots\cdots\cdots\cdots \\ \overline{G}_c^{I} = \overline{G}_c^{II} = \cdots = \overline{G}_c^{\phi} \end{array}\right\} \quad [c(\phi - 1) \text{ eqns.}] \tag{8.40}$$

Equation (8.40) contains a very important thermodynamic statement first conceived by J. W. Gibbs (1875), that $\overline{G}_i$ of component "$i$" is the same in all phases under equilibrium conditions.

## Phase Rule

Reconsider the preceding system of $\phi$ phases containing $c$ components. The total number of intensive variables which can define each phase is $c - 1$ composition variables, and pressure and temperature. For $\phi$ coexisting phases, therefore, the total number of intensive variables defining the system is

$$\text{Total number of variables} = 2 + \phi(c - 1) \tag{8.41}$$

where "2" refers to temperature and pressure of all phases. $\overline{G}_i$ of each component in each phase is a function of $P$, $T$, and $c - 1$ composition variables. According to (8.40), $\overline{G}_i$ of each component in one phase is equal to that in any other phase. Each line in (8.40) represents $\phi - 1$ independent equations, and therefore, $c$ lines represent $c(\phi - 1)$ independent equations; hence,

$$\text{The number of independent equations} = c(\phi - 1) \tag{8.42}$$

134    *Thermodynamics*

*The number of independent variables*, called the degrees of freedom, and denoted by $\Upsilon$, is equal to the total number of variables, minus the number of independent equations, i.e.: $2 + \phi(c-1) - c(\phi-1)$, or simply

$$\boxed{\text{Degrees of freedom} = \Upsilon = c - \phi + 2} \tag{8.43}$$

This equation is the formal statement of the well-known *phase rule* originally derived by J. W. Gibbs (1875). The degrees of freedom, $\Upsilon$ is sometimes called the *variance* and it represents the number of unrestricted variables of state which may be a set of variables out of $P$, $T$, and $c-1$ compositions. It must be emphasized that $\Upsilon$ can be zero or a positive number, and the negative values of $\Upsilon$ have never been observed. It should be remembered that the composition variables refer to each individual phase and not to the bulk composition of a heterogeneous system of two or more phases. The permissible number of composition restrictions could be all in one phase or in several phases; for example, fixing the mole fractions $x_i$ and $x_j$ of components $i$ and $j$ in one phase, or fixing $x_i$ in one phase and $x_j$ in another phase decreases the degrees of freedom by two.

It was stated that all the components are soluble in all phases in the derivation of the phase rule. However, in some cases the solubility of certain components in some phases is very small or practically zero. If, for example, the solubilities of $k_1$ components in Phase I, and $k_2$ components in Phase II are zero, the number of composition variables for Phase I decreases by $k_1$, and for Phase II, by $k_2$. Therefore, when $k_i$ components are insoluble in $\phi_j$ phases, the total number of variables in (8.41) decreases by $k_i \phi_j$ so that the right side of (8.41) becomes $2 + \phi(c-1) - k_i \phi_j$. The Gibbs energy $\mathcal{G}$, for a phase in which component "$i$" is insoluble, is independent of $n_i$; hence, $\overline{G}_i = \partial \mathcal{G}/\partial n_i = 0$. Consequently, the number of independent equations in (8.40) decreases by $\phi_j$ on each line of (8.40), and by $k_i \phi_j$ on all the lines; the phase rule then becomes

$$\Upsilon = 2 + \phi(c-1) - k_i \phi_j - [c(\phi-1) - k_i \phi_j] = c - \phi + 2$$

and thus the absence of some components in some of the phases does not alter the phase rule.

It is sometimes convenient to fix the pressure and decrease the degrees of freedom by one in dealing with condensed phases such as the substances with low vapor pressures. The phase rule then becomes

$$\boxed{\Upsilon = c - \phi + 1, \quad \text{(constant pressure)}} \tag{8.44}$$

A few examples illustrate some applications of the phase rule. In a single component system, for one phase, $\Upsilon = 1 - 1 + 2 = 2$, hence, pressure and temperature may vary independently, for two phases $\Upsilon = 1 - 2 + 2 = 1$, either pressure or temperature may vary independently; if pressure or temperature is fixed $\Upsilon = 0$, or the system is completely defined. If pressure and temperature are both fixed, $\Upsilon = -1$; therefore, either the phase rule is violated, or an unnecessarily large number of restrictions are imposed on the system. For three phases, $\Upsilon$ is zero and again the system is completely defined. If either temperature or pressure is varied one of the phases must disappear to permit this degree of freedom.

In a system consisting of 3 components, the maximum value of $\Upsilon$ is 4, since the minimum number of phases for any system is 1. When $\phi = 3$, then $\Upsilon$ is 2 and $\Upsilon$ may be chosen out of the following total number of variables; 2(press. and temp.) $+ 3(3 - 1)$(compos. var.) $= 8$. For example, in order to define the system, it is sufficient to fix two compositions as for example the mole fraction of component 1, in Phase I and the mole fraction of component 2 in Phase III.

## Other Definitions of $\overline{G}_i$

The partial Gibbs energy $\overline{G}_i$ of component $i$ is usually defined by $\overline{G}_i = (\partial \mathcal{G}/\partial n_i)_{P,T,n_1,n_2,\ldots}$, but there are four additional definitions. The energy of a closed system, $\mathcal{U}$ is a function of entropy and volume since $d\mathcal{U} = T\, d\mathcal{S} - P\, d\mathcal{V}$, and for an open system, the moles of each component are also variables of state; therefore,

$$\mathcal{U} = f(\mathcal{S}, \mathcal{V}, n_1, n_2, n_3, \ldots) \tag{8.45}$$

The total differential of energy is then

$$d\mathcal{U} = T\, d\mathcal{S} - P\, d\mathcal{V} + \left(\frac{\partial \mathcal{U}}{\partial n_1}\right)_{S,V,n'} \bullet dn_1 + \left(\frac{\partial \mathcal{U}}{\partial n_2}\right)_{S,V,n'} \bullet dn_2 + \cdots \tag{8.46}$$

where $n'$ means that the numbers of moles other than that inside the parentheses are regarded as constants. The total differential of $\mathcal{G} = \mathcal{U} + P\mathcal{V} - T\mathcal{S}$ is

$$d\mathcal{G} = d\mathcal{U} + P\, d\mathcal{V} + \mathcal{V}\, dP - T\, d\mathcal{S} - \mathcal{S}\, dT \tag{8.47}$$

Substitution of the right side of (8.46) for $d\mathcal{U}$ in (8.47) gives

$$d\mathcal{G} = \mathcal{V}\, dP - \mathcal{S}\, dT + \left(\frac{\partial \mathcal{U}}{\partial n_1}\right)_{S,V,n'} \bullet dn_1 + \left(\frac{\partial \mathcal{U}}{\partial n_2}\right)_{S,V,n'} \bullet dn_2 + \cdots \tag{8.48}$$

From this equation, it is evident that an alternative definition of $\overline{G}_1$ is given by

$$\overline{G}_1 = \left(\frac{\partial \mathcal{G}}{\partial n_1}\right)_{P,T,n'} = \left(\frac{\partial \mathcal{U}}{\partial n_1}\right)_{S,V,n'} \tag{8.49}$$

Likewise, starting with $\mathcal{H} = f(\mathcal{S}, P, n_1, \ldots)$, $\mathcal{S} = f(\mathcal{U}, \mathcal{V}, n_1, \ldots)$ and $\mathcal{A} = f(\mathcal{V}, T, n_1, \ldots)$, and following a similar procedure, the following additional definitions of $\overline{G}_1$ can be derived

$$\overline{G}_1 = \left(\frac{\partial \mathcal{H}}{\partial n_1}\right)_{S,P,n'} = -T\left(\frac{\partial \mathcal{S}}{\partial n_1}\right)_{E,V,n'} = \left(\frac{\partial \mathcal{A}}{\partial n_1}\right)_{V,T,n'} \tag{8.50}$$

The definition given by $\overline{G}_i = (\partial \mathcal{G}/\partial n_i)_{P,T,n'}$ is more convenient than the remaining definitions because it is a property obtained under conveniently attainable conditions of constant temperature and pressure. Therefore, the additional definitions given in this section are seldom used in thermodynamics, except the definitions based on $\mathcal{A}$ and $\mathcal{U}$.

136   Thermodynamics

## Useful Partial (Molar) Properties

Numerous partial properties, in addition to $\overline{G}_i$, are derivable from the appropriate thermodynamic functions. Some of the most useful of them are those in which $P$ and $T$ are the restricted variables, i.e., $\overline{G}_i = (\partial G/\partial n_1)_{P,T,n'}$. Thus for $\overline{H}_i$ and $\overline{S}_i$ one may write

$$\overline{H}_i = \left(\frac{\partial H}{\partial n_1}\right)_{P,T,n'}, \quad \text{and} \quad \overline{S}_i = \left(\frac{\partial S}{\partial n_i}\right)_{P,T,n'}$$

Substitutions of these relations into $(\partial G/\partial n_i) = (\partial H/\partial n_i) - T(\partial S/\partial n_i)$, where $P, T, n'$ are constant outside the parentheses, give

$$\boxed{\overline{G}_i = \overline{H}_i - T\overline{S}_i} \tag{8.51}$$

This is a very useful equation in thermodynamics. The total differential of $\overline{G}_i$ is $d\overline{G}_i = \overline{V}_i\,dP - \overline{S}_i\,dT$, where $\overline{V}_i = (\partial V/\partial n_i)_{P,T,n'}$.

## $\overline{G}_i$ and Criterion of Equilibrium

In general, for a finite process such as a chemical reaction where $\alpha$ moles of $A$, and $\beta$ moles of $B$ are transformed into $\lambda$ moles of $L$, $\eta$ moles of $M$, ..., i.e.,

$$\alpha A + \beta B \cdots = \lambda L + \eta M + \cdots \tag{8.52}$$

the change in the Gibbs energy is

$$\Delta G = \lambda \overline{G}_L + \eta \overline{G}_M + \cdots - (\alpha \overline{G}_A + \beta \overline{G}_B + \cdots) \tag{8.53}$$

and at equilibrium $\Delta G$ is equal to zero. It will be seen in Chapter XII that the physico-chemical equilibria of all types, e.g., phase transition, dissolution, precipitation, dissociation, association, synthesis, etc., may be written as reactions generalized by (8.52). Equation (8.53), set equal to zero, is then the most important relationship formulating the criterion of equilibrium at constant temperature and pressure, and constant mass (closed system). It will be discussed in detail in Chapter XII, that if $\Delta G$ is greater than zero, the reaction cannot proceed as written, but if $\Delta G$ is less than zero, the reaction may proceed spontaneously unless there are kinetic barriers that hinder the reaction.

## PROBLEMS

8.1  Show that $R\,d\ln P/d(1/T) = -\Delta_{vap}H°$, $R\ln P + RT(d\ln P/dT) = \Delta_{vap}S°$, and $-2RT\,d\ln P/dT + RT^2\,d^2\ln P/dT^2 = \Delta_{vap}C_p°$ for a liquid which vaporizes into an ideal gas.

8.2  Vapor pressures of solid and liquid benzene are as follows:

| $T$/K:   | 260.93 | 269.26 | 278.68 | 305.37 | 333.15 | 349.82 |
|----------|--------|--------|--------|--------|--------|--------|
| $P$/bar: | 0.0127 | 0.0242 | 0.0478 | 0.175  | 0.522  | 0.911  |

The triple point (solid-liquid-vapor point) is 278.68 K. Calculate: (a) the enthalpies of vaporization and sublimation, (b) the boiling point at 1 bar, and (c) the vapor pressure of supercooled liquid at 260.93 K, by assuming that the enthalpies of vaporization and sublimation are independent of temperature and the vapor is ideal. (d) Fit the vapor pressure of liquid into the Antoine equation and repeat (b) and (c). Compare the calculated results for the boiling point and the enthalpy of vaporization with the following experimental values: boiling point: 352.81 K; enthalpy of vaporization, 30,765 J mol$^{-1}$.

8.3 The densities of $\alpha$ and $\gamma$ iron are 7.571 and 7.633 g ml$^{-1}$ respectively at their transformation temperature of 910°C under 1 bar. The enthalpy of phase change, $\Delta_{tr}H°$, from $\alpha$ to $\gamma$ is 900 J mol$^{-1}$ at 910°C and 1665 J mol$^{-1}$ at 827°C. Assuming that (a) $\Delta_{tr}H°$ is independent of pressure but varies linearly with temperature, and (b) the value of $V_\alpha - V_\gamma$ is constant, calculate the pressure under which both forms of iron coexist at 827°C. (From L. S. Darken and R. D. Smith, Ind. Eng. Chem., 43, 1815, 1951.)

Ans.: 15,635 bars

8.4 When the vapor pressures of two pure liquids at two sets of temperatures are equal, i.e., $P^A$(at $T_\alpha$) = $P^B$(at $T_\beta$) and $P^{A'}$(at $T'_\alpha$) = $P^{B'}$(at $T'_\beta$), show that $T_\alpha/T_\beta = T'_\alpha/T'_\beta$ when the Trouton rule is applicable.

8.5 (a) Express $P$ as a linear function of temperature for the line $ED$ and find the volume and entropy changes for Ice V $\to$ Ice III at $-24.3$°C from $\Delta H(v \to III) = 67$ J mol$^{-1}$. (b) Assuming that $\Delta V$(vol.) for $v \to III$ is independent of $P$ and $T$, compute $\Delta H(v \to III)$ at $-17.0$°C and then use $\Delta H = 4,628$ J mol$^{-1}$ for III $\to$ Liquid at $-17.0$°C, to calculate $\Delta H$ for V $\to$ L at $-17.0$°C.

Ans.: (a) $\Delta V = 0.984$ ml mol$^{-1}$ (b) $\Delta H(v \to L) = 4,696$ J mol$^{-1}$.

8.6 The vapor pressure of liquid zinc is given by General Ref. (2) as follows:

$$\ln P(\text{bar}) = -(15,375/T) - 1.274 \ln T + 22.055$$

Calculate the enthalpy of vaporization at 1000 K.

Ans.: 117,243 J mol$^{-1}$

8.7 Three phases coexist in a system which has two degrees of freedom and one reaction among a number of species. Find the number of components and species.

Ans.: $c = 3$, species $= 4$

8.8 Three phases coexist at a fixed temperature in a system which has two degrees of freedom and two independent reactions among a number of species. Find the number of species and components.

Ans.: Species $= 6$, $c = 4$

8.9 Give the details of derivation of the phase rule for a system of three phases and three components and find (a) the degrees of freedom in this system if the mole fraction of a component is fixed in 2 phases, and (b) the degrees of freedom when one of the phases is pure component $i$.

8.10 Construct the phase diagram for graphite-diamond system from the information in the text for the temperature range of 0 to 2000 K. The portion of the diagram from 0 to 600 K should be schematic.

CHAPTER IX

# FUGACITY AND ACTIVITY

## Introduction

It was shown earlier that the molar Gibbs energy of an ideal gas can be obtained by integrating $dG = V\,dP = RT\,d\ln P$ at constant temperature; the result is

$$G = G°(T) + RT \ln P; \qquad [G°(T) = G°(P = 1, T)] \tag{9.1}$$

where $G$ is a function of both temperature and pressure but $G°$ is only a function of temperature at a fixed standard pressure of one bar (one atm in older publications). The value of $G$ cannot be determined experimentally but $G - G°$ can be obtained from $G - G° = RT \ln P$. The relationship $dG = RT\,d\ln P$ is not applicable to real gases, but a purely convenient function called the fugacity, $f$, was first introduced by G. N. Lewis in 1901 so that for real gases

$$dG = RT\,d\ln f, \qquad \text{(constant } T\text{)} \tag{9.2}$$

This equation is the formal definition of fugacity. Both $f$ and $P$ are expressed in the same units. Integration of (9.2) gives

$$G(P, T) - G(P_\circ, T) = RT \ln f - RT \ln P_\circ$$

where $P_\circ \to 0$, or $P_\circ$ is sufficiently low in order that $f$ can be set equal to $P_\circ$ as shown by the second term on the right side. Summation of the preceding equation and $G(P_\circ, T) = G°(P = 1, T) + RT \ln P_\circ$ from (9.1) gives

$$\boxed{G(P, T) = G°(P = 1, T) + RT \ln f} \tag{9.3}$$

where $G°(P = 1, T)$ is the standard Gibbs energy of the same gas as a hypothetically existing ideal gas at one bar. This is evident from $G°(P = 1, T) - G(P_\circ, T) = RT \ln(1/P_\circ)$, which is the change in $G$ for compressing the gas from $P_\circ$ where it is ideal to 1 bar by assuming that it follows the ideal behavior.

A real gas becomes ideal with decreasing pressure and in the limit $f \to P$, when $P_\circ = P \to 0$, and the state of $f \to P$ defines the reference state for the fugacity of a real gas. A term defined by $\phi = f/P$ for any pressure is called the fugacity coefficient.[*]

---

[*]O. A. Hougen, K. M. Watson, and R. A. Ragatz, "Chemical Process Principles Charts," John Wiley & Sons (1960); H. V. Kehiaian, Gen. Ref. (25).

## Fugacity of Pure Gases

For a real or ideal gas at constant temperature, $dG = V\,dP$ where $V$ is the real or actual molar volume, and from $dG = V\,dP$ and (9.2),

$$RT\,d\ln f = V\,dP, \quad \text{(constant } T\text{)} \tag{9.4}$$

Integration of the right side requires an equation of state from which $V$ can be expressed as a function of pressure. For this purpose, the virial equation of state, first proposed by Kammerlingh Onnes (1901) is useful. The virial equation in terms of pressure is

$$Z = PV/RT = 1 + (B_2/RT)P + \left[(B_3 - B_2^2)/R^2T^2\right]P^2 \tag{9.5}$$

[cf. (3.52a)]. $B_2$ and $B_3$ are functions of $T$ only. ($B_1 = RT$). Substitution of $V$ from (9.5) into $RT\,d\ln f = V\,dP$ gives

$$RT\,d\ln f - RT\,d\ln P = B_2\,dP + \left[(B_3 - B_2^2)/RT\right]P\,dP$$

Integration of the left side from $P_\circ = P \to 0$ to $P$ gives $RT\ln(f/P) - RT\ln(f_\circ/P_\circ)$, because, at $P_\circ$, $(f_\circ/P_\circ)$ is unity; hence,

$$RT\ln f = RT\ln P + B_2 P + 0.5\left(B_3 - B_2^2\right)P^2/RT \tag{9.6}$$

The virial coefficients $B_2, B_3, \ldots$ are related to the interactions in clusters of $2, 3, \ldots$ molecules respectively as signified by the subscripts. At about 50 bars of pressure at ordinary temperatures, or when the deviation from ideality is on the order of 2 percent, the terms beyond $B_2 P$ may be neglected, and then (9.6) becomes

$$RT\ln f = RT\ln P + B_2 P \tag{9.7}$$

Combination of (9.7) with (9.3) gives

$$G(P, T) = G^\circ(T) + RT\ln P + B_2 P \tag{9.8}$$

The second virial coefficient $B_2$, whose dimensions are volume per mole varies with temperature according to the empirical equation

$$B_2 = B_\circ - (\alpha/T) - (\beta/T^2) \tag{9.9}$$

where $B_\circ$, $\alpha$, and $\beta$ are purely empirical constants. When $P$ is in bars then $B_2$ has to be in $10 \times \text{cm}^3$. Thus, for Ar and $N_2$, we have

$$B_2 = 357.3 - \frac{128{,}500}{T} - \frac{7.55 \times 10^6}{T^2}; \quad \text{(Ar)} \tag{9.9a}$$

$$B_2 = 423.8 - \frac{108{,}900}{T} - \frac{9.25 \times 10^6}{T^2}; \quad (N_2) \tag{9.9b}$$

Equation (9.9b) yields 324.27 K for the Boyle temperature, at which $B_2$ is zero, and within the validity of this equation, usually a few bars, $PV = RT$(Boyle).

**Table 9.1** Fugacity of oxygen and hydrogen at 300 K.

|  | P/bar | f/bar* | Z | f ≈ PZ |
|---|---|---|---|---|
| Oxygen: | 10 | 10.001 | 0.9940 | 9.94 |
|  | 40 | 39.20 | 0.9772 | 39.09 |
|  | 70 | 67.6 | 0.9634 | 67.44 |
|  | 100 | 95.5 | 0.9540 | 95.40 |
| Hydrogen: | 10 | 10.1 | 1.005 | 10.05 |
|  | 40 | 41.0 | 1.023 | 40.92 |
|  | 70 | 73.0 | 1.041 | 72.9 |
|  | 100 | 106.3 | 1.060 | 106.0 |

*Computed from compressibility data from General Ref. (6).

For a mixture of gases, the fugacity $f_i$ of a component gas $i$ is closely related to its mole fraction $y_i$ and its fugacity $f_i(P)$ of pure $i$ at the pressure of mixture $P$ by

$$f_i = f_i(P) y_i \tag{9.10}$$

This relationship, which states that "the fugacity of a gas in a mixture is equal to the product of the fugacity of pure gas at the pressure of the mixture times its mole fraction" is known as the Lewis and Randall rule (1923) Gen. Ref. (22). Actual determinations of fugacities of some gaseous mixtures agree fairly well with this rule for moderate pressures. A more rigorous treatment of gas mixtures are in Gen. Refs. 15, 19, 22 and 30, and it is similar to the regular solutions in condensed states.

### Alternative Equations for Fugacity

The data on the compressibility factor $Z$ may often simplify the calculation of fugacity. Equation (9.5) with only the second virial coefficient gives

$$Z = 1 + B_2 P / RT \tag{9.11}$$

and substitution of this equation into (9.7) yields

$$\ln(f/P) = Z - 1 \tag{9.12}$$

If $f/P$ is not too far from unity, then $\ln(f/P) \approx (f/P) - 1$ and substitution of this approximation into (9.12) gives

$$f \approx PZ \tag{9.13}$$

It can be seen from Table 9.1 that the fugacity of oxygen and hydrogen up to 100 bars is adequately expressed by (9.13) at 300 K but as the temperature is decreased toward the critical point, additional virial coefficients are necessary for (9.11).

Other equations of state also yield the corresponding equations for the fugacity. For example the van der Waals equation can be solved for $P$ as $P = RT/(V - b) - a/V^2$ and $dP$ is expressed in terms of the right side as a function of volume and then substituted in $RT\, d\ln f = V\, dP$ to obtain

$$d\ln f = -\frac{V\, dV}{(V-b)^2} + \frac{2aV\, dV}{RTV^3}$$

The numerator in the first term on the right may be written as $(V - b)\, dV + b\, dV$ after which integration from $V_\circ \to \infty$ to $V$, corresponding to $P_\circ$ and $P$ respectively, can be carried out to obtain

$$\ln f\Big]_{P_\circ}^{P} = \ln\frac{V_\circ}{V-b} + \frac{b}{V-b} - \frac{2a}{RTV} + \frac{2a - bRT}{RTV_\circ}$$

Since $V_\circ$ at $P_\circ \to 0$ is very large, $V_\circ - b$ has been replaced by $V_\circ$. In addition, the last term is negligible since its denominator is very large. From $P_\circ V_\circ = RT$, $\ln V_\circ = \ln RT - \ln P_\circ$, and $-\ln P_\circ$ on the right side cancels out $-\ln f\, (\text{at } P_\circ) = -\ln P_\circ$ on the left side; the final result is

$$\ln f = \ln\frac{RT}{V-b} + \frac{b}{V-b} - \frac{2a}{RTV} \tag{9.14}$$

For the Berthelot equation,

$$PV = RT + \left[\frac{9RT_c}{128 P_c T}\left(1 - \frac{6T_c^2}{T^2}\right)\right]P$$

the result is even simpler because of the terms inside the square brackets are equal to $B_2$, as can be seen by comparison with (9.5); hence,

$$\ln f = \ln P + \left[\frac{9T_c}{128 P_c T}\left(1 - \frac{6T_c^2}{T^2}\right)\right]P \tag{9.15}$$

The fugacity can also be computed from the reduced equation of state. Appropriate charts for the fugacity coefficient $\phi = f/P$ versus the reduced pressure at various parametric values of reduced temperatures are available for a fairly close estimation of the fugacities, and for closer calculations, extensive tables are given in Gen. Ref. (19).

## Variation of Fugacity with Temperature

The temperature dependence of fugacity is given by the appropriate preceding equations such as (9.7) when the temperature variation of the virial coefficients is known. Alternatively, when the enthalpy and entropy effects are known, or they can be calculated, the effect of temperature on the fugacity may be obtained from $G - G^\circ = RT \ln f = H - H^\circ - T(S - S^\circ)$ rewritten in the form of

$$\ln f = \frac{(H - H^\circ)}{RT} - \frac{(S - S^\circ)}{R} \tag{9.16}$$

For a given moderate pressure and moderate range of temperature not close to the critical temperature $H - H°$ is very nearly constant, and so is $S - S°$, and the change in the value of $f$ from $T_1$ to $T_2$ is obtained from (9.16) as follows:

$$\ln \frac{f_2}{f_1} = \left(\frac{H - H°}{R}\right)\left(\frac{1}{T_2} - \frac{1}{T_1}\right); \quad \text{(constant } P\text{)} \tag{9.17}$$

where $f_1$ and $f_2$ are the fugacities corresponding to any pressure which is held constant when the gas is heated from $T_1$ to $T_2$. The enthalpy $H°$ refers to the gas at $P_o$ where the gas becomes ideal. The change in enthalpy, $H - H°$ is related to the Joule-Kelvin coefficient by $[\partial(H - H°)/\partial P]_T = -C_P \mu_{JK}$ since $\partial H°/\partial P$ is zero. For this reason and due to the fact that the change of enthalpy $(H - H°)$ refers to ideal gas $\rightarrow$ real gas, this quantity is sometimes called the Joule-Kelvin enthalpy change. In general, $H - H°$ is small and it decreases with decreasing $f$ and increasing temperature. The value of $H - H°$ is obtained from

$$H(P, T) - H°(P_o, T) = -\int_{P_o}^{P} C_P \mu_{JK} \, dP \tag{9.18}$$

Alternatively $dH = V \, dP + T \, dS$ and $(\partial H/\partial P)_T = V + T(\partial S/\partial P)_T$ but from one of the Maxwell relations (5.5), $(\partial S/\partial P)_T = -(\partial V/\partial T)_P$; hence,

$$H(P, T) - H°(P_o, T) = \int_{P_o}^{P} \left[V - T\left(\frac{\partial V}{\partial T}\right)_P\right] dP \tag{9.19}$$

where the quantities inside the square brackets can be obtained from an equation of state. The entropy equation corresponding to (9.19) can be derived by recalling that $dS = dH/T$. All the thermodynamic properties of non-reacting gases or gas mixtures can therefore be formulated when their equations of state are known.

## Definition of Activity

For a pure condensed phase, i.e., a liquid or solid phase, the Gibbs energy $G_i°$ is equal to the Gibbs energy of the vapor $G_i$ when the condensed phase is in equilibrium with the gas phase; hence,

$$G_i°(\text{cond}) = G_i(\text{vapor}) \equiv G_i°(\text{vapor}) + RT \ln f_i° \tag{9.20}$$

where $f°$ is the fugacity of vapor over the pure condensed phase. When the condensed phase is a solution containing the component "$i$," then $\overline{G}_i(\text{cond}) = \overline{G}_i(P, T, x_i \ldots)$, and the partial Gibbs energy of "$i$," is given by

$$\overline{G}_i(\text{cond}) = G_i°(\text{vapor}) + RT \ln f_i \tag{9.21}$$

where $G_i°(\text{vapor}) = G_i°(T)$ for an ideal gas, and $f_i$ refers to the fugacity of "$i$" over the solution, and $\overline{G}_i(\text{cond}) = \overline{G}_i(\text{vapor})$ at equilibrium. Subtraction of (9.20) from (9.21) gives

$$\boxed{\overline{G}_i(\text{cond}) - G_i°(\text{cond}) = RT \ln(f_i/f_i°)} \tag{9.22}$$

The quantity $f_i/f_i^\circ$, representing the ratio of the fugacity of a component over a solution to the fugacity over the pure component at the same temperature, is called the activity of component $i$. Thus, by definition, the activity of $i$, denoted by $a_i$, is

$$\boxed{a_i \equiv f_i/f_i^\circ \cong P_i Z_i (\text{at } P_i)/P_i^\circ Z_i^\circ} \tag{9.23}$$

With this definition the partial Gibbs energy $\overline{G}_i$ in (9.22) becomes

$$\boxed{\overline{G}_i(\text{cond}) = G_i^\circ(\text{cond}) + RT \ln a_i}; \qquad G_i^\circ(\text{cond}) = G_i^\circ(P, T) \tag{9.24}$$

The dimensionless quantity that relates the activity to the concentration is called the activity coefficient. If the concentration is expressed in terms of the mole fraction, the activity coefficient $\gamma_i$ is defined by

$$\boxed{a_i = \gamma_i x_i} \tag{9.25}$$

Frequently, in the pressure and temperature range of interest where the fugacity is very close to the pressure, the activity is very closely expressed by*

$$\boxed{a_i \cong P_i/P_i^\circ} \tag{9.26}$$

In a two-component condensed phase, the activities of components 1 and 2 are shown as functions of mole fractions in Figs. 9.1 and 9.2.

According to (9.23), the activity of a pure condensed phase is unity at one bar. The state of unit activity is called the standard state. In some cases, particularly in dilute solutions, it is convenient to use other standard states as will be seen later. All other standard states differ from that defined from (9.23) by a constant factor.

## Raoult's Law

A component in a solution obeys Raoult's law (F. M. Raoult, 1887) when its activity and mole fraction are equal; hence, from (9.25)

$$a_i = x_i, \quad \text{or} \quad \gamma_i = 1 \tag{9.27}$$

It has been observed experimentally that Raoult's Law is obeyed in the limiting case when $x_i$ approaches unity, i.e.,

$$\lim a_i = x_i, \quad \text{or} \quad \lim \gamma_i = 1, \qquad (\text{with } x_i \to 1) \tag{9.28}$$

The usefulness of these relationships is the fact that $\gamma_i$ is very close to unity in a finite range of composition approaching $x_i = 1$ because otherwise $\gamma_i = 1$ at $x_i = 1$ is trivial and simply defines the standard state. The broken lines in

---

*For measurement of activities, see Z.-C. Wang, X.-H. Zhang, Y.-Z. He, and Y.-H. Bao, J. Chem. Soc., Faraday Trans. 1, *84*, 4369 (1984); Z. C. Wang, Y. W. Tian, H. L. Yu, X. H. Zhang, and J. K. Zhou, Met. Trans., *25B*, 103 (1992); ibid., *23B*, 623 (1992). See also J. Chem. Thermod., *25*, 711 (1993); S. Howard, Met. Trans., *20B*, 845 (1989); M. J. Stickney, M. S. Chandrasekharaiah, K. A. Gingerich, and J. A. Speed, ibid., *22A*, 1937 (1991); S. Srikanth and K. T. Jacob, ibid., *22B*, 607 (1991); S. M. Howard and J. P. Hager, ibid., *9B*, 51 (1978); R. Chastel, M. Saito, and C. Bergman, J. Alloys and Compounds, *205*, 39 (1994).

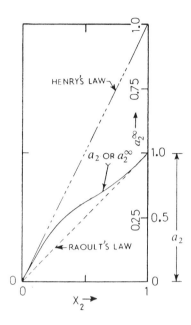

**Fig. 9.1** Activity of component 2 in a hypothetical binary solution. Deviation from Raoult's law is positive, and deviation from Henry's law is negative. Standard state for vertical scale outside is pure component 2. Standard state for $a_2^\infty$ based on Henry's law is hypothetical state at which $a_2^\infty = 1$; reference state for $a_2^\infty$ is infinitely dilute solution wherein $a^\infty \to x_2$ as $x_2$ approaches zero. Inner vertical scale is for $a_2^\infty$.

Figs. 9.1 and 9.2 represent Raoult's law, or the ideal behavior in a binary solution. An ideal solution is the analog of an ideal gas mixture and the "ideal behavior in solutions" and Raoult's law are frequently used as interchangeable terms.

### Henry's Law

The activity of a component $i$ becomes proportional to its mole fraction $x_i$, as $x_i$ approaches zero; therefore,

$$a_i = \gamma_i^\circ x_i, \quad \text{(with } x_i \to 0\text{)} \tag{9.29}$$

where $\gamma_i^\circ$ is the activity coefficient based on pure $i$ as the standard state. Combination of this equation with the definition of activity gives

$$a_i = f_i/f_i^\circ = \gamma_i^\circ x_i, \quad \text{or} \quad f_i = k_i x_i, \quad \left(\text{with } k_i = f_i^\circ \gamma_i^\circ\right) \tag{9.30}$$

This equation gives the modern statement of Henry's law. The last relationship becomes $P_i = k_i x_i$ when $f_i$ is equal to $P_i$, and this simple equation is the classical statement of Henry's law, i.e., the solubility of a sparingly soluble gas is proportional to its partial pressure over the solution. In thermodynamics this concept

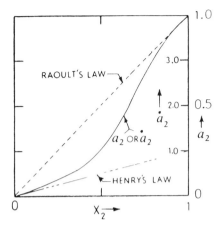

**Fig. 9.2** Activity of component 2 in a hypothetical binary solution. Deviation from Raoult's law is negative, from Henry's law, positive. Outer vertical scale is for $a_2$ with standard state as pure component 2, and inner vertical scale, for $a_2^\infty$ with reference state $x_2 \to 0$. In some unusual systems, curve for activity crosses diagonal broken straight line for Raoult's law, so that positive deviation from Raoult's law exists in certain range of composition, and negative deviation, in remaining range of composition.

is extended to sparing solubilities of any phase in any solvent provided that the dissolving substance does not change its molecular identity. The most useful form of this law in this text is obtained when the proportionality constant in (9.29) is arbitrarily set to unity purely for convenience; thus $a_i^\infty = x_i$ for $x_i \to 0$, but for higher concentrations,

$$a_i^\infty = \gamma_i^\infty x_i, \quad (\gamma_i^\infty \to 1 \text{ with } x_i \to 0) \tag{9.31}$$

where $\gamma_i^\infty$ is the activity coefficient with the reference state being the infinitely dilute solution and this activity scale is indicated by "$\infty$" over $a_i$. The activity scales based on Raoult's and Henry's laws may be summarized as follows:

| Basis of Activity | Definition | Standard State | Reference State |
|---|---|---|---|
| Raoult's law | $f_i/f_i^\circ = a_i = \gamma_i x_i$ | Pure "$i$"; i.e., $a_i = x_i = 1$ | Unnecessary |
| Henry's law | $a_i^\infty = \gamma_i^\infty x_i$ | $a_i^\infty = 1$, hypothetical state, usually nonexisting | $x_i \to 0$ |

The standard state is always the state corresponding to $a_i = 1$ or $a_i^\infty = 1$. In some texts the standard state is also called the reference state for the activity based on Raoult's law but such a term in this case is unnecessary and confusing.

All the solutes in metals dissolve as monatomic species so that when a diatomic gas, such as hydrogen, is dissolved in metals, the reaction and the activity are as follows:

$$0.5\, H_2(g) \rightarrow [H]; \qquad a_2^\infty = \gamma_2^\infty x_2/[P(H_2)]^{0.5},$$

$$\text{with } \gamma_2^\infty \rightarrow 1 \text{ as } x_2 = x_H \rightarrow 0 \qquad (9.9b)$$

This relationship, that the solubility of a diatomic gas, $x_2$, is proportional to the square root of its gas pressure, is known as the Sieverts law (1908) and will be discussed later in detail. Likewise if the solute in gaseous state is polyatomic, e.g., $As_4(g)$, then the activity* is related to the gas pressure, $P(As_4)$, by $a_2^\infty = \gamma_2^\infty \times x_2/[P(As_4)]^{0.25}$.

The partial Gibbs energy of component $i$, $\overline{G}_i$, is given by

$$\boxed{\overline{G}_i = G_i^\circ(P, T) + RT \ln \gamma_i x_i} \qquad (9.32)$$

$$\boxed{\overline{G}_i = G_i^\infty(P, T) + RT \ln \gamma_i^\infty x_i} \qquad (9.33)$$

where $G_i^\circ(P, T)$ and $G_i^\infty(P, T)$ are functions of both temperature and pressure. The partial Gibbs energy $\overline{G}_i$ is a function of temperature, pressure, and mole fraction $x_i$ in a two-component system and it is *independent* of the choice of standard states. The value of $\overline{G}_i$ becomes equal to $G_i^\circ$ or $G_i^\infty$ when $a_i$ or $a_i^\infty$ is set to unity and for this reason $G_i^\circ$ and $G_i^\infty$ are called the standard Gibbs energy. At a given value of $x_i$, the left sides of (9.32) and (9.33) are equal, and

$$G_i^\circ - G_i^\infty = -RT \ln \left(a_i/a_i^\infty\right) \qquad (9.34)$$

The left side of this equation is a constant at a constant temperature and pressure and the activities on the two scales are proportional to each other, i.e., $a_i/a_i^\infty$ is a constant independent of $x_i$ for constant $T$ and $P$. If the concentration scales for $a_1$ and $a_1^\infty$ are the mole fractions, then

$$\boxed{G_i^\circ - G_i^\infty = -RT \ln \left(\gamma_i/\gamma_i^\infty\right)} \qquad (9.35)$$

hence the ratio of activity coefficients is a constant for given values of $P$, $T$, and independent of $x_i$.

The laws of Raoult and Henry are obeyed by all substances in their respective limiting concentrations, therefore these laws are sometimes called the limiting laws of solutions. The range of composition in which each law is obeyed depends on the system and on the accuracy of measurements, but the important point is that the activity curves approach the lines for Raoult's and Henry's laws by becoming tangent to them where the slope of the activity curves for Raoult's law is unity and the slope for Henry's law is finite and nonzero as shown in Figs. 9.1 and 9.2. It will be seen later that if the dissociation or the combination of a solute "$i$" is not

---

*See for example D. C. Lynch, Met. Trans. B, *11B*, 623 (1980).

148   Thermodynamics

taken into consideration the slope for the activity at the limiting concentration of $x_i \to 0$ may be zero or infinity.

The activities and the concentrations in the preceding paragraphs are both expressed in terms of mole fraction but it is also common to use molality, molarity or weight fractions for the solutes dissolved in water or other liquids when convenience dictates the choice as will be seen in the succeeding chapters.

The activity coefficients may assume values greater or smaller than unity as deviations from the laws of Raoult and Henry are encountered. A positive deviation from Raoult's law occurs when $\gamma_i > 1$, and a negative deviation, when $\gamma_i < 1$. Likewise, a positive deviation from Henry's law is defined by $\gamma_i^\infty > 1$, and a negative deviation, by $\gamma_i^\infty < 1$.

The variation of activity with temperature, pressure and composition can be obtained from the functional dependence of $\overline{G}_i - G_i^\circ$ from (9.32), and $\overline{G}_i - G_i^\infty$ from (9.33) on these variables as will be discussed in the next chapter. It is sufficient to point out here that $\ln a_i$ from (9.32) may be written as

$$\ln a_i = \left[(\overline{H}_i - H_i^\circ)/RT\right] - (\overline{S}_i - S_i^\circ)/R \tag{9.36}$$

and the variation of $a_i$ with $T$, $P$ and $x_i$ is obtained when $\overline{H}_i - H_i^\circ$ and $\overline{S}_i - S_i^\circ$ are expressed as appropriate functions of $T$, $P$ and $x_i$ as will be seen in the succeeding chapters. It may be noted here that for a given composition, (9.36) generally requires that Raoult's law be approached with increasing temperature but there are notable exceptions such as the nicotine-water system in which the functional dependence of both terms in (9.36) play unusual roles. As a solution approaches the behavior required by Raoult's law, it is evident that the line for Henry's law also approaches the line for Raoult's law.

**Example 1**   The total vapor pressure, $P$, the liquid composition $x_2$ and the gas composition $y_2$ for $C_3H_6O$ in the binary system $C_6H_6$–$C_3H_6O$ (benzene-acetone) are listed on the first three columns in the following table:*

| $x_2$ | $y_2$ | $P$/bar | $a_1$ | $\gamma_1$ | $a_2$ | $\gamma_2$ |
|---|---|---|---|---|---|---|
| 0 | 0 | 0.29819 | 1 | 1 | 0 | — |
| 0.0470 | 0.1444 | 0.33429 | 0.9592 | 1.007 | 0.0706 | 1.502 |
| 0.0963 | 0.2574 | 0.36666 | 0.9131 | 1.010 | 0.1381 | 1.434 |
| 0.4759 | 0.6697 | 0.53293 | 0.5903 | 1.126 | 0.5221 | 1.097 |
| 1 | 1 | 0.68357 | 0 | — | 1 | 1 |

Calculate the activities and the activity coefficients of components.

**Solution**   The pressures are sufficiently low and the fugacity coefficients partially cancel out so that within a very good approximation, $a_i \cong P_i/P_i^\circ \cong y_i P/P_i^\circ$, where $P_i = y_i P$. Thus, for the third value from the top, $a_2 = 0.2574 \times 0.36666/0.68357 = 0.1381$, as listed. The activity coefficient is $\gamma_2 = a_2/x_2 =$

*I. Brown and F. Smith, Aust. J. Chem., *10*, 423 (1957); see also Gen. Ref. (14).

$0.1381/0.0963 = 1.434$; hence, the system has positive deviations from Raoult's law. More rigorous calculations by using the fugacities differ by about 2% on the average in the values of the activities, as given in the first edition of this book, probably about twice the experimental errors.

**Example 2** The solubility of carbon dioxide, $CO_2(g)$, in liquid ethanol ($C_2H_5OH$) under 1 bar of $CO_2$ is 0.01755 mole fraction at $-25°C$ and 0.00898 mole fraction at $0°C$.* Calculate (a) the Henry's law constant and (b) the solubility at 0.1, 0.4, and 4 bars of $CO_2$ by using the following compressibility data from Gen. Ref. (6):

| P/bar: | 0.1 | 0.4 | 1.0 | 4.0 |
|---|---|---|---|---|
| Z, at $-25°C$: | 0.9991 | 0.9964 | 0.9909 | 0.9615 |
| Z, at $0°C$: | 0.9993 | 0.9974 | 0.9935 | 0.9717 |

**Solution** (a) The constant $k_i$ for Henry's law from (9.40) is $k_i = f_i/x_i$. The fugacity $f_i$ of $CO_2$ at 1 bar is given very closely by $f_i = P_i Z_i = 0.9909$ bar at $-25°C$ and $f_i = 0.9935$ bar at $0°C$; hence $k_i = 0.9909/0.01755 = 56.46$ at $-25°C$ and $k_i = 0.9935/0.00898 = 110.6$ at $0°C$. Note that the inverse of $k_i$ is the solubility under a fugacity of 1 bar. (b) The constant $k_i$ is independent of pressure since the mole fraction of dissolved $CO_2$ is small and $k_i$ is expressed in terms of the fugacity $f_i$ of $CO_2$. The simple relationship $f_i \approx Z_i P_i$ is accurate enough within the accuracy of measurements in $x_i$. The solubilities computed from $x_i = f_i/k_i$ are as follows:

| P/bar: | 0.1 | 0.4 | 1.0 | 4.0 |
|---|---|---|---|---|
| $x_i$ at $-25°C$: | 0.00177 | 0.00706 | 0.01755 | 0.0681 |
| $x_i$ at $0°C$: | 0.000904 | 0.00361 | 0.00898 | 0.0351 |

At higher pressures and lower temperatures more accurate values of the fugacity than those calculated from $f_i \approx P_i Z_i$ are necessary for reliable values of $x_i$ provided that $x_i$ is sufficiently small so that Herny's law is obeyed. Modern and mathematically elegant definitions of Raoult's and Henry's laws are in Chapter XI.

## PROBLEMS

9.1   Determine the fugacities of $CO_2$ at various pressures from the following data at 320 K by using an equation for Z quadratic in P:

| P/bar: | 1.01 | 4.05 | 7.09 | 10.1 | 40.5 | 71.0 |
|---|---|---|---|---|---|---|
| Z: | 0.9960 | 0.9838 | 0.9714 | 0.9588 | 0.8206 | 0.6417 |

Compare the results with the approximate values from $f \approx ZP$.

*General Refs. (16–17); data of C. Bohr.

## 150    Thermodynamics

**9.2** Use $Z = 1 + 350P - 140,000P/T$ in (9.4) to derive in detail an equation for $\ln f$ and show that $f = P$ at the Boyle temperature.

**9.3** Calculate the change in Gibbs energy when one mole of $CO_2$ at 320 K and 40 bars is expanded isothermally to 4 bars of pressure. Use the data in Problem 9.1.

**9.4** Show that for a real gas $(\partial \ln f/\partial T)_P \approx (\partial \ln Z/\partial T)_P$ at moderate pressures and that as $P$ approaches zero this relationship becomes zero.

**9.5** Derive the following equations:
   (a) $[R\partial \ln f/\partial(1/T)]_P = H - H°$;
   (b) $S° - S = R \ln f + RT(\partial \ln f/\partial T)_P$;
   (c) $(C_P° - C_P)/RT = (T\partial^2 \ln f/\partial T^2) + 2\partial \ln f/\partial T$.

**9.6** In a binary system of $A$ and $B$, the two coexisting liquids $L_1$ and $L_2$ contain 0.90 and 0.15 mole fraction of $A$ respectively. Assuming that $A$ in $L_1$ and $B$ in $L_2$ obey Raoult's law, calculate the activities and the activity coefficients of both components in each phase.

**9.7** Plot the activity and logarithm of activity coefficient of Fe and Ni in liquid Fe–Ni system versus the mole fraction of Ni, $(x_2)$, by using the following data in Gen. Ref. (9):

| $x_2$: | 0.1 | 0.2 | 0.4 | 0.6 | 0.8 | 0.9 | |
|---|---|---|---|---|---|---|---|
| $a_2$: | 0.070 | 0.141 | 0.290 | 0.485 | 0.758 | 0.888 | (nickel) |
| $a_1$: | 0.898 | 0.796 | 0.590 | 0.346 | 0.119 | 0.047 | (iron) |

**9.8** The pressure of gaseous $CH_4$ is 0.34508 bar over pure liquid methane and 0.20340 bar over a liquid solution of 0.3935 mole fraction of $CH_4$ and 0.6065 mole fraction of $C_3H_8$ at 100 K. The vapor pressure of $C_3H_8$ is practically zero at 100 K. The second virial coefficient of $CH_4$ is given by
$B_2 (\text{cm}^3 \text{ mol}^{-1}) = 460 - (88,000/T)$
Calculate: (a) The fugacity of $CH_4$ as a function of pressure and temperature, (b) $H - H°$ and $S - S°$ in (9.29) and (c) the activity coefficient of $CH_4$ in the liquid solution and compare it with that calculated from $a_i \approx P_i/P_i°$ where $P_i$ is the pressure of $CH_4$ over the solution.*

**9.9** Derive the equations for $H - H°$ and $S - S°$ by using the fugacity equation from the van der Waals and Berthelot equations of state.

**9.10** Calculate the activity coefficient $\gamma_2°$, and the mole fraction $x_2$ of water in benzene and express $\ln \gamma_2°$ and $\ln x_2$ as functions of temperature from the following data†:

| Temperature, °C: | 31.96 | 29.95 | 22.95 | 17.94 |
|---|---|---|---|---|
| Weight fraction of $H_2O \times 10^6$: | 915.8 | 853.5 | 686.7 | 591.3 |

Assume that the solubility of benzene in water is negligibly small.
Partial ans.: $a_2 \cong 1$ throughout; $x_2 = 0.00395$; hence $\gamma_2 = 253.2$ at 31.96°C.

---

*A. J. B. Cutler and J. A. Morrison, Trans. Faraday Soc., 61, 429 (1965); see also Gen. Ref. (14).
†R. Karlsson, J. Chem. Eng. Data. 18, 290 (1973).

# CHAPTER X

# SOLUTIONS

## PART I  IDEAL SOLUTIONS

In this chapter the liquid and solid solutions of various substances will be considered in detail. The behavior of solutions is generally discussed in terms of the deviations from ideality. An ideal solution therefore plays an important role in thermodynamics of solutions.

An ideal solution, like an ideal gas mixture, is merely a hypothetical solution whose properties are approached but seldom encountered in real or existing solutions. Certain properties of real solutions always approach those of ideal solutions under limiting conditions. In the preceding chapter, an ideal solution was defined by

$$a_i = x_i \quad \text{or} \quad \gamma_i = 1; \quad [a_i = f_i/f_i^\circ \approx P_i/P_i^\circ] \tag{10.1}$$

where $a_i$, the activity of a component $i$, was defined by $f_i/f_i^\circ$, and $\gamma_i$ by $a_i/x_i$. The partial Gibbs energy $\overline{G}_i$ of component $i$ in an ideal solution is

$$\overline{G}_i = G_i^\circ(P, T) + RT \ln x_i \tag{10.2}$$

where $G_i^\circ$ is the standard molar Gibbs energy of the component in the pure state. From (10.2), $\partial \ln x_i/\partial T = -(\overline{H}_i - H_i^\circ)/RT^2$, and since the left side of this equation is zero because $x_i$ is independent of $T$, it is evident that $\overline{H}_i - H_i^\circ = 0$, or $\overline{H}_i = H_i^\circ$. It follows, therefore, that

$$\overline{G}_i - G_i^\circ = RT \ln x_i = -T(\overline{S}_i - S_i^\circ); \quad \overline{S}_i - S_i^\circ = -R \ln x_i \tag{10.3}$$

The molar Gibbs energy of a solution is given by $G = \sum x_i \overline{G}_i$ as shown in Chapter II, and the sum of the Gibbs energies of pure components prior to mixing is $\sum x_i G_i^\circ$; consequently, the molar Gibbs energy of mixing is given by $\Delta_{\text{mix}} G = G - \sum x_i G_i^\circ$. The term $\Delta_{\text{mix}} G$ is also appropriately called the Gibbs energy of formation of one mole of solution. Thus,

$$\Delta_{\text{mix}} G \equiv \sum_i x_i (\overline{G}_i - G_i^\circ) = RT \sum_i x_i \ln x_i = -T \sum_i x_i (\overline{S}_i - S_i^\circ) \tag{10.4}$$

where the last term without $-T$ is the entropy of mixing, i.e., $\Delta_{mix}S = \sum x_i(\overline{S}_i - S_i^\circ)$, as for an ideal gas mixture. In an ideal solution, there are no attractive and repulsive forces among the molecules and $\overline{H}_i - H_i^\circ$ is zero; hence, $\Delta_{mix}H$ is also zero. Similarly, from (10.3), $\partial(\overline{G}_i - G_i^\circ)/\partial P = \overline{V}_i - V_i^\circ = RT\, \partial \ln x_i/\partial P$ and since $x_i$ is independent of $P$ the right side is zero; hence, $\overline{V}_i - V_i^\circ$ is zero, and therefore,

$$\Delta_{mix}V \equiv \sum_i x_i(\overline{V}_i - V_i^\circ) = V - \sum_i x_i V_i^\circ = 0 \tag{10.5}$$

where $V = \sum x_i \overline{V}_i$ is the molar volume of the solution and $\sum x_i V_i^\circ$ is the sum of the molar volumes of pure components prior to mixing, and $\Delta_{mix}V$ is called the molar volume of mixing. It has been observed that the more nearly a solution is ideal, the smaller the value of the molar volume of mixing. The preceding properties of ideal solutions for a two-component (binary) system are shown in Fig. 10.1. It is important to observe that $\Delta_{mix}G$ is negative; hence the solution is thermodynamically more stable than the pure components. It is evident from $(\overline{S}_2 - S_2^\circ)/R = -\ln x_2$ that $\partial \ln x_2/\partial x_2 = 1$ as $x_2 \to 1$; hence the slope of $(\overline{S}_2 - S_2^\circ)/R$ versus $x_2$ is $-1$ at $x_2 \to 1$. Similarly, from $\Delta_{mix}S/R = -x_1 \ln x_1 - x_2 \ln x_2$, it can be shown that $\Delta_{mix}S/R$, as well as $\Delta_{mix}G/R$, approach the vertical axes at $x_2 = 0$ and $x_2 = 1$ with infinite slopes.

## Equilibrium Between an Ideal Solution and Its Vapor

The total pressure of vapor over an ideal solution and its composition are related to the composition of solution by simple relationships. Consider, for example, a binary system and let $x_1$ and $x_2$ represent the mole fractions of components 1 and 2 in solution, and $y_1$ and $y_2$ those in the vapor. For the solution

$$P_1 = x_1 P_1^\circ, \quad \text{and} \quad P_2 = x_2 P_2^\circ \tag{10.6}$$

For the ideal vapor, the total pressure $P$ is given by $P = P_1 + P_2$, or by

$$P = x_1 P_1^\circ + x_2 P_2^\circ \tag{10.7}$$

There is one independent composition variable in a binary system; hence, $x_1$ may be eliminated by using $x_1 = 1 - x_2$; thus (10.7) becomes

$$P = P_1^\circ + x_2(P_2^\circ - P_1^\circ) \tag{10.8}$$

At a constant temperature, $P_1^\circ$ and $P_2^\circ$ are constants and $P$ is a linear function of $x_2$. The total pressure $P$, and the partial pressures $P_1$ and $P_2$ are represented in Fig. 10.2. In the vapor phase, the ratio of the mole fractions $y_1$ and $y_2$ is equal to the ratio of partial pressures, i.e., $P_1/P_2 = y_1/y_2$, and the substitution of $P_1 = P_1^\circ x_1$ and $P_2 = P_2^\circ x_2$ into this ratio gives

$$\frac{y_1}{y_2} = \frac{P_1^\circ x_1}{P_2^\circ x_2} \tag{10.9}$$

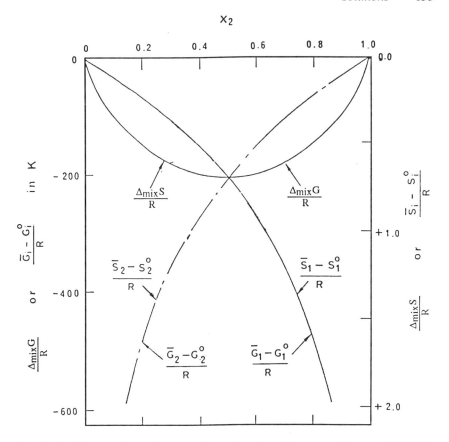

**Fig. 10.1** Thermodynamic properties of an ideal binary solution at 298.15 K. $\Delta_{mix}H$ and $\Delta_{mix}V$ are zero and independent of composition.

The substitution of $1 - y_2$ for $y_1$ and $1 - x_2$ for $x_1$ in this equation and then the elimination of $x_2$ by means of (10.8) yields

$$P = \frac{P_1^\circ P_2^\circ}{P_2^\circ - y_2(P_2^\circ - P_1^\circ)} \tag{10.10}$$

This relation gives the total vapor pressure as a function of the composition of the vapor phase at a given temperature. Although $P$ is a linear function of $x_2$ of the solution at a constant temperature, it is a hyperbolic function of $y_2$ of the vapor. The total pressure from (10.8) and (10.10) are shown in Fig. 10.3, where the line $P_1^\circ L' P_2^\circ$ is called the bubble point, and $P_1^\circ V' P_2^\circ$, the dew point. It can be seen that at a selected total pressure shown by the horizontal broken line $P'V'L'$, the composition of the liquid is $x_2 = \alpha$ and that of the vapor is $y_2 = \beta$ and that the vapor contains a higher concentration of the more volatile component 2 than the

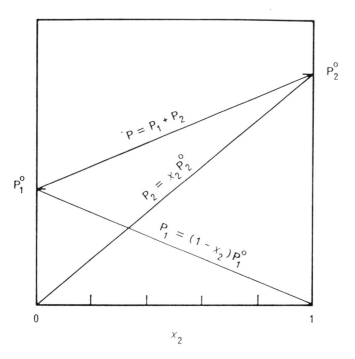

**Fig. 10.2** Vapor pressure of an ideal binary solution as a function of mole fraction $x_2$ of component 2 in liquid.

corresponding liquid. This statement can also be verified from (10.9); i.e., for $P_2^\circ > P_1^\circ$, it is evident that $P_1^\circ/P_2^\circ < 1$ and from (10.9), $y_1/y_2 < x_1/x_2$ and therefore the vapor phase is richer in the more volatile component 2 than the co-existing liquid phase. This conclusion is important in fractional distillation of solutions.

The relationship between $x_2$ and $y_2$ can be obtained by eliminating $P$ from (10.10) by means of (10.8) so that

$$y_2 = \frac{P_2^\circ x_2}{P_1^\circ + x_2(P_2^\circ - P_1^\circ)} \tag{10.11}$$

Similar equations can also be obtained for multi-component solutions by following an identical procedure.

The expressions for $\overline{G}_i$ in the gas and liquid phases are

$$\begin{aligned}\overline{G}_1(\text{gas}) &= G_i^\circ(\text{gas}) + RT \ln Py_1 \\ \overline{G}_1(\text{liquid}) &= G_i^\circ(\text{pure liq.}) + RT \ln x_1\end{aligned} \tag{10.12}$$

The left sides of these equations are equal as required by equilibrium; hence,

$$\begin{aligned}G_1^\circ(\text{gas}) - G_1^\circ(\text{pure liq.}) &= \Delta_{\text{vap}} G_1^\circ = \Delta_{\text{vap}} H_1^\circ - T\Delta_{\text{vap}} S_1^\circ \\ &= -RT \ln(Py_1/x_1)\end{aligned} \tag{10.13}$$

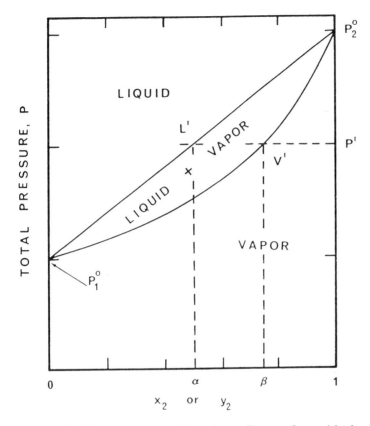

**Fig. 10.3** Constant temperature phase diagram for an ideal solution. $P_1^\circ L' P_2^\circ$ is called "bubble point," and $P_1^\circ V' P_2^\circ$, "dew point." At pressure $P'$, liquid $L'$ and vapor $V'$ coexist. Intersection of vertical line from $L'$ with horizontal axis gives composition of liquid, and that from $V'$, composition of vapor.

Similarly, for the second component,

$$\Delta_{vap} G_2^\circ = \Delta_{vap} H_2^\circ - T \Delta_{vap} S_2^\circ = -RT \ln(P y_2 / x_2) \qquad (10.14)$$

and subtraction of (10.14) from (10.13) gives, after rearrangement,

$$\ln \frac{y_2}{y_1} - \ln \frac{x_2}{x_1} = \frac{\Delta_{vap} H_1^\circ - \Delta_{vap} H_2^\circ}{RT} - \frac{\Delta_{vap} S_1^\circ - \Delta_{vap} S_2^\circ}{R} \qquad (10.15)$$

If component 2 is more volatile than 1, the boiling point of component 2 is lower than that of 1, and from Trouton's rule $\Delta_{vap} S_i^\circ \approx 88$ J mol$^{-1}$ K$^{-1}$, it is necessary that $\Delta_{vap} H_2^\circ < \Delta_{vap} H_1^\circ$; therefore, (10.15) simplifies into

$$\ln \frac{y_2}{y_1} - \ln \frac{x_2}{x_1} \approx \frac{\Delta_{vap} H_1^\circ - \Delta_{vap} H_2^\circ}{RT} \qquad (10.16)$$

156    Thermodynamics

and since $\Delta_{vap}H_1^\circ - \Delta_{vap}H_2^\circ > 0$, the right side increases with decreasing temperature, and the difference between the two logarithmic terms on the left also increases with decreasing temperature. Therefore, the ratio $y_2/y_1$ becomes larger than $x_2/x_1$, i.e., the vapor phase becomes relatively rich in the more volatile component at lower temperatures. This explains the advantage of distillation and purification at low temperatures.

## Constant Pressure Binary Equilibrium Diagrams

It was seen earlier that Fig. 10.3 is a binary diagram relating the composition of the gaseous and condensed phases at a *constant temperature*. However, at a constant pressure, and for convenience at 1 bar, the temperature-composition diagram is generally more useful. The line representing the temperature-composition relation for liquid in equilibrium with vapor can be obtained (i) by substituting the arbitrarily chosen constant total pressure of 1 bar in (10.13) and (10.14), (ii) eliminating $x_1$ and $y_1$ by using $x_1 = 1 - x_2$ and $y_1 = 1 - y_2$, (iii) solving (10.13) for $y_2$, and (iv) substituting the equation for $y_2$ in (iii), into (10.14); the result is

$$x_2 = \frac{1 - \exp[(-\Delta H_1^\circ + T\Delta S_1^\circ)/RT]}{\exp[(-\Delta H_2^\circ + T\Delta S_2^\circ)/RT] - \exp[(-\Delta H_1^\circ + T\Delta S_1^\circ)/RT]};$$

$(\Delta = \Delta_{vap})$; bubble point    (10.17)

This equation is represented in Fig. 10.4 by assuming that Trouton's rule is applicable and the boiling points for components 1 and 2 are 1000 and 1500 K respectively. Similarly, the temperature composition relation for the vapor can be obtained by eliminating $x_2$ instead of $y_2$ in the steps leading to the derivation of (10.17); the result is

$$y_2 = \frac{1 - \exp[(\Delta H_1^\circ - T\Delta S_1^\circ)/RT]}{\exp[(\Delta H_2^\circ - T\Delta S_2^\circ)/RT] - \exp[(\Delta H_1^\circ - T\Delta S_1^\circ)/RT]};$$

$(\Delta = \Delta_{vap})$; dew point    (10.18)

The line representing this equation with Trouton's rule is also plotted in Fig. 10.4 to complete the vapor-liquid phase diagram.

The foregoing relations are also applicable to ideal *solid solution–ideal gas* equilibria when $\Delta H^\circ$ and $\Delta S^\circ$ refer to sublimation and $x_i$ is the mole fraction in the solid solution. In fact any type of ideal phase equilibria may be represented by such equations, and the details of derivation are exactly the same as those for (10.17) and (10.18).

A frequently encountered equilibrium involves the coexistence of a solid and a liquid for which $\Delta_{vap}H_i^\circ$ and $\Delta_{vap}S_i^\circ$ in (10.17) and (10.18) refer to the enthalpy and entropy of *freezing*, and $x_i$ again refers to the liquid but $y_i$ refers to the solid. In addition, if it is assumed that the Richards rule is valid, i.e., $\Delta_{frz}H^\circ/RT_{frz} = \Delta_{frz}S^\circ/R \approx -1$, and $T_{1frz} = 400$ K and $T_{2frz} = 800$ K are substituted in (10.17)

Solutions 157

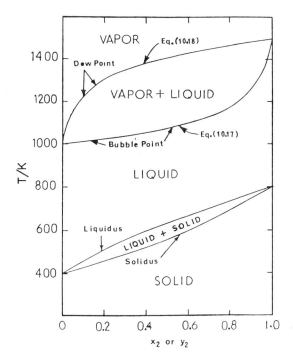

**Fig. 10.4** Vapor-liquid and liquid-solid equilibria in an ideal binary system. Melting points are 400 K(1), and 800 K(2), boiling points, 1000 K(1), and 1500 K(2) for components (1) and (2). $x_2$ is for liquid phase and $y_2$ for vapor, or for solid phase.

and (10.18), the lower set of curves in Fig. 10.4 for the liquid-solid phase diagram is obtained. The curves representing the limits of coexistence of the liquid and solid phases are called the liquidus and the solidus as shown in the figure. Some metallic systems with atomic sizes differing slightly from each other, e.g., Ag-Au, Ce-La, Ge-Si, and Nb-Ti, have liquidus and solidus curves closely represented by (10.17) and (10.18). (Phase diagrams are discussed in detail in Chapter XV.)

## Equilibria Between Pure Immiscible Solids and Ideal Liquid Solutions

Liquid solutions often separate into immiscible and fairly pure solid components upon freezing. For example, solutions of water and many organic and inorganic substances usually separate into pure solid components, with immeasurably small mutual solid solubilities. If the liquid solution is ideal and the solid phase is the relatively pure component $i$, then $y_i$ for the solid is very close to unity, then (10.13)

158    Thermodynamics

with $P = 1$ yields

$$RT \ln(x_i/y_i) = \Delta_{frz} G_i^\circ = \Delta_{frz} H_i^\circ - T \Delta_{frz} S_i^\circ,$$

$$[\ln(x_i/y_i) \approx \ln x_i \text{ for } y_i \to 1] \quad (10.19)$$

When $y_i < 0.95$, the left side of (10.19) is $RT \ln(x_i/y_i)$ and the approximate equality to $RT \ln x_i$ should not be used. As the temperature is lowered, the last term, $-T\Delta_{frz} S_i^\circ$, decreases and the liquid phase is depleted in component $i$, because nearly pure component $i$ precipitates out of the liquid upon cooling so that $x_i$ decreases with temperature.

It is possible to simplify (10.19) by substituting $\Delta_{frz} S_i^\circ = \Delta_{frz} H_i / T_{i,frz}$ by assuming that $\Delta_{frz} S_i^\circ$ and therefore $\Delta_{frz} H_i^\circ$ is independent of temperature; the result is

$$\ln \frac{x_i}{y_i} = \frac{\Delta_{frz} H_i^\circ}{R} \left( \frac{T_{i,frz} - T}{T T_{i,frz}} \right); \quad [\ln(x_i/y_i) \approx \ln x_i \text{ for } y_i \to 1] \quad (10.20)$$

For example, the system water (component 1)–sucrose (2) precipitates out pure ice upon freezing, and $\Delta_{frz} H^\circ / R = -722.79$ K, $T_{1,frz} = 273.15$ K and for $T = 268.15$ K, $x_1 = 0.9519$ and $y_1 = 1.00$; for $T = 263.15$ K, $x_1 = 0.9043$ and $y_1 = 1.00$.

Numerous binary systems form continuous solutions above the melting points of both components but as the solution is cooled, relatively pure solid phases, often consisting of nearly pure components, are precipitated out of the liquid solution. Equation (10.20) for $y_1 = 1$ and $y_2 = 1$ is then

$$\ln x_1 = \frac{\Delta_{frz} H_1^\circ}{R} \left( \frac{T_{1,frz} - T}{T T_{1,frz}} \right) \quad (10.21)$$

$$\ln x_2 = \frac{\Delta_{frz} H_2^\circ}{R} \left( \frac{T_{2,frz} - T}{T T_{2,frz}} \right) \quad (10.22)$$

For example, the liquidus curves calculated from these equations are shown by the unbroken lines for the Pb-Sb system in Fig. 10.5. The point of intersection $e$ is called the *eutectic point* which represents the simultaneous solution to (10.21) and (10.22). The agreement between the calculated curves and the experimental broken curves is good in this example, and this is frequently the case for simple eutectics. More precise calculations are in Chapter XV.

Equations (10.19) and (10.20) are useful in estimating the liquidus lines and the eutectic point if the standard enthalpies and entropies of fusion are known but the phase diagram is not known. Conversely, when the phase diagram is known in the vicinity of pure components, the standard enthalpies and entropies of fusion can be calculated if their values are not known from calorimetric measurements.

## Relative Positions of Liquidus and Solidus Lines

It can be seen from (10.19) that the relative positions of the liquidus and solidus lines near the pure components depend upon the relative compositions $x_i$ and $y_i$

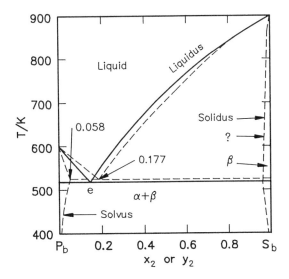

**Fig. 10.5** Pb-Sb equilibrium diagram. Solid curves, calculated; broken curves, actual. At eutectic point e, liquid and two solid solutions are in equilibrium. Sb is component 2. Broken curves are from General Ref. (18). (There are similar diagrams wherein gas occupies liquid region and two immiscible liquids occupy $\alpha + \beta$ region.) Note solidus and solvus curves.

of the two phases. In the vicinity of component $i$, if the liquid is richer in $i$ than the solid, $\ln(x_i/y_i)$ is positive, hence $T$ must be above the fusion point of the pure component, e.g., $T > T_{1,\text{frz}}$ in Fig. 10.4 since $\Delta_{\text{frz}} H_1^\circ$ is negative, i.e., the liquidus and solidus lines must slope upward. Conversely, when $x_i/y_i < 1$, or when the solid is richer in component $i$ than the liquid, then $\ln(x_i/y_i)$ in (10.19) is negative, and $T$ is below $T_{i,\text{frz}}$; e.g., the liquidus and solidus must slope downward away from pure component 2 in Fig. 10.4, and away from both pure components in Fig. 10.5. The fusion point of the solvent is then said to be *depressed*. When the curves are downward in the vicinity of the pure components and the liquidus curves from both sides intersect each other at a eutectic point, then the eutectic point is closer to the lower melting component; e.g., the point $e$ is closer to Pb than to Sb in Fig. 10.5. The limits of solid solubility are the solvus.

## Depression of Freezing Point

It was mentioned earlier that in a number of binary systems the solid phase separating from the liquid is virtually pure. The point at which the solid phase appears in a particular solution is called the freezing point. Actually the locus of such points represents the liquidus line, and the difference between the freezing point of the pure solvent and any point of the liquidus is called the depression of the freezing point. The relationship between the composition and the depression of the freezing point can be obtained from (10.20) with $i = 1$ as the solvent and

160   Thermodynamics

$i = 2$ as the solute for which $x_2$ is very small and typically $x_2 < 0.05$, and $y_1 \approx 1$; consequently

$$\ln x_1 = \frac{\Delta_{\text{frz}} H_1^\circ}{R}\left(\frac{\Delta T_{\text{frz}}}{TT_{1,\text{frz}}}\right) \tag{10.23}$$

where $T_{1,\text{frz}} - T = \Delta T_{\text{frz}}$ is the depression of the freezing point. If the solution is dilute, it is possible to write $TT_{1,\text{frz}} \approx T_{1,\text{frz}}^2$, and if $x_1$ is not far from unity, then $\ln x_1 \approx x_1 - 1 = -x_2$ from the Stirling approximation; hence, (10.23) becomes

$$x_2 \approx -\Delta_{\text{frz}} H_1^\circ \cdot \Delta T_{\text{frz}}/RT_{1,\text{frz}}^2; \qquad (y_1 \approx 1,\ \text{typically}\ x_2 < 0.05) \tag{10.24}$$

Thus in a dilute solution if the solvent freezes out as a pure solid, the mole fraction of solute and the depression of freezing point are proportional to each other. Equation (10.24) is also obeyed in a non-ideal solution, because $a_1 \to x_1$ as $x_1$ approaches unity, and $\ln x_1$ in (10.23) is very close to $-x_2$.

## Elevation of the Boiling Point

In a binary solution, if the solute is non-volatile, the total pressure over the solution is that of the solvent, and $y_1$ in (10.19) and (10.20) is unity. If the total pressure over the solution is one bar, the corresponding temperature is called the boiling point of the solution. Equation (10.23) and (10.24) are again applicable except that $\Delta_{\text{frz}} H_i^\circ$ must be replaced by the standard enthalpy of vaporization $\Delta_{\text{vap}} H_i^\circ$. It should be observed that $\Delta_{\text{frz}} H_i^\circ$ is negative, hence the freezing point is depressed but $\Delta_{\text{vap}} H_i^\circ$ is positive, hence the boiling point is elevated with increasing values of $x_2$. Equation (10.24) for the elevation of the boiling point is

$$x_2 \approx \Delta_{\text{vap}} H_1^\circ \cdot \Delta T_{\text{vap}}/RT_{1,\text{vap}}^2 \tag{10.25}$$

so that the elevation of the boiling point, $\Delta T_{\text{vap}}$ is positive as defined by $\Delta T_{\text{vap}} = T - T_{1,\text{vap}}$.

## Determination of Molecular Weights

The molecular weight of a substance can be determined by means of (10.24) from the depression of the freezing point. Let $M_1$ and $M_2$ represent the molecular weights of the solvent and the solute respectively and $W_1$ and $W_2$ their weights in a dilute solution. The molecular weight of the component 2, chosen as the solute is unknown but that of component 1 is known. The mole fraction of the solute is related to $W_1$ and $W_2$ by

$$x_2 = \frac{W_2/M_2}{(W_2/M_2) + (W_1/M_1)}$$

In a dilute solution, $W_1/M_1$ is much larger than $W_2/M_2$, therefore

$$x_2 \approx W_2 M_1/(W_1 M_2)$$

Solutions 161

**Table 10.1** Solubilities of simple gases at 25°C and 1 bar; listed solubility values are $10^5 x_2$.

| Solvent | He | $H_2$ | $N_2$ | CO | $O_2$ | Ar | $CH_4$ |
|---|---|---|---|---|---|---|---|
| Ideal | 16* | 79 | 99 | 126 | 130 | 158 | 345 |
| $CCl_4$ | — | 32.3 | 63.4 | 87 | 118 | — | 282 |
| $(C_2H_5)_2O$ | — | 54.5 | 124 | 167 | 195 | — | 447 |
| $(CH_3)_2CO$ | 10.7 | 22.8 | 58.4 | 84.3 | 91 | 90 | 220 |
| $N_2O_4$* | 10.1 | — | 65 | — | 101 | 97 | — |
| $N_2H_4$* | 0.51 | — | 0.71 | — | — | 1.22 | — |
| $H_2O$ | 0.7 | 1.5 | 1.2 | 1.8 | 2.3 | 2.5 | 2.4 |

*From E. T. Chang, N. A. Gokcen, and T. M. Poston, J. Phys. Chem., 72, 638 (1969); J. Spacecraft and Rockets, 6, 1177 (1969); remaining values from J. H. Hilderbrand and R. L. Scott, "The Solubility of Nonelectrolytes," Reinhold Publ. Co., (1950). See also L. R. Field, E. Wilhelm, and R. Battino, J. Chem. Thermod., 6, 237 (1974). H.-J. Hinz, editor, "Thermodynamic Data for Biochemistry and Biotechnology," Springer-Verlag (1986).

Substitution of this equation into (10.24) and rearrangement give

$$M_2 = -\frac{RT_{1,\text{frz}}^2 \cdot W_2 M_1}{\Delta_{\text{frz}} H_1^\circ \cdot \Delta T_{\text{frz}} \cdot W_1} \tag{10.26}$$

The unknown molecular weight $M_2$ can thus be obtained from the known and measured quantities on the right side of this equation. A similar relation can be obtained for the elevation of the boiling point from (10.25).

## Ideal Solubilities of Gases

At ordinary temperatures, gases are usually dissolved in liquids without chemical reactions. The cases involving reactions will be considered later in another chapter. To calculate the ideal solubility of a gas from its pressure, $P_2$, over a solution, Raoult's law may be rewritten as $x_2 = P_2/P_2^\circ$ where $P_2^\circ$ is the equilibrium pressure of the pure liquified gas. The value of $P_2^\circ$ is obtained from $R \ln P_2^\circ = (-\Delta_{\text{vap}} H_2^\circ / T) + \Delta_{\text{vap}} S_2^\circ$ by assuming that $\Delta_{\text{vap}} H_2^\circ$ is a constant and that this relationship is valid even if the temperature at which the calculated $P_2^\circ$ is above the critical point of the gas. If the equilibrium pressure of the gas over the solution is taken to be one bar, i.e., $P_2 = 1$, the corresponding value of $x_2$ is the solubility of the gas. The result for $x_2$ is then

$$x_2 = P_2/P_2^\circ = 1/P_2^\circ; \quad (P_2 = 1) \tag{10.27}$$

The order of magnitude of ideal solubilities computed from this equation for a number of gases in a number of solvents is correct as shown in Table 10.1. For water, the solubilities are about an order of magnitude lower. The vapor pressure of pure liquid gas, $P_2^\circ$, increases with increasing temperature; hence the ideal solubility at 1 bar decreases with increasing temperature as required by (10.27).

162    *Thermodynamics*

## PART II   REAL SOLUTIONS

### Definition of Real Solutions

Real or non-ideal solutions deviate from ideal behavior, i.e., the activity coefficient of components in non-ideal solutions deviate from unity. The attractive forces among the component molecules of a solution make $\gamma_i < 1$, and this is called negative deviation from Raoult's law, and conversely, repulsive forces make $\gamma_i > 1$, called positive deviation from Raoult's law. The activity $a_i$ is therefore smaller than $x_i$ for the negative deviation, and greater than $x_i$ for the positive deviation. The activities of components in a condensed phase are obtained conveniently from the vapor-condensed phase equilibria. Equilibria among liquids and solids are also useful in determining the activities but in general this procedure is not as accurate as the vapor-condensed phase equilibria. There are other methods of measurement of activities, particularly the electrochemical methods, as will be seen in Chapter XIV.

### Equilibrium Between a Real Solution and Its Vapor at Constant Temperature

In general, molecules of a component in the gaseous state are so much farther apart than in the solution that the vapor is nearly ideal and therefore the activity is usually obtained by $a_i = P_i/P_i^\circ$ rather than the exact relation $a_i = f_i/f_i^\circ$, because the fugacities are often not available. A vapor phase consisting of two components may condense with increasing pressure at a *given temperature* to form two immiscible liquids as shown in Fig. 10.5A. As the immiscibility tends to disappear, e.g., possibly by increasing temperatures, the liquid and vapor boundaries may form an azeotrope as shown later in Figs. 10.6–10.7. In another system, the vapor may form a nonideal miscible liquid phase as shown in Fig. 10.5B. The total vapor pressure $P$, and the compositions of the coexisting vapor and liquid phases determine the activities in liquid. For example, at $P = 11.00$ kPa (total pressure), $x_B = 0.20$, and $y_B = 0.90$, and $P_B^\circ = 18.12$ kPa; hence, the activity of component B is $a_B = P_B/P_B^\circ = P \times y_B/P_B^\circ = 11.00 \times 0.90/18.12 = 0.546$. It can be shown that, for component A, $a_A = 11.00 \times 0.10/1.43 = 0.77$ for $y_A = 0.10$ in the diagram, and for the same liquid composition.

The measurements of the vapor pressures of two typical azeotropic binary solutions taken from the classical work of Zavidski (1900) are shown in Figs. 10.6 and 10.7. The system carbon disulfide and acetone in Fig. 10.6 shows positive deviations from ideal behavior for both components. The uppermost curve represents the total pressure over the liquid, versus the composition of the liquid. The point $A$, for example, represents the composition of liquid and the sum of the partial pressures represented by $DB$ and $DC$. The pressure-composition curves for the liquid and the vapor are not given for the corresponding ideal solution but they are similar to those in Fig. 10.3. At the total pressure of $P^1$, the composition of liquid is $E$, and that of the coexisting vapor is $F$, hence the vapor is richer in

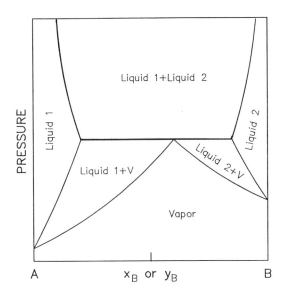

**Fig. 10.5A** Pressure-composition diagram for a hypothetical binary system exhibiting immiscibility in liquid state.

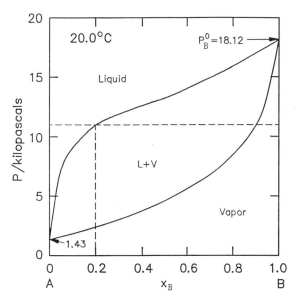

**Fig. 10.5B** Vapor-liquid equilibria in binary system $N_2H_4$ = A and $N_2H_2C_2H_6$ = B. This system is important as a fuel in space vehicles; from N. A. Gokcen and E. T. Chang, J. Phys. Chem., *72*, 2556 (1968).

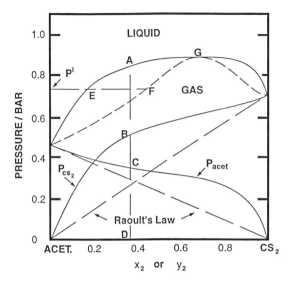

**Fig. 10.6** Partial and total vapor pressures of acetone-carbon disulfide system at 35.17°C. Top solid curve shows composition of liquid; above this curve only liquid exists; short dashed curve shows composition of gas; below this curve only gas phase exists, and between these curves gas and liquid phases coexist. Partial pressures of components versus liquid composition are shown by lower set of solid curves. Broken straight lines are for the partial pressures for Raoult's law. Carbon disulfide is component 2.

the more volatile carbon disulfide than the solution. Beyond the maximum point $G$ the reverse composition relationship prevails. At $G$, the composition of the liquid and vapor are identical and both curves are horizontally tangent to each other at their maximum points. The liquid then boils at a constant pressure and a constant composition; such a liquid is called an azeotropic solution. A binary system similar to that in Fig. 10.6 is shown in Fig. 10.6A*.

A maximum or a minimum in the total pressure curve requires that the composition of both phases be equal. The thermodynamic arguments requiring that the curves for the liquid and the vapor must touch each other at an extremum (maximum or minimum) are as follows. The Gibbs energy $\mathcal{G}$ is given by $\mathcal{G} = n_1 \overline{G}_1 + n_2 \overline{G}_2$ and also functionally by $\mathcal{G} = \mathcal{G}(P, T, n_1, n_2)$ where the boldface $\mathcal{G}$ refers to $n$ moles of mixture. The total differentials of these equations are

$$d\mathcal{G} = n_1 \, d\overline{G}_1 + n_2 \, d\overline{G}_2 + \overline{G}_1 \, dn_1 + \overline{G}_2 \, dn_2$$

$$d\mathcal{G} = \mathcal{V} \, dP - \mathcal{S} \, dT + \overline{G}_1 \, dn_1 + \overline{G}_2 \, dn_2$$

---

*L. H. Horsley, et al., "Azeotropic Data," Am. Chem. Soc., Washington, D.C. (1952); Part II, by L. H. Horsley and W. S. Tamplin (1962); Part III, edited by R. F. Gould (1973).

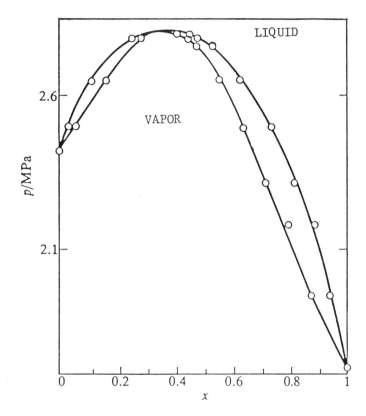

**Fig. 10.6A** Liquid-vapor equilibrium for $\{(1-x)CO_2 + xC_2H_6\}$ at 260.0 K. Adapted from A. Q. Clark and K. Stead, J. Chem. Thermod., **20**, 413 (1988).

These equations yield $\mathcal{V}\,dP - \mathcal{S}\,dT = n_1\,d\overline{G}_1 + n_2\,d\overline{G}_2$, where $\mathcal{V}$ and $\mathcal{S}$ refer to $n$ moles, and after dividing by $(n_1 + n_2)$ the following equation is obtained:

$$V\,dP - S\,dT = x_1\,d\overline{G}_1 + x_2\,d\overline{G}_2 \tag{10.28}$$

At a constant temperature, (10.28) for the liquid and for the coexisting gas may be written as

$$V(l)\,dP = x_1\,d\overline{G}_1 + x_2\,d\overline{G}_2; \qquad V(g)\,dP = y_1\,d\overline{G}_1 + y_2\,d\overline{G}_2 \tag{10.29}$$

where $\overline{G}_i$ in the gas-phase is identical with that in the liquid-phase, i.e., $\overline{G}_i(g) = \overline{G}_i(l)$ due to equilibrium between these phases. Subtraction of the first equation from the second, and division by $dx_2$ yields

$$[V(g) - V(l)]\frac{dP}{dx_2} = (y_1 - x_1)\frac{d\overline{G}_1}{dx_2} + (y_2 - x_2)\frac{d\overline{G}_2}{dx_2} \tag{10.30}$$

166    Thermodynamics

With the preceding equation, it is possible to prove the following theorems called the Gibbs-Konovalow theorems:*

***Theorem I***   If the curve for the liquid has a maximum or minimum (= extremum) at a point $(P, x_2)$, i.e., if $dP/dx_2$ is zero at $(P, x_2)$, it is necessary and sufficient that the compositions of both phases be equal.

***Proof***   $d\overline{G}_1/dx_2$ and $d\overline{G}_2/dx_2$ in (10.30) are finite as can be seen from the definition of $\overline{G}_i$ by $\overline{G}_i = G_i^\circ + RT \ln \gamma_i x_i$. In addition, $V(g) - V(l)$ is a large positive quantity. Since $dP/dx_2$ is zero at an extremum, the left side of (10.30) is zero; therefore it is necessary that $y_1 - x_1$ and $y_2 - x_2$ also be zero, hence, $y_1 = x_1$ and $y_2 = x_2$. The proof that this is also sufficient is very easy because if $y_1 = x_1$ and $y_2 = x_2$, it is evident from (10.30) that $dP/dx_2$ must be zero.

***Theorem II***   If $dP/dx_2$ is zero at $(P, x_2)$ it is necessary and sufficient that $dP/dy_2$ also be zero at $(P, x_2)$.

***Proof***   When $dP/dx_2$ is zero, $y_1 - x_1 = 0$ and $y_2 - x_2 = 0$ from (10.30). When (10.30) is multiplied by $dx_2/dy_2$, the left side becomes $[V(g) - V(l)] \, dP/dy_2$, and now $dP/dy_2$ must be zero because $y_1 - x_1$ and $y_2 - x_2$ are both zero, and this completes the proof. The proof that this is also sufficient is evident because when $dP/dy_2$ is zero $dP/dx_2$ is also zero as can be shown by following the same type of arguments in reverse. Theorems I and II may therefore be summarized by the statement that "*if the curve for the liquid, or for the vapor for a two-phase field has an extremum at a point, it is necessary and sufficient that the other curve also have an extremum at the same point.*" Both curves must therefore be horizontally tangent to each other. (A horizontal critical point is not an extremum here; cf. Fig. 15.3 in Chapter XV.)

At the azeotropic point, there is a simple relationship between the activity coefficient and the total pressure and the pressures of pure components, since the compositions of both phases are the same (see problem 10.6). In the event that there is no extremum in the total pressure curve for the liquid, or for the vapor, the slope must not change sign, and there must be no azeotropic point. The maxima and minima in phase diagrams involving condensed phases are not called azeotropic points, although in every respect they resemble such points, and obey Theorems I and II.

## Equilibrium Between a Real Solution and Its Vapor at Constant Pressure

Equilibrium between a real solution and its vapor at constant pressure is shown in Figs. 10.8 and 10.9. The vertical axis represents the boiling temperature under 1.013 bar of total pressure, and the horizontal axis is the mole fraction. At the

---

*General Refs. (15) and (22).

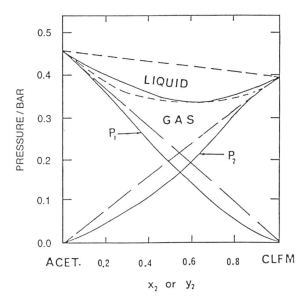

**Fig. 10.7** Partial and total pressures, acetone-chloroform system at 35.17°C. For solid, short- and long-broken lines see caption to Fig. 10.6. Top broken line is for total pressure if the solution were ideal. Component 1 is acetone, and component 2, chloroform, and $P_1$ and $P_2$ are their partial pressures.

azeotropic point, or at any similar point where a phase boundary curve separating a single phase field from a two-phase field meets with another such curve, their point of contact must be horizontal as shown in Figs. 10.6 and 10.7. The thermodynamic arguments that the two-curves must meet at an extremum are identical to those of Figs. 10.6 and 10.7. For this purpose, $V\,dP$ in (10.29) must be replaced by $-S\,dT$ so that the left side of (10.30) becomes $[S(l) - S(g)]\,dT/dx_2$ and for an extremum $dT/dx_2$ becomes zero.

The separation of an azeotropic solution into its components is not possible by using a distillation tower if the pressure is held at a value corresponding to the azeotropic point. It is possible, however, to distill at a higher or lower pressure when the azeotropic composition favorably shifts with pressure, or when the azeotropy disappears altogether as shown in Fig. 10.10 for the butanone-methanol system. The azeotropy usually disappears at sufficiently high pressures but in unusual cases such as in the methanol-acetone system, it disappears not only at high pressures but also at low pressures. When the azeotropy is absent, which is the case for the majority of binary systems, the curves for the phase boundaries do not change the algebraic sign of their slopes as shown in Fig. 10.4, but if they change, they become discontinuous at the eutectic, or other "—tectic" points as shown in Fig. 10.5, and discussed in Chapter XV. At the

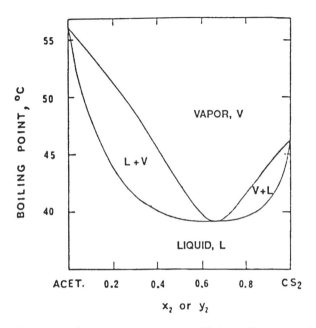

**Fig. 10.8** Constant-pressure equilibrium diagram at 1.013 bars for acetone–carbon disulfide system. See Fig. 10.6 for constant-temperature diagram of this system. Component 1 is acetone, and component 2, carbon disulfide.

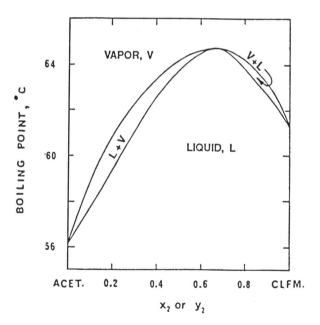

**Fig. 10.9** Constant-pressure equilibrium diagram at 1.013 bars for acetone–chloroform system. See Fig. 10.7 for constant-temperature diagram of this system. Component 1 is acetone, and component 2, chloroform.

Solutions 169

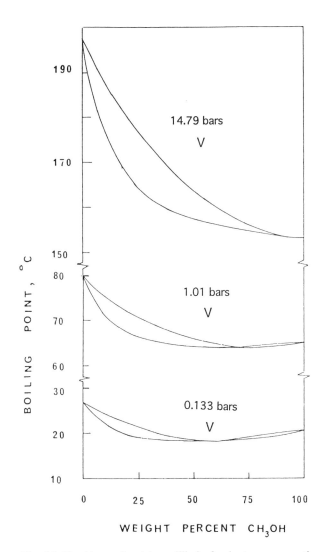

**Fig. 10.10** Vapor-liquid equilibria for butanone-methanol system at various pressures. Azeotropic point shifts to right with increasing pressure, and disappears. Pressure is in bars.

critical point, $dT/dx_2$ is also zero as well as $d^2T/dx_2^2$ as will be discussed in Chapter XV; it is sufficient to note here that the Gibbs-Konovalow theorem is applicable when there are two distinct phases in equilibrium, and $d^2T/dx_2^2$ is not zero so that the curves have an extremum point. At the critical point, two distinguishable phases become one indistinguishable phase as in Fig. 15.3 in Chapter XV.

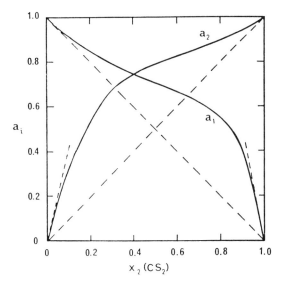

**Fig. 10.11** Activities in liquid acetone-carbon disulfide system at 35.17°C. Component 2 is carbon disulfide. Broken diagonal lines are for Raoult's law, and short broken lines tangent to the activity curves at $x_2 = 0$ and $x_2 = 1$ are for Henry's law.

## Variation of Activity and Activity Coefficients with Composition in Binary Solutions

The activity of the binary systems shown in Figs. 10.6 and 10.7 are represented in Figs. 10.11 and 10.12. The shape of one curve for the activity is related to the shape of the other by the Gibbs-Duhem equation, (2.27), i.e.,

$$x_1 \, d \ln a_1 + x_2 \, d \ln a_2 = 0, \quad \text{(constant } P \text{ and } T\text{)} \tag{10.31}$$

This equation determines the interrelationship of the activity curves. The corresponding relationship for the activity coefficients is obtained by substituting $\ln a_i = \ln x_i + \ln \gamma_i$ and observing that $x_1 \, d \ln x_1 + x_2 \, d \ln x_2 = dx_1 + dx_2 = 0$; the result is

$$x_1 \, d \ln \gamma_1 + x_2 \, d \ln \gamma_2 = 0, \quad \text{(constant } P \text{ and } T\text{)} \tag{10.32}$$

A simple analytical example for the relationship between the curves for $\gamma_1$ and $\gamma_2$ is as follows. Let $\ln \gamma_1$ and $\ln \gamma_2$ be expressed by

$$\ln \gamma_1 = A_1 x_2 + A_2 x_2^2 \tag{10.33}$$

$$\ln \gamma_2 = B_1 x_1 + B_2 x_1^2 \tag{10.34}$$

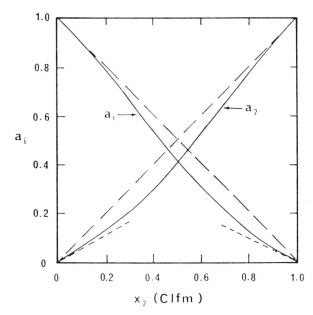

**Fig. 10.12** Activities in liquid acetone-chloroform system at 35.17°C. Component 2 is chloroform. Broken diagonal lines are for Raoult's law, and short broken lines, for Henry's law.

The boundary conditions that $\gamma_i = 1$ for $x_i = 1$ are satisfied by both equations, i.e., there are no constant terms on the right sides of these equations. Substitution of these equations into (10.32) gives,

$$A_1 x_1\, dx_2 + 2A_2 x_1 x_2\, dx_2 + B_1 x_2\, dx_1 + 2B_2 x_1 x_2\, dx_1 = 0$$

Elimination of $x_1$ and $dx_1$ by using $x_1 = 1 - x_2$ and $dx_1 = -dx_2$ yields

$$A_1 - (A_1 - 2A_2 + B_1 + 2B_2)x_2 - (2A_2 - 2B_2)x_2^2 = 0$$

This equation must be zero for all values of the independent variable $x_2$; therefore, $A_1 = 0$, $A_2 = B_2$ from the first and last terms, and $A_1 = -B_1 = 0$ from the coefficient of $x_2$; consequently, (10.33) and (10.34) become

$$\ln \gamma_1 = A_2 x_2^2 \tag{10.35}$$

$$\ln \gamma_2 = A_2 x_1^2 \tag{10.36}$$

For additional terms such as $2A_3 x_2^3$ in (10.33), and $2B_3 x_1^3$ in (10.34), it is easy to verify that $B_2 = A_2 + 3A_3$ and $2B_3 = -2A_3$, in addition to $A_1 = -B_1 = 0$.* For an ideal solution $\gamma_1 = \gamma_2 = 1$, and all the coefficients in a power series of the type given by (10.33) are zero. The solutions obeying (10.35) and (10.36) are called the regular solutions as will be seen in Chapter XI.

*S. Malanowski and A. Anderko, "Modeling Phase Equilibria," John Wiley & Sons (1992).

## Variation of Activities in Binary Solutions with Pressure and Temperature

From the definitional equation

$$\ln a_1 = (\overline{G}_1 - G_1^\circ)/(RT) = (\overline{H}_1 - H_1^\circ)/RT - (\overline{S}_1 - S_1^\circ)/R \qquad (10.37)$$

variation of the activity with pressure is obtained by

$$\left(\frac{\partial \ln a_1}{\partial P}\right)_{T,x_2} = \frac{\overline{V}_1 - V_1^\circ}{RT} \qquad (10.38)$$

At constant $x_2$, $d \ln a_1 = d \ln \gamma_1$; therefore,

$$\left(\frac{\partial \ln \gamma_1}{\partial P}\right)_{T,x_2} = \frac{\overline{V}_1 - V_1^\circ}{RT} \qquad (10.39)$$

Variation of the activity with pressure is of interest in high pressure phenomena. Integration of (10.39) requires experimental data on $\overline{V}_1 - V_1^\circ$ as a function of pressure.

The effect of temperature on the activity is expressed by

$$\overline{G}_1 - G_1^\circ = RT \ln a_1 = (\overline{H}_1 - H_1^\circ) - T(\overline{S}_1 - S_1^\circ) \qquad (10.40)$$

The preceding equations may be rewritten for component 2 by exchanging the subscript 1 for 2. The analytical equations for representing $\ln \gamma_1$ as functions of both temperature and composition will be given in Chapter XI.

## Dilute Solutions

It was shown in Chapter IX that in real dilute solutions the activity of a solute varies linearly with its mole fraction when its concentration becomes very small. Thus for the solute 2, the fugacity $f_2$ is related to the concentration $x_2$ in solution as follows:

$$\lim_{x_2 \to 0} f_2/f_2^\circ = \gamma_2^\circ x_2 = a_2; \qquad [\gamma_2 (x_2 \to 0) = \gamma_2^\circ] \qquad (10.41)$$

where $\gamma_2^\circ$ is independent of composition when $x_2$ is sufficiently low. Substitution of $k_2$ for $\gamma_2^\circ f_2^\circ$ gives $f_2 = k_2 x_2$, and Henry's law may be stated by

$$f_2 = k_2 x_2, \quad \text{or} \quad a_2 = \gamma_2^\circ x_2; \qquad (\gamma_2^\circ = \text{constant at each } T) \qquad (10.42)$$

where $f_2$ may often be replaced by its close equivalent $P_2$. In (10.42), the standard state for component 2 is the pure component 2, but if the reference state is taken to be the infinitely dilute solution of component 2 in component 1, then (10.42) becomes

$$a_2^\infty = f_2/k_2 = a_2/\gamma_2^\circ = x_2;$$

$$\left(\text{Here, } \gamma_2^\infty = 1 \text{ with } x_2 \to 0, \gamma_2^\infty \neq 1 \text{ with } x_2 > 0\right)$$

Deviations from Henry's law are then written as $a_2^\infty = \gamma_2^\infty x_2$, with the ratio of $\gamma_2^\infty/\gamma_2$ (at $x_2 > 0$) $= 1/\gamma_2^\circ$ (at $x_2 \to 0$), since the ratios of activity coefficients must be the same for a given system irrespective of the composition as shown in Chapter IX, (here, it is essential to refer to Figs. 9.1 and 9.2); [see (9.34)].

Deviations from Raoult's and Henry's laws become noticeable at relatively small concentration ranges. In general when the temperature is increased, the range at which these laws are applicable also increases, and the Henry's law $\gamma_i^\infty$ tends to approach unity. The upper limit at which either law is valid is not distinctly clear since the activity curves approach such behavior in a tangential manner. It is clear that the more accurately the activities are determined the smaller is the range at which Raoult's and Henry's laws are obeyed. These laws are called the limiting laws of solutions because they are valid in the limiting case with $x_1 \to 1$, or $x_2 \to 0$. The more precise statements of these laws are in Chapter XI.

## Molar, Partial Molar, and Excess Thermodynamic Properties of Solutions

The partial molar properties $\overline{G}_i - G_i^\circ$, $\overline{V}_i - V_i^\circ$, $\overline{H}_i - H_i^\circ$ and $\overline{S}_i - S_i^\circ$ are related to the measurable property $a_i$ by (10.37), (10.38), and (10.39). The molar properties are related to these equations by the following relationships for multicomponent systems.

$$\Delta_{\text{mix}} G = \sum_{i=1}^{c} x_i \left( \overline{G}_i - G_i^\circ \right) = RT \sum_{i=1}^{c} x_i \ln a_i$$

$$= RT \sum_{i=1}^{c} x_i \ln \gamma_i + RT \sum_{i=1}^{c} x_i \ln x_i \qquad (10.43)$$

$$\Delta_{\text{mix}} H = \sum_{i=1}^{c} x_i \left( \overline{H}_i - H_i^\circ \right) = -RT^2 \sum_{i=1}^{c} x_i \left( \frac{\partial \ln \gamma_i}{\partial T} \right)_{P,x_i} \qquad (10.44)$$

$$\Delta_{\text{mix}} S = \sum_{i=1}^{c} x_i \left( \overline{S}_i - S_i^\circ \right) = -RT \sum_{i=1}^{c} \left[ x_i \left( \frac{\partial \ln \gamma_i}{\partial T} \right)_{P,x_i} \right] - R \sum_{i=1}^{c} x_i \ln a_i$$

$$(10.45)$$

The derivatives at constant $P$ and $x_i$ refer to constant pressure and composition. A property such as $\Delta_{\text{mix}} G$ is called the molar Gibbs energy of mixing, or the molar Gibbs energy of formation of solution. The activity coefficient is a measure of deviation from ideality, therefore, the thermodynamic properties related to the activity coefficients are called *the excess thermodynamic properties*.* The excess molar Gibbs energy $G^E$, and the excess partial Gibbs energy $\overline{G}_i^E$, are defined by

$$G^E \equiv G(\text{real}) - G(\text{ideal}), \quad \text{and} \quad \overline{G}_i^E \equiv g_i \equiv \overline{G}_i(\text{real}) - \overline{G}_i(\text{ideal})$$

$$(10.46)$$

---

*Definition of excess properties was introduced by G. Scatchard [Gen. Ref. (22)].

where $g_i$ is used for $\overline{G}_i^E$ for simplicity and the real properties are the experimentally measured properties, and the ideal properties refer to the ideal solution in which the activity coefficients are assumed to be unity. The excess partial Gibbs energy $\overline{G}_i^E$ is therefore obtained by subtracting $\overline{G}_i(\text{ideal}) = G_i^\circ + RT \ln x_i$ from $\overline{G}_i = G_i^\circ + RT \ln a_1$; thus

$$\overline{G}_i^E \equiv g_i \equiv [\overline{G}_i - G_i^\circ] - [\overline{G}_i(\text{ideal}) - G_i^\circ]$$
$$= RT \ln a_i - RT \ln x_i = RT \ln \gamma_i \qquad (10.47)$$

and likewise,

$$\overline{H}_i^E \equiv h_i \equiv \overline{H}_i - H_i^\circ = -RT^2 \left(\frac{\partial \ln \gamma_i}{\partial T}\right)_{P, x_i} \qquad (10.48)$$

where $H_i^E = h_i$ is defined by $h_i = \overline{H}_i - \overline{H}_i(\text{ideal})$ with $\overline{H}_i = \overline{H}_i(\text{real})$, and in (10.48), $H_i^\circ$ is equal to $\overline{H}_i(\text{ideal})$ because, for an ideal solution, $\overline{H}_i(\text{ideal}) - H_i^\circ = 0$. The excess partial molar entropy may be obtained from (10.47) and (10.48) substituted in $\overline{G}_i^E = \overline{H}_i^E - T\overline{S}_i^E \equiv h_i - Ts_i$; thus,

$$\overline{S}_i^E \equiv s_i \equiv -R \ln \gamma_i - RT\left(\frac{\partial \ln \gamma_i}{\partial T}\right)_{P, x_i} \qquad (10.49)$$

The corresponding excess molar properties are obtained by substituting the foregoing equations into the functional form of $G_i^E = \sum_{i=1}^{c} x_i \overline{G}_i^E$; i.e.,

$$G^E = RT \sum_{i=1}^{c} x_i \ln \gamma_i \qquad (10.50)$$

$$H^E = -RT^2 \sum_{i=1}^{c} x_i \frac{\partial \ln \gamma_i}{\partial T} \qquad (10.51)$$

$$S^E = -R \sum_{i=1}^{c} x_i \ln \gamma_i - RT \sum_{i=1}^{c} x_i \frac{\partial \ln \gamma_i}{\partial T} \qquad (10.52)$$

**Example 1** The excess molar and partial Gibbs energy for the carbon tetrachloride-benzene system, obtained from vapor-liquid equilibria,* are expressed by

$$g_1/R = x_2^2[59.757 - 0.0692T + 0.855(x_2 - x_1)] \qquad (10.53)$$

$$G^E/R = x_1 x_2 (59.757 - 0.0692T + 0.855 x_2) \qquad (10.54)$$

where the subscript 1 is for $CCl_4$, and 2, for $C_6H_6$. The measurements were made in the range of 30 to 70°C and the activities were computed from $f_i/f_i^\circ$ over each solution, and then $G^E$ and $g_1$ were fitted empirically into the foregoing equations. Compute $G^E$, $H^E$, $S^E$, $g_i$, $h_i$, and $s_i$ ($i = 1$ or 2) for this system at $x_2 = 0.3$ and 300 K. Compute also Henry's law constants $\gamma_i^\circ$.

*G. Scatchard and L. B. Ticknor, J. Am. Chem. Soc., 74, 3724 (1952).

***Solution*** The equation for $g_2$ can be easily obtained by substituting (10.53) and (10.54) in $G^E = x_1 g_1 + x_2 g_2$ and solving for $g_2/R$; the result is

$$g_2/R = x_1^2(59.757 - 0.0692T + 1.71x_2) \tag{10.55}$$

The remaining functions are

$$\frac{\partial(G^E/RT)}{\partial(1/T)} = H^E/R = x_1 x_2(59.757 + 0.855 x_2)$$

$$\partial(G^E/R)/\partial T = -S^E/R = -0.0692 x_1 x_2$$

$$\frac{\partial(g_1/RT)}{\partial(1/T)} = h_1/R = x_2^2[59.757 + 0.855(x_2 - x_1)]$$

$$h_2/R = x_1^2(59.757 + 1.71x_2)$$

$$\partial(g_1/R)/\partial T = -s_1/R = -0.0692 x_2^2$$

$$s_2/R = 0.0692 x_1^2$$

The values of all these equations at $x_2 = 0.3$ and 300 K are listed as follows:

| $G^E/R$ | $g_1/R$ | $g_2/R$ | $H^E$ | $h_1/R$ | $h_2/R$ | $S^E/R$ | $s_1/R$ | $s_2/R$ |
|---|---|---|---|---|---|---|---|---|
| 8.3 | 3.5 | 19.4 | 12.6 | 5.3 | 29.5 | 0.0145 | 0.00623 | 0.0339 |

The value of $\gamma_i^\circ$ is obtained from (10.53), which gives $g_1/R = 39.85$ at 300 K and $x_1 = 0$, and from $g_1/R = T \ln \gamma_1^\circ$, $\gamma_1^\circ$ is 1.142, and it is easily seen from (10.55) that $\gamma_2^\circ = 1.139$.

***Example 2*** The molar, excess molar, partial, and excess partial properties of liquid Al-Cu solutions, summarized by P. D. Desai (1970) in General Ref. (9) are plotted in Fig. 10.13, where $x_2$ is the mole fraction of copper.* It will be shown in the next chapter that $\overline{G}_1 - G_1^\circ$, $\overline{G}_2 - G_2^\circ$ and $\Delta_{mix} G$ coincide where $\partial \Delta_{mix} G / \partial x_2$ is zero. The excess thermodynamic properties also follow the same pattern. It will be seen in Chapter XI that it is best to fit the data for $\gamma_i$ with a Margules-type equation, from which all the related thermodynamic equations can readily be derived.

---

*For various systems see Gen. Refs. (9–14); for methods of activity measurements see the following: Gen. Ref. (4); Y. Maa, A. Mikula, Y. A. Chang, and W. Schuster, Met. Trans., *13A*, 1115 (1982); J. K. Gibson, L. Brewer, and K. A. Gingerich, ibid., *15A*, 2075 (1984); S. N. Sinha, H. Y. Sohn, and M. Nagamori, ibid., *15B*, 441, (1984); L. Timberg, J. M. Toguri, and T. Azakami, ibid., *12B*, 275 (1981); C. Bergman, R. Chastel, and R. Castanet, J. Phase Equil., *13*, 113 (1992); M. R. Baren and N. A. Gokcen, in "Advances in Sulfide Smelting," edited by H. Y. Sohn, D. B. George, and A. D. Zunkel, Conference Proceedings, TMS-AIME, p. 41 (1983); J. N. Pratt, Met. Trans., *21A*, 1223 (1990); S. Howard, ibid., *20B*, 845 (1989); M. J. Stickney, M. S. Chandrasekharaiah, K. A. Gingerich, and J. A. Speed, ibid., *22A*, 1937 (1991); S. Srikanth and K. T. Jacob, ibid., *22B*, 607 (1991); S. M. Howard and J. P. Hager, ibid., *9B*, 51 (1978); R. Chastel, M. Saito, and C. Bergman, J. Alloys and Compounds, *205*, 39 (1994); Z. C. Wang, Y. W. Tian, H. L. Yu, X. H. Zhang, and J. K. Zhou, Met. Trans., *25B*, 103 (1992); ibid., *23B*, 623 (1992). See also J. Chem. Thermod., *25*, 711 (1993); E. Kato, J. Mass Spect. Soc. Japan, *41* (6), 297 (1993) and other papers in this issue.

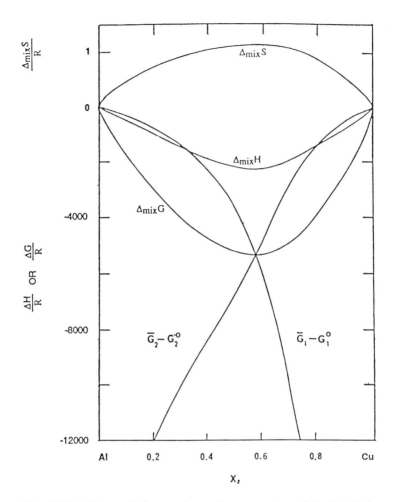

**Fig. 10.13** Selected thermodynamic properties of liquid Al-Cu system at 1373 K. Subscript 1 is for Al, and 2, for Cu. Curves for $\Delta_{mix}G$, $\overline{G}_1 - G_1^\circ$ and $\overline{G}_2 - G_2^\circ$ coincide where $\Delta_{mix}G$ is minimum, (cf. Fig. 10.1). Lower left scale is for $\Delta_{mix}H$, $\Delta_{mix}G$ and $\overline{G}_i - G_i^\circ$. From A. Yazawa and K. Itagaki; remaining values are largely based on data by T. C. Wilder, and by H. Mitani and H. Nagai as summarized in General Ref. 9. See also J. L. Murray, Intern. Met. Reviews, 30, 211 (1985).

## PROBLEMS

10.1 Show that for an ideal solution $\overline{C}_{p,i} = C^\circ_{p,i}$, $\Delta_{mix}C_p = 0$, $\partial^2(\overline{G}_2 - G^\circ_i)/\partial T^2 = 0$ and $\partial^2(\overline{G}_i - G^\circ_i)/\partial T \partial x_i = R/x_i$.

10.2 Calculate the boiling point and the freezing point of water containing 1 molal solution of $C_6H_{12}O_{11}$ at 1.0133 bars. For water $\Delta_{vap}H^\circ/R = 4890$, and $\Delta_{fus}H^\circ/R = 722.79$.

*Ans.*: $\Delta_{boil}T = 0.504$, and $\Delta_{frz}T = 1.827$ K

10.3 What are the values of $\Delta_{mix}G$, $\Delta_{mix}S$, $\overline{S}_1 - S^\circ_1$, $\overline{G}_1 - G^\circ_1$ for a real solution when $x_1$ approaches zero?

*Partial ans.*: $\overline{G}_1 - G^\circ_1 = -\infty$, and $\overline{S}_1 - S^\circ_1 = \infty$

10.4 The vapor pressure of liquid argon is 633 bars at 25°C; calculate the ideal solubility of argon in acetone and in cyclohexane at 25°C and 2 bars. The actual solubility at 1 bar is about $9 \times 10^{-4}$ mole fraction for acetone and $15 \times 10^{-4}$ for cyclohexane; calculate $\gamma^\circ_2$ for argon in each solution.

*Partial ans.*: $\gamma^\circ_2$ (in acetone) = 1.755

10.5 A solution of water containing 0.01 weight fraction of an unknown compound lowers the freezing point by 0.24°C. Calculate the molecular weight of the compound (a) assuming that it does not dissociate and (b) dissociates completely into three ionic species. $\Delta_{fus}H^\circ = 6009.5$ J for one mole of water. Note that each ionized species acts as an independent molecule in lowering the freezing point.

*Ans.*: (a) 78.3, (b) 234.9

10.6 Show that (a) in a binary system consisting of one azeotrope similar to Figs. 10.9 and 10.10 at a constant temperature, $\gamma_1 = P/P^\circ_1$ and $\gamma_2 = P/P^\circ_2$, where $P$ is the total pressure, $P^\circ_1$ and $P^\circ_2$ are the vapor pressures of the pure components, and (b) $\gamma_1$ and $\gamma_2$ are larger than unity when the total pressure-composition diagram has a maximum and they are smaller than unity when it has a minimum.

10.7 Derive the following relationship at the azeotropic point in a binary system under one bar of total pressure, (cf. Problem 10.6 and for $P = 1$, $a_1 = x_1/P^\circ_1$):

$$G^\circ_1(\text{gas}) - G^\circ_1(\text{liq.}) = +RT \ln \gamma_1$$

10.8 The following sets of equations are proposed to express the activity coefficients in two binary solutions:

Set 1:

$\ln \gamma_1 = \alpha_1 + \beta_1 x_2^{3/2}$

$\ln \gamma_2 = \alpha_2 + \beta_2 x_1^{3/2}$

Set 2:

$\ln \gamma_1 = 2A_3 x_2^3 + 3A_4 x_2^4$

$\ln \gamma_2 = 2A_4 x_1^3 + 3B_4 x_1^4$

Prove that all the coefficients are zero.

10.9 The values of silver gas pressure over liquid Ag-Au alloys at 1350 K are as follows:

| $P_{Ag}$/Pa: | 3.140 | 2.393 | 1.583 | 1.212 | 0.882 | 0.361 |
|---|---|---|---|---|---|---|
| $x_{Au}$: | 0.0 | 0.20 | 0.40 | 0.50 | 0.60 | 0.80 |

Calculate $a_{Ag}$, $\overline{G}_{Ag} - G^\circ_{Ag}$ and $g_{Ag}$.

*Partial ans.*: $g_{Ag}$ is very closely expressed by $g_{Ag} = x^2_{Au}(-8.444 - 6.312 x_{Ag})$.

10.10 The activities in the acetone-carbon disulfide system from the vapor pressure measurements at 35.17°C are as follows:

| $x_2$: | 0.0624 | 0.1330 | 0.2761 | 0.453 | 0.6161 | 0.828 | 0.935 | 0.9692 |
|---|---|---|---|---|---|---|---|---|
| $a_2$: | 0.216 | 0.403 | 0.631 | 0.770 | 0.835 | 0.908 | 0.960 | 0.970 |
| $a_1$: | 0.963 | 0.896 | 0.800 | 0.720 | 0.656 | 0.524 | 0.318 | 0.1803 |

Subscript 1 is for acetone and 2, for carbon disulfide. Calculate and tabulate the values of $\overline{G}_i - G_i^\circ$, $g_i$ and $\Delta_{mix}G = x_1(\overline{G}_1 - G_1^\circ) + x_2(\overline{G}_2 - G_2^\circ)$.

10.11 An empirical equation for the excess Gibbs energy of liquid Ag-Cu alloys is $G^E/R = -(2{,}100 + 0.511\,T)x_1 x_2$. ($x_1$ is for Ag, $x_2$ is for Cu). Derive the equations for $H^E$, $S^E$, $g_1$ and $g_2$, and compute $\gamma_1$ and $\gamma_2$ at 1550 K and $x_2 = 0.5$.

*Partial ans.*: $H^E/R = -2{,}100\,x_1 x_2$, $\gamma_1 = \gamma_2 = 0.627$

10.12 $G^E = -4{,}098 + 0.793\,T$ J mol$^{-1}$ for liquid Fe-Ni alloys at the equiatomic composition.* Assuming that $G^E$ varies symmetrically with composition as in Prob. 10.11, calculate $H^E$, $S^E$, $g_i$, $h_i$, and $s_i$ at 1800 K and $x_1 = 0.50$.

*Partial ans.*: $H^E = -4{,}098$.

10.13 The partial excess enthalpy $h_i$, (which is also equal to the partial enthalpy), in liquid Mn-Ni alloys is $h_1 = -8{,}000\,x_2^2 - 48{,}400\,x_2^3$ for Mn, and $h_2 = -80{,}600\,x_1^2 + 48{,}400\,x_1^3$ for Ni in J mol$^{-1}$. ($i = 1 = $ Mn, and $2 = $ Ni).† Express $H^E$ as a function of composition and calculate $H^E$ at the equiatomic composition.

*Partial ans.*: $H^E = -11{,}075$ at $x_2 = 0.5$.

10.14 The values of $G^E$ at 1743 for the Mn-Ni system in the preceding problem is expressed by $G^E \approx -43{,}200 x_1 x_2$ J mol$^{-1}$ (constant $T$). Use $H^E$ and $h_i$ in Problem (10.13) and express $S^E$, $g_i$ and $s_i$ as functions of $x_1$ and $x_2$, and compute $s_i$ at $x_2 = 0.5$. [Retain only the term with 59.757 in (10.54) to obtain (10.53) with one term and thus observe the derivation of $g_i$ from $G^E = A_{12} x_1 x_2$].

*Partial ans.*: $s_1 = +1.6$, and $s_2 = -1.9$ J mol$^{-1}$K$^{-1}$.

---

*L. J. Swartzendruber, V. P. Itkin, and C. B. Alcock, J. Phase. Equil., *12*, 288 (1991).
†N. A. Gokcen, ibid., p. 313.

CHAPTER XI

# PARTIAL (MOLAR) PROPERTIES

## Introduction*

The partial molar properties, more briefly and properly, *the partial properties*, are obtained from the corresponding extensive properties expressed as functions of $P$, $T$, and the moles, $n_1, n_2, \ldots$ (cf. Chapter II):

$$\mathcal{G} = \mathcal{G}(P, T, n_1, n_2, \ldots) \tag{11.1}$$

Note that $\mathcal{G}$ is any thermodynamic property in Chapter II, and the Gibbs energy, in the succeeding chapters. The extensive property $\mathcal{G}$ can also be expressed as a function of other sets of independent variables but the set of variables in (11.1) is very convenient, and therefore, generally preferred to other sets. The function $\mathcal{G}$ is a homogeneous function of the first degree in $n_1, n_2, \ldots$, and if $\mathcal{G}$ is divided by any one of the moles, $n_i$, or the total moles, $n$, it becomes a function zeroth degree in the moles; thus,

$$G = \frac{\mathcal{G}}{\sum_{i}^{c} n_i} = G(P, T, x_1, x_2, \ldots, x_{c-1}) \tag{11.2}$$

where $G$ is an intensive property which is called a *molar property* (we retain "molar" for G). The term molal is also used in some texts in preference to molar. The term "molar" in this chapter should not be confused with the "molar" referring to the concentration of a solute in one liter of solution. The number of independent composition variables is $c - 1$ because $\sum x_i = 1$, from which one composition variable can be eliminated.

It was seen in the preceding chapter that for an ideal solution, partial properties are expressed by

$$\overline{V}_i = V_i^\circ; \quad \overline{H}_i = H_i^\circ; \quad \overline{S}_i = S_i^\circ - R \ln x_i; \quad \overline{G}_i = G_i^\circ + RT \ln x_i \tag{11.3}$$

where the superscript (°) refers to the standard state, and $G_i^\circ = G_i^\circ(T)$ is a function of $T$. The extensive properties such as the volume, enthalpy, entropy, and Gibbs

---

*This chapter is for advanced readers. Its omission beyond (11.23) will not affect understanding the remaining chapters.

energy of an ideal solution are given by

$$\left.\begin{array}{l}\mathcal{V} = n_1\overline{V}_1 + n_2\overline{V}_2 + \cdots = n_1 V_1^\circ + n_2 V_2^\circ + \cdots \\ \mathcal{H} = n_1\overline{H}_1 + n_2\overline{H}_2 + \cdots = n_1 H_1^\circ + n_2 H_2^\circ + \cdots \\ \mathcal{S} = n_1 S_1^\circ - n_1 R \ln x_1 + n_2 S_2^\circ - n_2 R \ln x_2 + \cdots \\ \mathcal{G} = n_1 \overline{G}_1 + n_2 \overline{G}_2 + \cdots \\ \phantom{\mathcal{G}} = n_1 G_1^\circ + n_1 RT \ln x_1 + n_2 G_2^\circ + n_2 RT \ln x_2 + \cdots \end{array}\right\} \quad (11.4)$$

The changes in volume, enthalpy, entropy and Gibbs energy of an ideal solution upon mixing can be obtained from (11.4); thus,

$$\Delta_{\text{mix}}\mathcal{V} = \mathcal{V} - (n_1 V_1^\circ + n_2 V_2^\circ + \cdots) = 0; \qquad \Delta_{\text{mix}}\mathcal{H} = 0$$

$$\Delta_{\text{mix}}\mathcal{S} = \mathcal{S} - (n_1 S_1^\circ + n_2 S_2^\circ + \cdots) = -R(n_1 \ln x_1 + n_2 \ln x_2 + \cdots) \quad (11.5)$$

$$\Delta_{\text{mix}}\mathcal{G} = + RT(n_1 \ln x_1 + n_2 \ln x_2 + \cdots)$$

In real solutions there are deviations from these relationships, and the corresponding functions require experimental measurements of $\mathcal{V}$, $\Delta_{\text{mix}}\mathcal{H}$ and the activities. This chapter deals with the principles of determination and correlation of partial properties from such measurements.

## Partial (Molar) Properties of Binary Systems

It was shown in Chapter II that the partial property* $\overline{G}_1$ of component 1 in a binary system is related to $\mathcal{G}$ by

$$\overline{G}_1 = \left(\frac{\partial \mathcal{G}}{\partial n_1}\right)_{P,T,n_2} \quad (11.6)$$

and $\overline{G}_2$ can be obtained similarly or by interchanging the subscripts 1 and 2. In general it is convenient to treat a measured thermodynamic property in terms of $G$ in (11.2), which refers to one mole of a phase, instead of the extensive property $\mathcal{G}$. The partial properties may then be evaluated by substituting $\mathcal{G} = (n_1 + n_2)G$ from (11.2) into (11.6) and differentiating with respect to $n_1$ at constant $P$, $T$, and $n_2$; the result is

$$\overline{G}_1 = G + (n_1 + n_2)\left(\frac{\partial G}{\partial n_1}\right)_{P,T,n_2} \quad (11.7)$$

Differentiation of $x_1 = n_1/(n_1 + n_2)$ at constant $n_2$ gives

$$dx_1 = \frac{n_2\, dn_1}{(n_1 + n_2)^2} = \frac{x_2\, dn_1}{n_1 + n_2}$$

---

*Again, the term "molar" for $\overline{G}_1$ is redundant, because, in (11.6), $\partial \mathcal{G}$ is divided by the molar term $\partial n_1$.

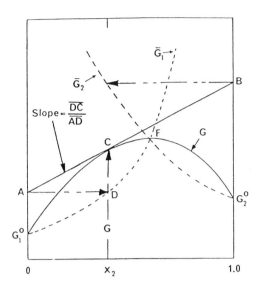

**Fig. 11.1** Method of intercepts. Solid curve is for molar property $G = G(x_2)$, and dotted curves are for $\overline{G}_1$ and $\overline{G}_2$. $G$ and $\overline{G}_1$ may be Gibbs energy, enthalpy, entropy, heat capacity, volume, or other thermodynamic properties. $G$, $\overline{G}_1$ and $\overline{G}_2$ coincide where $\partial G/\partial x_2 = 0$.

Substitution of $x_2/dx_1$ for $(n_1 + n_2)/dn_1$, from this equation, into (11.7) gives

$$\overline{G}_1 = G + x_2 \left( \frac{\partial G}{\partial x_1} \right)_{P,T,n_2}$$

Since $G$ is a function of one composition variable, e.g., $x_1$, it is independent of $n_2$; therefore, the restriction required by the constancy of $n_2$ is unnecessary; thus

$$\boxed{\overline{G}_1 = G + (1 - x_1)\left( \frac{\partial G}{\partial x_1} \right)_{P,T}} \tag{11.8}$$

where $1 - x_1$ has been substituted for $x_2$ for symmetry in subscripts. Likewise for component 2,

$$\boxed{\overline{G}_2 = G + (1 - x_2)\left( \frac{\partial G}{\partial x_2} \right)_{P,T}}; \quad \boxed{\Delta\overline{G}_2 = \Delta G + (1 - x_2)\left( \frac{\partial \Delta G}{\partial x_2} \right)} \tag{11.9}$$

which can be obtained by replacing the subscripts 1 in (11.8) by 2. Equations (11.7)–(11.9) are also valid for $\Delta G$, yielding $\Delta\overline{G}_i$ as shown by the second equation in (11.9). If $G$ is expressed graphically, as shown in Fig. 11.1, it is possible to evaluate $\overline{G}_1$ and $\overline{G}_2$ roughly by the method of intercepts as follows. Let $AB$

represent a tangent at any point $C$ to the curve representing the molar property $G$ of a phase. For the line $AB$

$$\text{Slope} = \frac{\partial G}{\partial x_2} = \frac{\overline{DC}}{AD} = \frac{\overline{DC}}{x_2}$$

from which $x_2(\partial G/\partial x_2) = \overline{DC}$ or $(1-x_1)(\partial G/\partial x_1) = -\overline{DC}$ since $dx_2 = -dx_1$. Substitution of this relation into (11.8) gives

$$\overline{G}_1 = G - DC = \overline{x_2 C} - \overline{DC} = \overline{x_2 D} = \overline{0A}$$

Consequently *the intercept on the vertical axis at $x_1 = 1$ of the tangent to the curve for $G$ at a given value of $x_2$ represents the value of the partial property $\overline{G}_1$ of component* 1. The intercept $A$ transposed on the corresponding composition, $x_2$, is $D$, the locus of such points is the broken line $G_1^\circ D F$ representing the partial property $\overline{G}_1$. Likewise, the intercept $B$ is the value of $\overline{G}_2$ at $x_2$ and the line $G_2^\circ F \overline{G}_2$ represents $\overline{G}_2$ as a function of composition. At the maximum point, $F$, on the curve for $G$, the slope is zero, hence $G$, $\overline{G}_1$ and $\overline{G}_2$ are equal as required by (11.8) and (11.9), i.e., all three curves intersect one another at $F$. Graphical evaluation of $\overline{G}_i$ from $G$ is undesirable because it is possible to fit the data for $G$ with an appropriate power series in $x_1$ or $x_2$, and substitute the resulting equation for $G$ into (11.8) and (11.9) to obtain $\overline{G}_1$ and $\overline{G}_2$. It is sufficient here to consider the excess Gibbs energy defined by $G^E = G - G(\text{ideal})$ to illustrate this procedure because all other excess properties may be fitted in an identical manner.

## Excess Gibbs Energy; Binary Systems

The excess partial Gibbs energy was defined by (10.47), which is

$$\boxed{\overline{G}_i^E \equiv \overline{G}_i - \overline{G}_i(\text{ideal}) \equiv \left[\overline{G}_i - G_i^\circ\right] - \left[\overline{G}_i(\text{ideal}) - G_i^\circ\right]} \qquad (11.10)$$

The use of $\Delta$ before $\overline{G}_i^E$ is redundant because the superscript "$E$" signifies $\Delta$. The properties $\overline{G}_i$ and $\overline{G}_i(\text{ideal})$ are not measurable but $\overline{G}_i - G_i^\circ$, called the relative partial Gibbs energy of $i$, is measurable. Note that $\Delta_{\text{mix}} \overline{G}_i = \overline{G}_i - G_i^\circ$ refers to the process "Pure $i \to i$(in solution)," and, hereafter, we use $\Delta$ for $\Delta_{\text{mix}}$ for brevity in notation in this chapter. The terms in brackets in (11.10) are given by

$$\boxed{\Delta \overline{G}_i = \overline{G}_i - G_i^\circ = RT \ln a_i} \qquad (\Delta = \Delta_{\text{mix}}) \qquad (11.11)$$

$$\boxed{\Delta \overline{G}_i(\text{ideal}) = \overline{G}_i(\text{ideal}) - G_i^\circ = RT \ln x_i} \qquad (11.12)$$

Substitution of these equations in (11.10) gives

$$\boxed{\overline{G}_i^E = \Delta \overline{G}_i - \Delta \overline{G}_i(\text{ideal}) = RT \ln a_i - RT \ln x_i} \qquad (11.13)$$

Since $a_i/x_i = \gamma_i$, (11.13) assumes the following useful form:

$$\boxed{\overline{G}_i^E = RT \ln \gamma_i} \qquad (11.14)$$

For a binary system, $i$ is either 1 or 2. Combination of (11.14) with the definitional relationship

$$\boxed{G^E = x_1 \overline{G}_1^E + x_2 \overline{G}_2^E} \qquad (11.15)$$

yields the following important equation for a binary system:

$$\boxed{G^E = RT(x_1 \ln \gamma_1 + x_2 \ln \gamma_2)} \qquad (11.16)$$

We also use (10.43) and $\Delta G(\text{ideal}) = RT(x_1 \ln x_1 + x_2 \ln x_2)$ to derive another form of (11.16):

$$G^E = \Delta G - \Delta G(\text{ideal}); \quad (\Delta = \Delta_{\text{mix}}) \qquad (11.16a)$$

Note that (11.13) and (11.16a) have a parallel forms. For simplicity in notation, we replace $\overline{G}_i^E$ with $g_i$, and write

$$\boxed{g_i \equiv \overline{G}_i^E = RT \ln \gamma_i, \quad \text{and} \quad G^E = x_1 g_1 + x_2 g_2} \qquad (11.17)$$

Equations (11.10) to (11.16a) are applicable to any other property. Thus, the ideal enthalpy of mixing is zero, i.e.,

$$\Delta H(\text{ideal}) = H - x_1 H_1^\circ - x_2 H_2^\circ = 0 \qquad (11.18)$$

Hence, $\overline{H}_i^E$ and $H_i^E$ are given by

$$\overline{H}_i^E \equiv h_i = \overline{H}_i - H_i^\circ \quad \text{and} \quad H^E = H - (x_1 H_1^\circ + x_2 H_2^\circ) = \Delta H \qquad (11.19)$$

and likewise, $\overline{V}_i^E(\text{ideal})$ and $\Delta V(\text{ideal})$ are zero, and

$$\overline{V}_i^E = \overline{V}_i - V_i^\circ \quad \text{and} \quad V^E = V - (x_1 V_1^\circ + x_2 V_2^\circ) \qquad (11.20)$$

## Representation of $G^E$

*Binary Systems* Numerous power series have been suggested for representation of $G^E$ as functions of composition. All $g_i$ for a given alloy must obey the Gibbs-Duhem relation, $\sum x_i \, dg_i = 0$, and $G^E$ must be zero for any $x_i = 1$ from (11.16) and should contain only the mathematically independent terms so that they can be evaluated by using a set of experimental data. For a binary system the total number of independent terms for a given maximum power, $e$, is $e - 1$. Out of the many equations developed in the past, we propose the following Margules-type[*]

---

[*] E. Hala, J. Pick, V. Fried, and O. Vilim, "Vapor-Liquid Equilibrium, Second Edition," translated by G. Standart, Pergamon Press (1967); see also J. Tomiska, CALHAD, Vol. 4, p. 63 (1980); N. A. Gokcen and Z. Moser, J. Phase Equilibria, Vol. 14, p. 288 (1993).

equation (1895):

$$G^E = x_1 x_2 (A_{21} x_1 + A_{12} x_2) + x_1^2 x_2^2 (B_{21} x_1 + B_{12} x_2)$$
$$+ x_1^3 x_2^3 (C_{21} x_1 + C_{12} x_2) + \cdots \tag{11.21}$$

Substitution of this equation in (11.8) yields

$$g_1 = x_2^2 \big[ A_{12} + 2x_1(A_{21} - A_{12} + B_{12}) + 3x_1^2(B_{21} - 2B_{12} + C_{12})$$
$$+ 4x_1^3(B_{12} - B_{21} + C_{21} - 3C_{12}) + 5x_1^4(3C_{12} - 2C_{21})$$
$$+ 6x_1^5(C_{21} - C_{12}) + \cdots \big] \tag{11.22}$$

Setting $A_{12} = A_{21}$, $B_{12} = B_{21}$, or $C_{12} = C_{21}$ would reduce the order of each term in (11.21) by one since $x_1 + x_2 = 1$; e.g., $(A_{12} x_1 + A_{21} x_2)$ is first order but for $A_{12} = A_{21}$, it is zeroth order, and equal to $A_{12}$. Interchanging the subscript 1 and 2 does not change (11.21), and the same operation in (11.22) yields the equation for $g_2$:

$$g_2 = x_1^2 \big[ A_{21} + 2x_2(A_{12} - A_{21} + B_{21}) + 3x_2^2(B_{12} - 2B_{21} + C_{21})$$
$$+ 4x_2^3(B_{21} - B_{12} + C_{12} - 3C_{21}) + 5x_2^4(3C_{21} - 2C_{12})$$
$$+ 6x_2^5(C_{12} - C_{21}) + \cdots \big] \tag{11.23}$$

The values of $g_1 = RT \ln \gamma_1$ can be determined by various methods, e.g., by vapor pressure measurement, or by galvanic cells to be discussed in Chapter XIV. Then the values of $g_1$ can be substituted in (11.22) to determine its parameters, preferably by the least squares method. The resulting values of the parameters can be used to obtain $G^E$ and $g_2$ of (11.21) and (11.23).

Prior to extensive use of computers, $g_2$ was obtained from $g_1$ by graphical integration of the Gibbs-Duhem relation, i.e.,

$$g_2 = -\int_{x_2=1}^{x_2} \left( \frac{x_1}{x_2} \right) dg_1$$

Difficulties arose as $x_2$ approached zero and $x_1/x_2$ tended to infinity, and other difficulties and uncertainties were encountered when $x_1/x_2$ was plotted versus $g_1$ to obtain graphical integration of this equation. It is therefore evident, without further discussion, that the graphical integration method should be avoided. Simple personal computer programs are available for solving $n$ parameters in (11.22) by using a set of data for $g_1$.

**Ternary Systems** For a ternary system, the following basic <u>fourth order</u> Margules equation can be used with conveniently designated coefficients:

$$G^E = x_1 x_2 (A_{21} x_1 + A_{12} x_2) + B_{12} x_1^2 x_2^2 + x_1 x_3 (A_{31} x_1 + A_{13} x_3)$$
$$+ B_{13} x_1^2 x_3^2 + x_2 x_3 (A_{32} x_2 + A_{23} x_3) + B_{23} x_2^2 x_3^2 \tag{11.24}$$
$$+ x_1 x_2 x_3 (A_{21} + A_{13} + A_{32} + J_1 x_1 + J_2 x_2 + J_3 x_3)$$

The first set of three terms in this equation is for the binary component systems, obtained from (11.21), and these terms contain 9 adjustable parameters. To make these terms fourth order, it is necessary to set $B_{ij} = B_{ji}$ and $C_{ij} = C_{ji} = 0$ in (11.21). For a fifth order equation, $B_{ij}$ is not equal to $B_{ji}$ but $C_{ij} = C_{ji} = 0$, and the additional independent ternary terms, i.e. the additional terms inside the last set of parentheses are $E_{12}x_1x_2$, $E_{13}x_1x_3$, and $E_{23}x_2x_3$. Observe that for a third order equation, the ternary term is obtained by setting $J_1 = J_2 = J_3 = J$, and since $x_1 + x_2 + x_3 = 1$, there would be only one independent adjustable ternary parameter, $J$, and for $J = 0$, there is no independent adjustable ternary parameter. Thus, if we let the order of (11.24) be $\varepsilon$, we obtain the number of independent parameters in the ternary Margules equation as follows:

$$\begin{array}{lcccccl}
\varepsilon: & 1 & 2 & 3 & 4 & 5 \cdots & \varepsilon \\
\text{Binary terms:} & 0 & 3 & 6 & 9 & 12 \cdots & 3(\varepsilon - 1) \\
\text{Ternary terms:} & 0 & 0 & 1 & 3 & 6 \cdots & (\varepsilon - 1)(\varepsilon - 2)/2
\end{array} \qquad (11.25)$$

The derivation of $g_i$ from $G^E$ requires a general equation applicable to multicomponent systems. For this purpose, it is sufficient to write $G^E = G^E(n_1 + n_2 + n_3 + n_4)$ for a quaternary system first, and then differentiate $G^E$ with respect to $n_1$ to obtain

$$g_1 = \left(\frac{\partial G^E}{\partial n_1}\right)_{P,T,n_1'} = G^E + (n_1 + n_2 + n_3 + n_4)\left(\frac{\partial G^E}{\partial n_1}\right)_{P,T,n_1'} \qquad (11.26)$$

where the subscript $n_1'$ means that all $n$'s except $n_1$ are held constant. The partial differential of $x_1$ is

$$(dx_1)_{n_1'} = d\left(\frac{n_1}{n_1 + n_2 + n_3 + n_4}\right) = \frac{(n_2 + n_3 + n_4)(dn_1)_{n_1'}}{(n_1 + n_2 + n_3 + n_4)^2}$$

$$= \frac{(1 - x_1)(dn_1)_{n_1'}}{n_1 + n_2 + n_3 + n_4}$$

The variable $x_1$ is an intensive property that may be written as $x_1 = (n_1/n_3)/[(n_1/n_3) + (n_2/n_3) + 1 + (n_4/n_3)]$ where the restriction of $n_1'$, or the constancy of $n_2$, $n_3$, and $n_4$, is simply equivalent to the constancy of ratios $r = x_2/x_3$ and $s = x_3/x_4$ because holding $n_3$ as a constant does not affect either $x_1$ or $n_1/n_3$ in view of the fact that all the possible values of $n_1/n_3$ can be obtained by varying $n_1$ irrespective of $n_3$. The substitution of $(1 - x_1)/dx_1$ from the preceding equation into (11.26) gives

$$\boxed{g_1 = G^E + (1 - x_1)\left(\frac{\partial G^E}{\partial x_1}\right)_{P,T,x_2/x_3,x_3/x_4}} \qquad (11.27)$$

It is immediately clear that the additional restrictions in (11.27) for a multicomponent system are $x_4/x_5$, $x_5/x_6, \ldots$, corresponding to the additional components.

186   Thermodynamics

The analytical equation for $\overline{G}_1^E$ for a *ternary* system may now be obtained by substituting (11.24) into (11.27) and carrying out the differentiation with respect to $x_1$. Care must be exercised in this procedure because complications arise in obtaining the correct derivations unless the sum of the mole fractions is written as $x_1 + x_2 + x_3 = x_1 + x_2(r+1)/r = 1$, and $x_1 + (r+1)x_3 = 1$, from $x_2/x_3 = r$; then from $dx_1 + [(r+1)/r]\,dx_2 = 0$, and from $dx_1 + (r+1)\,dx_3 = 0$, the derivatives are correctly written as

$$(\partial x_2/\partial x_1)_r = -x_2/(x_2 + x_3), \quad \text{and} \quad (\partial x_3/\partial x_1)_r = -x_3/(x_2 + x_3)$$

These equations may be rewritten in the following convenient forms:

$$(1 - x_1)(\partial x_2/\partial x_1)_r = -x_2, \quad \text{and} \quad (1 - x_1)(\partial x_3/\partial x_1)_r = -x_3 \qquad (11.28)$$

The second one of these equations can be obtained from the first by replacing the subscript 2 with 3. It is now possible to return to the derivation of $g_1$ from (11.24) and (11.27). The result is

$$\begin{aligned}
g_1 =\ & x_2^2[A_{12} + 2x_1(A_{21} - A_{12} + B_{12}) - 3B_{12}x_1^2] \\
& + x_3^2[A_{13} + 2x_1(A_{31} - A_{13} + B_{13}) - 3B_{13}x_1^2] \\
& + x_2x_3[A_{21} + A_{13} - A_{32} + 2x_1(A_{31} - A_{13}) + 2x_3(A_{32} - A_{23}) \\
& - 3B_{23}x_2x_3 + J_1x_1(2 - 3x_1) + J_2x_2(1 - 3x_1) + J_3x_3(1 - 3x_1)]
\end{aligned}$$

(11.29)

The results for $g_2$ and $g_3$ can be obtained by rotating the subscripts as $1 \to 2 \to 3 \to 1$, because this rotation leaves (11.24) unchanged. Thus, we obtain the following equations:

$$\begin{aligned}
g_2 =\ & x_3^2[A_{23} + 2x_2(A_{32} - A_{23} + B_{23}) - 3B_{23}x_2^2] \\
& + x_1^2[A_{21} + 2x_2(A_{12} - A_{21} + B_{12}) - 3B_{12}x_2^2] \\
& + x_3x_1[A_{32} + A_{21} - A_{13} + 2x_2(A_{12} - A_{21}) \\
& + 2x_1(A_{13} - A_{31}) - 3B_{13}x_3x_1 + J_2x_2(2 - 3x_2) \\
& + J_3x_3(1 - 3x_2) + J_1x_1(1 - 3x_2)]; \quad (B_{ij} = B_{ji})
\end{aligned}$$

(11.30)

$$\begin{aligned}
g_3 =\ & x_1^2[A_{31} + 2x_3(A_{13} - A_{31} + B_{13}) - 3B_{13}x_3^2] \\
& + x_2^2[A_{32} + 2x_3(A_{23} - A_{32} + B_{23}) - 3B_{23}x_3^2] \\
& + x_1x_2[A_{13} + A_{32} - A_{21} + 2x_3(A_{23} - A_{32}) \\
& + 2x_2(A_{21} - A_{12}) - 3B_{12}x_1x_2 + J_3x_3(2 - 3x_3) \\
& + J_1x_1(1 - 3x_3) + J_2x_2(1 - 3x_3)]; \quad (B_{ij} = B_{ji})
\end{aligned}$$

(11.31)

It may be desirable to check the consistency of (11.24) with (11.29)–(11.31), or each term in these equations. This requires an additional thermodynamic relationship. For this purpose, we differentiate $G^E = x_1g_1 + x_2g_2 + x_3g_3$, and use the Gibbs-Duhem relation $x_1 \, dg_1 + x_2 \, dg_2 + x_3 \, dg_3 = 0$ to get $dG^E = g_1 \, dx_1 + g_2 \, dx_2 + g_3 \, dx_3$. At a fixed value of $x_3$, $dx_3 = 0$, and $dx_1 = -dx_2$, and then the equation for $dG^E$ becomes $dG^E = -g_1 \, dx_2 + g_2 \, dx_2$; this equation yields

$$g_2 = g_1 + \left(\frac{\partial G^E}{\partial x_2}\right)_{x_3} \tag{11.32}$$

The rotation of subscripts yields additional similar equations. For example, we can use the term $x_1^2 x_2 x_3 J_1$ in (11.24) to show that $x_1 x_2 x_3 J_1 (2 - 3x_1)$ in (11.29) plus $(\partial x_1^2 x_2 x_3 J_1 / \partial x_2)_{x_3}$ give $J_1 x_1^2 x_3 (1 - 3x_2)$ in (11.30).

Consider the usual type of measurements for a ternary system in which one of the components, labelled as component 1, lends itself to convenient and accurate determination of $g_1$, e.g., the vapor pressure of $P_1$ over the system is accurately measurable from which $g_1 = RT \ln P_1 / (x_1 P_1^*)$ where $P_1^*$ refers to the pure component. With adequate data, well spaced in composition, all the coefficients in (11.29) can be determined on a computer. These coefficients can then be used to calculate $g_2$, $g_3$, and $G^E$, all from the measurement of $g_1$ for one component.

The foregoing equations, or their higher order forms, should be used to correlate and obtain $g_i = RT \ln \gamma_i (i = 1, 2, 3)$ and $G^E$ for ternary systems. The quaternary systems require the binary and ternary terms, and the quaternary terms containing $x_1 x_2 x_3 x_4(\ldots)$.

Graphical integration methods for the Gibbs-Duhem relation were used to obtain $g_i$ ($i \neq 1$) from the measured values of $g_1$ prior to the advent of computers. Such methods are obsolete and highly inaccurate, particularly for ternary and multicomponent systems.

## Alternative Equations

Most of the analytical equations for $G^E$ are the infinite series satisfying the Gibbs-Duhem relation, (2.27), and the usual boundary condition that $G^E = 0$ for $x_i = 1$ when each standard state is the pure component in the same physical state as the solution. The most important of these equations are presented here for completeness.

The Bale-Pelton equation* is limited to the binary systems, but represents an interesting aspect in thermodynamics of solutions. In (11.22), the coefficients vary somewhat with the numbers of terms selected to represent a given set of data, i.e., the coefficients (parameters) in (11.22) are correlated. To overcome this problem, Bale and Pelton have used the Legendre Polynomials, $P_n(x_2)$. Their equations for $g_1$ of a binary system 1-2 and $P_n(x_2)$ are as follows:

$$g_1 = x_2^2 \sum_{n=0}^{n'} A_n \cdot P_n(x_2) \tag{11.33}$$

$$P_0(x_2) = 1; \quad P_n(x_2) = (2n - 1)(2x_2 - 1) P_{n-1}(x_2) / n \tag{11.34}$$

---

*C. W. Bale and A. D. Pelton, Met. Trans., Vol. 5, p. 2323 (1974); Canadian Metallurgical Quarterly, Vol. 14, p. 213 (1975); A. D. Pelton and C. W. Bale, Met. Trans., 17A, 1057 (1986).

From the generating function for $P_n(x_2)$, and for $n'$ up to 3, $g_1$ is given by

$$g_1 = x_2^2\big[A_0 \cdot (1) + A_1 \cdot (2x_2 - 1) + A_2 \cdot \big(6x_2^2 - 6x_2 + 1\big) \\ + A_3 \cdot \big(20x_2^3 - 30x_2^2 + 12x_2 - 1\big)\big] \quad (11.35)$$

where the four terms in four sets of parentheses are the zeroth to the third order Legendre polynomials. The reason that the values of various $A_n$ calculated from experimental data are not correlated is that the polynomials are orthogonal, i.e., for the integral from $x_2 = 0$ to $x_2 = 1$, the following integral is zero:

$$\int_{x_2=0}^{x_2=1} P_n(x_2) \cdot P_m(x_2)\, dx_2 = 0; \qquad (m \neq n) \quad (11.36)$$

For example, $P_0(x_2) \cdot P_1(x_2) = 2x_2 - 1$, which upon integration yields $[x_2^2 - x_2] = 0$, as required by (11.36). The coefficients $A_n$ for a given set of data are independent of the number of terms selected for (11.33), provided that there are an adequate number of well-spaced data points. However, for a set of data of usually encountered accuracy, the value of each $A_n$ depends slightly on the number of terms selected for (11.33). For ternary and multicomponent systems, Legendre polynomials have not been developed, and it becomes practical to resort to (11.24) or (11.29) for their constituent binary systems.

Equations by van Laar, Wohl, and Scatchard and Hamer* express $G^E$ as functions of variables called the effective volume fractions often defined by $z_i = x_i V_i^\circ /(x_1 V_1^\circ + x_2 V_2^\circ + x_3 V_3^\circ)$ with $i = 1, 2,$ and 3. The functional forms of these equations are similar to (11.24); e.g., the equation by Wohl for $\varepsilon = 3$ is

$$G^E = \big[x_1 + \big(x_2 V_2^\circ / V_1^\circ\big) + \big(x_3 V_3^\circ / V_1^\circ\big)\big]\{z_1 z_2\big[\big(A_{21} z_1 V_1^\circ / V_2^\circ\big) + A_{12} z_2\big] \\ + z_1 z_3\big[\big(A_{31} z_1 V_1^\circ / V_3^\circ\big) + A_{13} z_3\big] + z_2 z_3\big[\big(A_{32} z_2 V_1^\circ / V_3^\circ\big) \\ + \big(A_{23} z_3 V_1^\circ / V_2^\circ\big)\big] + z_1 z_2 z_3\big[\big(A_{21} V_1^\circ / V_2^\circ\big) + A_{13} \\ + \big(A_{32} V_1^\circ / V_3^\circ\big) - C\big]\}; \qquad z_i = \frac{x_i V_i^\circ}{\sum x_i V_i^\circ} \quad (11.37)$$

The ratio $V_i^\circ / V_j^\circ$ is also replaced by empirical effective volume ratio $q_i/q_j$. Such equations do not offer a great tangible practical advantage although from a theoretical point of view the molecular volume fraction in solution is assumed to be more significant than the mole fraction. For equal molar volumes $x_i = z_i$ and all such equations become basically the same as (11.24). The van Laar equation for $\varepsilon = 2$, represents the contribution of the pairwise interactions of the molecules to $G^E$ so that

$$G^E = \frac{A_{21} x_1 x_2 + A_{31} x_1 x_3 + A_{23}\big(V_3^\circ / V_1^\circ\big) x_2 x_3}{x_1 + x_2\big(V_2^\circ / V_1^\circ\big) + x_3\big(V_3^\circ / V_1^\circ\big)} \quad (11.38)$$

*See the footnote on the next page.

While the original equation of van Laar contains $b_i/b_j$ of van der Waals constant, it is more appropriate to use the molar ratios (for details, see Hala, et al.*) The equation derived by Scatchard and Hamer* for a binary system is

$$G^E = \left(x_1 + x_2 V_2^\circ/V_1^\circ\right) z_1 z_2 \left(A_{21} z_1 V_1^\circ/V_2^\circ + A_{12} z_2 - D_{12} z_1 z_2\right) \quad (11.39)$$

This equation becomes $G^E = x_1 x_2 (A_{21} x_1 + A_{12} x_2 - D_{12} x_1 x_2)$ when $V_1^\circ = V_2^\circ$.

Alternative equations presented in this section have no great advantage over (11.24), and in fact the use of only one equation, i.e, (11.24), would facilitate communication among the thermodynamicists of solutions. The selected equation should have an adequate number of terms for judicious representation of a set of data within the accuracy of measurements, and satisfy the Gibbs-Duhem relation for $g_i$. An equation such as (11.24) that satisfies the boundary conditions of $G^E = 0$ for each $x_i = 1$, automatically satisfies the Gibbs-Duhem relation when the standard state of unit activity for component $i$ is the pure component $i$.

## Regular Solutions

The concept of regular solutions has been the subject of numerous publications. It is therefore appropriate to discuss it in sufficient detail as a special subject. The quadratic form of (11.21) for a binary system assumes the following simple form

$$G^E = A_{12} x_1 x_2 \quad (11.40)$$

A treatment of the binary system 1-2 obeying (11.40) can therefore be extended to include the multicomponent systems limited to quadratic terms, i.e.,

$$G^E = A_{12} x_1 x_2 + A_{13} x_1 x_3 + A_{23} x_2 x_3 + \cdots \quad (11.41)$$

This equation, to be considered later, can be obtained from (11.24) by setting $B_{ij} = 0$, $J_i = 0$, and $A_{ij} = A_{ji}$. Equation (11.40) has been derived on a theoretical basis[†] by van Laar and Lorenz, Hertzfeld and Heitler, and Hildebrand, and a solution obeying (11.40) was named a regular solution by Hildebrand. The theoretical basis for (11.40) is very simple as will be seen later, and involves pairwise interactions of molecules having equal sizes. The excess partial Gibbs energies are obtained by substituting (11.40) into (11.8) and (11.9); thus,

$$g_1 = A_{12} x_2^2 \quad \text{and} \quad g_2 = A_{12} x_1^2; \quad [A_{12} = A_{21}] \quad (11.42)$$

---

*For a summary of these equations, and extensive references to original papers see E. Hala, J. Pick, V. Fried, and O. Vilim, "Vapor-Liquid Equilibrium," Second edn., transl. by G. Standart, Pergamon Press (1967) and S. Malanowski and A. Anderko "Modeling Phase Equilibria," John Wiley & Sons (1992). See also N. A. Gokcen, J. Phase Equil., 15, 147 (1994), and N. A. Gokcen and Z. Moser, ibid., 14, 288 (1993) for other references.

[†]For summaries see C. Wagner, "Thermodynamics of Alloys," Addison-Wesley Press (1952); I. Prigogine, A. Bellemans, and V. Mathot, "The Molecular Theory of Solutions," Interscience Publishers (1957); J. S. Rowlinson, "Liquid and Liquid Mixtures," Butterworths (1959); Gen. Ref. (15).

190    Thermodynamics

The corresponding equations for $\Delta G = G^E + \Delta G(\text{ideal})$ and $\Delta \overline{G}_i = g_i + \Delta \overline{G}_i(\text{ideal})$ are

$$\Delta G = A_{12}x_1x_2 + RT(x_1 \ln x_1 + x_2 \ln x_2) \tag{11.43}$$

$$\left.\begin{array}{l}\Delta \overline{G}_1 = RT \ln a_1 = A_{12}x_2^2 + RT \ln x_1 \\ \Delta \overline{G}_2 = RT \ln a_2 = A_{12}x_1^2 + RT \ln x_2\end{array}\right\} \tag{11.44}$$

All these equations are symmetric because interchanging the subscripts 1 and 2 to $x$ does not alter (11.40) and (11.43), and interchanging 1 and 2 (with $A_{12} = A_{21}$) converts one equation into another in (11.42) and in (11.44). The value of $A_{12}$ is a constant assumed to be independent of both temperature and composition at ordinary pressures.

Differentiation of $\Delta G/T$ from (11.43) with respect to $1/T$ gives

$$\frac{\partial(\Delta G/T)}{\partial(1/T)} = \Delta H = A_{12}x_1x_2 \tag{11.45}$$

and $\Delta G$ with respect to $T$ gives

$$\partial \Delta G/\partial T = -\Delta S = R(x_1 \ln x_1 + x_2 \ln x_2) \tag{11.46}$$

therefore, for a regular solution

$$G^E = H^E = \Delta H = A_{12}x_1x_2, \quad \text{and} \quad S^E = 0 \tag{11.47}$$

hence the distribution of molecules in a regular solution is random and deviation from ideality is entirely due to $\Delta H$. A single experimental value of $\Delta H$ or $G^E$ is sufficient to determine the numerical form of all the equations for a regular solution. Some liquid solutions such as Ar-$O_2$, $C_6H_6$-$CCl_4$, $C_6H_6$-Cyclohexane, Au-Cu closely follow the regular behavior.

The regular solution equations are valid at wider concentration ranges than the simpler equations for Raoult's and Henry's laws, but in the intermediate ranges the regular behavior is seldom obeyed; consequently additional terms are necessary in (11.40) to express $G^E$ as a satisfactory function of composition; cf. (11.21).

## Maximum, Minimum, and Critical Points in (11.43)

Equation (11.43) for $\Delta G$ of mixing and the activity of component 2, $a_2$, are shown as functions of $x_2$ in Fig. 11.2. The ordinate for (11.43) is the dimensionless quantity $\Delta G/RT$ in the upper figure so that

$$\Delta G/RT = \alpha x_1 x_2 + x_1 \ln x_1 + x_2 \ln x_2; \quad (\alpha = A_{12}/RT) \tag{11.43a}$$

where $\alpha = A_{12}/RT$, and $\alpha^*$ varies inversely with temperature; hence the solution tends to ideality with increasing temperature.

---

*To name $\alpha = A_{12}/RT$, or $g_1/x_2^2$ of (11.22) as the "alpha function" is redundant, and should be avoided. Also, naming the term with the first set of parentheses in (11.21), i.e., the cubic Margules equation, as the "sub-regular solution" is redundant and without a theoretical basis.

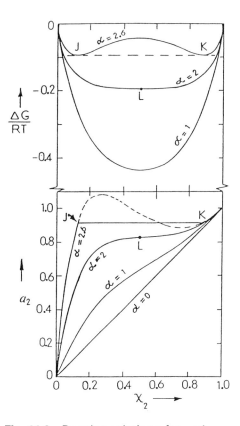

**Fig. 11.2** Regular solutions for various values of $\alpha$. Upper curves are for $\Delta G/RT$. Broken horizontal line is tangent at $J$ and $K$ to curve for $\Delta G/RT$, with $\alpha = 2.6$. $L$ is critical point. Lower curves are for activity $a_2$ of component 2; activity $a_1$ is symmetric to $a_2$ about vertical line through $x_2 = 0.5$.

The curves for (11.43) or (11.43a), and those for more complex equations may have maximum, minimum, and critical points. We differentiate (11.43a) after eliminating $x_1$ by using $x_1 = 1 - x_2$ in order to characterize these points as follows:

$$\frac{1}{RT}\left(\frac{\partial \Delta G}{\partial x_2}\right)_T = \alpha(1 - 2x_2) + \ln\left(\frac{x_2}{1-x_2}\right) \quad (11.48)$$

$$\frac{1}{RT}\left(\frac{\partial^2 \Delta G}{\partial x_2^2}\right)_T = -2\alpha + \frac{1}{x_2(1-x_2)}; \quad (A_{12} = \alpha RT) \quad (11.49)$$

(1) For any value of $\alpha$, (11.48) is zero for $x_2 = 0.5$, and since for $\alpha < 2$, (11.49) is positive, then, (11.43a) has a minimum point for each value of $\alpha < 2$ at $x_2 = 0.5$, as shown for $\alpha = 1$ in Fig. 11.2. For $\alpha = 0$, the solution is ideal,

and the corresponding curve is similar to that for $\alpha = 1$, but located lower. Negative deviations occur in $a_2$ when $\alpha$ is negative (not shown in Fig. 11.2).

(2) For $\alpha = 2$, $[A_{12} = 2RT]$, (11.48) and (11.49) both become zero at $x_2 = 0.5$ as shown in the figure; hence, this is the critical point $T_c$, given by $T_c = A_{12}/2R$. At $x_2 = 0.5$, $G^E = A_{12}x_1x_2 = A_{12}/4$, and $A_{12} = 4G^E$, and $T_c$ is also given by $T_c = 2G^E$(at $x_2 = 0.5)/R$.

(3) The curves for $\Delta G/RT$ exhibit one maximum at $x_2 = 0.5$, and two minima at two different values of $x_2$ for $\alpha > 2$. Thus, for $\alpha = 2.6$, the minimum points are at $J$ with $x_2 = 0.124$ and at $K$ with $x_2 = 0.876$ with $J$ and $K$ located symmetrically with respect to each other. In the range of composition between $J$ and $K$ the solution separates into two phases where the composition of each phase is constant but their relative proportions vary from $J$ to $K$. The Gibbs energy of mixing for a *single phase* between $J$ and $K$ is higher than that for two phases and therefore the single phase is unstable. For example at $x_2 = 0.5$, $\Delta G/RT = 2.6 \times 0.5^2 + \ln 0.5 = -0.0431$ for the single phase but for two phases, noting that each phase at $x_2 = 0.5$ is 0.5 mole, then $\Delta G/RT = [0.5\Delta G(\text{phase at }J) + 0.5\Delta G(\text{phase at }K)]/RT = -0.0924$. The tangent to the curve at $J$ and $K$ is horizontal and the intercepts of the tangent are given by $\Delta \overline{G}_1/RT = \Delta \overline{G}_2/RT = \Delta G/RT$, with $\Delta \overline{G}_1$, $\Delta \overline{G}_2$, and $\Delta G$ calculated either at $x_2 = 0.124$ or at $x_2 = 0.876$. The activity $a_2$ varies correspondingly as shown by the lower set of curves.

## Spinodal Points

A regular binary solution for which $\alpha > 2$, $(A_{12} > 2RT)$, there are two nonhorizontal inflection points at $x_2 \neq 0.5$ for which (11.49) is zero, but (11.48) is not. These inflection points are called the spinodal points, or the spinodes, and they are obtained by setting (11.49) to zero, from which we derive $2\alpha = 1/x_1x_2$, and

$$T(\text{spinodal}) = \frac{2A_{12}}{R}x_2(1-x_2) \tag{11.50}$$

where $x_2$ here is the spinodal composition. For example, when $\alpha = 2.6$, as shown in Fig. 11.2, the spinodal composition is given by $2 \times 2.6 = 1/[x_2(1-x_2)]$, which is satisfied by $x_2 = 0.2598$ and $x_2 = 0.7402$. Spinodal points are important in kinetic theories of nucleation and growth of new phases from a homogeneous single phase upon cooling.*

## Theoretical Derivation of (11.40)

There are four main assumptions used in the theoretical derivation of the regular solution equation: (1) the molecules of both components have equal volumes, (2) the number of nearest molecules surrounding any molecule is the same in the pure components and in the solution, (3) the solution is random so that $S^E$ is

---

*J. W. Cahn and J. E. Hilliard, J. Chem. Phys., *31*, 688 (1959); J. E. Hilliard, in "Phase Transformations," H. I. Aaronson, ed. ASM, p. 497 (1970); J. J. Hoyt, Acta Met. Mat., *38*, 227 (1990).

zero, and (4) $A_{12}$ or $\Delta H$ of mixing is independent of temperature. The statistical derivation of (11.40) with these assumptions is very simple. Consider that all the molecules in a solid or liquid solution are observed instantaneously so that each molecule occupies a fixed site or position. Each molecule has $\bar{z}$ neighbors and $\bar{z}$ is called the coordination number,* which is four for the tetrahedral crystals such as silicon and diamond, 8 for the body-centered cubic, and 12 for the close-packed hexagonal and the face-centered cubic crystals. The probability that a molecule of component 1 occupies a selected site is proportional to its mole fraction $x_1$, and the probability that another molecule of component 1 can become a neighbor in a given direction is $x_1 x_1$, or $x_1^2$. Likewise, for a pair of 1 and 2, the probability is $x_1 x_2$. If the same procedure is repeated by starting with a molecule of 2, then the probabilities become $x_2^2$ and $x_2 x_1$. The sum of all probabilities is unity as required in statistics; thus $x_1^2 + x_1 x_2 + x_2 x_1 + x_2^2 = (x_1 + x_2)^2 = 1$. There are $N_A \bar{z}/2$ bonds in one mole of solution, $N_A$ being the Avogadro number, and these bonds are distributed among the pairs of molecules in proportion to their probabilities in one direction. Therefore, there are $N_A \bar{z} x_1^2 / 2$ bonds of 1-1, $N_A \bar{z} x_2^2 / 2$ bonds of 2-2, and $N_A \bar{z} x_1 x_2$ bonds of 1-2. The reader may construct a one-dimensional crystal, e.g., two types of beads on a string, with $\bar{z} = 2$, to understand the types of bonds and to count their numbers when the beads are randomly strung. The energy associated with each bond $i$-$j$ is designated by $e_{ij}$, and the total energies corresponding to the foregoing types of bonds are $N_A \bar{z} x_1^2 e_{11}/2$, $N_A \bar{z} x_2^2 e_{22}/2$, and $N_A \bar{z} x_1 x_2 e_{12}$. The bonds and energies of $x_1$ moles of 1, and $x_2$ moles of 2 prior to mixing are $N_A \bar{z} x_1 e_{11}/2$, and $N_A \bar{z} x_2 e_{22}/2$. The enthalpy of mixing $\Delta H = H^E$ is the difference between these two sets, i.e.,

$$H^E = (N_A \bar{z}/2)\left(x_1^2 e_{11} + 2 x_1 x_2 e_{12} + x_2^2 e_{22} - x_1 e_{11} - x_2 e_{22}\right)$$
$$= (N_A \bar{z}/2)(2 e_{12} - e_{11} - e_{22}) x_1 x_2 = N_A W_{12} x_1 x_2 \qquad (11.51)$$

where $W_{12} = (\bar{z}/2)(2 e_{12} - e_{11} - e_{22})$ is called the molecular exchange energy. Since the molecules are assumed to be random as in an ideal solution, $S^E$ is zero, and from $H^E = G^E$, the foregoing equation becomes $G^E = N_A W_{12} x_1 x_2 = A_{12} x_1 x_2$, where $N_A W_{12} = A_{12}$ [cf. (11.40)].

The simple assumption that $S^E$ is zero can be removed by further statistical treatments which yield the first approximation to the regular solutions. A treatment by Gokcen and Chang,[†] based on systematic enumerations of configurations gives the following equation for the first approximation to the regular solutions:

$$G^E = RT x_1 \ln \frac{\beta + x_1 - x_2}{x_1(\beta + 1)} + RT x_2 \ln \frac{\beta + x_2 - x_1}{x_2(\beta + 1)} \qquad (11.52)$$

The parameter $\beta$ is given by

$$\beta^2 = 1 - 4 x_1 x_2 [1 - \exp(N_A W_{12}/RT)] \qquad (11.53)$$

---

*See for example, C. A. Wert and R. M. Thomson, "Physics of Solids," McGraw-Hill Book Co. (1970); P. Haasen, "Physical Metallurgy," translated by J. Mordike, Cambridge Univ. Press (1986).
[†]N. A. Gokcen and E. T. Chang, J. Chem. Phys., 55. 2279 (1971). See also Gen. Ref. (15).

**Table 11.1** Experimental and calculated values of $H^E$ from first approximation to regular solutions. $G^E$ and $H^E$ are in J (mol solution)$^{-1}$ at $x_1 = x_2 = 0.5$.

| Mixture | $T/K$ | $G^E$ | $H^E$(exper.) | $H^E$(calc.) |
|---|---|---|---|---|
| $N_2$-$O_2$ | 70 | 43 | 66 | 41 |
| $C_6H_6$-$CCl_4$ | 298.15 | 82 | 110 | 80 |
| $CCl_4$-$CH_3I$ | 298.15 | 205 | 280 | 196 |
| K-Na | 298.15 | 670 | 523 | 548 |
| Cd-Mg | 923 | −4644 | −5611 | −5464 |
| Cd-Zn | 800 | 2000 | 2092 | 1582 |
| Bi-Pb | 700 | −1255 | −1109 | −1364 |
| In-Pb | 673 | 586 | 962 | 552 |
| Sb-Zn | 823 | −3138 | −3473 | −3611 |

It is convenient to evaluate $N_A W_{12}$ by taking the experimental value of $G^E$ at $x_1 = x_2 = 0.5$ so that

$$G^E = RT \ln[2\beta/(\beta + 1)] \tag{11.54}$$

and at $x_1 = 0.5$, $\beta^2$ is identical with $\exp(N_A W_{12}/RT)$. The equation for $H^E$ is readily obtained from (11.53) by differentiation of $G^E/T$ with respect to $1/T$; the result is

$$H^E = 2N_A W_{12} x_1 x_2/(\beta + 1) \tag{11.51a}$$

It can be seen from (11.52) and (11.51a) and from $TS^E = H^E - G^E$ that $S^E$ is not zero whereas for the zeroth approximation of (11.47), $S^E$ is zero. Comparison of (11.52) with $G^E = x_1 g_1 + x_2 g_2$ shows that the coefficient of $x_1$ in (11.52) is $g_1$ and that of $x_2$, $g_2$. Equations (11.52) and (11.51a) have a fair degree of success in correlating $G^E$ and $H^E$ for spherical and symmetrical molecules as shown in Table 11.1. In fact the correlation is good for one out of two liquid metallic solutions for which adequate data are available. It may be noted that for $\beta = 1$ or $W_{12} = 0$, $G^E$ and $H^E$ are both zero and this constitutes a redefinition of an ideal solution. As an example, consider the last system, Sb-Zn, in Table 11.1. From the experimental value of $G^E$, $-750/823R = -0.45859 = \ln[2\beta/(\beta+1)]$; therefore, $\beta = 0.4622$, and $N_A W_{12}/RT = -1.5435$, or $N_A W_{12} = -10{,}560$ J mol$^{-1}$. The result for $H^E$ from this value of $N_A W_{12}$ and from (11.51a) is $-3{,}611$, which is in good agreement with the experimental value of $-3473$ with an estimated error of $\pm 800$ J mol$^{-1}$. It is difficult to evaluate $e_{ij}$ for each solution, and in addition $\bar{z}$ is not well-defined for liquids; therefore $W_{12}$ in (11.51), combining $e_{ij}$ and $\bar{z}$, is often taken to be a parameter that can be evaluated from a single experimental value of $H^E$ or $G^E$. Attempts to determine $W_{12}$ theoretically have not succeeded in realistic representations of $H^E$ and $G^E$.

## Effect of Temperature on $G^E$ and $g_i$

The coefficients in all the preceding equations are functions of $P$ and $T$, and at a given constant pressure, usually one bar, each constant $A_{ij}$, as well others in (11.24), may be written as a linear function of $T$, i.e.,

$$A_{ij} = \overset{\cdot}{\overline{A}}_{ij} - \overset{\cdot}{\overline{B}}_{ij} T \tag{11.55}$$

The constants $\overset{\cdot}{\overline{A}}_{ij}$ and $\overset{\cdot}{\overline{B}}_{ij}$ are independent of temperature when the excess partial heat capacity $\overline{C}^E_{p,i}$ is zero and they are related to the excess partial enthalpy $h_i$ and entropy $s_i$ respectively through the following thermodynamic equations:

$$\frac{\partial(g_i/T)}{\partial T} = -\frac{h_i}{T^2}, \qquad \frac{\partial g_i}{\partial T} = -s_i, \qquad g_i = h_i - T s_i \tag{11.56}$$

Similar equations may also be written for the excess property $G^E$. The assumption that $\overline{C}^E_{p,i} = 0$, known as Kopp's rule, is valid within experimental errors for a large number of mixtures. When $\overline{C}^E_{p,i}$ is a non-zero constant $\overline{C}_i$, (11.55) becomes

$$A_{ij} = \overset{\cdot}{\overline{A}}_{ij} - (\overline{B}_{ij} + \overline{C}_i \ln T)T \approx \overset{\cdot}{\overline{A}}_{ij} - \overset{\cdot}{\overline{B}}_{ij} T \tag{11.57}$$

The approximate equality $\approx$ holds particularly well at high temperatures and at a moderate range of temperature where $\ln T$ does not sensibly vary and merely adds a nearly constant quantity to $\overline{B}_{ij}$. When $\overline{C}^E_{p,i} = \overline{C}_i + \overline{C}'_i T$ a term involving $T^2$ is added to (11.57) but the contribution of such a term is very small and the excess partial heat capacity is very seldom known to such a degree of accuracy.[†] Therefore, (11.55) is quite adequate for expressing the temperature dependence of $G^E$ and the related thermodynamic properties. Equation (11.55) requires a minimum of two sets of data for $G^E$ at two temperatures for determining two values of $A_{ij}$, after which $\overset{\cdot}{\overline{A}}_{ij}$ and $\overset{\cdot}{\overline{B}}_{ij}$ can be calculated.

*Example 1* The separation process for bismuth from lead is technologically important. For this reason, numerous investigations have been carried out on thermodynamic properties of liquid Bi-Pb (Bi = 1; Pb = 2). The excess molar enthalpy of formation (or mixing), $H^E$, has been determined calorimetrically by Kleppa, Wittig and Huber, and Itagaki and Yazawa.[‡] The excess partial gibbs energy of Pb, $g_2$, has been measured by using galvanic cells, more recently by

---

[*] T. Tanaka, N. A. Gokcen, and Z. Morita, Z. Metallkunde, *81*, 49 and 349 (1990); T. Tanaka, N. A. Gokcen, D. Neuschutz, P. J. Spencer, and Z. Morita, Steel research, *62*, 385 (1991); T. Tanaka, N. Imai, A. Kiyose, T. Iida, and Z. Morita, Z. Metallkunde, *82*, 836 (1991); see also T. Tanaka et al., Z. Metallkunde, *84*, 100 (1993).
[†] C. Bergman and K. L. Komarek, CALPHAD, *9*, 1 (1985).
[‡] The list of all references are cited in an assessment of all data by N. A. Gokcen, J. Phase Equilibria, *13*, 21 (1992).

196    Thermodynamics

Ptak, Moser, Sugimoto, Codron and their collaborators, among others cited in a recent assessment. (loc. cit.) An equation has been derived by Gokcen (see footnote ‡ cited on the previous page) for $g_2$ as a power series in $x_1$, which can be rewritten in the form of (11.23) as follows:

$$g_2 = x_1^2[-3400 - 0.994T + 2x_2(-4050 + 1.128T)$$
$$- 3x_2^2(-4050 + 1.405T)] \qquad (11.58)$$

where $g_2$ is in J/(atom mol of liquid alloy), or (J/g. atom). Comparison with (11.23) shows that $B_{12} = B_{21}$, and therefore, $B_{21} = -4050 + 1.405T$; $A_{21} = -3400 - 0.994T$, and $A_{12} = -3400 - 1.271T$. Since $g_2 = h_2 - Ts_2$ as in (11.56), the constant terms in (11.58) represent $h_2$, and the terms that are the coefficients of $T$ represent $s_2$. These parameters, ($A_{21}$, $A_{12}$, and $B_{21}$), can be substituted in (11.21) and (11.22) to obtain the equations for $G^E$ and $g_1$.

**Example 2**  Thermodynamic calculations for the binary system pyridine-toluene ($C_5H_5N$-$C_7H_8$) are summarized in Fig. 11.3, where $x_2$ is the mole fraction of toluene. We write the equation in Fig. 11.3 for $G^E$ first, and then rewrite it in the cubic form of (11.21) after substituting $a_i = \alpha_i$ to avoid confusion with the activity, and further, we use $\alpha_1 = \alpha_1(x_1 + x_2)$, and thus obtain

$$G^E/RT = x_1x_2[\alpha_1 + \alpha_2(x_1 - x_2)] = x_1x_2[(\alpha_1 + \alpha_2)x_1 + (\alpha_1 - \alpha_2)x_2]$$
$$= x_1x_2(A'_{21}x_1 + A'_{12}x_2); \quad A'_{21} = \alpha_1 + \alpha_2; \quad A'_{12} = \alpha_1 - \alpha_2 \qquad (11.59)$$

where $A'_{ij} = A_{ij}/RT$ of (11.21). The numerical result for (11.59) is

$$G^E/RT = x_1x_2(0.27999x_1 + 0.30733x_2) \qquad (11.59a)$$

The temperature is 373.15 K, therefore, at the equimolecular composition, $G^E = 227.8$ J, as indicated in the figure. The relationship in Fig. 11.3 is the Redlich-Kister equation in terms of $(x_1 - x_2)$, which will not be used in this text.

**Example 3**  The system ethyl alcohol (1)-methylcyclopentane (2)-n-hexane (3), labelled respectively as (1), (2), and (3) after each component, has been investigated under 101325 Pa of pressure by Kaes and Weber.* The samples of gas and liquid phases have been analyzed by gas chromatography, calibrated with known prepared mixtures. The results consist of 57 measurements, about one third of which are listed in Table 11.2. The activity coefficients have been calculated from $\gamma_i = y_iP/x_iP_i^\circ$ as well as from the equation of state for the gas phase accounting for deviations from the ideal gas behavior but the difference is very small, and on the average less than the experimental errors. The vapor pressures of pure components are given by the Antoine equation, i.e.,

$$\log P_i^\circ (\text{Pa}) = A_i'' - [B_i''/(t + C_i'')] \qquad (11.60)$$

*G. L. Kaes and J. H. Weber, J. Chem. Eng. Data, 7, 344 (1962).

| Components: | 1. Pyridine, $C_5H_5N$ [110-86-1] | Author(s): | Rogalski, M.; Bylicki, A. (Instytut Chemii Fizycznej, Polska Akademia Nauk, Kasprzaka 44/52, 01-224 Warszawa, Poland) |
|---|---|---|---|
| | 2. Toluene, $C_7H_8$ [108-88-3] | Edited by: | Oracz, P. (Wydzial Chemii, Uniwersytet Warszawski, ul. Pasteura 1, 02-093 Warszawa, Poland) |
| State: | Binary system, single-phase liquid; pure components, both liquid | | |
| Variables: | $G^E$, molar excess Gibbs energy | | |
| | $\mu_i^E$, partial molar excess Gibbs energy of component $i$ | | |
| | $x_i$, mole fraction of component $i$ | | |
| Parameters: | $T$, temperature | SOURCE OF DATA | |
| Constants: | $P$, pressure | Rogalski, M.; Bylicki, A. (Institute of Physical Chemistry, Polish Academy of Sciences, Warsaw, Poland); FIRST PUBLISHED RESULTS | |
| Method: | Calculation from direct experimental liquid-vapor equilibrium pressure, $P_{exp}$ data at variable $x_i$ and constant $T$; ref. 1 | | |

### CALCULATED VALUES

Notes: $P$, satn. pressure ($< 25 \cdot 10^3$ Pa)

| $T$/K | 373.15 | | |
|---|---|---|---|
| $x_1$ | $G^E$ | $\mu_1^E$ | $\mu_2^E$ |
| | J mol$^{-1}$ | | |
| 0.00 | 0.0 | 953.5 | 0.0 |
| 0.05 | 45.1 | 852.9 | 2.6 |
| 0.10 | 85.1 | 758.6 | 10.2 |
| 0.15 | 120.0 | 670.5 | 22.8 |
| 0.20 | 149.8 | 588.5 | 40.2 |
| 0.25 | 174.8 | 512.5 | 62.2 |
| 0.30 | 194.9 | 442.3 | 88.9 |
| 0.35 | 210.2 | 377.8 | 119.9 |
| 0.40 | 220.7 | 318.8 | 155.3 |
| 0.45 | 226.5 | 265.3 | 194.8 |
| 0.50 | 227.8 | 217.2 | 238.4 |
| 0.55 | 224.4 | 174.2 | 285.9 |
| 0.60 | 216.6 | 136.3 | 337.2 |
| 0.65 | 204.4 | 103.3 | 392.1 |
| 0.70 | 187.8 | 75.1 | 450.6 |
| 0.75 | 166.9 | 51.6 | 512.5 |
| 0.80 | 141.7 | 32.7 | 577.7 |
| 0.85 | 112.4 | 18.2 | 646.0 |
| 0.90 | 78.9 | 8.0 | 717.4 |
| 0.95 | 41.5 | 2.0 | 791.6 |
| 1.00 | 0.0 | 0.0 | 868.7 |

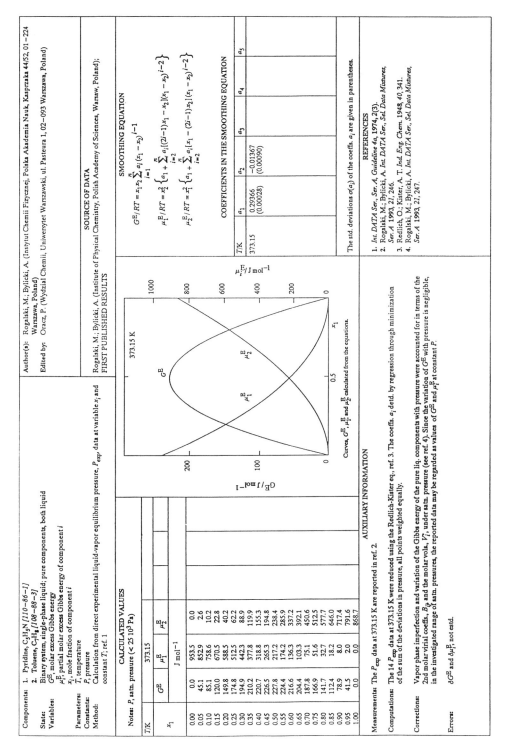

### SMOOTHING EQUATION

$$G^E/RT = x_1 x_2 \sum_{i=1}^{n} a_i (x_1 - x_2)^{i-1}$$

$$\mu_1^E/RT = x_2^2 \left\{ a_1 + \sum_{i=2}^{n} a_i [(2i-1)x_1 - x_2](x_1 - x_2)^{i-2} \right\}$$

$$\mu_2^E/RT = x_1^2 \left\{ a_1 + \sum_{i=2}^{n} a_i [x_1 - (2i-1)x_2](x_1 - x_2)^{i-2} \right\}$$

### COEFFICIENTS IN THE SMOOTHING EQUATION

| $T$/K | $a_1$ | $a_2$ | $a_3$ | $a_4$ | $a_5$ |
|---|---|---|---|---|---|
| 373.15 | 0.29366 (0.00028) | −0.01367 (0.00090) | | | |

The std. deviations $\sigma(a_i)$ of the coeffs. $a_i$ are given in parentheses.

### REFERENCES

1. *Int. DATA Ser., Ser. A Guideline 4a*, 1974, 2(3).
2. Rogalski, M.; Bylicki, A. *Int. DATA Ser., Sel. Data Mixtures, Ser. A* 1993, 21, 246.
3. Redlich, O.; Kister, A. T. *Ind. Eng. Chem.* 1948, 40, 341.
4. Rogalski, M.; Bylicki, A. *Int. DATA Ser., Sel. Data Mixtures, Ser. A* 1993, 21, 247.

### AUXILIARY INFORMATION

| Measurements: | The $P_{exp}$ data at 373.15 K are reported in ref. 2 |
|---|---|
| Computations: | The 14 $P_{exp}$ data at 373.15 K were reduced using the Redlich-Kister eq., ref. 3. The coeffs. $a_i$ detd. by regression through minimization of the sum of the deviations in pressure, all points weighted equally. |
| Corrections: | Vapor phase imperfection and variation of the Gibbs energy of the pure liq. components with pressure were accounted for in terms of the 2nd molar virial coeffs, $B_{ij}$ and the molar vols, $V_i^L$ under satn. pressure (see ref. 4). Since the variation of $G^E$ with pressure is negligible, in the investigated range of satn. pressures, the reported data may be regarded as values of $G^E$ and $\mu_i^E$ at constant $P$. |
| Errors: | $\delta G^E$ and $\delta \mu_i^E$, not estd. |

**Fig. 11.3** Thermodynamic properties of $C_5H_5N$-$C_7H_8$ System; Gen. Ref. (14). Courtesy of K. N. Marsh and R. C. Wilhoit. ($\mu_i^E$ here = $g_i$ in this book).

198  Thermodynamics

**Table 11.2**  Vapor-liquid equilibrium data for ethyl alcohol (1)-methyl cyclopentane (2) and n-hexane (3) at 101325 Pa; G. L. Kaes and J. H. Weber, loc. cit.

| Temp., °C | $x_1$ | $x_2$ | $y_1$ | $y_2$ | $\gamma_1$ | $\gamma_2$ | $\gamma_3$ |
|---|---|---|---|---|---|---|---|
| 62.5 | 0.050 | 0.839 | 0.269 | 0.633 | 10.48 | 1.01 | 1.07 |
| 60.1 | 0.416 | 0.485 | 0.364 | 0.519 | 1.88 | 1.56 | 1.56 |
| 61.1 | 0.731 | 0.184 | 0.415 | 0.371 | 1.17 | 2.84 | 3.20 |
| 60.0 | 0.616 | 0.191 | 0.397 | 0.287 | 1.39 | 2.19 | 2.17 |
| 60.3 | 0.151 | 0.634 | 0.320 | 0.496 | 4.52 | 1.13 | 1.12 |
| 59.8 | 0.215 | 0.475 | 0.333 | 0.382 | 3.36 | 1.18 | 1.23 |
| 59.3 | 0.544 | 0.189 | 0.369 | 0.247 | 1.51 | 1.95 | 1.95 |
| 59.2 | 0.405 | 0.200 | 0.349 | 0.198 | 1.92 | 1.48 | 1.57 |
| 60.3 | 0.098 | 0.480 | 0.300 | 0.346 | 6.49 | 1.04 | 1.10 |
| 58.9 | 0.300 | 0.208 | 0.339 | 0.176 | 2.56 | 1.28 | 1.36 |
| 59.3 | 0.165 | 0.222 | 0.315 | 0.172 | 4.26 | 1.16 | 1.14 |
| 59.1 | 0.170 | 0.118 | 0.322 | 0.085 | 4.25 | 1.08 | 1.14 |
| 64.0 | 0.016 | 0.163 | 0.156 | 0.131 | 17.42 | 1.03 | 1.01 |
| 65.4 | 0.015 | 0.815 | 0.165 | 0.682 | 18.58 | 1.02 | 1.01 |
| 60.3 | 0.305 | 0.645 | 0.358 | 0.591 | 2.49 | 1.32 | 1.38 |
| 67.1 | 0.919 | 0.039 | 0.578 | 0.187 | 0.99 | 5.51 | 5.94 |
| 58.9 | 0.164 | 0.063 | 0.303 | 0.046 | 4.18 | 1.10 | 1.16 |
| 59.9 | 0.336 | 0.500 | 0.345 | 0.480 | 2.22 | 1.41 | 1.42 |
| 59.8 | 0.266 | 0.518 | 0.337 | 0.451 | 2.76 | 1.28 | 1.31 |
| 60.2 | 0.328 | 0.588 | 0.349 | 0.565 | 2.27 | 1.39 | 1.36 |

where $t$ is in °C. The values of the constants are given by Hala, Wichterle, Polak and Boublik* as follows:

| Component | $A_i^{''}$ | $B_i^{''}$ | $C_i^{''}$ |
|---|---|---|---|
| (1) | 10.28780 | 1623.220 | 228.980 |
| (2) | 8.98773 | 1186.059 | 226.042 |
| (3) | 9.00266 | 1171.530 | 224.366 |

At 59.8°C and $x_1 = 0.266$, on the second row from the bottom of Table 11.2, $y_1$ is 0.337; $P_1^\circ$ from the Antoine equation is 46,435 Pa; therefore the activity coefficient of ethyl alcohol is $\gamma_1 = 101325 \times 0.337/(0.266 \times 46435) = 2.76$, as listed on

---

*E. Hala, E. Wichterle, J. Polak, and T. Boublik, "Vapor-Liquid Equilibrium Data at Normal Pressures," Pergamon Press (1968). See also J. Prausnitz, T. Anderson, E. Grens, C. Eckert, R. Hsieh, and J. O'Connell, "Computer Calculations for Multicomponent Vapor-Liquid and Liquid-Liquid Equilibria," Prentice-Hall (1980).

the sixth column. Hala, et al. have fitted these data into a fourth order Margules equation [cf. (11.29)]. They have obtained twelve constants, 3 for each of the three binary systems and 3 for the ternary system, by computer programming of the data. However, the constants for the binary system 2-3, ($A_{32}$, $A_{23}$, and $B_{23}$) are set to zero because this system is very nearly ideal. The form of the Margules equation used by Hala et al., for $G^E$ is identical with (11.24), except $G^E/(2.3026RT)$ is substituted for $G^E$ of (11.24) so that their parameters, e.g. $A'_{ij}$, are our parameters $A_{ij}/(2.3026RT)$. Their parameters are as follows:

$A'_{12} = 1.1397$, $\quad A'_{23} = 0$, $\quad A'_{31} = 0.9361$, $\quad J'_1 = -0.8407$

$A'_{21} = 0.8360$, $\quad A'_{32} = 0$, $\quad A'_{13} = 1.1487$, $\quad J'_2 = +0.3219$

$B'_{12} = -0.4825$, $\quad B'_{23} = 0$, $\quad B'_{13} = -0.5483$, $\quad J'_3 = +0.8477$

These values can be substituted in (11.29) to obtain

$$\frac{g_1}{2.3026RT} = \log \gamma_1 = x_2^2[1.1397 - 1.5724x_1 + 1.4475x_1^2]$$
$$+ x_3^2[1.1487 - 1.5218x_1 + 1.6449x_1^2]$$
$$+ x_2x_3[1.9847 - 0.4252x_1 - 0.8407x_1(2 - 3x_1)$$
$$+ 0.3219x_2(1 - 3x_1) + 0.8477x_3(1 - 3x_1)] \quad (11.61)$$

The value of $\gamma_1$ calculated from (11.61) at $x_1 = 0.266$, $x_2 = 0.518$ and $x_3 = 0.216$ is 2.808 and this is very close to 2.76, i.e., within 1.7 percent; (cf. the second row from the bottom in Table 11.2). It is difficult to obtain better than about 3 percent agreement between $\gamma_i$ from the analytical equations and $\gamma_i$ from the experiments because only a fourth order ($\varepsilon = 4$) Margules equation is used.

Experimental difficulties exist in determination of thermodynamic properties at high temperatures, particularly for alloys of metals. The optimum procedure appears to be (1) direct measurement of activities as by vapor pressure and e.m.f. techniques, for which a number of references were cited in Chapter X, followed by (2) direct calorimetry as exemplified by the references cited* on this page.

## Equations with Henrian Reference States

The standard states in all the preceding equations of this chapter is Raoultian, i.e., $\gamma_i = 1$ for $x_i \to 1$. Therefore, $G^E$ and $g_i$ are both zero for $x_i = 1$ of each

---

*C. Colinet, A. Pasturel, and K. H. J. Buschow, Met. Trans., *18A*, 903 (1987); L. Topor and O. J. Kleppa, ibid., *19A*, 1827 (1988); see also J. Phase Equil. *15*, 240 (1994); K. Fitzner, W.-G. Jung, O. J. Kleppa, Met. Trans., *22A*, 1103 (1991); D. El Allam, M. Gaune-Escard, J.-P. Bros, and E. Hayer, ibid., *23B*, 39 (1992); Z. Moser, M. Zakulski, Z. Panek, M. Kucharski, and L. Zabdyr, ibid., *21B*, 707 (1990); W. Dokko and R. G. Bautista, ibid., *11B*, 511 (1980); S. Hassan, J. Agren, M. Gaune-Escard, and J. P. Bros, ibid., *21A*, 1877 (1990); Y. Yin and B. B. Argent, J. Phase Equil., *14*, 588 (1993); H. Feufel, M. Krishnaiah, F. Sommer, and B. Predel, ibid., *15*, 303, (1994); R. Haddad, M. Gaune-Escard, and J. P Bros., ibid., *15*, 310 (1994); R. Castanet, ibid., p. 339; R. Ferro, G. Borzone, N. Parody, and G. Cacciamani, ibid., *15*, 317 (1994); J. C. Gachon and J. Hertz, CALPHAD, *7(1)*, 1 (1983); I. Arpshofen, R. Luck, B. Predel, and J. F. Smith, J. Phase Equil., *12*, 141 (1991).

component $i$. When the reference state is Henrian, emphasized in this section with superscript $\infty$, i.e., $\gamma_i^\infty = 1$ for $x_i \to 0$, it was shown by (9.35) that

$$RT \ln \gamma_i^\infty - RT \ln \gamma_i = g_i^\infty - g_i = G_i^\circ - G_i^\infty = C \tag{11.62}$$

where $C$ is a constant at each temperature, and $g_i^\infty = RT \ln \gamma_i^\infty$. Henry's law is particularly convenient for the solutes of limited solubilities. We designate the solvent as component 1, and the solute as 2, and for simplicity, we consider $g_2 = A_{21} x_1^2$ of (11.42), and then extend the results to multicomponent systems requiring more complex equations, such as (11.24). At $x_1 = 1$, $g_2 = A_{21}$, and $g_2^\infty = 0$ because $\gamma_2^\infty = 1$; consequently, $C$ in (11.62) is equal to $-A_{21}$. With this constant, the rearrangement of (11.62) yields

$$g_2^\infty = -A_{21} + A_{21} x_1^2 = -A_{21} + g_2 \tag{11.63}$$

Here, the relationship $g_2^\infty = -A_{21} + g_2$ is also valid if (11.30) is used for $g_2$. Further, when the Henrian scale is also used for a third dilute component, then the exchange of subscript 2 in (11.63) for subscript 3 yields

$$g_3^\infty = -A_{31} + g_3 \tag{11.64}$$

The resulting equation for $G^{E\infty} = x_1 g_1 + x_2 g_2^\infty + x_3 g_3^\infty$ is

$$G^{E\infty} = -A_{21} x_2 - A_{31} x_3 + G^E \tag{11.65}$$

The equation for the ternary regular solutions (11.41) was obtained by substituting $B_{ij} = J_i = 0$ and $A_{ij} = A_{ji}$ in (11.24), and the same substitution in (11.29)–(11.31) now yields the corresponding excess partial Gibbs energies. The resulting equations can be substituted in (11.63)–(11.65), with $g_1$ to complete the following set:

$$G^{E\infty} = -A_{21} x_2 - A_{31} x_3 + A_{21} x_1 x_2 + A_{31} x_1 x_3 + A_{32} x_2 x_3 \tag{11.66}$$

$$g_1 = A_{21} x_2^2 + A_{31} x_3^2 + x_2 x_3 (A_{21} + A_{31} - A_{32}) \tag{11.67}$$

$$g_2^\infty = -A_{21} + A_{21} x_1^2 + A_{32} x_3^2 + x_1 x_3 (A_{21} + A_{32} - A_{31}) \tag{11.68}$$

$$g_3^\infty = -A_{31} + A_{31} x_1^2 + A_{32} x_2^2 + x_1 x_2 (A_{32} + A_{31} - A_{21}) \tag{11.69}$$

These equations are useful for dilute solutions.

## Wagner Interaction Parameters

Wagner[*] derived $g_i^\infty$ for sparingly soluble components by using the linear Taylor series. Each term in the series was expressed as $[\partial g_i^\infty / \partial x_j] x_j$, where the coefficient of $x_j$, $\varepsilon_{ij} = \partial g_i / \partial x_j$, was called the Wagner interaction parameter. For a dilute ternary solution, there are two such linear terms corresponding to two independent variables. The Wagner equation for a ternary system can be derived much more

---

[*]For a full discussion, see N. A. Gokcen, J. Phase Equilibria, *15*, 147 (1994), and *16*, 8 (1995).

conveniently by substituting $1 - x_2 - x_3$ for $x_1$ in (11.68) and (11.69), and then retaining only the linear terms; thus, we obtain,

$$g_2^\infty = -2A_{21}x_2 + (A_{23} - A_{21} - A_{31})x_3; \quad (x_1 \to 1) \quad (11.70)$$

$$g_3^\infty = -2A_{31}x_3 + (A_{23} - A_{31} - A_{21})x_2; \quad (x_1 \to 1) \quad (11.71)$$

Here, $\partial g_2^\infty/\partial x_2 = -2A_{21}$, and $\partial g_2^\infty/\partial x_3 = \partial g_3^\infty/\partial x_2 = A_{23} - A_{21} - A_{31}$. It is necessary to stress that these equations are valid for dilute solutions and that the substitution is made for $x_1$ to obtain the equations with the independent variables $x_2$ and $x_3$.

The Wagner equation has been the subject of numerous publications because of its simple linear form. The value of $-2A_{21}$ is determined by plotting $g_2^\infty = RT \ln \gamma_2^\infty$ vs. $x_2$ in the dilute binary system 1-2, and the slope of the nearly linear portion at $x_2 \to 0$ yields $-2A_{21}$. The coefficient of $x_3$ in (11.70) is best obtained by plotting $g_2^\infty + 2A_{21}x_2$ vs. $x_3$ upon adding small amounts of component 3 in dilute binary alloy 1-2; the slope is then the coefficient of $x_3$ in (11.70). The coefficient $-2A_{31}$ in (11.71) is determined for the binary system 1-3 in the same way as $-2A_{21}$ is determined for the binary system 1-2.

The quadratic terms are neglected in deriving (11.70) and (11.71); hence this procedure makes $g_1$ of (11.66) zero, i.e., the solvent is nearly ideal Raoultian when $x_1$ is not far from unity. It was shown in Chapter X that such linear equations for $g_i$ violate the Gibbs-Duhem relation, i.e., the linear term in (10.33) must be zero so that it could assume the correct form given by (10.35). Further, when one of the solutes, e.g., 2, yields reliable values of $g_2^\infty$, then (11.70) can determine only two coefficients whereas (11.68) can determine all of the three parameters;[†] therefore (11.68) and (11.69) should be used instead of (11.70) and (11.71) whenever feasible, particularly when the mole fractions of solutes are larger than 0.03.

## PROBLEMS

11.1 Show that for a binary system at constant pressure and temperature

(a) $\overline{G}_2 - \overline{G}_1 = \partial G/\partial x_2$, (b) $\partial(G/x_2)/\partial x_2 = -\overline{G}_1/x_2^2$, and

(c) $\partial \overline{G}_2/\partial x_2 = x_1 \partial^2 G/\partial x_2^2$

11.2 Determine the values of the unknown coefficients in the following equations for a binary system and write the corresponding equation for the excess molar

---

[*]M. Ohtani and N. A. Gokcen, Trans. Metal. Soc. AIME, *218*, 533 (1960); A. K. Jena and M. B. Bever, ibid., *245*, 1035 (1969); F. Neumann and H. Schenck, Archiv für das Eisenhütt., *30*, 477 (1959).
[†]N. A. Gokcen, High Temp. Science, *15*, 293 (1982); Q. Han, Y. Dong, X. Feng, C. Xiang, and S. Yang, Met. Trans., *16B*, 785 (1985); "Steelmaking Data Sourcebook," Japan Society for the Promotion of Science, 19th Comm. on Steelmaking, Gordon & Breach Science Publishers (1988).

Gibbs energy $G^E$.

$$\ln \gamma_1 = A_2 x_2^2 - 8x_2^3 + 3x_2^4$$
$$\ln \gamma_2 = x_1^2 + B_3 x_1^3 + B_4 x_1^4$$

*Partial ans.: $A_2 = 7$, $B_3 = 0$ and $B_4 = 3$*

11.3 Plot the molar Gibbs energy of mixing $\Delta G = A_2 x_1 x_2 + x_1 RT \ln x_1 + x_2 RT \ln x_2$ versus $x_2$ at 25°C for the following values of $A_2/RT$: 0, +2.5, and −2.5. Compute the maximum, minimum and the inflection points for $A_2/RT = +2.5$.

11.4 Derive $G^E$ and $g_2$ for the binary system of components 2 and 3 from (11.24) for $\varepsilon = 2$ and $\varepsilon = 3$.

11.5 Derive $g_1$, $g_2$ and $g_3$ for a ternary system by substituting (11.41) in (11.27) for $g_1$, $g_2$ and $g_3$.

11.6 Express $S^E$ of the first approximation to the regular solutions and compute the values of $G^E/RT$, $H^E/RT$, and $S^E/RT$ for $\beta = 2.0$ and $3x_1 = x_2$ for a binary system.

11.7 Use (11.21) with $\varepsilon = 3$, eliminate $x_1$ and solve for $A_{21}$ and $A_{12}$ at the critical point for $\Delta G$ versus $x_2$ when $x_2$ is taken to be (a) 0.5 and (b) 0.6.

*Partial ans.: (a) $A_{12} = A_{21} = 2RT$*

11.8 Derive the equation similar to (11.61) for log $\gamma_2$ and compute the numerical value of $\gamma_2$ for $x_1 = 0.409$ and $x_2 = 0.400$ on the twelfth row in Table 11.2. Show that log $\gamma_2$ is zero for all values of $x_2$ and $x_3$ when $x_1$ is zero.

11.9 Express $G^E$ for the Bi-Pb system by using the parameters in (11.58) and compute $G^E$ at $x_2 = 0.5$.

11.10 Calculate log $\gamma_1$ in (11.61) at the ternary compostitions ($x_1 = 0.6$, $x_2 = 0.2$), ($x_1 = 0.2$, $x_2 = 0.6$), and ($x_1 = 0.3$, $x_2 = 0.2$) and fit the results to the regular ternary equation; compare the results from this equation for log $\gamma_1$ with those from (11.61) at $x_1 = x_2 = x_3$ and at $x_1 = 2x_2 = 2x_3$.

11.11 Calculate $g_2 (= \mu^E)$ in Fig. 11.3 at $x_2 = 0.3$, 0.5 and 0.7 from the listed equation and check the results with the tabulated values.

11.12 Check the term $J_3 x_1 x_2 x_3$ in (11.30) by using (11.24) and (11.32).

11.13 Determine the value of $A_{12}$ in terms of $RT$ in Problem 11.7 for $A_{21} = 2A_{12}$ when there is a spinodal point at $x_2 = 0.6$.

*Ans.: $A_{12} = -1.736 RT$*

CHAPTER XII

# GIBBS ENERGY CHANGE OF REACTIONS

## Introduction

A chemical reaction can be represented by

$$\nu_1 A_1 + \nu_2 A_2 + \cdots = \nu_I A_I + \nu_{II} A_{II} + \cdots \tag{12.1}$$

where $\nu_1, \nu_2, \ldots$ are the numbers of moles of the reactants $A_1, A_2, \ldots$, and $\nu_I, \nu_{II}, \ldots$ those of the products $A_I, A_{II}, \ldots$. The coefficients $\nu$ with various subscripts are called the stoichiometric coefficients to which positive values are assigned in this book. The Gibbs energies of the reactants and the products are represented by $\overline{G}_i = G_i^\circ(P, T) + RT \ln a_i$ if $i$ is a condensed phase and $\overline{G}_i = G_i^\circ(T) + RT \ln f_i$ if $i$ is a gas, and the Gibbs energy change for Reaction (12.1) is

$$\Delta_r G = (\nu_I \overline{G}_I + \nu_{II} \overline{G}_{II} + \cdots) - (\nu_1 \overline{G}_1 + \nu_2 \overline{G}_2 + \cdots) \tag{12.2}$$

It is important to remember that $\Delta_r$ refers to the sum of the thermodynamic properties of the products minus those of the reactants throughout this chapter. Although (12.2) is perfectly general and without restriction, the temperature and pressure of the reactants and the products are held constant during the reaction so that the changes in thermodynamic properties are solely those accompanying the chemical reaction. The substitution of $\overline{G}_i = G_i^\circ + RT \ln a_i$ into (12.2) for condensed reactants and products gives

$$\Delta_r G = \left(\nu_I G_I^\circ + \nu_I RT \ln a_I + \nu_{II} G_{II}^\circ + \nu_{II} RT \ln a_{II} + \cdots\right)$$
$$- \left(\nu_1 G_1^\circ + \nu_1 RT \ln a_1 + \nu_2 G_2^\circ + \nu_2 RT \ln a_2 + \cdots\right) \tag{12.3}$$

Separating the logarithmic terms from others and noting that

$$\Delta_r G^\circ = \left(\nu_I G_I^\circ + \nu_{II} G_{II}^\circ + \cdots\right) - \left(\nu_1 G_1^\circ + \nu_2 G_2^\circ + \cdots\right) \tag{12.4}$$

transforms (12.3) into

$$\boxed{\Delta_r G = \Delta_r G^\circ + RT \ln\left(\frac{a_I^{\nu_I} \cdot a_{II}^{\nu_{II}} \cdots}{a_1^{\nu_1} \cdot a_2^{\nu_2} \cdots}\right) = \Delta_r G^\circ + RT \ln J} \tag{12.5}$$

where

$$J = \left(\frac{a_I^{\nu_I} \cdot a_{II}^{\nu_{II}} \cdots}{a_1^{\nu_1} \cdot a_2^{\nu_2} \cdots}\right) \quad (12.6)$$

HEREAFTER, WE SHALL USE $\Delta$ FOR $\Delta_r$ IN THIS CHAPTER, EXCEPT FOR EMPHASIS. The quantity $J$ is related to the activities in phases of arbitrarily chosen nonequilibrium concentrations. The change in the Gibbs energy, $\Delta G$ corresponds to Reaction (12.1) in which $\nu_1$ moles of $A_1$ and $\nu_2$ moles of $A_2$, etc., from every large amounts of their phases with fixed compositions are reacted at constant pressure and temperature to form $\nu_I$ moles of $A_I$ and $\nu_{II}$ moles of $A_{II}$, etc., in their phases of fixed compositions at the same pressure and temperature, and $\Delta G°$ represents the change in standard Gibbs energy when $\nu_1$ moles of $A_1$, etc., in their standard states are reacted at the same temperature to form $\nu_I$ moles of $A_I$, etc., in their standard states as evident from (12.4).

At equilibrium $\Delta G$ is zero, hence (12.5) becomes

$$0 = \Delta G° + RT \ln \left(\frac{a_I^{\nu_I} \cdot a_{II}^{\nu_{II}} \cdots}{a_1^{\nu_1} \cdot a_2^{\nu_2} \cdots}\right)_{\text{equil.}} ; \quad (\Delta \equiv \Delta_r) \quad (12.7)$$

The ratio in parentheses in this equation, i.e.,

$$\left(\frac{a_I^{\nu_I} \cdot a_{II}^{\nu_{II}} \cdots}{a_1^{\nu_1} \cdot a_2^{\nu_2} \cdots}\right)_{\text{equil.}} = K_p \quad (12.8)$$

is a constant at a constant temperature and (12.8) is the formal definition of the equilibrium constant $K_p$ of Reaction (12.1). Substitution of $K_p$ into (12.8) and rearrangement gives

$$\Delta G° = -RT \ln K_p = \Delta H° - T\Delta S° \quad (12.9)$$

This equation signifies that when the reactants in their standard states are transformed into the products in their standard states the accompanying change in the standard Gibbs energy, $\Delta G°$, is equal to $-RT \ln K_p$. For gaseous reactants and products the activity $a_i$ is replaced by the fugacity.

The equilibrium constant for ideal gas mixtures is very simple, i.e.,

$$K_p = \frac{P_I^{\nu_I} \cdot P_{II}^{\nu_{II}} \cdots}{P_1^{\nu_1} \cdot P_2^{\nu_2} \cdots} \quad (12.10)$$

Equation (12.10) is often used in the absence of data for fugacities and moderate total pressures and sufficiently high temperatures where the gas mixtures are nearly ideal.

## Feasibility of Chemical Reactions

Combination of (12.9) with (12.5) gives

$$\Delta G = -RT \ln K_p + RT \ln J = RT \ln(J/K_p) \quad (12.11)$$

Equation (12.11), first derived explicitly by J. H. van't Hoff (1886) is called the reaction isotherm. If the value of $J$ is greater than $K_p$, $J > K_p$, then $\Delta G > 0$; hence the reaction cannot proceed from the left to the right as shown by (12.1). The reverse reaction, however, is possible since this is equivalent to multiplying (12.1) and (12.11) by $-1$, hence $\Delta G_{\text{forward}} = -\Delta G_{\text{reverse}}$ and $\Delta G_{\text{reverse}} < 0$. Conversely if $J < K_p$, then $\Delta G < 0$ and (12.1) is possible or a spontaneous reaction is possible since *the accompanying change in the Gibbs energy is negative*. When $J$ is equal to $K_p$, $\Delta G = 0$, then equilibrium prevails. In summary,

$$\begin{aligned} J > K_p \quad &\text{or} \quad \Delta G > 0 \quad \text{Reaction is not possible} \\ J = K_p \quad &\text{or} \quad \Delta G = 0 \quad \text{Reaction is at equilibrium} \\ J < K_p \quad &\text{or} \quad \Delta G < 0 \quad \text{Reaction is possible} \end{aligned} \quad (12.12)$$

Thermodynamics can predict whether a reaction is possible or impossible, but when it is possible there might be kinetic barriers that could make a detectable rate of reaction impossible.

If the arbitrarily chosen activities are unity, i.e., the reactants and the products are in their standard states, then the change in the Gibbs energy $\Delta G$ in (12.2) is the same as the change in the standard Gibbs energy $\Delta G°$, and then $J$ is unity from (12.6). Under these conditions a spontaneous reaction is feasible when $\Delta G°$ is negative, the reaction is impossible when $\Delta G°$ is positive, and when $\Delta G°$ is zero, the reaction is at equilibrium. It must be remembered that negative or positive values of the standard Gibbs energy change, $\Delta G°$, alone cannot predict whether a reaction is feasible or not unless the arbitrary activities or fugacities of the reactants and the products in $J$ are specified.

**Example 1** Equilibrium pressures of $CO_2$ over solid $CaCO_3$ and $CaO$ are 0.445 and 2.045 bars at 1100 and 1200 K respectively.* Assuming that the actual pressure of $CO_2$ is 1 bar, calculate the values of $K_p$, $J$, $\Delta G$, and $\Delta G°$ at each temperature for

$$CaCO_3(s) = CaO(s) + CO_2(g)$$

and show whether the reaction is feasible or impossible.

**Solution** The equilibrium constant is the same as the equilibrium pressure of carbon dioxide because the activities of solid $CaCO_3$ and $CaO$ are unity, i.e., $K_p = P_{CO_2}$. The actual pressure is 1 bar, hence $J = 1$. Since the reactants and the products are in their standard states, $\Delta G$ in (12.2) is the same as $\Delta G°$, i.e., $\Delta G = \Delta G°$, and $\Delta G°$ at 1100 K is

$$\Delta G° = -RT \ln K_p = -1100 R \ln 0.445 = 7,405 \text{ J}$$

Since $\Delta G = \Delta G° > 0$ the reaction is impossible at 1 bar of $CO_2$. At 1200 K,

$$\Delta G° = -1200 \, R \ln 2.045 = -7,138 \text{ J}$$

*See General Refs. (1) and (5).

and evidently $\Delta G° < 0$; therefore the reaction is possible. At 1151 K, the equilibrium pressure of $CO_2$ is also 1 bar and $K_p = J$ and $\Delta G = \Delta G° = 0$, and equilibrium prevails because $\Delta G$ is zero.

**Example 2** The standard Gibbs energy changes at 1400 K for the combustion of $H_2$ and CO from Gen. Ref. (1) are as follows:

(I)  $H_2(g) + 1/2 O_2(g) = H_2O(g)$,   $\Delta G°_I = -170{,}089$ J
(II) $CO(g) + 1/2 O_2(g) = CO_2(g)$,   $\Delta G°_{II} = -161{,}091$ J

(a) Calculate the equilibrium constant at 1400 K for the reaction

(III) $CO(g) + H_2O(g) = CO_2(g) + H_2(g)$

(b) What is the composition of the gas mixture when 0.4 mole of CO and 1 mole of $H_2O$ are introduced into a chamber at 1400 K and 1 bar? What is $\Delta G$ at start?

**Solution** (a) Subtraction of Reaction I from II yields III, and subtraction of $\Delta G°_I$ from $\Delta G°_{II}$ yields $\Delta G°_{III}$, from which $K_p$ is obtained; thus,

$$\Delta G°_{III} = \Delta G°_{II} - \Delta G°_I = 8998 \text{ J}$$

$$\ln K_p = -\frac{\Delta G°}{RT} = \frac{-8998}{8.3145 \times 1400} = -0.773$$

$$K_p = \frac{P_{CO_2} \cdot P_{H_2}}{P_{CO} \cdot P_{H_2O}} = 0.462$$

Here, $P_i$ is written instead of fugacity $f_i$ because at high temperatures and moderate pressures the gas mixture is ideal.

(b) Let $x$ be the number of moles of CO reacted completely according to Reaction III. After the attainment of equilibrium, it follows from stoichiometry that the number of moles $n_i$ of gas species are

$$n_{CO} = 0.4 - x, \qquad n_{H_2O} = 1 - x, \qquad n_{CO_2} = x, \qquad n_{H_2} = x$$

and the total number of moles is 1.4; the mole fractions are therefore obtained by dividing each number of moles by 1.4. Since the total pressure is one bar, the mole fractions also represent the partial pressures; thus,

$$P_{CO} = \frac{0.4 - x}{1.4}, \qquad P_{H_2O} = \frac{1 - x}{1.4}, \qquad P_{CO_2} = \frac{x}{1.4}, \qquad P_{H_2} = \frac{x}{1.4} \qquad (12.13)$$

and

$$K_p = 0.462 = \frac{x^2}{(0.4 - x)(1 - x)}$$

from which

$$1.1645 x^2 + 1.4 x - 0.4 = 0$$

Gibbs Energy Change of Reactions   207

where $x$ without a subscript is an unknown, not a mole fraction. The values of $x$ satisfying this equation are 0.2384 and $-2.240$. The negative value is not acceptable. The mole fractions in the gas mixture from (12.13) are $x_{CO} = 0.1154$, $x_{H_2O} = 0.5440$, $x_{CO_2} = 0.1703$, $x_{H_2} = 0.1703$. At the beginning, when CO and $H_2O$ are introduced, $J$ is zero, or an infinitesimally small positive number; hence, $\Delta G = -\infty$ and the reaction has the utmost driving force to proceed.

**Example 3*** A gas mixture initially consists of 0.50 mole of CO, 0.18 mole of $H_2$, and 0.32 mole of $H_2O$ at 360 K and 1.5 bars. It is heated to 1200 K at constant volume to attain equilibrium by reacting as follows:

$$H_2O(g) + CO(g) = CO_2(g) + H_2(g)$$
$$0.32 - x \quad 0.50 - x \quad x \quad 0.18 + x$$

where $x$ is the mole of completely reacting $H_2O$. Compute the composition and the final pressure of gas mixture at 1200 K, given $\Delta G° = 3,146$ J for the reaction from Gen. Ref. (1); see also Appendix II.

**Solution** The occurrence of reaction itself does not change the pressure; therefore, the total moles before and after the reaction remain the same, and in this case 1 mole. The equilibrium constant can be written, as in Example 2, in terms of mole fractions as follows:

$$K_p = \frac{x(0.18 + x)}{(0.32 - x)(0.50 - x)}$$

where the quantities under each species in the reaction also represent the mole fractions. The value of the equilibrium constant from $\Delta G° = 3,146 = -1200R \ln K_p$, is $K_p = 0.7296$. Expansion of $K_p$ as a quadratic equation yields

$$0.3706x^2 + 1.0667x - 0.16 = 0$$

The acceptable root of this equation is 0.1429; the other root is $-3.0212$, which is unacceptable, because, it signifies that the amount of $H_2$ formed is negative! The final pressure $P_f$ from $P_f/1.5 = 1200/360$ is 5.0 bars.

**Example 4** The following reactions and their $\Delta_r G°$ at 1600 K are from Gen. Ref. (2):

(I) $\quad 0.5S_2(g) + O_2(g) = SO_2(g); \quad\quad \Delta_r G° = -58,554$ cal mol$^{-1}$
$\quad\quad\quad x \quad\quad\quad y \quad\quad z$

(II) $\quad 0.5S_2(g) + 1.5O_2 = SO_3(g); \quad\quad \Delta_r G° = -46,602$ cal mol$^{-1}$
$\quad\quad\quad x \quad\quad\quad y \quad\quad w$

The standard pressure is 1 atm [=1.01325 bar in Gen. Ref. (2)]. The symbols $x, y, z$, and $w$ are the moles of each species at equilibrium. We use the preceding

---

*Adapted from Y. K. Rao, "Stoichiometry and Thermodynamics of Metallurgical Processes," Cambridge University Press, Cambridge (1985); p. 328.

208    Thermodynamics

equations to compute the gas composition when 0.20 mole of $SO_2$ is fed through one tube and 1.00 mole of $O_2$ is fed through another tube into one chamber at 1600 K, assuming that the species of sulfur compounds other than $S_2$, $SO_2$, and $SO_3$ are negligible, and the final pressure is 1 atm. [This is an example for the units used in Gen. Ref. (2)].

**Solution**   The total pressure is 1.0 atm; hence, the mole fractions can be used instead of the partial pressures in writing the equilibrium constants. We let $N = x + y + z + w$ for brevity and then express the equilibrium constants $K_I$ and $K_{II}$ as follows:

$$K_I = \frac{z\sqrt{N}}{y\sqrt{x}}; \qquad K_{II}\frac{wN}{y^{1.5}\sqrt{x}}; \qquad (N = x + y + z + w)$$

For the first reaction we have $\Delta_r G° = -58{,}554 = -1.987216 \times 1600 \times \ln K_I$, from which $K_I = 9.9517 \times 10^7$, and likewise, for the second reaction, $\Delta_r G° = -46{,}602 = -1.987216 \times 1600 \times \ln K_{II}$, and $K_{II} = 2.3194 \times 10^6$. [$R = 1.987216$ cal mol$^{-1}$K$^{-1}$] Next, we need sulfur and oxygen material balances (atomic balances) to obtain two more equations:

Sulfur $= 0.20 = 2x + z + w$.

Oxygen $= 0.20 \times 2 + 1.00 \times 2 = 2y + 2z + 3w$

These equations, and those for $K_I$ and $K_{II}$ provide four independent equations with four unknowns, which can be solved to obtain the values of the unknowns. Computer programs are available for this purpose. However, it is evident from the values of the equilibrium constants that $x$ for $S_2$ is very small, since $y$ is not small, and $x$ in the denominators as a factor can be eliminated by dividing $K_{II}$ by $K_I$ and setting the values of $x$ in the remaining terms to zero, and solving the resulting equation by successive approximation. This procedure is equivalent to subtracting Reaction (I) from (II) to obtain

(III)   $SO_2 + 0.5 O_2 = SO_3;$     $\Delta_r G° = 11{,}952$ cal mol$^{-1}$
          $z$        $y$       $w$        $K_{III} = K_{II}/K_I = 0.0233064$
                                          $= w \times N^{0.5}/[z \times y^{0.5}]$

The equilibrium constant $K_{III}$ provides one equation. Sulfur and oxygen balances without $x$ provide two more equations, from which $w = 2 - 2y$, and $z = 2y - 1.8$. The elimination of $w$ and $z$ from $K_{III}$ by using these equations and the rearrangement of result yield

$$(2 - 2y)(0.2 + y)^{0.5} - 0.0233064(2y - 1.8)y^{0.5} = 0$$

where $N = 0.2 + y$ with very small $x$. Successive approximations yield $y = 0.997917$ from which $w = 0.004166$ and $z = 0.195834$. Next, it is possible to solve for $x$ by substituting the foregoing values of $y, w, z$ in $K_I$. The result is $x \simeq 4.66 \times 10^{-18}$. [Generally, partial pressures or fugacities must be used to express $K_p$ for all gas reactions].

## Equilibria in Real Gas Mixtures

The fugacities of each constituent in a real gas mixture approach their partial pressures in the limit as the total pressure approaches zero; the equilibrium constant then assumes the form expressed by (12.10). A useful form of the equilibrium constant for real gas mixtures is obtained by substituting for the fugacity, $f_i$, of $i$, the product of the total pressure, the mole fraction and the fugacity coefficient $\phi_i$, i.e. $f_i = \phi_i x_i P$; thus

$$K_p = \left( \frac{x_I^{\nu_I} \cdot x_{II}^{\nu_{II}} \cdots}{x_1^{\nu_1} \cdot x_2^{\nu_2} \cdots} \right) \left( \frac{\phi_I^{\nu_I} \cdot \phi_{II}^{\nu_{II}} \cdots}{\phi_1^{\nu_1} \cdot \phi_2^{\nu_2} \cdots} \right) P^{\Delta \nu} \tag{12.14}$$

where $\Delta \nu = (\nu_I + \nu_{II} + \cdots) - (\nu_1 + \nu_2 + \cdots)$. The first and the last factors in (12.14) are represented by

$$K_p' = \left( \frac{x_I^{\nu_I} \cdot x_{II}^{\nu_{II}} \cdots}{x_1^{\nu_1} \cdot x_2^{\nu_2} \cdots} \right) P^{\Delta \nu} \tag{12.15}$$

The second quotient in (12.14) is

$$\Gamma = \left( \frac{\phi_I^{\nu_I} \cdot \phi_{II}^{\nu_{II}} \cdots}{\phi_1^{\nu_1} \cdot \phi_2^{\nu_2} \cdots} \right) \tag{12.16}$$

Thus (12.14) becomes

$$K_p = K_p' \Gamma \tag{12.17}$$

A useful approximation can be obtained by using (9.10) if the fugacities of pure gases are known. Determination of $\Gamma$ requires data on the fugacities of each constituent at various pressures, temperatures and compositions. Such data are not always available. The alternative method consists of determining the values of $K_p'$ at various pressures and compositions and then obtaining $\Gamma$ from the ratio $K_p/K_p'$ where the equilibrium constant $K_p$ is the value of $K_p'$ when the pressure is sufficiently low. A frequently used extrapolation consists of plotting $\ln K_p'$ versus $P$ and extrapolating the results linearly to zero pressure where the curvature in the plot is very small.

*Example* The following reaction

$$0.5 N_2 + 1.5 H_2 = NH_3$$

has been investigated at 475°C by equilibrating gas mixtures initially containing 0.25 mole fraction of $N_2$ and 0.75 mole fraction of $H_2$. The gas mixtures were sampled properly and analyzed for ammonia. From the following mole fractions of ammonia,* calculate $K_p$, $K_p'$, and $\Gamma$:

| $x_{NH_3}$: | 0.0165 | 0.0468 | 0.133 | 0.480 | 0.640 |
|---|---|---|---|---|---|
| $P$/bars: | 10 | 30 | 100 | 600 | 1000 |

---

*Based on original work cited in Gen. Refs. (1) and (2).

**Solution** Two moles of initial gas mixture, taken as a convenient basis for calculation, contain 0.5 mole of $N_2$ and 1.5 moles of $H_2$. The reaction at equilibrium is

$$0.5N_2 + 1.5H_2 = NH_3$$
$$(0.5 - 0.5x) \quad (1.5 - 1.5x) \quad (x); \quad \text{(Total moles} = 2 - x)$$

where $(0.5 - 0.5x)$ and $(1.5 - 1.5x)$ represent the remaining numbers of moles of $N_2$ and $H_2$ respectively after the formation of $x$ moles of $NH_3$. The mole fraction of ammonia at 10 bars is expressed by

$$0.0165 = x/(2 - x)$$

where $2 - x$ is the total number of moles of all species at equilibrium. This equation gives $x = 0.03246$, hence the mole fractions of $N_2$ and $H_2$ and $NH_3$ are

$$x_{N_2} = (0.5 - 0.5x)/(2 - x) = 0.2459;$$

$$x_{H_2} = (1.5 - 1.5x)/(2 - x) = 0.7376;$$

$$x_{NH_3} = x/(2 - x) = 0.0165.$$

Substitution of these values in (12.15) at $P = 10$ bars gives

$$K'_p = 0.0165/[(10)(0.2459)^{0.5}(0.7376)^{1.5}] = 0.00525$$

The factor $P^{-1} = 10^{-1}$ shows clearly that the ammonia yield $x_{NH_3}$ increases with increasing pressure. The values of $K'_p$ calculated in this manner are as follows:

| $P$/bars: | 10 | 30 | 100 | 600 | 1000 |
|---|---|---|---|---|---|
| $K'_p \times 10^3$: | 5.25 | 5.29 | 5.45 | 9.11 | 15.21 |
| $\Gamma$: | | 1.00 | 0.992 | 0.963 | 0.576 | 0.345 |

Within the accuracy of results, $K'_p$ at 10 bars is assumed to be virtually identical with $K_p$. The values of $\Gamma$, obtained from $K_p/K'_p$, are also listed above for the corresponding pressures. It is seen that $\Gamma$ does not differ greatly from unity even at 100 bars, but at 1000 bars, it is considerably smaller than unity. When $\Gamma$ decreases $K'_p$ increases, hence the ammonia yield also increases. This example illustrates that if the reaction decreases the volume as it proceeds from the left to the right, or, stated differently, if for the gaseous species of reaction $\Delta n = (\Sigma n_i \text{ of products}) - (\Sigma n_i \text{ of reactants})$ is negative, then an increase in externally applied pressure on the gas mixture can make the reaction proceed from the left to the right even when, unlike the ammonia reaction, $\Gamma$ increases somewhat with pressure. Thus, the ammonia yield increases eightfold for a pressure increase of tenfold from 10 to 100 bars when the change in $\Gamma$ is negligibly small. In addition, from Gen. Refs. (1)–(3), in the vicinity of 475°C, $K_p$ for $NH_3$ varies according to

$$\Delta G° = -RT \ln K_p = -53{,}160 + 114.72T \text{ J}$$

$$R \ln K_p = \frac{+53{,}160}{T} - 114.72$$

Thus, the reaction is exothermic, [cf. (12.9)] and $K_p$ increases with decreasing temperature. In summary (i) if the reaction decreases the volume, the applied external pressure promotes the reaction and (ii) if the reaction is exothermic, cooling promotes the reaction. These observations are the basis of the Le Chatelier principle (1884) that may be stated as follows: If a reaction under equilibrium in a closed system is subjected to external stresses, such as pressurization, cooling, and heating, the reaction will seek a new equilibrium to relieve partially the applied external stresses. In addition, for open systems, if a reaction product or reactant is removed, the reaction will proceed to replenish the removed species. The Le Chatelier principle is therefore a consequence of the second law of thermodynamics.

## Equilibria Involving Condensed Phases

The activities in the condensed phases and the fugacities in the gaseous phase for a reaction involving such phases must be known in order to evaluate the equilibrium constant. If one or more reactants and the products are dissolved in condensed phases to a limited extent, the reference state for the activity is usually chosen as the infinitely dilute solution, but usually the standard state is taken to be the pure component. Similar remarks are also valid if the reaction takes place in a homogeneous liquid phase, or a solid phase. Equilibria involving more than one phase are called heterogeneous chemical equilibria. The following example illustrates the important aspects of this type of chemical equilibria.

*Example* The activity of zinc in the liquid Cu-Zn system is $a_{Zn} = 0.657$ at $x_{Zn} = 0.70$ and 1200 K as determined from electromotive force measurements described in Chapter XIV, and pressure measurements.*Calculate molar ratio of $CO/CO_2$ over the liquid alloy and over the pure zinc, assuming that zinc is reactive but not Cu. The required reaction and the equilibrium constant are given as follows:

$$Zn(\text{in liq. soln.}) + CO_2(g) = ZnO(s) + CO(g);$$

$$K_p = \frac{P_{CO}}{a_{Zn} \cdot P_{CO_2}} = 91.44$$

The vapor pressure of pure zinc is 1.236 bars at 1200 K.

*Solution* For the alloy, $K_p \cdot a_{Zn} = 91.44 \times 0.657 = 60.08 = P_{CO}/P_{CO_2}$, which is also the molar ratio of CO to $CO_2$. For pure Zn, the activity is unity and $P_{CO}/P_{CO_2} = 91.44$.

The calculation of partial pressure of each gas requires the total pressure, and the partial pressure of zinc over the alloy. The pressure of Zn(g), $P_{Zn}^\circ$ over pure zinc is 1.236 bars and $P_{Zn} = 1.236 \times 0.657 = 0.812$ bar over the solution, from the definition of activity by $a_{Zn} = P_{Zn}(\text{over soln.})/P_{Zn}^\circ(\text{Pure Zn})$. For example, at

---

*See data by O. J. Kleppa and C. E. Thalmayer, and by D. B. Downie, as summarized in General Ref. (9).

2 bars of total pressure, $P_{Zn} = 0.812$, $P_{CO} = 1.169$ and $P_{CO_2} = 0.0190$ over *the alloy*; and $P_{Zn}^\circ = 1.236$, $P_{CO} = 0.7515$ and $P_{CO_2} = 0.0125$ over the pure zinc.

## Determination of Standard Gibbs Energy Changes

In general there are four methods by which the standard Gibbs energy changes of reactions can be determined. Each method has its advantages and limitations as will be described in the succeeding sections. Details of each method may often involve complex instrumentation and interpretation beyond the scope of this book; therefore, it is sufficient here to outline the basic principles for determining the standard Gibbs energy changes.*

## Method I—Determination of $\Delta G°$ from Equilibrium Constant

The standard Gibbs energy change can be obtained directly from experimental measurements of the equilibrium constant. This method requires the attainment of true equilibrium by allowing sufficient time for the reaction to proceed with and without catalysts, and by approaching equilibrium from both sides, i.e., by making the reaction proceed in the direction of the products and then in the direction of the reactants. The equilibrium concentrations of all species must be determined from analysis of representative samples.

It is usually difficult to analyze gas mixtures in which a reaction is at equilibrium because if the gas sample is at a different temperature and pressure than that of the equilibrated gases, the reversal of reaction may not be avoided. The composition of the equilibrium mixture can often be obtained from the change in pressure when the reaction occurs at a constant volume and the numbers of moles of the reactants are not equal to the numbers of moles of the products, and therefore the reaction causes a change in pressure. When the reaction does not cause a change in pressure, it is desirable to correlate such properties as conductivity, spectral properties, etc., with the composition. In some cases it is possible to bleed a minute amount of gas through a very small orifice into a mass spectrometer to determine the equilibrium composition. The procedure is rapid and the reversal of the chemical reaction in the sample is often negligibly small.

Consider, for example, when one mole of pure methane in a sealed chamber at 0.334 bar and 300 K, is heated to 800 K at constant volume in the presence of graphite. The pressure, measured by means of a capillary manometer attached to the chamber is 1.545 bars. If there were no dissociation, the pressure at 800 K would have been directly proportional to the temperature in K, i.e., $P = 800 \times 0.334/300 = 0.891$ bar. However, owing to the change in the number of moles of gaseous species after dissociation, the pressure is higher than 0.891 bar. Thus from the complete dissociation of $x$ mole of methane out of one mole, there are

---

*For details, see N. A. Gokcen, R. V. Mrazek, and L. B. Pankratz, U.S. Bureau of Mines IC 8853, pp. 438 (1981); see also "Experimental Thermochemistry," cited after (3.85).

$1-x$, $x$, and $2x$ moles of $CH_4$, C, and $H_2$ respectively at equilibrium, i.e.,

$$CH_4(g) = C(graph) + 2H_2(g)$$
$$(1-x) \qquad (x) \qquad (2x)$$

hence the total number of gaseous moles is $(1-x) + 2x = 1 + x$. The presence of some graphite other than that produced by the reaction may assure that amorphous carbon is not deposited; nevertheless, it is necessary to analyze deposited carbon by diffraction to be certain that it is graphite. From the observed pressure and from the ideal gas law, $1.545V = (1+x)800R$, and from the initial conditions, $0.334V = 300R$. Division of these equations side by side and rearrangement of the result gives

$$1.545 = (1+x)0.891, \quad \text{and} \quad x = 0.734$$

The equilibrium constant is therefore,

$$K_p = \frac{\left(\frac{2x}{1+x}1.545\right)^2}{\left(\frac{1-x}{1+x}1.545\right)} = 1.545\frac{(0.847)^2}{0.1534} = 7.226$$

If, by increasing and decreasing the temperature and returning to 800 K, the equilibrium is shifted to either direction, and the same value of $K_p$ is obtained again at 800 K, then it is certain that $K_p$ is the correct equilibrium constant. The problem is not this simple in practice because it is also necessary to ascertain whether or not other hydrocarbons are created by side reactions, which interfere with pressure measurement.

Solid compounds, particularly sulfides, carbides, nitrides, borides, and oxides deviate from stoichiometry with increasing temperature. Therefore a reaction such as $FeS_x + H_2(g) = FeS_{x-1} + H_2S(g)$, does not have $K_p = P_{H_2S}/P_{H_2}$ as its equilibrium constant unless $x$ is (i) independent of the ratio of $H_2S/H_2$ and (ii) independent of the temperature. Otherwise $FeS_x$ is a phase in the binary iron-sulfur system, and each ratio of $H_2S/H_2$ is in equilibrium with a corresponding value of $x$. In fact, solid FeS dissolves S in solid state above 1000°C, so that $x$ is larger than one at sufficiently high values of the ratio $H_2S/H_2$ or sulfur pressure $S_2(g)$; and in liquid $FeS_x$ above 1190°C, $x$ has a wide range of values around one depending on the pressure of $S_2(g)$ over the liquid.* The equilibrium constant then refers to

$$S(\text{in Fe-S}) + H_2(g) = H_2S(g), \qquad K_p = \frac{P_{H_2S}}{P_{H_2}a_S}$$

---

*For a summary of Fe-S diagram and original references see O. Kubaschewski, "Iron-Binary Phase Diagrams," Springer Verlag (1982). For methods of equilibrium pressure measurements, see Gen. Refs. (21) and (23).

In fact the activity of sulfur in the system Fe-S and other metal-sulfur systems is investigated by equilibration with mixtures of $H_2 + H_2S$ or by measuring the pressure of $S_2$ over the metal-sulfur system, i.e., S(in Fe-S) → $0.5\, S_2(g)$. Likewise, $CaCl_2$ is a compound at ordinary temperatures but at high temperatures and high pressures $CaCl_2$ dissolves either chlorine or calcium depending on the preponderance of either element; therefore, $CaCl_2$ is simply a one-phase region of the Ca-Cl phase diagram, where the concept of a true compound need not be invoked.

## Method II—Thermal Data

The standard Gibbs energy change can be obtained from thermal, or calorimetric determinations of (a) $\Delta H_T^\circ$ of reaction at a convenient temperature $T$, (b) enthalpies of phase transformation, and (c) $H_T^\circ - H_0^\circ$, of each reactant and product from 0 K to as high a temperature as possible. From the two latter measurements the enthalpy of reaction can be obtained at any temperature. The measurements of heat capacity and enthalpies of transition of each substance permit calculation of the standard entropy. The standard change of entropy of a reaction as in (12.1), $\Delta S_T^\circ$ at any temperature, is obtained from

$$\Delta S_T^\circ = (\nu_{\text{I}} S^\circ(\text{I}) + \nu_{\text{II}} S^\circ(\text{II}) + \cdots) - (\nu_1 S^\circ(1) + \nu_2 S^\circ(2) + \cdots) \quad (12.18)$$

Thus from the thermally determined values of $\Delta H_T^\circ$ and $\Delta S_T^\circ$, as functions of temperature, $\Delta G_T^\circ$ can be obtained from

$$\boxed{\Delta G_T^\circ = \Delta H_T^\circ - T \Delta S_T^\circ} \quad (12.19)$$

## Method III—Electromotive Force (emf) Method

It will be seen in Chapter XIV that the change in the standard Gibbs energy of certain reactions can be obtained from the emf of appropriate cells. A brief reference to this method is therefore sufficient in this chapter.

The change in the Gibbs energy is identical with the reversible *net* work $W'$. If the reactants and the products in a reaction, which occurs in a reversible cell, are in their standard states, then

$$\Delta G^\circ = \Delta W'_{\text{rev}} = -zFE^\circ \quad (12.20)$$

where $z$ is the number of faradays of electricity, $F$ is the Faraday constant 96,485.3 coulombus per gram equivalent, hence $zF$ is the amount of electricity in coulombs, and $E^\circ$ is the standard emf in volts. The reversibility is attained by counterbalancing the cell with an external opposing source of emf so that the measured potential of the cell is a maximum and the current, a minimum. A slight increase or decrease in the external emf would therefore change the direction of chemical reaction. It is necessary to be sure that the observed value of $E^\circ$ refers to the reaction under investigation and not to any side reactions.

## Method IV—Spectroscopic Data and Mechanics of Molecules

Thermodynamic properties of gases and gaseous reactions can be obtained with a high degree of precision from the analysis and interpretation of spectroscopic data by means of statistical mechanics. An appropriate interpretation of mechanics of molecules and of interactions with various photons is necessary for this purpose. The accuracy of resulting data cannot often be attained by any other method. Although the method of calculation, based on certain postulates, is somewhat involved and is beyond the realm of classical thermodynamics, the tabulated results can readily be utilized in obtaining the changes in standard Gibbs energy of gaseous reactions.

## Thermodynamic Equations

Thermodynamic equations, i.e., "thermodynamic models" in computer jargon, for expressing various properties above 298 K are based on an empirical equation for $C_p^\circ$ first used by Maier and Kelly* and later expanded by JANAF as follows:

$$C_p^\circ = \alpha + 2\beta T + 6\lambda T^2 - 2\epsilon T^{-2} \tag{12.21}$$

This equation was cited earlier as (3.71); its determination for condensed phases requires calorimetric data. Substitution of this equation in $dH^\circ = C_p^\circ dT$ and integration as in (3.72), gives the relative enthalpy, also called the enthalpy function (3.73). This equation is repeated here for convenience as follows:

$$H_T^\circ - H_{298}^\circ = \alpha T + \beta T^2 + 2\lambda T^3 + 2\epsilon T^{-1} - I \tag{12.22}$$

where the subscript 298 is the abbreviation of 298.15. The integration constant $I$ is obtained by substituting $T = 298.15$ K and solving for $I$. The use of "heat content" for $H_T^\circ - H_{298}^\circ$ is to be discouraged, because, heat is a form of energy, not a properly. The term "heat" appears only in "heat capacity", which is a traditional inheritance. The subscript $T$ in $H_T^\circ$, and later in $S_T^\circ$, is for emphasis when necessary.

We substitute (12.21) in $dS^\circ = C_p^\circ dT/T$, and integrate it to obtain

$$S_T^\circ = S_\theta^\circ + \alpha \ln T + 2\beta T + 3\lambda T^2 + \epsilon T^{-2} \tag{12.23}$$

where $S_\theta^\circ$ is an integration constant that can be determined with the experimental value of $S_{298}^\circ$ so that

$$S_\theta^\circ = S_{298}^\circ - \alpha \ln 298.15 - 2\beta 298.15 - 3\lambda(298.15)^2 - \epsilon(298.15)^{-2}; \tag{12.24}$$

The values of $C_p^\circ$ for condensed phases, obtained by low temperature calorimetry and fitted to a Sommerfeld and Debye-type equation for $C_p^\circ$, can be substituted in (5.3) to obtain $S_{298}^\circ$ i.e.,

$$S_{298}^\circ - S_0^\circ = S_{298}^\circ = \int_0^{298.15} dS^\circ = \int_0^{298.15} C_p^\circ dT/T \tag{12.25}$$

where the lower integration limit for the left side, $S_0^\circ$, is zero by the third law.

*See Ref. 151 in Gen. Ref. (2). For JANAF, see Gen. Ref. (1).

The standard changes in entropy and enthalpy for chemical reactions can be expressed by using the preceding equations, after which the standard change in Gibbs energy can be formulated. The resulting equations are identical with those for vaporization, because the vaporization can also be written as a reaction. Therefore, we rewrite (8.16) and (8.17) for a reaction as follows by using (12.22) and (12.23):

$$\Delta H_T^\circ = \Delta H_\theta^\circ + \Delta\alpha T + \Delta\beta T^2 + 2\Delta\lambda T^3 + 2\Delta\epsilon T^{-1} \quad (12.26)$$

$$\Delta S_T^\circ = \Delta S_\theta^\circ + \Delta\alpha \ln T + 2\Delta\beta T + 3\Delta\lambda T^2 + \Delta\epsilon T^{-2} \quad (12.27)$$

Determination of $\Delta H_\theta^\circ = \Delta H_{298}^\circ - \Delta I$ requires the experimental value of $\Delta H_{298}^\circ$ for the reaction. Substitution of (12.26) and (12.27) in (12.9) gives an equation equivalent to (8.18); thus,

$$\Delta G_T^\circ = -RT \ln K_p = \Delta H_\theta^\circ - \Delta\alpha T \ln T$$
$$- \Delta\beta T^2 - \Delta\lambda T^3 - \left(\Delta S_\theta^\circ - \Delta\alpha\right)T + \Delta\epsilon T^{-1} \quad (12.28)$$

The foregoing equations have been used in various assessments, and tabulation* of thermodynamic data.

## Tabulation of Thermodynamic Data

Available thermodynamic data have been assessed, correlated, tabulated, and usually computerized by various research groups. Each system of tabulation has its own assessed data and units. Gen. Refs. (1) and (3) use the SI units with 1 bar (= 0.1 MPa) for the standard state pressure, but Gen. Refs. (2) and (7) use calories, (1 cal = 4.1840 J), and one atmosphere (1 atm = 1.01325 bars). However, most of the tables, e.g. those in Gen. Refs. (1), (2), (5), and (7), have very nearly the same format. Typical tables for aluminum in its STANDARD STATE (stable under a standard pressure) from Gen. Refs. (1) and (2) are shown in Tables 12.1 and 12.2 respectively. Table 12.1 is for the "REFERENCE STATE", a term that signifies "STABLE STANDARD STATE FOR THE ELEMENT" to differentiate it from the table for gaseous Al(g) in Table 12.3, which is also in its STANDARD STATE of 1 bar, but Al(g) is not stable below its boiling point in the temperature range in which Al(g) is listed in Table 12.3 as will be seen later. The "reference state" is used only for a pure element in its stable standard state, not for a compound. The STANDARD STATE IS ALWAYS DESIGNATED BY "°". The first column in Table 12.1 contains the temperature in K in multiples of 100 K before and after 298.15 K, except at a first order phase transition, where the exact temperature appears twice, e.g. the melting point of Al, 933.450 in Table 12.1. The second column is the standard heat capacity, $C_p^\circ$, and the third column is the standard entropy, $S^\circ$. The column for $S^\circ$ up to 298.15 has been calculated by using an equation quite similar to (12.25) for the solid.

---

*I. Ansara and B. Sundman in "Computer Handling and Dissemination of Data," edited by P. S. Glaeser Elsevier (1987); Gen. Ref. (1) and (2).

## Gibbs Energy Change of Reactions

**Table 12.1** Thermodynamic properties of Al in its standard reference state. Reproduced from Gen. Ref. (1).

| | ($J\,K^{-1}\,mol^{-1}$)* | | | ($kJ\,mol^{-1}$)* | | | |
|---|---|---|---|---|---|---|---|
| $T/K$ | $C_p^\circ$ | $S^\circ$ | $-[G^\circ-H^\circ(T_r)]/T$ | $H^\circ-H^\circ(T_r)$ | $\Delta_f H^\circ$ | $\Delta_f G^\circ$ | $Log\,K_f$ |
| 0 | 0. | 0. | Infinite | −4.539 | 0. | 0. | 0. |
| 100 | 12.997 | 6.987 | 47.543 | −4.056 | 0. | 0. | 0. |
| 200 | 21.338 | 19.144 | 30.413 | −2.254 | 0. | 0. | 0. |
| 298.15 | 24.209 | 28.275 | 28.275 | 0. | 0. | 0. | 0. |
| 300 | 24.247 | 28.425 | 28.276 | 0.045 | 0. | 0. | 0. |
| 400 | 25.784 | 35.630 | 29.248 | 2.553 | 0. | 0. | 0. |
| 500 | 26.842 | 41.501 | 31.129 | 5.186 | 0. | 0. | 0. |
| 600 | 27.886 | 46.485 | 33.283 | 7.921 | 0. | 0. | 0. |
| 700 | 29.100 | 50.872 | 35.488 | 10.769 | 0. | 0. | 0. |
| 800 | 30.562 | 54.850 | 37.663 | 13.749 | 0. | 0. | 0. |
| 900 | 32.308 | 58.548 | 39.780 | 16.890 | 0. | 0. | 0. |
| 933.450 | 32.959 | 59.738 | 40.474 | 17.982 | Crystal ↔ Liquid | | |
| 933.450 | 31.751 | 71.213 | 40.474 | 28.693 | Transition | | |
| 1000 | 31.751 | 73.400 | 42.594 | 30.806 | 0. | 0. | 0. |
| 1100 | 31.751 | 76.426 | 45.534 | 33.981 | 0. | 0. | 0. |
| 1200 | 31.751 | 79.189 | 48.225 | 37.156 | 0. | 0. | 0. |
| 1300 | 31.751 | 81.730 | 50.706 | 40.331 | 0. | 0. | 0. |
| 1400 | 31.751 | 84.083 | 53.007 | 43.506 | 0. | 0. | 0. |
| 1500 | 31.751 | 86.273 | 55.153 | 46.681 | 0. | 0. | 0. |
| 1600 | 31.751 | 88.323 | 57.162 | 49.856 | 0. | 0. | 0. |
| 1700 | 31.751 | 90.247 | 59.052 | 53.031 | 0. | 0. | 0. |
| 1800 | 31.751 | 92.062 | 60.836 | 56.207 | 0. | 0. | 0. |
| 1900 | 31.751 | 93.779 | 62.525 | 59.382 | 0. | 0. | 0. |
| 2000 | 31.751 | 95.408 | 64.129 | 62.557 | 0. | 0. | 0. |
| 2100 | 31.751 | 96.957 | 65.656 | 65.732 | 0. | 0. | 0. |
| 2200 | 31.751 | 98.434 | 67.112 | 68.907 | 0. | 0. | 0. |
| 2300 | 31.751 | 99.845 | 68.505 | 72.082 | 0. | 0. | 0. |
| 2400 | 31.751 | 101.196 | 69.839 | 75.257 | 0. | 0. | 0. |
| 2500 | 31.751 | 102.493 | 71.120 | 78.432 | 0. | 0. | 0. |
| 2600 | 31.751 | 103.738 | 72.350 | 81.607 | 0. | 0. | 0. |
| 2700 | 31.751 | 104.936 | 73.535 | 84.782 | 0. | 0. | 0. |
| 2790.812 | 31.751 | 105.986 | 74.574 | 87.665 | Liquid ↔ Ideal Gas | | |
| 2790.812 | 20.795 | 211.333 | 74.574 | 381.667 | Fugacity = 1 bar | | |
| 2800 | 20.795 | 211.401 | 75.023 | 381.858 | 0. | 0. | 0. |
| 2900 | 20.796 | 212.131 | 79.738 | 383.938 | 0. | 0. | 0. |
| 3000 | 20.798 | 212.836 | 84.163 | 386.017 | 0. | 0. | 0. |
| 3100 | 20.800 | 213.518 | 88.325 | 388.097 | 0. | 0. | 0. |
| 3200 | 20.804 | 214.178 | 92.248 | 390.178 | 0. | 0. | 0. |

*Continued*

218    Thermodynamics

**Table 12.1**  Continued.

| | ($J\ K^{-1}\ mol^{-1}$)* | | | ($kJ\ mol^{-1}$)* | | | |
|---|---|---|---|---|---|---|---|
| T/K | $C_p°$ | $S°$ | $-[G°-H°(T_r)]/T$ | $H°-H°(T_r)$ | $\Delta_f H°$ | $\Delta_f G°$ | Log $K_f$ |
| 3300 | 20.808 | 214.818 | 95.952 | 392.258 | 0. | 0. | 0. |
| 3400 | 20.815 | 215.440 | 99.458 | 394.339 | 0. | 0. | 0. |
| 3500 | 20.823 | 216.043 | 102.780 | 396.421 | 0. | 0. | 0. |
| 3600 | 20.833 | 216.630 | 105.934 | 398.504 | 0. | 0. | 0. |
| 3700 | 20.846 | 217.201 | 108.934 | 400.588 | 0. | 0. | 0. |
| 3800 | 20.862 | 217.757 | 111.790 | 402.673 | 0. | 0. | 0. |
| 3900 | 20.881 | 218.299 | 114.515 | 404.760 | 0. | 0. | 0. |
| 4000 | 20.904 | 218.828 | 117.116 | 406.849 | 0. | 0. | 0. |
| 4100 | 20.932 | 219.345 | 119.603 | 408.941 | 0. | 0. | 0. |
| 4200 | 20.964 | 219.849 | 121.984 | 411.036 | 0. | 0. | 0. |
| 4300 | 21.002 | 220.343 | 124.265 | 413.134 | 0. | 0. | 0. |
| 4400 | 21.046 | 220.826 | 126.455 | 415.236 | 0. | 0. | 0. |
| 4500 | 21.088 | 221.299 | 128.557 | 417.341 | 0. | 0. | 0. |
| 4600 | 21.143 | 221.763 | 130.578 | 419.452 | 0. | 0. | 0. |
| 4700 | 21.206 | 222.219 | 132.523 | 421.570 | 0. | 0. | 0. |
| 4800 | 21.276 | 222.666 | 134.396 | 423.694 | 0. | 0. | 0. |
| 4900 | 21.352 | 223.105 | 136.202 | 425.824 | 0. | 0. | 0. |
| 5000 | 21.439 | 223.537 | 137.945 | 427.964 | 0. | 0. | 0. |
| 5100 | 21.535 | 223.963 | 139.627 | 430.112 | 0. | 0. | 0. |
| 5200 | 21.641 | 224.382 | 141.253 | 432.271 | 0. | 0. | 0. |
| 5300 | 21.757 | 224.795 | 142.825 | 434.441 | 0. | 0. | 0. |
| 5400 | 21.884 | 225.203 | 144.347 | 436.623 | 0. | 0. | 0. |
| 5500 | 22.021 | 225.606 | 145.821 | 438.818 | 0. | 0. | 0. |
| 5600 | 22.170 | 226.004 | 147.249 | 441.027 | 0. | 0. | 0. |
| 5700 | 22.330 | 226.398 | 148.634 | 443.252 | 0. | 0. | 0. |
| 5800 | 22.496 | 226.787 | 149.978 | 445.491 | 0. | 0. | 0. |
| 5900 | 22.680 | 227.173 | 151.284 | 447.749 | 0. | 0. | 0. |
| 6000 | 22.836 | 227.552 | 152.551 | 450.005 | 0. | 0. | 0. |

Previous: June 1979 (1 atm)                                   Current: June 1983 (1 bar)

*Enthalpy Reference Temperature = $T_r$ = 298.15 K; Standard State Pressure = $p°$ = 0.1 MPa

The equation for $C_p°$ of solid in the footnote to Table 12.2 has three terms of (12.21), i.e. $\alpha = 4.592, 2\beta = 0.003454$ and $2\epsilon = -17{,}400$. These values substituted with $S_{298}° = 6.776$ in (12.23) give

$$S° = -20.319 + 4.592 \ln T + 0.003454 T - 8{,}700 T^{-2};$$

cal/mol · K    (12.29)

This equation gives 14.301 at 933.61 K, very close to 14.303, calculated with a multiple-term equation on a computer. The enthalpy of melting is given as

**Table 12.2** Thermodynamic properties of Al(c,l) in its standard state. Reproduced from Gen. Ref. (2).*

| T, K | $C_p°$ | $S°$ | $-(G° - H°_{298})/T$ | $H° - H°_{298}$, kcal/mol |
|---|---|---|---|---|
| | (cal/mol·K) | | | |
| 298.15 | 5.820 | 6.776 | 6.776 | 0 |
| 300 | 5.830 | 6.812 | 6.776 | .011 |
| 400 | 6.120 | 8.530 | 7.007 | .609 |
| 500 | 6.380 | 9.922 | 7.456 | 1.233 |
| 600 | 6.660 | 11.110 | 7.970 | 1.884 |
| 700 | 7.000 | 12.161 | 8.495 | 2.566 |
| 800 | 7.390 | 13.120 | 9.014 | 3.285 |
| 900 | 7.770 | 14.014 | 9.520 | 4.045 |
| 933.61 | 8.060 | 14.303 | 9.687 | 4.310 |
| 933.61 | 7.590 | 17.067 | 9.687 | 6.890 |
| 1000 | 7.590 | 17.589 | 10.194 | 7.395 |
| 1100 | 7.590 | 18.312 | 10.899 | 8.154 |
| 1200 | 7.590 | 18.973 | 11.545 | 8.913 |
| 1300 | 7.590 | 19.580 | 12.140 | 9.672 |
| 1400 | 7.590 | 20.143 | 12.692 | 10.431 |
| 1500 | 7.590 | 20.666 | 13.206 | 11.190 |
| 1600 | 7.590 | 21.156 | 13.688 | 11.949 |
| 1700 | 7.590 | 21.616 | 14.141 | 12.708 |
| 1800 | 7.590 | 22.050 | 14.568 | 13.467 |
| 1900[†] | 7.590 | 22.461 | 14.974 | 14.226 |
| 2000 | 7.590 | 22.850 | 15.358 | 14.985 |
| 2100 | 7.590 | 23.220 | 15.723 | 15.744 |
| 2200 | 7.590 | 23.573 | 16.072 | 16.503 |
| 2300 | 7.590 | 23.911 | 16.406 | 17.262 |
| 2400 | 7.590 | 24.234 | 16.725 | 18.021 |
| 2500 | 7.590 | 24.544 | 17.032 | 18.780 |

*The term, "reference state," is not used in Gen. Ref. (2).
[†]Data above 1800 K extrapolated.
*Phase changes*: 933.61 K, melting point of Al; $\Delta H° = 2.580$ kcal/mol.
2798 K, boiling point of Al; $\Delta H° = 70.1$ kcal/mol.
*Heat capacity (cal/mol·K) and enthalpy (kcal/mol) equations:*

298.15 – 933.61 K:  $C_p° = 4.592 + 3.454 \times 10^{-3}T + 0.174 \times 10^5 T^{-2}$.
$H° - H°_{298} = 4.592 \times 10^{-3}T + 1.727 \times 10^{-6}T^2$
$- 0.174 \times 10^2 T^{-1} - 1.464$

933.61 – 2500 K:  $C_p° = 7.590$
$H° - H°_{298} = 7.590 \times 10^{-3}T - 0.196$

*Sources*: Entropy at 298 K from CODATA (*36*). Other data from Hultgren (*99*) who extrapolated above 1800 K by assuming constant heat capacity. Data corrected to IPTS-68.

## Thermodynamics

**Table 12.3** Thermodynamic properties of Al(g). Reproduced from Gen. Ref. (1).

| | (J K$^{-1}$ mol$^{-1}$) | | | (kJ mol$^{-1}$) | | | |
|---|---|---|---|---|---|---|---|
| T/K | $C_p^\circ$ | $S^\circ$ | $-[G^\circ - H^\circ(T_r)]/T$ | $H^\circ - H^\circ(T_r)$ | $\Delta_f H^\circ$ | $\Delta_f G^\circ$ | Log $K_f$ |
| 0 | 0. | 0. | Infinite | −6.919 | 327.320 | 327.320 | Infinite |
| 100 | 25.192 | 139.619 | 184.197 | −4.458 | 329.297 | 316.034 | −165.079 |
| 200 | 22.133 | 155.883 | 166.528 | −2.129 | 329.824 | 302.476 | −78.999 |
| 250 | 21.650 | 160.764 | 164.907 | −1.036 | 329.804 | 295.639 | −61.770 |
| 298.15 | 21.390 | 164.553 | 164.553 | 0. | 329.699 | 289.068 | −50.643 |
| 300 | 21.383 | 164.686 | 164.554 | 0.040 | 329.694 | 288.816 | −50.287 |
| 350 | 21.221 | 167.969 | 164.813 | 1.104 | 329.524 | 282.016 | −42.089 |
| 400 | 21.117 | 170.795 | 165.388 | 2.163 | 329.309 | 275.243 | −35.943 |
| 450 | 21.046 | 173.278 | 166.130 | 3.217 | 329.060 | 268.499 | −31.167 |
| 500 | 20.995 | 175.492 | 166.957 | 4.268 | 328.781 | 261.785 | −27.348 |
| 600 | 20.930 | 179.314 | 168.708 | 6.363 | 328.141 | 248.444 | −21.629 |
| 700 | 20.891 | 182.537 | 170.460 | 8.454 | 327.385 | 235.219 | −17.552 |
| 800 | 20.866 | 185.325 | 172.147 | 10.542 | 326.492 | 222.112 | −14.502 |
| 900 | 20.849 | 187.781 | 173.751 | 12.628 | 325.436 | 209.126 | −12.137 |
| 1000 | 20.836 | 189.977 | 175.266 | 14.712 | 313.605 | 197.027 | −10.292 |
| 1100 | 20.827 | 191.963 | 176.695 | 16.795 | 312.513 | 185.422 | −8.805 |
| 1200 | 20.821 | 193.775 | 178.044 | 18.877 | 311.420 | 173.917 | −7.570 |
| 1300 | 20.816 | 195.441 | 179.319 | 20.959 | 310.327 | 162.503 | −6.529 |
| 1400 | 20.811 | 196.984 | 180.526 | 23.041 | 309.233 | 151.172 | −5.640 |
| 1500 | 20.808 | 198.419 | 181.672 | 25.122 | 308.139 | 139.921 | −4.872 |
| 1600 | 20.805 | 199.762 | 182.761 | 27.202 | 307.045 | 128.742 | −4.203 |
| 1700 | 20.803 | 201.023 | 183.798 | 29.283 | 305.950 | 117.631 | −3.614 |
| 1800 | 20.801 | 202.212 | 184.789 | 31.363 | 304.856 | 106.585 | −3.093 |
| 1900 | 20.800 | 203.337 | 185.736 | 33.443 | 303.760 | 95.600 | −2.628 |
| 2000 | 20.798 | 204.404 | 186.643 | 35.523 | 302.665 | 84.673 | −2.211 |
| 2100 | 20.797 | 205.419 | 187.513 | 37.603 | 301.570 | 73.800 | −1.836 |
| 2200 | 20.796 | 206.386 | 188.349 | 39.682 | 300.475 | 62.979 | −1.495 |
| 2300 | 20.795 | 207.311 | 189.153 | 41.762 | 299.379 | 52.209 | −1.186 |
| 2400 | 20.795 | 208.196 | 189.928 | 43.841 | 298.284 | 41.486 | −0.903 |
| 2500 | 20.794 | 209.044 | 190.676 | 45.921 | 297.188 | 30.808 | −0.644 |
| 2600 | 20.794 | 209.860 | 191.398 | 48.000 | 296.092 | 20.175 | −0.405 |
| 2700 | 20.794 | 210.645 | 192.097 | 50.080 | 294.997 | 9.583 | −0.185 |
| 2790.812 | 20.795 | 211.333 | 192.712 | 51.968 | Fugacity = 1 bar | | |
| 2800 | 20.795 | 211.401 | 192.773 | 52.159 | 0. | 0. | 0. |
| 2900 | 20.796 | 212.131 | 193.428 | 54.239 | 0. | 0. | 0. |
| 3000 | 20.798 | 212.836 | 194.063 | 56.318 | 0. | 0. | 0. |
| 3100 | 20.800 | 213.518 | 194.680 | 58.398 | 0. | 0. | 0. |
| 3200 | 20.804 | 214.178 | 195.279 | 60.478 | 0. | 0. | 0. |

*Continued*

**Table 12.3**  *Continued.*

| | | | | | | | |
|---|---|---|---|---|---|---|---|
| 3300 | 20.808 | 214.818 | 195.861 | 62.559 | 0. | 0. | 0. |
| 3400 | 20.815 | 215.440 | 196.428 | 64.640 | 0. | 0. | 0. |
| 3500 | 20.823 | 216.043 | 196.980 | 66.722 | 0. | 0. | 0. |
| 3600 | 20.833 | 216.630 | 197.518 | 68.805 | 0. | 0. | 0. |
| 3700 | 20.846 | 217.201 | 198.042 | 70.889 | 0. | 0. | 0. |
| 3800 | 20.862 | 217.757 | 198.553 | 72.974 | 0. | 0. | 0. |
| 3900 | 20.881 | 218.299 | 199.053 | 75.061 | 0. | 0. | 0. |
| 4000 | 20.904 | 218.828 | 199.541 | 77.150 | 0. | 0. | 0. |
| 4100 | 20.932 | 219.345 | 200.017 | 79.242 | 0. | 0. | 0. |
| 4200 | 20.964 | 219.849 | 200.484 | 81.337 | 0. | 0. | 0. |
| 4300 | 21.002 | 220.343 | 200.940 | 83.435 | 0. | 0. | 0. |
| 4400 | 21.046 | 220.826 | 201.386 | 85.537 | 0. | 0. | 0. |
| 4500 | 21.088 | 221.299 | 201.823 | 87.642 | 0. | 0. | 0. |
| 4600 | 21.143 | 221.763 | 202.252 | 89.753 | 0. | 0. | 0. |
| 4700 | 21.206 | 222.219 | 202.672 | 91.870 | 0. | 0. | 0. |
| 4800 | 21.276 | 222.666 | 203.084 | 93.995 | 0. | 0. | 0. |
| 4900 | 21.352 | 223.105 | 203.488 | 96.125 | 0. | 0. | 0. |
| 5000 | 21.439 | 223.537 | 203.885 | 98.264 | 0. | 0. | 0. |
| 5100 | 21.535 | 223.963 | 204.274 | 100.413 | 0. | 0. | 0. |
| 5200 | 21.641 | 224.382 | 204.657 | 102.572 | 0. | 0. | 0. |
| 5300 | 21.757 | 224.795 | 205.033 | 104.742 | 0. | 0. | 0. |
| 5400 | 21.884 | 225.203 | 205.403 | 106.924 | 0. | 0. | 0. |
| 5500 | 22.021 | 225.606 | 205.766 | 109.119 | 0. | 0. | 0. |
| 5600 | 22.170 | 226.004 | 206.124 | 111.328 | 0. | 0. | 0. |
| 5700 | 22.330 | 226.398 | 206.476 | 113.553 | 0. | 0. | 0. |
| 5800 | 22.496 | 226.787 | 206.823 | 115.792 | 0. | 0. | 0. |
| 5900 | 22.680 | 227.173 | 207.165 | 118.050 | 0. | 0. | 0. |
| 6000 | 22.836 | 227.552 | 207.501 | 120.306 | 0. | 0. | 0. |
| Previous: June 1979 (1 atm) | | | | | | Current: June 1983 (1 bar) | |

*Enthalpy Reference Temperature $= T_r = 298.15$ K;   Standard State Pressure $= p° = 0.1$ MPa

2,580 cal/mol in the footnote, so that the entropy of melting is $2{,}580/933.61 = 2.7635$, which added to 14.303, gives 17.067 cal/mol·K, listed for the liquid at 933.61 K. The remaining values are based on $C_p° = 7.590$ in the following equation with the lower integration limit of 933.61 K:

$$S° = -34.841 + 7.590 \ln T; \qquad \text{cal/mol} \cdot \text{K} \tag{12.30}$$

At 2500 K, this equation gives $S° = 25.544$, as in the table.

Next, consider the fifth column in Table 12.2 and return to the fourth column later. The equation for $H° - H_{298}°(c)$ for the solid is given in the footnote wherein $I = 1{,}464$ cal/mol. This equation gives 4,311 cal/mol versus 4,310 in the table, but in kcal/mol. The addition of the enthalpy of melting (fusion), $\Delta_{\text{fus}} H° = 2{,}580$ cal/mol, gives $H_{933.61}°(l) - H_{298}°(c) = 6{,}890$ for the liquid at 933.61 K, relative to

222    Thermodynamics

solid at 298.15 K. Above the melting point, $C_p^\circ = 7.590$, and integration of $C_p \, dT$ from 933.61 to $T$ yields

$$H^\circ(l) - H_{933.61}^\circ(l) = 7.590T - 7{,}086; \quad \text{cal/mol} \tag{12.31}$$

The addition of $H_{933.61}^\circ(l) - H_{298}^\circ(c) = 6{,}890$ gives the enthalpy relative to the solid (c) at 298.15 K as the last equation in the table. This equation replicates the values of enthalpy relative to the solid at 298.15 K. The property $H^\circ - H_{298}^\circ$, the relative enthalpy, is also called the ENTHALPY FUNCTION.

The fourth column is for the Gibbs energy function $(G_T^\circ - H_{298}^\circ)/T$ often abbreviated as $Gef$. Older tables use $F^\circ$ for the standard Gibbs energy $G^\circ$. The values of $Gef$ are always negative, therefore, $-Gef$ is tabulated to avoid the repetition of the negative sign before each listed value. The Gibbs energy function is calculated from the entropy and the enthalpy by

$$\left(G_T^\circ - H_{298}^\circ\right)/T \equiv -S_T^\circ + \left(H_T^\circ - H_{298}^\circ\right)/T \tag{12.32}$$

It is important to recall that in computing $Gef$ from $H_T^\circ - H_{298}^\circ$, the fifth column must be multiplied by 1000 for consistency in units.

Next, we consider Table 12.3 from Gen. Ref. (1) at 1 bar of standard pressure, and Table 12.4 at 1 atm ($=1.01325$ bar) of standard pressure, both for the ideal gas Al(g). Gaseous Al is UNSTABLE at 1 bar below its boiling point of 2790.812 K, but such tables are useful in computation. We examine Table 12.4 in greater detail because it contains the equations for thermodynamic properties, but otherwise it is quite similar to Table 12.3. The second column gives the values of $C_p^\circ$ for Al(g), which are very close to $5R/2 = 4.98$, and given by the equation in the footnote. The third column for entropy, corresponding to (12.23) is

$$S^\circ = 11.081 + 4.966 \ln T - 6{,}550 T^{-2}; \quad \text{cal/mol} \cdot \text{K} \tag{12.33}$$

where the constant 11.081 has been obtained by using $S_{298} = 39.302$ and 298.15 in (12.24). At 1000 K, this equation yields $S^\circ = 45.378$ in agreement with the table. The *enthalpy equation*, relative to the gas at 298.15 K, $[(H_{298}^\circ(g)]$, is already given in the footnote. At 1000 K, this equation gives $H_{1000}^\circ(g) - H_{298}^\circ(g) = 3{,}516$ cal/mol, as listed on the fifth column of the table. Note that the enthalpy symbols both refer to the gas in Table 12.4, whereas in Tables 12.1 and 12.2, the reference state tables, they are $H^\circ(c, l, \text{or } g) - H_{298}^\circ(c)$. The fourth column contains $-Gef$ by using the third and fifth columns (in 12.32). Thus at 1000 K, $-Gef = 45.378 - 3{,}516/1000 = 41.862$, as listed in Table 12.4. $Gef$ is a slowly varying function whose use will be illustrated later with examples.

The sixth and seventh columns in Table 12.4 with $\Delta$ are for the formation reaction $[Al(c, l)] = Al(g)$ at the head of the table. The values of $\Delta H_f^\circ$ [$\Delta_f H_T^\circ$ in this book] are obtained from the following identity:

$$\Delta_f H_T^\circ \equiv \Delta_f H_{298}^\circ + \left[H_T^\circ(g) - H_{298}^\circ(g)\right] - \left[H_T^\circ(c, l) - H_{298}^\circ(c)\right] \tag{12.34}$$

where $\Delta_f H_{298} = 78{,}800$ cal/mol is the experimental value at 298.15 K, the values of the relative enthalpy for the gas are from Table 12.4, and those for the crystal are

## Gibbs Energy Change of Reactions

**Table 12.4** Thermodynamic properties of Al(g). Reproduced from Gen. Ref. (2). Aluminum (ideal monatomic gas); [Formation: Al(c, l) = Al(g)].

| | (cal/mol·K) | | | | (kcal/mol) | | |
|---|---|---|---|---|---|---|---|
| $T, K$ | $C_p^\circ$ | $S^\circ$ | $-(G^\circ - H_{298}^\circ)/T$ | $H^\circ - H_{298}^\circ$ | $\Delta H_f^\circ$ | $\Delta G_f^\circ$ | Log $Kf$ |
| 298.15 | 5.112 | 39.302 | 39.302 | 0 | 78.800 | 69.102 | −50.653 |
| 300 | 5.111 | 39.335 | 39.305 | .009 | 78.798 | 69.041 | −50.296 |
| 400 | 5.047 | 40.794 | 39.501 | .517 | 78.708 | 65.802 | −35.952 |
| 500 | 5.018 | 41.917 | 39.877 | 1.020 | 78.587 | 62.589 | −27.357 |
| 600 | 5.002 | 42.830 | 40.295 | 1.521 | 78.437 | 59.405 | −21.638 |
| 700 | 4.993 | 43.600 | 40.713 | 2.021 | 78.255 | 56.248 | −17.561 |
| 800 | 4.987 | 44.266 | 41.116 | 2.520 | 78.035 | 53.118 | −14.511 |
| 900 | 4.983 | 44.854 | 41.501 | 3.018 | 77.773 | 50.017 | −12.146 |
| 933.61 | 4.982 | 45.037 | 41.625 | 3.185 | 77.675 | 48.982 | −11.466 |
| 933.61 | 4.982 | 45.037 | 41.625 | 3.185 | 75.095 | 48.982 | −11.466 |
| 1000 | 4.980 | 45.378 | 41.862 | 3.516 | 74.921 | 47.132 | −10.301 |
| 1100 | 4.978 | 45.853 | 42.204 | 4.014 | 74.660 | 44.365 | −8.814 |
| 1200 | 4.976 | 46.286 | 42.526 | 4.512 | 74.399 | 41.623 | −7.581 |
| 1300 | 4.975 | 46.684 | 42.831 | 5.009 | 74.137 | 38.902 | −6.540 |
| 1400 | 4.974 | 47.053 | 43.119 | 5.507 | 73.876 | 36.202 | −5.651 |
| 1500 | 4.973 | 47.396 | 43.393 | 6.004 | 73.614 | 33.519 | −4.884 |
| 1600 | 4.972 | 47.717 | 43.654 | 6.501 | 73.352 | 30.854 | −4.214 |
| 1700 | 4.972 | 48.018 | 43.901 | 6.999 | 73.091 | 28.208 | −3.626 |
| 1800 | 4.972 | 48.302 | 44.138 | 7.496 | 72.829 | 25.575 | −3.105 |
| 1900 | 4.971 | 48.571 | 44.364 | 7.993 | 72.567 | 22.958 | −2.641 |
| 2000 | 4.971 | 48.826 | 44.581 | 8.490 | 72.305 | 20.353 | −2.224 |

*Phase change*: 933.61 K, melting point of Al; $\Delta H^\circ = 2.580$ kcal/mol.
*Heat capacity (cal/mol·K) and enthalpy (kcal/mol) equations*:

298.15 − 2000 K: $\quad C_p^\circ = 4.966 + 0.131 \times 10^5 T^{-2}$

$\qquad\qquad\qquad\quad H^\circ - H_{298}^\circ = 4.966 \times 10^{-3}T - 0.131 \times 10^2 T^{-1} - 1.437$

*Formation equations (kcal/mol)*:

298.15 − 933.61 K: $\quad \Delta H_f^\circ = 78.828 + 0.374 \times 10^{-3}T - 1.727 \times 10^{-6}T^2 + 4.300 T^{-1}$
$\qquad\qquad\qquad\qquad \Delta G_f^\circ = 78.828 - 0.374 \times 10^{-3}T \ln T + 1.727 \times 10^{-6}T^2$
$\qquad\qquad\qquad\qquad\qquad + 2.150 T^{-1} - 31.027 \times 10^{-3}T$

933.61 − 2000 K: $\quad \Delta H_f^\circ = 77.560 - 2.624 \times 10^{-3}T - 13.100 T^{-1}$
$\qquad\qquad\qquad\quad \Delta G_f^\circ = 77.560 + 2.624 \times 10^{-3}T \ln T - 6.550 T^{-1} - 48.550 \times 10^{-3}T$

*Sources*: Enthalpy of formation at 298 K from CODATA (*36*). All other data from Hultgren (*99*).

from Table 12.2. Thus, for 1000 K, we have $\Delta H_{f,1000}^\circ = 78,800 + 3516 - 7,395 = 74,921$ cal/mol, which is in Table 12.4. Substitution of the equations for the relative enthalpies of gas and condensed phases from Tables 12.4 and 12.2 in (12.34) yield the equations for the two standard enthalpies of formation equations for the two temperature ranges in the footnotes to Table 12.4. The standard change in entropy is obtained by subtracting the entropy in the third column of Table 12.2 from that of Table 12.4 for each temperature, e.g., for 1000 K, $\Delta Sf°_{1000} = 45.378 - 17.589 = 27.789$ cal/mol·K. We substitute the foregoing values of $\Delta Hf°$ and $\Delta Sf°$ in $\Delta Gf° = \Delta Hf°_{1000} - T\Delta Sf°_{1000} = 74{,}921 - 1000 \times 27.789 = 47{,}132$ cal/mol, as in Table 12.4. Likewise, the corresponding equations for $\Delta Gf°$ can be obtained from the equations for $\Delta Hf°_T$ and $\Delta Sf°_T$, and these equations for $\Delta Gf°$ are in Table 12.4. The last column represents $\text{Log } Kf = -\Delta Gf°_T/(2.303 \times RT)$ wherein $Kf$ is the equilibrium constant of formation reaction, and in this case, it is equal to the equilibrium pressure of Al(g) over the condensed phase; at 1000 K, $Kf = 5.006 \times 10^{-11}$ atm $= 5.072 \times 10^{-11}$ bar. Therefore, the conversion of $\Delta Gf°$ into Joules and the standard state of 1 bar requires the conversion of the listed values by multiplication with 4.184 and the addition of $-RT \ln 1.01325$; the result for 1000 K is $197{,}200 - 109 = 197{,}091$ J/mol (very close to 197,027 J mol$^{-1}$ in Table 12.3). The values of $S°$ are equal to the listed values multiplied by 4.184 plus $R \ln 1.01325$; for 1000 K, we have $S° = 45.378 \times 4.184 + R \ln 1.01325 = 189.971$ J/mol·K (nearly identical with 189.977 in Table 12.3). This conversion is based on the fact that 1 bar is lower in pressure than 1 atm, hence the entropy must increase. If we rewrite (5.8) as $S°(1 \text{ bar}) - S°$ (1 atm $= 1.01325$ bars) $= R \ln(1.01325/1) = 0.1094$, then it becomes evident that $S°(1 \text{ bar})$ is larger than $S°(1 \text{ atm})$ by $R \ln(1.01325)$. $\Delta Hf°_{1000}$ is independent of moderate changes in pressure, and it is $74{,}921 \times 4.184 = 313{,}469$ J/mol (313,605 in Table 12.3 due to different processes of assessment). These values are in Table 12.5, together with other values for 1000 to 1400 K for comparison with Table 12.3.

For the CONDENSED PHASES as those in Table 12.2, the conversion to J and 1 bar system of units would simply require multiplication by 4.184.

There are also the tables for liquid phases for convenience in certain types of computer calculations, e.g. Al(l) from 0–3000 K in Gen. Ref. (1), but such tables are not in Gen. Ref. (2).*

## Use of Tables

The use of tables in calculating $\Delta_r H°$, $\Delta_r S°$, and $\Delta_r G°$ for a reaction at a desired temperature is very simple, particularly when the temperature corresponds to the tabulated values. If the temperature lies between two consecutive tabulated values, a linear interpolation, or if the accuracy of data justifies it, a quadratic interpolation by fitting the adjoining three values into an appropriate equation is necessary. Let $A$, $B$, and $C$ be the successive values at temperatures such as 1100, 1200, and 1300 K respectively; a value $D$ midway between 1100 and 1200 K is obtained by quadratic fitting from

$$D = 0.375A + 0.75B - 0.125C \tag{12.35}$$

---

*For organic substances, see J. B. Pedley, R. D. Naylor, and S. P. Kirby, "Thermochemical Data of Organic Compounds," Chapman and Hall (1986). See also Gen. Ref. (7).

## Gibbs Energy Change of Reactions 225

**Table 12.5** Thermodynamic properties of Al(g);* adapted from Table 12.4.

| T, K | (J/mol·K) | | | (J/mol) | | |
|---|---|---|---|---|---|---|
| | $C_p^\circ$ | $S^\circ$ | $-(G^\circ - H_{298}^\circ)/T$ | $H^\circ - H_{298}^\circ$ | $\Delta H_f^\circ$ | $\Delta G_f^\circ$ |
| 1000 | 20.836 | 189.971 | 175.260 | 14,711 | 313,469 | 197,091 |
| 1100 | 20.828 | 191.958 | 176.691 | 16,795 | 312,377 | 185,503 |
| 1200 | 20.820 | 193.770 | 178.038 | 18,878 | 311,285 | 174,019 |
| 1300 | 20.815 | 195.435 | 179.314 | 20,958 | 310,189 | 162,624 |
| 1400 | 20.811 | 196.979 | 180.519 | 23,041 | 309,097 | 151,316 |

*Aluminum (ideal monatomic gas); [Formation: Al(c, l) = Al(g)]; standard pressure = 1 bar

For a reaction, such as (12.1), $\Delta_r H_T^\circ$ for any temperature is obtained as follows:

$$\Delta_r H_T^\circ = \nu_\mathrm{I} \cdot \Delta_f H_T^\circ(A_\mathrm{I}) + \nu_\mathrm{II} \cdot \Delta_f H_T^\circ(A_\mathrm{II}) + \cdots$$
$$-\nu_1 \cdot \Delta_f H_T^\circ(A_1) - \nu_2 \cdot \Delta_f H_T^\circ(A_2) + \cdots \quad (12.36)$$

The equations for $\Delta_r G_T^\circ$ and $\Delta_r S_T^\circ$ are identical in form. As a simple example, consider the following reaction at 1100–1300 K; for 1100 K, the results are

$$\text{Al(g)} + 3\text{HF(g)} = \text{AlF}_3(\text{c}) + 1.5\text{H}_2(\text{g}) \quad (12.37)$$
$$\downarrow \quad\quad \downarrow \quad\quad\quad\quad \downarrow \quad\quad\quad \downarrow$$

$\Delta_r H^\circ = -312,377 + 3 \times 274,870 - 1,509,488 + 1.5 \times 00 = -997,255$ J
$\Delta_r G^\circ = -185,503 + 3 \times 278,830 - 1,223,996 + 1.5 \times 00 = -573,009$ J

where the values are obtained from Tables 12.5–12.8. The formation properties of the stable elements are obviously zero as indicated for $H_2$. These values are written under each species with the stoichiometric coefficients and added algebraically to obtain the standard enthalpy and Gibbs energy of reaction at 1100 K. The value of $\Delta_r Gef$, i.e. the change in Gibbs energy function, at 1100 K is $\Delta_r Gef = 176.691 + 3 \times 190.647 - 118.707 - 1.5 \times 147.550 = 408.600$ J/K, where the sequence of numbers corresponds to (12.37). Similar calculations for 1200 and 1300 K are listed in Table 12.9. If we wish to compute the value of $\Delta_r G^\circ$ for 1150 K, a linear interpolation gives $-553,804$ J, whereas (12.35) would yield $\Delta_r G^\circ = -0.375 \times 573,009 - 0.75 \times 534,598 + 0.125 \times 496,485 = -553,766$. While these two values are closer than the experimental errors, at much lower temperatures, the use of (12.35) is recommended. The *Gef* varies slowly with temperature, and it can be used for linear interpolation of $\Delta_f G^\circ$. *Gef* is also useful in experimental work as will be seen later.

**Example 1** Calculate $\Delta_r H^\circ$ at 298.15 for Reaction (12.37), given $\Delta_r H_{1100}^\circ = -997,255$ J and $\Delta_r G_{1100}^\circ = -573,009$ J, both at 1100 K by using (a) the enthalpy functions, and (b) the Gibbs energy functions in Tables 12.5–12.8.

**Table 12.6** Thermodynamic properties of HF(g); adapted from Gen. Ref. (3).

| T, K | (J/(mol·K)) | | | J/mol | | |
|---|---|---|---|---|---|---|
| | $C_p^\circ$ | $S^\circ$ | $-(G^\circ - H_{298}^\circ)/T$ | $H^\circ - H_{298}^\circ$ | $\Delta H_f^\circ$ | $\Delta G_f^\circ$ |
| 1000.00 | 30.142 | 209.285 | 188.636 | 20,649 | −274,532 | −278,455 |
| 1100.00 | 30.516 | 212.178 | 190.647 | 23,684 | −274,870 | −278,830 |
| 1200.00 | 30.934 | 214.850 | 192.554 | 26,756 | −275,202 | −279,176 |
| 1300.00 | 31.384 | 217.344 | 194.366 | 29,872 | −275,519 | −279,494 |
| 1400.00 | 31.832 | 219.686 | 196.091 | 33,033 | −275,822 | −279,788 |

**Table 12.7** Thermodynamic properties of AlF$_3$(c); adapted from gen. Ref. (3).

| T, K | (J/(mol·K)) | | | J/mol | | |
|---|---|---|---|---|---|---|
| | $C_p^\circ$ | $S^\circ$ | $-(G^\circ - H_{298}^\circ)/T$ | $H^\circ - H_{298}^\circ$ | $\Delta H_f^\circ$ | $\Delta G_f^\circ$ |
| 1000.00 | 100.831 | 179.487 | 112.139 | 67,348 | −1,510,890 | −1,250,016 |
| 1100.00 | 101.893 | 189.147 | 118.707 | 77,485 | −1,509,488 | −1,223,996 |
| 1200.00 | 102.917 | 198.057 | 124.953 | 87,725 | −1,508,014 | −1,198,107 |
| 1300.00 | 103.915 | 206.334 | 130.898 | 98,067 | −1,506,467 | −1,172,343 |
| 1400.00 | 104.894 | 214.071 | 136.566 | 108,508 | −1,504,845 | −1,146,702 |

**Table 12.8** Thermodynamic properties of H$_2$(g); adapted from Gen. Ref. (3).

| T, K | (J/(mol·K)) | | | (J/mol) |
|---|---|---|---|---|
| | $C_p^\circ$ | $S^\circ$ | $-(G^\circ - H_{298}^\circ)/T$ | $H^\circ - H_{298}^\circ$ |
| 1000.00 | 30.205 | 166.216 | 145.536 | 20,680 |
| 1100.00 | 30.579 | 169.112 | 147.550 | 23,719 |
| 1200.00 | 30.992 | 171.790 | 149.460 | 26,797 |
| 1300.00 | 31.424 | 174.288 | 151.275 | 29,918 |
| 1400.00 | 31.863 | 176.633 | 153.003 | 33,082 |

**Table 12.9** Thermodynamic calculations for Reaction (12.37).

| T, K | $\Delta_r H^\circ$, J | $\Delta_r G^\circ$, J | $\Delta_r Gef$, J/K |
|---|---|---|---|
| 1100 | −997,255 | −573,009 | 408.600 |
| 1200 | −993,693 | −534,598 | 406.557 |
| 1300 | −990,099 | −496,485 | 404.602 |

**Solution** (a) For Reaction (12.37), we can write the following identity, which is another form of (3.89):

$$\sum (H° - H°_{298})_{products} - \sum (H° - H°_{298})_{reactants} \equiv \Delta_r H°_T - \Delta_r H°_{298}$$
(12.38)

The left side of this equation is $77,485(AlF_3) + 1.5 \times 23,719(H_2) - 16,795(Al) - 3 \times 23,684(HF) = 25,217$ J. We subtract this value from $\Delta_r H°_{1100} = -997,255$ to obtain $\Delta_r H°_{298} = -1,022,472$ J. [Complete tables down to 298.15 K yield directly $\Delta_r H°_{298} = -1,510,400 \ (AlF_3) + 3 \times 272,546(HF) - 329,699(Al) = -1,022,461$, within 11 J.]

(b) The definition of *Gef*, and its application to (12.37) give

$$\Delta_r [Gef] \equiv [\Delta_r G° - \Delta H°_{298}]/T = 408.600 = [-573,009 - \Delta H°_{298}]/1100$$

where 408.600 was calculated from Tables 12.5–12.8 earlier in the text. Solution of this equation yields $\Delta_r H°_{298} = -1,022,469$ J.

**Example 2** Calculate $\Delta_r S°$ for Reaction (12.37) at 1200 K from Table 12.9, and compare the result with that from Tables 12.5–12.8 by directly using the values of $S°$ for 1200 K.

**Solution** Substitution of $\Delta_r H°_{1200} = -993,693$, and $\Delta_r G°_{1200} = -534,598$ in (12.9) leads to

$$\Delta_r G° = \Delta_r H° - T\Delta_r S° = -534,598 = -993,693 - 1200 \Delta_r S° \quad (12.39)$$

from which $\Delta_r S°_{1200} = -382.579$ J/mol·K at 1200 K. From Tables 12.5–12.8, the algebraic summation of entropies gives $\Delta_r S°_{1200} = 98.057 \ (AlF_3) + 1.5 \times 171.790(H_2) - 193.770(Al) - 3 \times 214.850(HF) = -382.578$ J/mol·K at 1200 K.

**Example 3** Calculate $\Delta_r G°$, $\Delta_r H°$, and $\Delta_r S°$ at 1100 K for Reaction (12.37) by using Table 12.1 for liquid Al as the reactant instead of Al(g).

**Solution** The results for $\Delta_r G°$ and $\Delta_r H°$ are as follows:

$$\Delta_r G°_{1100} = -1,223,996(AlF_3) + 3 \times 278,830(HF) = -378,506 \text{ J}$$

$$\Delta_r H°_{1100} = -1,509,488(AlF_3) + 3 \times 274,870(HF) = -684,878 \text{ J}.$$

Observe that Al(l) is in its stable standard state for which the standard Gibbs energy and enthalpy of formation are zero. The result for the standard change in entropy is

$$\Delta_r S°_{1100} = 189.147(AlF_3) + 1.5 \times 169.112(H_2) - 76.426(Al)$$
$$- 3 \times 212.178(HF) = -270.145 \text{ J/mol·K}$$

The values of $\Delta_r G°$ and $\Delta_r H°$ substituted in $1100\Delta_r S° = -\Delta_r G° + \Delta_r H° = -297,372$ give $\Delta_r S° = -270.338$ J/mol·K, within 0.193.

**Example 4** Pure $Cu_2O$ and pure $Cu_2S$, each in its own container, are placed in a sealed evacuated chamber and then heated to 1000 K to attain equilibrium by gas-phase reactions. After equilibrium, some excess pure $Cu_2O$ and $Cu_2S$ remain. Assume that the gas phase contains $S_2$, SO, $SO_2$ and $SO_3$ as the major gas species. The values of $\Delta_f G°$ in J mol$^{-1}$ from Gen. Refs. (1) and (2) at 1000 K are as follows:

|  | $Cu_2O$(s) | $Cu_2S$(s) | SO(g) | $SO_2$(g) | $SO_3$(g) |
|---|---|---|---|---|---|
| $-\Delta_f G°_{1000}$: | 95,517 | 92,692 | 64,316 | 288,725 | 293,639 |

Calculate the partial pressures of gases in the chamber at 1000 K.

**Solution** The reactions involving all the species are

$$2\,Cu_2S(s) = 4\,Cu(s) + S_2(g); \quad \Delta_r G° = +185,384 \text{ J}$$
$$Cu_2O(s) + Cu_2S(s) = 4\,Cu(s) + SO(g); \quad \Delta_r G° = +123,893 \text{ J}$$
$$2\,Cu_2O(s) + Cu_2S(s) = 6\,Cu(s) + SO_2(g); \quad \Delta_r G° = -4,999 \text{ J}$$
$$3\,Cu_2O(s) + Cu_2S(s) = SO_3(g); \quad \Delta_r G° = +85,604 \text{ J}$$

The first reaction gives $\Delta_r G° = 185,384 = -1000R \ln K_p$; $K_p = P(S_2) = 2.07 \times 10^{-10}$ bar. The second reaction gives $\Delta_r G° = 123,893 = -RT \ln P(SO)$; $P(SO) = 3.38 \times 10^{-7}$. Likewise, the third reaction gives $P(SO_2) = 1.8244$ bars, and the fourth reaction, $P(SO_3) = 3.78 \times 10^{-5}$ bar. Other sulfur gas species of very small pressures also exist but ignored in these computations. Additional reactions could have taken place if the solids were mixed. (For a similar problem, see Y. K. Rao, p. 385, loc. cit.).

## Other Thermodynamic Tables and Compilations

Various other thermodynamic tables and compilations are available [Gen. Refs. (5), (7), (10)–(12)]. It is essential that the introductory chapters and methods of tabulation be read prior to their utilization. A useful compilation for minerals is in the U.S. Geological Survey Bulletin [Gen. Ref. (5)] by Robie and Hemingway. Table 12.10 for $CaCO_3$ is taken from this Bulletin to be published. The format is quite similar to that in Gen. Ref. (1), and requires no further comments. Table 12.11 for acetone and isopropyl alcohol is from a compilation of data for organic substances by Stull, Westrum and Sinke [Gen. Ref. (7)]. It is identical in form with Table 12.2.

New data and compilations appear in physical, chemical and engineering journals, particularly in Journal of Chemical Thermodynamics, High Temperature Science, Journal of Chemical and Engineering Data, and Journal of Physical and Chemical Reference Data. [See also the compilation in Gen. Ref. (4) by Kubaschewski, Alcock and Spencer.] Thermodynamic data for metals and alloys

**Table 12.10** Thermodynamic properties of calcium carbonate.* Reproduced from Gen. Ref. (5).

| | $(J \cdot mol^{-1} \cdot K^{-1})^\dagger$ | | | | $(kJ \cdot mol^{-1})$ | | |
|---|---|---|---|---|---|---|---|
| Temp., K | $C_p^\circ$ | $S_T^\circ$ | $\frac{(H_T^\circ - H_{298}^\circ)}{T}$ | $\frac{-(G_T^\circ - H_{298}^\circ)}{T}$ | $\Delta_f H^\circ$ | $\Delta_f G^\circ$ | Log $K_f$ |
| 298.15 | 83.47 | 91.71 | 0.00 | 91.71 | −1207.4 | −1128.5 | 197.70 |
| 300 | 83.82 | 92.23 | 0.52 | 91.71 | −1207.4 | −1128.0 | 196.39 |
| 400 | 97.00 | 118.36 | 23.18 | 95.19 | −1206.3 | −1101.6 | 143.86 |
| 500 | 104.55 | 140.88 | 38.75 | 102.13 | −1204.9 | −1075.6 | 112.37 |
| 600 | 109.88 | 160.43 | 50.18 | 110.25 | −1203.5 | −1049.9 | 91.40 |
| 700 | 114.16 | 177.70 | 59.02 | 118.68 | −1202.0 | −1024.4 | 76.44 |
| 800 | 117.88 | 193.19 | 66.15 | 127.04 | −1201.6 | −999.0 | 65.23 |
| 900 | 121.28 | 207.28 | 72.09 | 135.18 | −1200.3 | −973.8 | 56.52 |
| 1000 | 124.48 | 220.22 | 77.17 | 143.05 | −1199.1 | −948.7 | 49.55 |
| 1100 | 127.55 | 232.23 | 81.61 | 150.62 | −1198.1 | −923.7 | 43.86 |
| 1200 | 130.53 | 243.46 | 85.57 | 157.89 | −1204.9 | −898.2 | 39.10 |

\* Calcite; Formula wt. 100.087; CaCO$_3$: Rhombohedral crystals 298.15 to 1200 K.
† *Phase changes*:

| | | | |
|---|---|---|---|
| Melting T | K | Boiling T | K |
| $\Delta_{fus} H^\circ$ | kJ | $\Delta_{vap} H^\circ$ | kJ |
| | | Molar Vol. | 3.693 J · bar$^{-1}$ |
| $H_{298}^\circ - H_0^\circ$ | 14.48 kJ | | 36.93 cm$^3$ |
| A = −1.200E + 03; | B = 2.518E − 01; | C = −3.04E + 05 | 10/14/92 |

are in Gen. Refs. (9)–(13), and for organic substances, in Gen. Ref. (14). See also "Thermodynamic Data for Biochemistry and Biotechnology", edited by H.-J. Hinz, Springer-Verlag (1986).

## Use of Tabular Data in Experimental Work

Parts of compiled data are very useful in experimental work for determination of enthalpies of formation or reaction. For this purpose, experimental determination of equilibrium constants, $K_p$, is necessary, for which $\Delta G^\circ = -RT \ln K_p$. There are two methods by which the related standard enthalpy change can be derived, the second law method, and the third law method.

The second law method does not require a knowledge of some or all of the entropies of the species from 0 K to high temperatures for a given reaction. It requires at least two, and in practice many values of $K_p$ in a sufficiently wide range of temperature. The resulting data for $K_p$ are then used in

$$\ln K_p = -\Delta G^\circ / RT = -(\Delta H^\circ / RT) + (\Delta S^\circ / R) \tag{12.40}$$

to express the variation of $K_p$ with temperature. In this equation, $\Delta H^\circ$, hence $\Delta S^\circ$, is assumed to be independent of temperature, because (i) the range of temperature to which $K_p$ can be measured accurately is usually narrow and (ii) the errors in

## 230  Thermodynamics

**Table 12.11** Thermodynamic properties of acetone and isopropyl alcohol in idea gas state (1 atm). From Stull, Westrum and Sinke, Gen. Ref. (7).

| $T°K$ | (cal/(mol°K)) | | | (kcal/mol) | | | |
|---|---|---|---|---|---|---|---|
| | $C_p°$ | $S°$ | $-(G°-H_{298}°)/T$ | $H°-H_{298}°$ | $\Delta H_f°$ | $\Delta G_f°$ | $\text{Log } K_p$ |
| No. 554, Acetone, $C_3H_6O$; (Ideal Gas State); Mol. Wt. 58.078 | | | | | | | |
| 298 | 17.90 | 70.49 | 70.49 | 0.00 | −52.00 | −36.58 | 26.811 |
| 300 | 17.97 | 70.61 | 70.50 | 0.04 | −52.02 | −36.48 | 26.577 |
| 400 | 22.00 | 76.33 | 71.25 | 2.04 | −53.20 | −31.12 | 17.001 |
| 500 | 25.89 | 81.67 | 72.80 | 4.44 | −54.22 | −25.48 | 11.135 |
| 600 | 29.34 | 86.70 | 74.70 | 7.20 | −55.07 | −19.65 | 7.156 |
| 700 | 32.34 | 91.45 | 76.76 | 10.29 | −55.74 | −13.69 | 4.273 |
| 800 | 34.93 | 95.94 | 78.88 | 13.65 | −56.28 | −7.64 | 2.088 |
| 900 | 37.19 | 100.19 | 81.01 | 17.26 | −56.67 | −1.54 | 0.373 |
| 1000 | 39.15 | 104.21 | 83.13 | 21.08 | −56.93 | 4.61 | −1.007 |
| No. 520, Isopropyl Alcohol, $C_3H_8O$; (Ideal Gas State); Mol. Wt. 60.094 | | | | | | | |
| 298 | 21.21 | 74.07 | 74.07 | 0.00 | −65.15 | −41.49 | 30.411 |
| 300 | 21.31 | 74.21 | 74.08 | 0.04 | −65.18 | −41.34 | 30.118 |
| 400 | 26.78 | 81.09 | 74.98 | 2.45 | −66.65 | −33.17 | 18.120 |
| 500 | 31.89 | 87.64 | 76.86 | 5.40 | −67.82 | −24.66 | 10.776 |
| 600 | 35.76 | 93.81 | 79.18 | 8.78 | −68.74 | −15.94 | 5.804 |
| 700 | 39.21 | 99.58 | 81.68 | 12.53 | −69.46 | −7.07 | 2.208 |
| 800 | 42.13 | 105.01 | 84.26 | 16.61 | −69.99 | 1.87 | −0.511 |
| 900 | 44.63 | 110.12 | 86.86 | 20.95 | −70.36 | 10.88 | −2.642 |
| 1000 | 46.82 | 114.94 | 89.43 | 25.52 | −70.58 | 19.93 | −4.355 |

$K_p$ very seldom indicate the dependence of $\Delta H°$ on temperature. A plot of $\ln K_p$ versus $1/T$, yields a straight line whose slope is $-\Delta H°/R$ for the reaction. The equation for the straight line is obtained by the method of least squares, often with a standard deviation in $\ln K_p$. The value of $\Delta H°$ from the slope refers to the temperature corresponding to the average of all values of $1/T$, although in the majority of old publications either the average of all recorded temperatures, or the simple average of extreme temperatures has been taken. Extrapolation of $\Delta H_T°$ to 298.15 K is then carried out by (12.38) if the enthalpy functions are known from $T$ down to 298.15 K. The standard entropy change $\Delta S_T°$ from (12.40), which also refers to the average of all $1/T$, can be extrapolated to 298.15 K by

$$\Delta S_T° = \Delta S_{298}° + \Delta\left(S_T° - S_{298}°\right) \tag{12.41}$$

If the entropy of one of the species of the reaction at 298.15 K is not known from the third law but the entropies of the remaining species are known, then the unknown entropy can be computed from the entropy of reaction $\Delta S_{298}°$ obtained from (12.41). The unknown standard entropy determined in this manner is the second law standard entropy.

An alternative method is expressing $\Delta G°$ as a function of temperature as in (12.28), and then rearranging it to obtain

$$-\frac{\Delta H_\theta°}{T} + \left(\Delta S_\theta° - \Delta\alpha\right) = R\ln K_p - \Delta\alpha \ln T - \Delta\beta T - \Delta\lambda T^2 + \Delta\epsilon T^{-2}$$

(12.42)

The right side of this equation is designated by sigma, $\Sigma$; therefore,

$$\Sigma = R\ln K_p - \Delta\alpha \ln T - \Delta\beta T - \Delta\lambda T^2 + \Delta\epsilon T^{-2}$$

(12.43)

Equations (12.42) and (12.43) are identical in form with (8.18) to (8.18a). A plot of $\Sigma$ versus $1/T$, called the sigma plot, yields a straight line since $\Delta H_\theta°$ and $(\Delta S_\theta° - \Delta\alpha)$ are both constants. The plot should be made by the method of least squares, and from such a plot, the analytical equation for $\Sigma$ is

$$\Sigma = -\frac{\Delta H_\theta°}{T} + \left(\Delta S_\theta° - \Delta\alpha\right)$$

(12.44)

The values of $\Delta H_\theta°$ and $(\Delta S_\theta° - \Delta\alpha)$ substituted in (12.26)–(12.28) yield $\Delta H_T°$, $\Delta S_T°$ and $\Delta G_T°$ as functions of temperature, hence $\Delta H_{298}°$. The enthalpy of reaction $\Delta H_{298}°$ obtained by the preceding method is called the second law method by sigma plot. It is superior to the simple second law method, and it should be used whenever data for $C_p°$ are available above 298 K. At temperatures above 1000 K, the scattering in the data for $K_p$ is high, and therefore, either method is satisfactory, and generally the simple second law method described by (12.40) is preferred.

The use of heat capacity equations assumes that there are no phase transitions among the reaction species; however, if there are phase transitions, the values of thermodynamic quantities from the single phase tables have to be added to $\Delta H_{298}°$ and $\Delta S_{298}°$ if these values are expected to refer to the stable substances at 298.15 K.

The third law method is considerably simpler than the second law method, and requires the Gibbs energy function, *Gef*, which signifies that the heat capacity, entropy, and enthalpy function, for each species of reaction are known. From the definition of *Gef*, it is evident that

$$\Delta_r Gef \equiv \Delta_r \left(\frac{G_T° - H_{298}°}{T}\right) \equiv \frac{\Delta_r G_T°}{T} - \frac{\Delta_r H_{298}°}{T}$$

$$= -R\ln K_p - \frac{\Delta_r H_{298}°}{T}$$

(12.45)

where $\Delta G° = -RT\ln K_p$ is used to obtain the last equality. This equation shows that, in principle, a single value of $K_p$ is sufficient to determine the value of $\Delta H_{298}°$, though, additional values at various temperatures are desirable. An important advantage of the third law method is that a simple arithmetic averaging of $\Delta H_{298}$ is sufficient to obtain the desired value. Examples of the foregoing methods will now be presented.

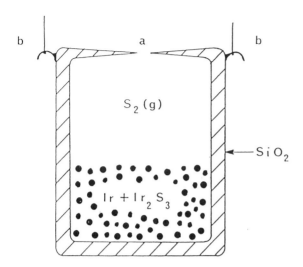

**Fig. 12.1** Silica Knudsen effusion cell. Orfice area is a; powdered Ir(s) + $Ir_2S_3$(s) mixture is on bottom; $S_2$(g) is above solid mixture and effuses out of orifice a. Cell is suspended from recording microbalance by b.

*Example 5* The equilibrium pressure $P$ of $S_2$(g) over Ir(s) + $Ir_2S_3$(s) at various temperatures has been measured* by the effusion technique described in numerous publications.† Pure $Ir_2S_3$(s), or a mixture of Ir(s) and $Ir_2S_3$ is placed in a quartz Knudsen cell having a circular orifice with sharp edges, as shown in Fig. 12.1, and suspended from a microbalance to measure the rate of weight loss **W** at a constant temperature. The rate of weight loss **W** in g sec$^{-1}$ substituted in the following equation, derived from the kinetic theory of gases, yields the equilibrium pressure $P$ of $S_2$:

$$P(\text{atm}) = 0.022557 \frac{\mathbf{W}}{\mathbf{a}} \left(\frac{T}{M}\right)^{0.5} ; \quad (1 \text{ atm} = 1.01325 \text{ bar}) \quad (12.46)$$

In this equation, **a** is the orifice area in cm$^2$, $T$, the temperature in K, and $M$ the molecular weight of the effusing gas, i.e., 64.13 for $S_2$. The logarithm of the sulfur pressure versus $1/T$ is plotted in Fig. 12.2 and fitted by the method of least squares into the following equation:

$$\log P(\text{atm}) = -(13{,}367/T) + 9.2133 \quad (12.47)$$

*E. T. Chang and N. A. Gokcen, High Temp. Sci., *4*, 432, (1972).
†See Gen. Ref. (21); S. Fujishiro and N. A. Gokcen, J. Electrochem. Soc., *109*, 835 (1962); W. G. O'Brien, H. J. Jensen, R. N. Benedict, and R. G. Bautista, Met. Trans., *7B*, 671 (1976); R. Chastel, M. Saito, and C. Bergman, J. Alloys and Compounds, *205*, 39 (1994).

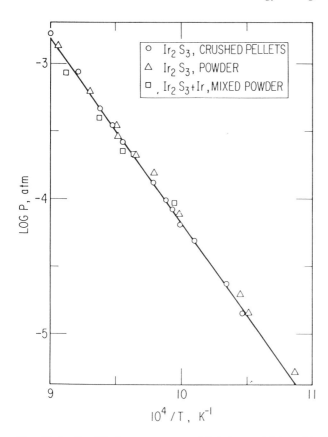

**Fig. 12.2** Equilibrium pressure of $S_2(g)$ over $Ir(c)$ and $Ir_2S_3(c)$ at various temperatures.

The chemical reaction investigated and its equilibrium constant are as follows:

$$2Ir(c) + 1.5S_2(g) = Ir_2S_3(c) \qquad K_p = P^{-1.5} \qquad (12.48)$$

The equation for $\Delta G°$ can be obtained by substituting (12.47) into $\Delta G° = 1.5 \times 4.5757 \times T \times \log P$, with the result that

$$\Delta G° = -91{,}745 + 63.236T \qquad \text{cal mol}^{-1} \text{ of } Ir_2S_3 \qquad (12.49)$$

where, $\Delta H°_{1025} = -91{,}745$ cal mol$^{-1}$, is called the second law enthalpy for Reaction (12.48). This value refers to 1025 K because the average value of $1/T$ for all the points in Fig. 12.2 is equal to $1/1025$. It is customary to reduce $\Delta H°_{1025}$ to 298.15 K with the enthalpy function $\Delta(H°_T - H°_{298})$ for Reaction (12.48) listed in Table 12.12 in the last row, obtained from the preceding columns for $H°_T - H°_{298}$; thus for example, $\Delta(H°_T - H°_{298})$ for 1000 K is calculated from $21{,}838 - 1.5 \times 5{,}975 - 2 \times 4{,}522 = 3{,}832$. A linear interpolation of the enthalpy function in

**Table 12.12** Values of $S_T^\circ$, $H_T^\circ - H_{298}^\circ$ for Ir(s), $S_2$(g), and $Ir_2S_3$(s) and $\Delta(H_T^\circ - H_{298}^\circ)$ for Reaction (12.48). Last row is obtained from columns for $H_T^\circ - H_{298}^\circ$.

|  | $S_T^\circ$ cal mol$^{-1}$ K$^{-1}$ | | | | $H_T^\circ - H_{298}^\circ$ cal mol$^{-1}$ | | |
| --- | --- | --- | --- | --- | --- | --- | --- |
| $T$, K | 298.15 | 900 | 1000 | 1100 | 900 | 1000 | 1100 |
| Ir(S) | 8.48 | 15.423 | 16.158 | 16.842 | 3,840 | 4,522 | 5,240 |
| $S_2$(g) | 54.510 | 63.758 | 64.687 | 65.531 | 5,093 | 5,975 | 6,860 |
| $Ir_2S_3$(s) | — | — | — | — | 18.411 | 21,838 | 25,371 |

| $T$, K | | 900 | 1000 | 1100 |
| --- | --- | --- | --- | --- |
| $\Delta(H_T^\circ - H_{298}^\circ)$/cal mol$^{-1}$ | | 3,092 | 3,832 | 4,601 |

the last row to 1025 K gives $\Delta(H_{1025}^\circ - H_{298}^\circ) = 4{,}024$ cal mol$^{-1}$, and from this equation, $4{,}024 = \Delta H_{1025}^\circ - \Delta H_{298}^\circ = -91{,}745 - \Delta H_{298}^\circ$; hence $\Delta H_{298}^\circ = -95{,}769$ cal mol$^{-1}$ for the second law enthalpy of reaction.

Combination of the following reactions and their $\Delta H_{298}^\circ$ yield the standard enthalpy of formation of $Ir_2S_3$(s), $\Delta_f H_{298}^\circ$, from its elements in their stable standard states, i.e.,

$$2Ir(s) + 1.5S_2(g) = Ir_2S_3(s) \qquad \Delta H_{298}^\circ = -95{,}769 \text{ cal mol}^{-1}$$
$$3S(\text{rhombic}) = 1.5S_2(g) \qquad \Delta H_{298}^\circ = +46{,}260 \text{ cal mol}^{-1}$$

$$2Ir(s) + 3S(\text{rhombic}) = Ir_2S_3(s) \qquad \Delta_f H_{298}^\circ = -49{,}509 \text{ cal mol}^{-1}$$

where $\Delta_f H_{298}^\circ = -49{,}509$ cal mol$^{-1}$ is the standard enthalpy of formation of $Ir_2S_3$ at 298.15 K. For the second reaction, $\Delta H_{298}^\circ = +46{,}260$ cal mol$^{-1}$ has been obtained from the single phase tables, and this shows the usefulness of such tables.

The *second law entropy of reaction* is $-63.236$ cal mol$^{-1}$ K$^{-1}$ in (12.49) at 1025 K. It can be extrapolated from 1025 to 298.15 K by using $\Delta(S_T^\circ - S_{298}^\circ)$ for Reaction (12.48) and by following an identical procedure. For $Ir_2S_3$(s), $S_{298}^\circ$ has never been measured; however, $S_T^\circ - S_{298}^\circ$ can be calculated from its heat capacity, $C_p^\circ = 24.3 + 0.0105\,T$, valid above 298 K. Thus, integration of $(C_p^\circ/T)\,dT$ gives

$$S_{1025}^\circ - S_{298}^\circ = 24.3 \ln(1025/298.15) + 0.0105(1025 - 298.15)$$
$$= 37.639 \text{ cal mol}^{-1} \text{ K}^{-1}; \qquad Ir_2S_3$$

For $1.5S_2$(g), $1.5(S_{1025}^\circ - S_{298}^\circ) = 15.582$, and for 2 Ir(s), $2(S_{1025}^\circ - S_{298}^\circ) = 15.698$ as linearly interpolated from Table 12.12. The value of $\Delta(S_{1025}^\circ - S_{298}^\circ)$ for Reaction (12.48) is then

$$\Delta(S_{1025}^\circ - S_{298}^\circ) = 37.639 - 15.582 - 15.698$$
$$= 6.359 \text{ cal(mole of } Ir_2S_3)^{-1} \text{ K}^{-1}$$

**Table 12.13** Equilibrium total pressure, $P_e$ and $\Delta H^\circ_{298}$ for $2\text{AlN(s)} = 2\text{Al(g)} + \text{N}_2\text{(g)}$.

| T, K | $P_e \times 10^6$/atm | $\Delta H^\circ_{298}$/kcal |
|---|---|---|
| 1776 | 66 | 308.1 |
| 1818 | 148 | 306.4 |
| 1887 | 406 | 306.5 |
| 1919 | 567 | 307.8 |
| 1972 | 1120 | 308.0 |
| Av. of 29 data : | | 307.3 ± 0.7 |

Substitution of $\Delta S^\circ_{1025} = -63.236$ cal mol$^{-1}$ K$^{-1}$ from (12.49) into the following identity

$$\Delta\left(S^\circ_{1025} - S^\circ_{298}\right) \equiv \Delta S^\circ_{1025} - \Delta S^\circ_{298} = 6.359 \text{ cal mol}^{-1} \text{ K}^{-1}$$

yields $\Delta S^\circ_{298} = -69.595$ cal(mole of Ir$_2$S$_3$)$^{-1}$ K$^{-1}$. The result for $S^\circ_{298}$ is related to the measured values of the entropies of 2 Ir(s) and 1.5 S$_2$(g) and the unknown entropy of Ir$_2$S$_3$(s) by

$$\Delta S^\circ_{298} = -69.595 = S^\circ_{298}(\text{Ir}_2\text{S}_3, \text{s}) - 1.5 S^\circ_{298}(\text{S}_2, \text{g}) - 2 S^\circ_{298}(\text{Ir}, \text{s})$$

and the substitution of $1.5 S^\circ_{298} = 81.765$ for $1.5 S_2$(g) and $2 S^\circ_{298} = 16.96$ for 2Ir(s) gives $S^\circ_{298}(\text{Ir}_2\text{S}_3) = 29.13$ cal mol$^{-1}$ K$^{-1}$, and this is the second law entropy of Ir$_2$S$_3$(s). The value of the standard entropy for solid Ir$_2$S$_3$ at 298.15 has never been determined from the third law because the heat capacity from 0 to 298.15 K has not been measured; therefore $S^\circ_{298} = 29.13$ cal mol$^{-1}$ K$^{-1}$ remains to be verified by direct measurements from 0 to 298.15 K. In fact, for this reason, $\Delta_f H^\circ$ of Ir$_2$S$_3$ cannot be calculated by the third law method. The foregoing values can readily be converted into units with J and bar.

*Example 6* The equilibrium *total pressure* $P_e$ of Al(g) and N$_2$(g) over AlN(s) has been measured by Hildenbrand and Hall,[*] by using the torsion effusion technique, similar to the simple Knudsen effusion method. Five experimental data from this investigation, out of 29 measurements in the range of 1776 to 1972 K, have been selected as examples and listed in Table 12.13. The reaction investigated is

$$2 \text{AlN(s)} = 2 \text{Al(g)} + \text{N}_2\text{(g)} \tag{12.50}$$

Mass Spectrometric analysis of the gases over AlN(s), reported by Schissel and Williams[†] shows that only Al(g) and N$_2$(g) are present. The partial pressures of

---

[*]D. L. Hildenbrand and W. F. Hall, J. Phys. Chem., 67, 888 (1963).
[†]P. O. Schissel and W. S. Williams, Bull. Am. Phys. Soc. [4], 4, 139 (1959).

Al(g) and $N_2$(g) are $P(Al) = 2P_e/3$ and $P(N_2) = P_e/3$. The equilibrium constant is therefore expressed by

$$K_p = 4P_e^3/27 = 0.148 P_e^3 \tag{12.51}$$

The Gibbs energy function for Al(g), $N_2$(g), and AlN(s) from the earlier edition of Gen. Ref. (1), yield the following values of $\Delta Gef$ for Reaction (I):

| $T$, K: | 1700 | 1800 | 1900 | 2000 |
|---|---|---|---|---|
| $-\Delta Gef$, cal K$^{-1}$: | 112.466 | 112.254 | 112.049 | 111.849 |

For example, for the first experiment at 1776 K in Table 12.13, $\Delta H_{298}^\circ$ obtained from (12.38) is

$$\Delta H_{298}^\circ = -1776 R \ln(0.148 \times 66^3 \times 10^{-18}) + 112.305 \times 1776$$
$$= 308{,}113 \text{ cal} = 308.1 \text{ kcal}$$

where $-\Delta Gef = 112.305$ cal K$^{-1}$ has been obtained by interpolation to 1776 K. The average of 29 values of $\Delta H_{298}^\circ$ gives the third law value of 307.3 kcal. The second law method gives 309.5 kcal according to Hildenbrand and Hall. The third law value is preferable since the Gibbs energy functions are accurately known. Combination of 307.3 kcal with $\Delta H_{298}^\circ = -156.0$ kcal for 2 Al(g) = 2 Al(s) from Tables 12.2 and 12.4 gives $\Delta H_{298}^\circ = 151.3$ kcal for 2 AlN = 2 Al(s) + $N_2$(g) for two moles for AlN, hence the standard molar enthalpy of formation of AlN is $\Delta_f H_{298}^\circ = -75.65$ kcal mol$^{-1}$. The accepted value of $-76.0$ kcal mol$^{-1}$ is the average of two calorimetric values by Neugebauer and Margrave, and Mah, King, Weller and Christensen as noted in Gen. Refs. (1) and (2).

*Example 7* Equilibria in the gas-phase dissociation of isopropyl alcohol (2-propanol) into acetone and hydrogen, i.e.,

$$C_3H_7OH(g) = (CH_3)_2CO(g) + H_2(g) \tag{12.52}$$
(2-Propanol) = (Acetone) + (Hydrogen)
(Species 1)   (Species 2)   (Species 3)

have been investigated by Kolb and Burwell in the range of 416.7 to 491.6 K by starting either with excess acetone and hydrogen, or with excess 2-propanol in the presence of copper as a catalyst [see Gen. Ref. (7)]. The equilibrium constant $K_p$ for this reaction is $K_p = P x_2 x_3 / x_1$ where $P$ is in atm, $x_1$, $x_2$, and $x_3$ are the mole fractions of the species 2-propanol, acetone, and hydrogen respectively [$K_p = 0.124$ at 416.7 K, 0.525 at 455.7 K, and 1.570 at 491.6 K]. The equilibrium constant is independent of pressure because deviations from ideality in gaseous acetone and 2-propanol are in the same direction, and therefore the fugacity coefficients cancel each other at law pressures. We consider the sigma plot of obtaining $\Delta H^\circ$ of reaction. For this purpose, we fit the values of $C_p^\circ$ for each substance into a quadratic equation as follows:

2-propanol: $C_p^\circ = 2.74 + 0.0673T - 18.0 \times 10^{-6}T^2$
Acetone: $C_p^\circ = 5.04 + 0.0452T - 7.0 \times 10^{-6}T^2$
Hydrogen: $C_p^\circ = 6.14 + 0.0037T - 4.0 \times 10^{-6}T^2$

The equation for $\Delta C_p^\circ$ corresponding to the chemical reaction is

$$\Delta C_p^\circ = 8.44 - 0.0184T + 7.0 \times 10^{-6}T^2 \text{ cal mol}^{-1} \text{ K}^{-1} \tag{12.53}$$

Substitution of this equation into (12.42) yields

$$-\frac{\Delta H_\theta^\circ}{T} + \left(\Delta S_\theta^\circ - \Delta \alpha\right) = \Sigma = R \ln K_p - 8.44 \ln T$$
$$+ 0.0092T - 1.167 \times 10^{-6}T^2$$

The values of $\Sigma$ versus $1/T$ are represented in Fig. 12.3. The straight line in this figure, obtained by the method of least squares is represented by

$$\Sigma = -\frac{11{,}858}{T} - 22.978$$

where $\Delta H_\theta^\circ = 11{,}858$ cal mol$^{-1}$. Substitution of the right side of this equation for $\Sigma$ in the preceding equation and solution for $\Delta G^\circ = -RT \ln K_p$ yields

$$\Delta G^\circ = 11{,}858 + 22.978T - 8.44T \ln T + 0.0092T^2 - 1.167 \times 10^{-6}T^3;$$

The equation for $\Delta H^\circ$ is given by (12.26), and since $\Delta H_\theta^\circ$ is 11,858 cal mol$^{-1}$, it is evident that

$$\Delta H_T^\circ = 11{,}858 + 8.44T - 0.0092T^2 + 2.333 \times 10^{-6}T^3$$

From $\partial(\Delta G^\circ)/\partial T = -\Delta S^\circ$, the corresponding equation for $\Delta S^\circ$ is

$$\Delta S_T^\circ = -14.538 + 8.44 \ln T - 0.0184T + 3.500 \times 10^{-6}T^2$$

The values of $\Delta H^\circ$ and $\Delta S^\circ$ at 298.15 from these equations are

$$\Delta H_{298}^\circ = 13{,}618 \text{ cal mol}^{-1}; \quad \Delta S_{298}^\circ = 28.375 \text{ cal mol}^{-1} \text{ K}^{-1} (\pm 0.743)$$

The standard entropy change for the reaction, $\Delta S_{298}^\circ$, can be obtained from the direct integrals of $\Delta C_p^\circ/T$ from 0 to 298.15 K, listed in Table 11 for species 1 and 2, and in Gen. Ref. (1) for species 3; the result is

$$\Delta S_{298}^\circ = 27.631 \text{ cal mol}^{-1} \text{ K}^{-1} \quad \text{(direct determination)}$$

The discrepancy of 0.743 with 28.375 cal mol$^{-1}$ K$^{-1}$ by sigma plot is too large, and requires further experimental data either on $\Delta C_p^\circ$ or on $K_p$ or both. It is probable that side reactions might affect the values of $K_p$ as is often the case in organic reactions.

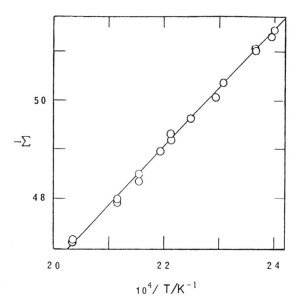

**Fig. 12.3** Sigma plot for gaseous reaction 2-Propanol = Acetone + Hydrogen, from Gen. Ref. (7).

## Complex Equilibria

Calculations of the equilibrium concentrations of the species in one reaction, formed from the given amounts of initial species, can be carried out in a simple manner for one reaction as was shown for $CO(g) + H_2O(g) = CO_2(g) + H_2(g)$, $1.5H_2(g) + 0.5N_2(g) = NH_3(g)$, and $CH_4(g) = C(gr.) + 2H_2$. When there are two or more simultaneous reactions, the calculations become complex and often necessitate a computer. The procedure requires the mass balance or atomic balance and the equilibrium constants to solve for all the species, formed from the starting species, after the attainment of equilibrium. As a simple example, assume that *two moles of graphite* and *one mole of $H_2O$* are injected into a closed chamber to form CO, $H_2$, and $CO_2$ at 1200 K and one bar of pressure as follows [Gen Ref. (1)]:

(I) $\quad C(graph) + H_2O(g) = CO(g) + H_2(g), \quad K_p(I) = 38.38$
$\qquad (v) \qquad\quad (w) \qquad\quad (x) \qquad (y)$

(II) $\quad CO(g) + H_2O(g) = CO_2(g) + H_2(g), \quad K_p(II) = 0.7296$
$\qquad (x) \qquad\quad (w) \qquad\quad (z) \qquad (y)$

The standard Gibbs energy change $\Delta_r G°(I)$ for the first reaction is obtained by writing $\Delta_r G°(I) = \Delta_f G°(CO) - \Delta_f G°(H_2O) = -217,819(CO) + 181,425(H_2O) = -36,394$ J. Similarly, for the second reaction, $\Delta_r G°(II) = -396,098(CO_2) +$

$181,425(H_2O) + 217,819(CO) = 3,146$ J. From $\Delta_r G°(I) = -1200R \ln K_p(I) = -36,394$ J, $K_p(I)$ is 38.38, and likewise, from $\Delta_r G°(II)$, $K_p(II)$ is 0.7296. There are five species and the number of moles of each species is written in parentheses under each formula. The atomic balance for carbon, hydrogen, and oxygen yields

$2 = v + x + z$ (carbon);  $\quad 2 = 2w + 2y$ (hydrogen);

$1 = w + x + 2z$ (oxygen)

where the left side is the initial number of atoms of each element prior to any reaction and the right side, the number of atoms of each element in all the species after the reactions attain equilibrium. These three equations for mass balance and the two equilibrium constant are sufficient to solve for the five unknowns. From $v = 2 - x - z$, $w = 1 - y$, and $2z = y - x$, the reactions may be rewritten as

(I) $\quad$ C $\quad + \quad H_2O \quad = \quad$ CO $\quad + \quad H_2$
$\quad\quad (2 - x - z) \quad (1 - y) \quad\quad (x) \quad\quad (y)$

(II) $\quad\quad\quad$ CO $+ \quad H_2O \quad = \quad CO_2 \quad + \quad H_2$
$\quad\quad\quad\quad (x) \quad\quad (1 - y) \quad\quad (y - x)/2 \quad (y)$

The total moles of gaseous species is

$1 - y + x + y + (y - x)/2 = 1 + (x + y)/2$

The partial pressure of each species is the same as its mole fraction because the total pressure is one bar; hence, the equilibrium constants are expressed by

$$K_p(I) = \frac{xy}{(1-y)[1 + (x+y)/2]} = 38.38$$

$$K_p(II) = \frac{y(y-x)/2}{x(1-y)} = 0.7296$$

Simplification and rearrangement of these equations give

$y^2 + 1.05211xy + y - x - 2 = 0; \quad\quad y^2 + 0.4592xy - 1.4592x = 0$

The second equation yields

$x = y^2/(1.4592 - 0.4592y)$

Substitution of this equation in the first equation gives

$y^3 + 4.0101y - 4.9222 = 0$

This equation can be solved by successive approximation because inspection shows that $y$ is very close to unity, and in fact for $y = 1$, the left side is 0.0879 instead of zero, and for $y = 0.990$ and 0.985, the left side is +0.0181 and −0.0166 respectively and the linear interpolation between these values yields $y = 0.9874$

240    *Thermodynamics*

in four decimal places. The value of $x$ is therefore 0.9694, and the results for all the species in moles are as follows:

[C, $(v)$]: 1.0216;   [H$_2$O, $(w)$]: 0.0126;   [CO, $(x)$]: 0.9694;

[H$_2$, $(y)$]: 0.9874;   [CO$_2$, $(z)$]: 0.0090

In complex reactions involving combustion, rocket propulsion, etc., it is necessary to have special computer programs for calculating the number of moles of all species;* such task become increasingly complex at high temperatures largely because of the occurrence of many simultaneous reactions. However, for temperatures encountered in processing materials and minerals, usually below 2000 K, the task is not unduly complex. A computer program with a data bank for personal computers are presented in Appendix III for calculating equilibria involving multispecies. It contains all the data in Gen. Ref. (2).

## Generalized Reactions and Their Equilibrium Constants

Chemical reactions, vaporization, solubility of one phase in another, distribution of a component between two condensed phases, and association and dissociation of molecules in a single phase have been treated as different subjects in various publications, based on traditional development of thermodynamics. Considerable simplicity is gained if all these processes are written as physico-chemical reactions and for each such reaction an appropriate equilibrium constant is written. The chemical reactions, which convert one set of species into another set were considered earlier, hence they need not be repeated here; some of the typical examples for the remaining processes are as follows.

*Vaporization*   An element or a compound may vaporize without change in molecular species either from a pure condensed phase or from a solution. If the condensed phase is pure, then the equilibrium constant is the same as the vapor pressure or the fugacity $f°$ of the vapor. For example,

$$\text{H}_2\text{O}(\text{pure l or c, or solution}) = \text{H}_2\text{O}(g) \qquad K_p = f° = f/a \qquad (12.54)$$

where the second equality is for the condensed solution, and if $a$ is defined by $f/f°$, it is evident that $K_p$ over the solution is also equal to $f°$.

A condensed phase may vaporize into two or more species as in the case of sodium, i.e.,

$$\text{Na}(c, l, \text{ or soln}) = \text{Na}(g) \qquad K_p = P_{\text{Na}}/a_{\text{Na}} \qquad (12.55)$$

$$2\,\text{Na}(c, l, \text{ or soln}) = \text{Na}_2(g) \qquad K_p = P_{\text{Na}_2}/a_{\text{Na}}^2 \qquad (12.56)$$

---

*W. T. Thompson, A. D. Pelton, and C. W. Bale, CALPHAD, 7(2), 113 (1993); (F*A*C*T Program); W. T. Thompson, F. Ajersch, and G. Eriksson, Ed., Gen. Ref. (29); G. Eriksson and K. Hack, Met. Trans., 21B, 1013 (1990); (ChemSage & SOLGASMIX Programs); L. V. Gurvich, IVANTHERMO, CRC Press (1992); B. Sundman, B. Jansson, and J.-O. Andersson, CALPHAD, 9(2), 153 (1985); (ThermoCalc Program).

where $P_i$ is used instead of $f_i$ because for most metals the accuracy of experimental data seldom justifies the use of fugacity. Liquid sodium contains Na and Na$_2$, and the definition of activity in sodium alloys becomes difficult when they contain more than one species. The experimental methods available for measuring the concentrations of various species in liquids are not yet sufficiently reliable. Therefore, the condensed phase is usually taken to be the predominant species, and in the foregoing example, simply the monatomic liquid sodium. With this convention, the activity of sodium in a solution is defined by $a_{Na} = P_{Na}/P^\circ_{Na} = (P_{Na_2}/P^\circ_{Na_2})^{0.5}$.

Some substances such as graphite may vaporize into as many as 30 gaseous species, and compounds such as AlCl$_3$ may sublime into a monomer and a dimer in gaseous state, and with increasing temperatures, dissociate into several molecular and atomic species.

*Dissolution of Gases, Liquids and Solids without Dissociation* In general, all substances dissolve in condensed phases under ambient conditions without dissociation, e.g.,

$$N_2(g) = N_2(\text{in liq. } H_2O); \qquad K_P = x_{N_2}/f_{N_2} \qquad (12.57)^*$$

$$S_8(c, \text{rhombic}) = S_8(\text{in liq. } CS_2); \qquad K_p = x_{S_8} \qquad (12.58)$$

For the first reaction, the equilibrium constant is the mole fraction of dissolved nitrogen divided by the fugacity of nitrogen because $x_{N_2}$ is small and Henry's law is obeyed. The activity coefficient $\gamma^\infty_{N_2}$ based on Henry's law can be determined from

$$K_p = \left(\frac{x_{N_2}}{f_{N_2}}\right)_{x_{N_2} \to 0} = \frac{\gamma^\infty_{N_2} x_{N_2}}{f_{N_2}} \qquad (12.59)$$

where $K_p$ is the value of the first ratio when $x_{N_2}$ or $f_{N_2}$ is very low. A precise method of obtaining $K_p$ is to plot $\ln(x_{N_2}/f_{N_2})$ versus $x_{N_2}$ and to extrapolate the curve to $x_{N_2} \to 0$ where $\ln(x_{N_2}/f_{N_2})$ is identical with $\ln K_p$.

*Dissolution by Dissociation and by Polymerization* Diatomic gases dissolve as monatomic species in metals and metallic solutions as discovered first by Sieverts (1909), e.g.,

$$0.5\, N_2(g) = N(\text{in metals and alloys}) \qquad K_p = x_N/P^{0.5}_{N_2} \qquad (12.60)$$

This equation is a formal statement of Sieverts' law. The solubilities of diatomic gases are therefore proportional to the square root of the partial pressure. While $K_p$ in (12.60) is a constant within a finite range of concentration at a given temperature, $K_{p,2}$ for the following assumed reaction is not; thus

$$N_2(g) = N_2 \text{ (in metals)} \qquad K_{p,2} = x_{N_2}/P_{N_2} = \frac{(x_N/2)/P^{0.5}_{N_2}}{P^{0.5}_{N_2}}$$

---

*See General Ref. (12); R. A. Pierotti, J. Phys. Chem., 69, 281 (1961); R. Battino, F. D. Evans, W. F. Danforth, and E. Wilhelm, J. Chem. Thermod., 3, 743 (1971); N. A. Gokcen, Faraday Trans. I, 69, 438 (1973).

where

$$K_{p,2} = K_p/2P_{N_2}^{0.5}, \qquad \text{with } x_{N_2} \approx x_N/2$$

and $K_{p,2}$ approaches infinity as $P_{N_2}$ approaches zero. Therefore, a plot of $x_N$ versus $P_{N_2}^{0.5}$ is a straight line whose slope is $K_p$, but a plot of $x_{N_2}$ versus $P_{N_2}$ is not linear and its slope becomes infinity as $P_{N_2}$ approaches zero as shown for liquid iron in Fig. 12.4. An additional example of different type is provided by[*] $H_2O(g) = H_2(g) + O$ (in metals), which shows that the dissolved oxygen is O, not $O_2$, because the equilibrium constant is $K_p = x_O P_{H_2}/P_{H_2O}$.

As the concentration of oxygen increases in a metal, or as other components are added to the system, $K_p$ must be written as $K_p = x_O \gamma_O^\infty P_{H_2}/P_{H_2O}$ where $\gamma_O^\infty$ is the activity coefficient of oxygen, for which the reference state is the infinitely dilute binary oxygen-metal system.

The reason for the dissolution of diatomic gases as monatomic species is that the diatomic species are too large to be accommodated by the space among the molecules of metals and alloys.

If a substance, in particular a gas $A$ polymerizes as it dissolves in a condensed phase, i.e.,

$$nA(g) = A_n(\text{in soln}); \qquad K_p = x_n/P^n \qquad (12.61)$$

where the equilibrium constant requires $P^n$ in the denominator. Such a process is seldom encountered. It is common, however, that $A$ can associate with the solvent molecules. In such an event the reaction is $A(g) + mB = AB_m(\text{in soln})$, where $B$ is the solvent.

*Distribution of a Component between Two or more Immiscible Components* A component such as iodine dissolves in two virtually immiscible liquids such as water and carbon disulfide and its concentration $x_2$ (in $H_2O$) divided by $y_2$ (in $CS_2$) is a constant at a given temperature for a finite limited range of iodine concentration. The ratio[†] is given by $K_p = x_2/y_2 = 0.00052$ up to about $y_2 = 0.02$ at 25°C, but above this concentration $K_p$ must be written as $K_p = x_2/(\gamma_2^\infty y_2) = 0.00052$ where $\gamma_2^\infty$ is the activity coefficient of iodine in $CS_2$ on Henry's law scale. The mole fraction $x_2$, is so small that the activity coefficient of iodine in water is unity. In some immiscible systems, as the concentration of the solute increases, immiscibility disappears, e.g., liquid Ag and Fe are immiscible but when the concentration of a third element, e.g., silicon, is increased beyond a certain point, the system becomes a one-phase ternary system.

---

[*]E. S. Tankins, N. A. Gokcen, and G. R. Belton, Trans. Metall. Soc. AIME, *230*, 820 (1964); T. B. Flanagan and J. F. Lynch, Metall. Trans. *6A*, 243 (1975); W. V. Venal and G. Geiger, ibid., *4*, 2567 (1973).

[†]A. A. Jaworskin; see J. H. Hildebrand and R. L. Scott, "The Solubility of Nonelectrolytes," Third Ed., Reinhold Publ. Co., New York (1950), p. 208; T. B. Flanagan and J. F. Lynch, Metall. Trans. *6A*, 243 (1975); W. V. Venal and G. Geiger, ibid., *4*, 2567 (1973).

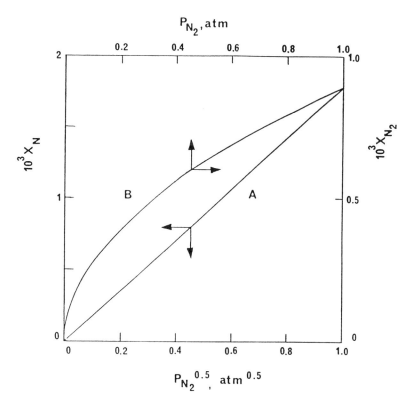

**Fig. 12.4** Solubility of $N_2(g)$ in liquid iron at 1880 K. Pressure of $N_2(g)$ in experiments range from about 0.01 to 1 atm. Adherence of data to straight line A confirms Sieverts' law, i.e. $K_p = x_N/P_{N_2}^{0.5} = 0.00177$ is constant. Curve B is for $x_{N_2}$ versus $P_{N_2}$ and its slope for $P \to 0$ approaches infinity because $K_{p,2} = x_{N_2}/P_{N_2} = x_N/2P = K_p/2P^{0.5}$. Arrows indicate sets of coordinates. From an assessment by H. A. Wriedt, N. A. Gokcen and R. H. Nafziger, Bull. Alloy Phase Diag., 8, 355 (1987) largely the data by R.D. Pehlke et al.

When a component such as silicon coexists in liquid and solid states, a second component such as carbon, is distributed between the liquid and solid phases. The distribution coefficient $K_p$ is given* by

$$K_p = y_2(\text{C in solid Si})/x_2(\text{C in liquid Si}) \approx 0.07 \qquad (12.62)$$

where $K_p$, is called the segregation coefficient in semiconductor technology. The smaller is $K_p$, the more effective is the purification of silicon by fractional freezing, which is similar to fractional distillation. Purification of condensed phases by fractional freezing is called *zone refining*.[†]

---

*T. Nozaki, Y. Yatsurugi, and N. Akiyama, J. Electrochem. Soc., *117*, 1566 (1970).
[†]W. G. Pfann, "Zone Melting," John Wiley and Sons, Inc., New York (1958).

244    Thermodynamics

The foregoing examples make it clear that any physico-chemical process may be written as a thermodynamic process for which an equilibrium constant can be written, and from the equilibrium constant, all the useful thermodynamic properties can be obtained. Therefore, $\Delta G° = \Delta H° - T\Delta S° = -RT \ln$ (equilibrium constant).

## PROBLEMS

(See tables in this chapter and in Appendix II for required thermodynamic data.)

12.1  The equilibrium constant for C(graph) + 2H$_2$(g) = CH$_4$(g) is 0.316 at 900 K. Calculate (a) the mole fraction of CH$_4$ at 1, 50, and 100 bars, (b) the standard Gibbs energy change, $\Delta_r G°$, and (c) the Gibbs energy change $\Delta_r G$ when the system consists of graphite and an equimolecular mixture of H$_2$, Ar, and CH$_4$ at one bar.
    *Partial ans.:* (a) $P(H_2) = 11.1$ and $P(CH_4) = 38.9$ bar at 50 bars,
(b) $\Delta_r G° = 8621$ J (c) $\Delta_r G = 16,842$ J.

12.2  The standard Gibbs energies of formation of solid BaO and BaO$_2$ are $-444,261$ and $-446,350$ J mol$^{-1}$ at 1100 K. Calculate (a) the dissociation pressure of BaO$_2$ when it is reduced to BaO, and (b) the dissociation pressure of BaO into barium and oxygen. Which one of these dissociations are likely to occur in air at one bar?
    *Ans.:* (a) 0.633 bar, (b) BaO$_2$ would associate in air, not BaO.

12.3  The vapor pressure of a pure supercooled liquid is 3,333 Pa at 300 K and that of the stable solid, 2,666.4 Pa at 300 K. Calculate the standard Gibbs energy changes of vaporization, sublimation and freezing of the supercooled liquid.

12.4  Calculate the equilibrium constants and gas compositions at 10.133 bars of pressure for the following separate gaseous reactions at 700°C from the tables in Appendix II:

    (I)   CO(g) + 0.5O$_2$(g) = CO$_2$(g)
    (II)  H$_2$(g) + CO$_2$(g) = CO(g) + H$_2$O(g)

Initial amounts of reactants are 1 mole of CO and 0.25 mole of O$_2$ for (I), and 1 mole H$_2$ and 0.5 mole of CO$_2$ for (II), before any reaction product is formed. Find the number of moles of each species at equilibrium.
    *Ans.:* (I) 0.50 mol of CO and 0.50 mol of CO$_2$ (II) H$_2$: 0.703, CO$_2$: 0.203,
CO: 0.297, H$_2$O: 0.297 mol.

12.5  Calculate the moles of each species formed at 600 K and one bar by starting with one mole of methyl chloride and one mole of water and from the following gaseous reactions and their equilibrium constants:

    (I)   CH$_3$Cl(g) + H$_2$O(g) = CH$_3$OH(g) + HCl(g),    $K_p$(I) = 0.00153
    (II)  2CH$_3$OH(g) = (CH$_3$)$_2$O(g) + H$_2$O(g),    $K_p$(II) = 10.6

    *Partial ans.:* number of moles of (CH$_3$)$_2$O is 0.0094

12.6  Two moles of ammonia and one mole of carbon dioxide are fed into a furnace chamber at 450°C where the following reactions attain equilibrium under one bar:

    (I)   2NH$_3$ = N$_2$ + 3H$_3$,    $K_p$(I) = 2.36 × 10$^6$
    (II)  H$_2$ + CO$_2$ = CO + H$_2$O,    $K_p$(II) = 0.135

Derive an equation from which the composition of gas mixture can be computed.

12.7 The standard Gibbs energy of formation of liquid water at 25°C is $-237{,}187$ J/mol$^{-1}$. The vapor pressure of water is 3,168 Pa. Calculate (a) the standard Gibbs energy of formation of steam at 25°C, (b) the standard Gibbs energy of vaporization at 25°C. Note that $\Delta_f G°$ of water is the same as $\Delta_f G$ of steam at 3,168 Pa. Standard pressure is 1 bar.

Ans.: (a) $\Delta_f G° = -228{,}629$ J mol$^{-1}$, (b) $\Delta_{vap} G° = 8{,}558$ J mol$^{-1}$

12.8 One mole of $C_2H_4$ and three moles of $H_2$ react at 1000 K and one bar of pressure as follows:

(I)   $C_2H_4(g) + 2H_2(g) = 2CH_4(g)$     $\Delta_r G°_{1000}/1000R = -9.652$
(II)  $C_2H_4(g) + H_2(g) = C_2H_6(g)$     $\Delta_r G°_{1000}/1000R = +1.158$

Assuming there are no other concurrent reactions, calculate the moles of each species at equilibrium.

12.9 Calculate $\Delta G°$, $\Delta H°$, and $\Delta S°$ at 550 K for the following reactions from the tabular values in Appendix and in this chapter:

$$Al(g) + 1.5Cl_2(g) = AlCl_3(g)$$

$$2Al(s) + 1.5O_2(g) = Al_2Cl_3(s)$$

$$C(s) + 2H_2(g) = CH_4(g)$$

12.10 Calculate $\Delta_r H°_{298}$ by the simple second law method for Reaction (12.52) by using $K_p = 0.151$ at 422.0 K and $K_p = 1.58$ at 491.6 K.

12.11 The solubility of oxygen in liquid water at 1.0133 bar is given by*

$$\log x_2[O_2 \text{ in } H_2O(l)] = (4271.1/T) + 28.1918 \log T - 88.72463.$$

Express $\Delta G°$, $\Delta H°$, $\Delta S°$, $\Delta C°_p$, and $\Sigma$ for $O_2(g) = O_2$ (in soln).

12.12 One mole of $CH_4$ is burned with 12 moles of air and the reaction products are maintained at 1200 K and 1.0 bar (air = 20% $O_2$ and 80% $N_2$). The gas analysis indicates $H_2$, $N_2$, CO, $CO_2$, and $H_2O$ as the major species in the combustion products. No carbon was deposited in the chamber. Calculate the composition of gas at 1200 K and 1.0 bar.

12.13 Recalculate Example 20—Part 2 in Gen. Ref. (2), Bull. 677, by considering only $S_1$, $S_2$, and $S_3$ as the significant gas species at 2000 K and 1 bar.

Ans.: $S_1 = 0.00316$, $S_2 = 0.9916$, and $S_3 = 0.00524$ bar.

12.14 Equilibrium oxygen pressures over the coexisting pure CuO and $Cu_2O$ are 0.0299 bar at 1200 K, and 0.2206 bar at 1300 K. Express $\Delta_r G°$ per mole of $O_2$ formed for the dissociation of CuO into $Cu_2O$ and $O_2$ as a linear function of $T$, and compute the $P(O_2)$ in bars at 1150 and 1250 K. [Based on Gen. Ref. (2)]. (A good method for the preparation of $Cu_2O$ is to pass pure Ar over a thin layer of powdered CuO at 1200 K). For a similar problem, see Y. K. Rao, p. 343, loc. cit.,

Partial ans.: $\Delta_r G° = 259{,}216 - 186.830T$; J (mol $O_2$)$^{-1}$.

---

*N. A. Gokcen and E. T. Chang, Denki Kagaku, J. Japan Electrochem. Soc., *43*, 232 (1975).

12.15 The solubility of $N_2$ in liquid Fe(l) in terms of nitrogen mole fraction $x_N$ and pressure $P(N_2)$ in bars is expressed by[†]

$$0.5N_2(g) = [N]; \qquad \log\{[x_N/[P(N_2)]^{0.5}\} = -2.618 - (255.4/T)$$

Derive the standard Gibbs energy of dissolution $\Delta G^\infty$ of nitrogen and calculate $x_N$ at 1550°C and 1.0 bar, and also [N] in wt%.

*Partial ans.:* 0.050 wt% [N].

12.16 Calculate the theoretical concentrations of dissolved Al and O in initially pure liquid Fe at 1900 K in equilibrium with pure $Al_2O_3(s)$. *Hint:* $Al_2O_3$ dissolves as $Al_2O_3(s) = 2[Al] + 3[O]$, where [ ] indicate the dissolved elements. Use the first table in Appendix II for $Al_2O_3(s)$.

*Ans.:* $x[Al] = 1.074 \times 10^{-6}$; $x[O] = 1.611 \times 10^{-6}$.

---

[†]H. A. Wriedt, N. A. Gokcen, and R .H. Nafziger, Bull. Alloy Phase Diag., *8*, 355 (1988); H. Wada and R. D. Pehlke, Met. Trans. B, *12*, 333 (1981).

CHAPTER XIII

# SOLUTIONS OF ELECTROLYTES

## Introduction

Electrolytes dissociate into ions when dissolved in water, liquid ammonia, dioxane, alcohols and a number of other solvents according to the Arrhenius theory. These ions act as individual species, not as independent components, in changing the properties of solvents, such as the depression of freezing point, the lowering of vapor pressure, etc. Solutions of electrolytes play a dominant role in production of chemicals, in purification and electrowinning of metals, and in manufacturing galvanic cells and storage batteries. Numerous biological and geological processes occur in electrolytic solutions. While various solids and liquids are solvents for electrolytes, the most important and unique solvent is water.

Electrolytes are divided into two groups, (1) strong electrolytes, which consist almost entirely of free ions in dilute solutions and, (2) weak electrolytes, which dissociate partially in dilute solutions. For example, acetic acid, $CH_3COOH$, is a weak electrolyte that dissociates to the extent of 0.09 fraction of total acid at 0.02 molal concentration in water; but HCl, NaBr, and $HNO_3$ are strong electrolytes that are nearly completely dissociated at the same concentration. On the other hand acetic acid is a strong electrolyte in ammonia. When an electrolyte dissociates to such a large extent that the undissociated fraction is immeasurably small at low molalities then it is fruitful to consider only the ions as the sole existing species and label the electrolyte as a *strong electrolyte*. At moderate and high concentrations for strong electrolytes, any association of ions as the molecule of the electrolyte is then considered to be a part of phenomena contributing to the changes in the activity coefficients of ions. It is thermodynamically fruitless and experimentally impossible with the existing techniques, except in a few favorable cases, to deal with the undissociated fraction of strong electrolytes even in moderate ranges of concentration because ionization is nearly complete.

The composition of an electrolyte in a solution is usually expressed in terms of molality, $m$, i.e., the number of moles of electrolyte or its ions in *one kilogram of solvent*. The molar ratio of an electrolyte or an ion to the solvent is then $m/55.508$ when the solvent is water. The composition is also expressed in molarity, $c$, i.e., the number of moles in 1000 cubic centimeters of solution. Conversion of $c$ into $m$ requires the density of solution.

## Thermodynamics

The charge number of an ion is denoted by $z_+$ or $z_-$; the value of $z_+$ is a positive integer, and that of $z_-$, a negative integer. For $K^+$, $Ca^{++}$, and $Fe^{3+}$ the charge number $z_+$ is $+1$, $+2$, and $+3$ respectively, and for $F^-$, and $SO_4^{--}$, and $PO_4^{3-}$, $z_-$ is $-1$, $-2$, and $-3$ respectively. In a solution containing $i$ positive ions, or cations, of molalities $m_{i+}$ and $j$ negative ions, or anions, of molalities and $m_{j-}$ the sum of the positive and negative charges is zero; i.e.,

$$\sum_{i=1}^{i} z_{i+} m_{i+} + \sum_{j=1}^{j} z_{j-} m_{j-} = 0; \qquad (z_{i+} > 0, \ z_{j-} < 0) \tag{13.1}$$

Equation (13.1) expresses the *principle of electrical neutrality* and imposes a restriction on the ionic species so that the number of ionic components is *one fewer* than the number of ionic species. For example, $(K^+, Na^+, F^-)$ contribute two components because when the concentrations of $K^+$ and $Na^+$ are fixed, that of $F^-$ is also fixed but $(K^+, Na^+, F^-, Cl^-)$ contribute three components because when the concentrations of $K^+$ and $Na^+$ are fixed the ratio of $F^-$ to $Cl^-$ can be varied. Therefore $C^{(i)}$ ionic species plus the solvent constitute $C^{(i)}$ independent components, not $C^{(i)} + 1$, and consequently, $C^{(i)} - 1$ independent composition variables.

A large portion of this chapter is devoted to strong electrolytes in aqueous solutions.* The weak electrolytes are discussed in a section at the end of this chapter. It is therefore not necessary to repeat the term *strong* when strong electrolytes are mentioned but the term *weak* will be repeated to avoid confusion when reference is made to weak electrolytes.

### Activity and Activity Coefficient

The activities, and the activity coefficients of individual ions are not measurable with existing experimental techniques, but a mean activity $a_\pm$ and the corresponding activity coefficient $\gamma_\pm$ of both ions of an electrolyte are measurable properties. The derivation of equations defining $a_\pm$ and $\gamma_\pm$ is sufficient to demonstrate this point. It is convenient first to consider a 1:1 electrolyte, such as HCl, which dissociates into two univalent ions and then to generalize the results to any type of electrolyte. The Gibbs-Duhem relation for a solution containing $H^+$ and $Cl^-$ of HCl in water is

$$55.508 \, d \ln a_1 + m_+ \, d \ln a_+ + m_- \, d \ln a_- = 0 \tag{13.2}$$

where $m = m_+ = m_-$ are the molalities, $a_1$ is the activity of water, $a_1 = (P_1 Z_1 / P_1^\circ Z_1^\circ)$ from (9.23), and $a_+$ and $a_-$ are the activities of the cation and

---

*For nonaqueous electrolytes, see G. J. Janz, Ed., "Molten Salts Handbook," Academic Press, New York (1967); G. J. Janz and R. P. T. Tomkins, and contributors (D. A. Aikens, J. Ambrose, D. N. Bennion, J. N. Butler, K. Doblhoffer, R. J. Gillespie, R. J. Jasinski, P. V. Johnson, A. A. Pilla, H. V. Venkatasetty, and N. P. Yao), "Nonaqueous Electrolytes Handbook," Academic Press, Vol. *I* (1972), *II* (1973).

Solutions of Electrolytes    249

the anion respectively. The equality $m = m_+ = m_-$ simplifies (13.2) so that

$$55.508 \, d \ln a_1 + m \, d \ln a_+ \cdot a_- = 0; \quad \text{(1:1 electrolyte)} \tag{13.3}$$

Since the activity of solvent, $a_1$, can be determined, e.g., by vapor pressure or depression of freezing point measurements, $a_+ \cdot a_-$ can be determined from (13.3). The measurable property $a_+ \cdot a_-$ is designated for brevity by the square of the mean activity $a_\pm$; hence, by definition

$$a_\pm^2 = a_+ \cdot a_- \tag{13.4}$$

The mean activity $a_\pm$ is therefore the geometrical mean of $a_+$ and $a_-$ since $a_\pm = \overline{(a_+ \cdot a_-)^{0.5}}$. The reason for this definition is that $a_+$ and $a_-$ are not individually measurable, and in addition $a_\pm$, $a_+$, and $a_-$ are expressed by the same units, such as $m$. Equation (13.4) must therefore be regarded as a definition simply for brevity. Dividing $a_+$ by $m_+$ gives the activity coefficient $\gamma_+$, i.e., $a_+/m_+ = \gamma_+$, and likewise, $a_-/m_- = \gamma_-$. If, again for brevity, the mean ionic molality $m_\pm$ is defined by $m_\pm^2 = m_+ \cdot m_- = m^2$, then (13.4) becomes

$$a_\pm^2/m_\pm^2 = \gamma_\pm^2 = \gamma_+ \cdot \gamma_-; \quad (m_\pm^2 = m_+ \cdot m_- = m^2);$$
$$\text{(1:1 electrolyte)} \tag{13.5}$$

where $\gamma_\pm$ is the mean activity coefficient.

The same general argument summarized in (13.2) to (13.5) can be used to show that the measurable mean activity of *any electrolyte* depends on the ionization reaction

$$A_{\nu_+} B_{\nu_-} = \nu_+ A^{z+} + \nu_- B^{z-}; \quad (\nu_+ > 0; \nu_- > 0) \tag{13.6}$$

where $\nu_+$ and $\nu_-$ are the stoichiometric coefficients, which are always positive by convention. If the molality of the electrolyte is $m$, then

$$m_+ = \nu_+ m = \text{molality of cation}; \quad m_- = \nu_- m = \text{molality of anion} \tag{13.7}$$

The substitution of these equations into (13.2) gives

$$55.508 \, d \ln a_1 + \nu_+ m \, d \ln a_+ + \nu_- m \, d \ln a_- = 0 \tag{13.8}$$

where $a_+$ refers to $a_{z+}$ but for brevity $z$ is omitted only from the subscript to the activity and the activity coefficient, e.g., $a_{z+} = a_+$ since there is no confusion in such a notation. Equation (13.8) may be rewritten as

$$55.508 \, d \ln a_1 + m \, d \ln a_+^{\nu_+} \cdot a_-^{\nu_-} = 0 \tag{13.9}$$

It is now possible to define a mean activity $a_\pm$ by

$$a_\pm^\nu = a_+^{\nu_+} \cdot a_-^{\nu_-}; \quad (\nu = \nu_+ + \nu_-) \tag{13.10}$$

The activity coefficients are defined by $a_i/m_i$ for each ion; hence

$$(a_\pm/m_\pm)^\nu = (a_+/m_+)^{\nu_+}(a_-/m_-)^{\nu_-} = \gamma_+^{\nu_+} \cdot \gamma_-^{\nu_-} \equiv \gamma_\pm^\nu \tag{13.11}$$

The numerators on both sides of the first equality in (13.11) are equal from (13.10); hence the denominators must also be equal so that,

$$m_\pm^\nu = m_+^{\nu_+} \cdot m_-^{\nu_-} = (\nu_+)^{\nu_+} \cdot (\nu_-)^{\nu_-} m^\nu; \qquad (m_\pm = \text{mean ionic molality}) \tag{13.12}$$

and the mean ionic activity is therefore

$$a_\pm = \left(\gamma_+^{\nu_+} \cdot \gamma_-^{\nu_-} \cdot \nu_+^{\nu_+} \cdot \nu_-^{\nu_-}\right)^{1/\nu} m = \left(\nu_+^{\nu_+} \cdot \nu_-^{\nu_-}\right)^{1/\nu} \gamma_\pm m \tag{13.13}$$

For example, for $Ba_3(PO_4)_2$, $\nu_+ = 3$, $\nu_- = 2$, and $\nu = 5$; hence, $m_\pm = (3^{0.6})\cdot(2^{0.4})m = 2.55085$ m. With the preceding relationships and definitions, the general form of the Gibbs-Duhem relation (13.9) for an electrolyte in water becomes

$$55.508\, d\ln a_1 + \nu m\, d\ln a_\pm = 0 \tag{13.14}$$

The activity coefficient of water is $\gamma_1 = a_1/x_1$, where the mole fraction of water is given by

$$x_1 = 55.508/(55.508 + \nu_+ m + \nu_- m) \tag{13.15}$$

## Debye-Hückel Theory

The activity coefficient of an electrolyte, $\gamma_i$, is a thermodynamic property that can be evaluated by experimental measurements. The Debye-Hückel theory* (1923) however, gives the activity coefficients of electrolytes in a very dilute concentration range without experimental measurements of thermodynamic nature. Although the theory has a number of limitations, it constitutes an important foundation in the physical chemistry of electrolytes and accounts for the deviation from Henry's law at concentrations well below 0.001 m, whereas for nonelectrolytes Henry's law is obeyed at much higher concentrations, i.e., roughly 0.5 to 3 m. This behavior is due to the difference in the interaction of a pair of neutral molecules, and that of a pair of electrically charged anion and cation. The interaction energy between two nonionized molecules in proportional to $\mathbf{d}_i^{-6}$ according to the so-called London forces, whereas the interaction energy between two oppositely charged ions is proportional to $\mathbf{d}_i^{-1}$ according to Coulomb's law of attraction, when $d_i$ is the distance between a pair of molecules or ions. It is therefore not surprising that appreciable deviations from Henry's law occur at concentrations as low as 0.0002 m. Consequently, deviation of $\gamma_i$ from unity is entirely due to the coulombic attraction of ions according to the Debye-Hückel theory. For an ion, the partial Gibbs energy is given by $\overline{G}_i = G_i^\infty + RT \ln a_i = G_i^\infty + RT \ln \gamma_i + RT \ln m_i$.

---

*(I) H. S. Harned and B. B. Owen, "The Physical Chemistry of Electrolytic Solutions," Reinhold Publ. Corp. (1958); (II) R. A. Robinson and R. H. Stokes, "Electrolyte Solutions," Butterworths (1959), (III) I. D. Zaytsev and G. S. Aseyev, "Aqueous Solutions of Electrolytes," translated by M. A. Lazarev and V. R. Sorochenko, CRC Press (1992). Most of the numerical examples in Chapters XIII and XIV are from these references with the citation as Ref. (I), Ref. (II), and Ref. (III).

Solutions of Electrolytes    251

This relationship can be separated into $G_i^\infty + RT \ln m_i$, which is the contribution to $\overline{G}_i$ when the uncharged ion obeys Henry's law, and the remaining term, which is due to the electrostatic interaction, $\overline{G}_i^{el}$. The equation for $\overline{G}_i^{el}$ is therefore

$$\overline{G}_i^{el} = RT \ln \gamma_i \tag{13.16}$$

which gives the relationship between $\overline{G}_i^{el}$ of an ion $i$ due to electrostatic forces, and the activity coefficient $\gamma_i$. Interaction between the ions of the same charge or between the ion and the solvent molecules do not contribute to $\gamma_i$ in very-dilute ionic solutions.

The derivation of the equation for $\overline{G}_i^{el}$ is in the realm of statistical mechanics; therefore, only a brief outline of the theory and the final equation is sufficient in this text. An ion $i$ with a charge of $z_i e$, where $z_i$ is the charge number is attracted by the remaining ions of opposite charge through the solvent acting as a dielectric, and having a dielectric constant designated by $D$. The property $D$ is a dimensionless quantity defined by the electric capacitance of a condenser separated by a dielectric medium divided by the capacitance when the condenser is in a perfect vacuum. Since ions of opposite charges are attracted to each other, the probability of finding an atmosphere of oppositely charged ions around the selected ion $i$ is overwhelmingly larger than that of finding ions of the same charge. The charge density decreases according to the Boltzmann distribution law as assumed by Debye and Hückel. The effects of $D$, the coulombic energy of attraction, and ionic concentration of solution lead to

$$\overline{G}_i^{el} = -\frac{z_i^2 e^2 N_A \kappa}{8\pi e_\circ D(1 + \kappa \mathring{a})} \tag{13.17}$$

where $N_A$ is Avogadro's number, $\mathring{a}$ the distance of closest approach between ions of opposite charges, $e_\circ$, permittivity of perfect vacuum, and $\kappa$ is given by

$$\kappa^2 = \frac{e^2 N_A^2 \rho_1}{e_\circ D R T} \sum m_i z_i^2 \tag{13.18}$$

where $R$ is the gas constant, $\rho_1$ the density of solvent in kg per cubic meter.

The ionic strength* for any solution is defined by

$$I = 0.5 \sum m_i z_i^2 = \text{Ionic strength} \tag{13.19}$$

For a single electrolyte in solution, (13.1) requires that $m_- z_- = -m_+ z_+$, and with $m_+ = m\nu_+$, the ionic strength becomes

$$I = 0.5(m_+ z_+^2 + m_- z_-^2) = 0.5(m_+ z_+^2 - m_+ z_+ z_-)$$
$$= 0.5 m \nu_+ z_+ (z_+ - z_-); \quad (z_- < 0 < z_+) \tag{13.19a}$$

The substitution of the ionic strength and $e = 1.6021 \times 10^{-19}$ Coulomb, $N_A = 6.02214 \times 10^{23}$ molecules per mole, $e_\circ = 8.85419 \times 10^{-12}$ Coulomb meter$^{-1}$ volt,

---

*See General Ref. (20). Definition of ionic strength, empirically introduced by Lewis and Randall, was justified two years later on theoretical grounds by Debye and Hückel, in 1923.

and $R = 8.31451$ J mol$^{-1}$ K$^{-1}$ in (13.18) and then in (13.17) gives

$$\kappa^2 = \frac{2e^2 N_A^2 \rho_1}{e_\circ DRT} I = 2.5289 \times 10^{20} \frac{\rho_1 I}{DT} \tag{13.20}$$

$$\ln \gamma_i = -\frac{z_i^2 e^2 N_A \kappa}{8\pi e_\circ DRT(1+\kappa \mathring{a})} = -\frac{132,853 \, z_i^2 (\rho_1 I)^{0.5}}{(DT)^{1.5}(1+\kappa \mathring{a})} \tag{13.21}$$

The dimension of $\kappa$ is in $m^{-1}$; thus, $\kappa \mathring{a}$ in the denominator is dimensionless. The mean activity coefficient $\gamma_\pm$ from (13.11) is given by $\ln \gamma_\pm = (\nu_+/\nu) \ln \gamma_+ + (\nu_-/\nu) \ln \gamma_-$; hence, the summation of (13.20) for $i$ for all cations and all anions according to the requirement for $\ln \gamma_\pm$, with $\nu_+ z_+ = -\nu_- z_-$, gives

$$\ln \gamma_\pm = \frac{z_+ z_- 132,853(\rho_1 I)^{0.5}}{(DT)^{1.5}(1+\kappa \mathring{a})} \tag{13.22}$$

where the principle of electrical neutrality is used to eliminate the stoichiometric coefficients $\nu_+$ and $\nu_-$. Equation (13.22) is called the Debye-Hückel equation. For water at 25°C, $D = 78.54$, $\rho_1 = 997.08$ kg/m$^3$ and (13.22) becomes

$$\ln \gamma_\pm = \frac{z_+ z_- 1.171 I^{0.5}}{1 + I^{0.5}}, \quad (25°C; \text{ solvent: } H_2O) \tag{13.23}$$

where $\kappa \mathring{a} = I^{0.5}$ as suggested by Guggenheim* because realistic values of the interionic distance $\mathring{a}$ do not satisfy the experimental data. This is equivalent to setting $\mathring{a}$ equal to $3 \times 10^{-10}$ m. Actually the theory is strictly correct when the solution is dilute and the second term in the denominator of (13.23) is negligible; the resulting equation is called the Debye-Hückel limiting equation, or limiting law, i.e.,

$$\ln \gamma_\pm = z_+ z_- 1.171 I^{0.5}, \quad (25°C, H_2O); \quad (I^{0.5} < 0.001) \tag{13.24}$$

This equation has been tested repeatedly and found to be correct for various values of $z_+ \cdot z_-$, and for the same electrolyte dissolved in solvents having various values of the dielectric constant $D$. Remarkable aspects of the Debye-Hückel equation are that (1) $\gamma_\pm$ can be calculated in the limiting case indirectly by the physically measurable properties and (2) the limiting slope of $\ln \gamma_\pm$ versus $I^{0.5}$ is given by the coefficient of $I^{0.5}$ in (13.24). Equation (13.22), as well as (13.23), is usually written as

$$\ln \gamma_\pm = \frac{z_+ z_- \alpha I^{0.5}}{1 + I^{0.5}}, \tag{13.25}$$

with $\alpha$ as a convenient single quantity defined by

$$\alpha = 132,853 \rho_1^{0.5}/(DT)^{1.5} \tag{13.26}$$

---

*E. A. Guggenheim, Phil. Mag., *19*, 588 (1935). Some authors use $1 + \beta I^{0.5}$ in (13.23) with $\beta$ as an adjustable empirical parameter for each type of solution.

Solutions of Electrolytes    253

**Table 13.1** Values of $D$, $\rho_1$, and $\alpha$ for water at various temperatures; $D$ is dimensionless, and $\rho_1$ is in kg m$^{-3}$; $\alpha$ is defined by (13.26).

| °C: | 0 | 10 | 20 | 25 | 30 | 50 | 75 | 100 |
|---|---|---|---|---|---|---|---|---|
| $D$: | 88.15 | 84.15 | 80.36 | 78.54 | 76.77 | 70.10 | 62.62 | 55.90 |
| $\rho_1$: | 999.840 | 999.700 | 998.206 | 997.048 | 995.650 | 988.038 | 974.844 | 958.357 |
| $\alpha$: | 1.125 | 1.142 | 1.161 | 1.171 | 1.181 | 1.225 | 1.289 | 1.365 |

**Table 13.2** Mean activity coefficient $\gamma_\pm$ of HCl, NaCl, KOH, Na$_2$SO$_4$, and LaCl$_3$. First column is molality $m$ of electrolytes, second column is for 0°C, and remaining columns are for 25°C.*

| | $\gamma_\pm$, 0°C | $\gamma_\pm$, 25°C | | | | | |
|---|---|---|---|---|---|---|---|
| $m$ | HCl | HCl | NaCl | KCl | KOH | Na$_2$SO$_4$ | LaCl$_3$ |
| 0.0005 | 0.9756 | 0.9751 | — | — | — | — | 0.839 |
| 0.001 | 0.9668 | 0.9653 | — | — | — | — | 0.788 |
| 0.002 | 0.9541 | 0.9523 | — | — | — | — | — |
| 0.005 | 0.9303 | 0.9285 | 0.9283 | 0.927 | — | — | 0.637 |
| 0.01 | 0.9065 | 0.9048 | 0.9032 | 0.902 | — | — | 0.559 |
| 0.02 | 0.8774 | 0.8755 | 0.8724 | — | — | — | 0.484 |
| 0.05 | 0.8346 | 0.8304 | 0.8215 | 0.816 | 0.824 | 0.529 | 0.392 |
| 0.1 | 0.8027 | 0.7964 | 0.7784 | 0.770 | 0.790 | 0.445 | 0.336 |
| 0.2 | 0.7756 | 0.7667 | 0.732 | 0.719 | 0.757 | 0.365 | 0.293 |
| 0.5 | 0.7761 | 0.7571 | 0.679 | 0.652 | 0.728 | 0.268 | 0.285 |
| 1.0 | 0.8419 | 0.8090 | 0.656 | 0.607 | 0.756 | 0.204 | 0.366 |
| 1.5 | 0.9452 | 0.8962 | 0.656 | 0.585 | 0.814 | 0.180 | 0.554 |
| 2.0 | 1.078 | 1.009 | 0.670 | 0.577 | 0.888 | 0.158 | — |
| 3.0 | 1.452 | 1.316 | 0.719 | 0.574 | 1.081 | 0.137 | — |
| 4.0 | 2.006 | 1.762 | 0.791 | 0.581 | 1.352 | — | — |

*Adapted from Harned and Owen, Ref. (I), Refs. (II), (III), and W. H. Hamer and Y. C. Wu, J. Phys. and Chem. Ref. Data, *1*, 1047 (1972).

The values of $D$ by Wyman and Ingalls,[†] and the density of water from Gen. Ref. (25) are listed in Table 13.1. These values, substituted in (13.26), yield the values of $\alpha$ on the last row of this table.

Precise data for a number of electrolytes in water are listed in Table 13.2. Figure 13.1, based on the data in this table, shows that $(\ln \gamma_\pm)/z_+ z_-$ versus $I^{0.5}$ for four different types of electrolytes in water approach the same slope of $\alpha = 1.171$ as predicted by (13.24). The deviation from (13.24) is approximately the same for the same value of $I$ in the range of $I \leq 0.04$ for different electrolytes.[‡] As

---
[†]J. Wyman, Jr. and E. N. Ingalls, J. Am. Chem. Soc., *60*, 1182 (1938); Felix Franks, Ed.,"Water: A Comprehensive Treatise," Vol. *1* (1972), and Vol. *3* (1973).
[‡]P. Delahay and C. W. Tobias, "Advances in Electrochemistry and Electrochemical Engineering," 9 vols. (1961–1973).

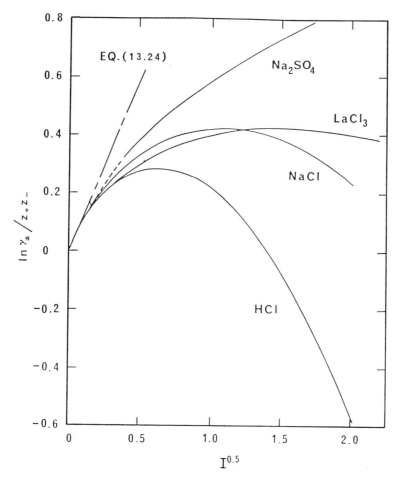

**Fig. 13.1** Mean ionic activity coefficients of some strong electrolytes and Debye-Hückel limiting law, (13.24), at 25°C. Data are listed in Table 13.2.

the concentration of a single electrolyte increases in each solution, $\gamma_\pm$ decreases first because the ionic attraction increases as the oppositely charged ions get closer to each other, but at considerably higher concentrations, e.g., above $m = 0.5$ for HCl in Table 13.2, highly complex phenomena lead to repulsive forces among the ions, and association of ions with each other occurs to increase $\gamma_\pm$. It is evident that the range of validity of (13.24) is very low, i.e., up to $I = 10^{-4}$, but it is important that the slopes of the curves are finite and identical, and that without (13.24) it would be impossible to extrapolate $\ln \gamma_\pm$ to infinite dilution to establish a reference state for $\gamma_\pm$, which must approach unity as the ionic strength approaches zero.

Solutions of Electrolytes    255

The preceding equations account for the effects of more than one electrolyte through the ionic strength $I$. For example, the value of $\gamma_\pm$ for a solution containing $m = 0.005$ HCl alone is the same as that for a solution containing 0.002 m HCl plus 0.003 m NaCl, because the ionic strength is the same in both cases and the value calculated from (13.23) is in very good agreement with the experimental value. As electrolytes of different valence types are mixed, e.g., HCl and CaCl$_2$, the concentration range where (13.23) gives good results becomes significantly lower. For about 0.001 to 0.01 m, for both single and mixed electrolytes, an additional linear term in $m$ is necessary for a precise correlation of $\ln \gamma_\pm$ and $m$:

$$\ln \gamma_\pm = \frac{z_+ z_- \alpha I^{0.5}}{1 + I^{0.5}} + \beta m \qquad (13.26a)$$

**Example 1**  The ionic strength $I$ of $m$ molal solution of LaCl$_3$ in water at 25°C is 0.84. Calculate the values of $m$, $m_+$, $m_-$, $m_\pm$, and $x_1$. Estimate $\gamma_\pm$ from Table 13.2, observing the behavior of LaCl$_3$ in Fig. 13.1.

**Solution**  The stoichiometric coefficients are $\nu_+ = 1$, and $\nu_- = 3$; hence, $m_+ = m$, and $m_- = 3m$. Also, $z_+ = 3$, and $z_- = -1$ so that from (13.19), $I = 0.5[m \times 3^2 + 3m \times (-1)^2] = 6m = 0.84$, from which $m = m_+ = 0.14$, and $m_- = 0.42$. The substitution in (13.12) gives $m_\pm = (3)^{3/4} \times 0.14 = 0.31913$. Further, (13.19) yields $x_1 = 55.508/(55.508 + m + 3m) = 0.9900$. The behavior of $\ln \gamma_\pm$ in Fig. 13.1 is nearly linear in the vicinity of $I = 0.84$; accordingly, the linear interpolation of $\ln \gamma_\pm$ between $m = 0.1$ to 0.2 to $m = 0.14$ gives $\ln \gamma_\pm = \ln 0.293 + 0.6[\ln 0.336 - \ln 0.293] = -1.14542$, from which $\gamma_\pm = 0.318$; a linear interpolation of the values of $\gamma_\pm$ in Table 13.2 yields 0.319.

**Example 2**  One kg of water contains $u$ moles of YCl$_3$ and $v$ moles of NaCl at 25°C. The ionic strength of solution of 0.010, and the sum of molalities of cations is 0.004. Calculate $u$, $v$, $m'$ (for NaCl), and $\gamma'_\pm$ (for NaCl) by using (13.23).

**Solution**  We use the symbols without prime for YCl$_3$, and with prime for NaCl: $\nu_+ = 1$, $\nu_- = 3$, $\nu = 4$, $m_+ = u$, $m_- = 3u$, $z_+ = 3$, $z = -1$; $\nu'_+ = \nu'_- = 1$, $\nu' = 2$, $m'_+ = m'_- = v$, $z'_+ = -z'_- = 1$. From (13.19), $I = 0.5(m_+ \times 3^2 + m_- + m'_+ + m'_-) = 0.5(9u + 3u + 2v) = 6u + v = 0.010$. The total moles of cations is $u + v = 0.004$. These two simultaneous equations yield $u = 0.0012$ and $v = 0.0028$; therefore, $m_+ = 0.0012$, $m_- = 0.0036$, and $m'_+ = m'_- = 0.0028$. The total chloride ion concentration is $0.0028 + 0.0036 = 0.0064$, and $m'_\pm = (0.0028 \times 0.0064)^{0.5} = 0.00423$. The value of $\gamma'_\pm$ is given by

$$\ln \gamma'_\pm = \frac{-1.171(0.01)^{0.5}}{1 + (0.01)^{0.5}} = -0.10645; \qquad \gamma'_\pm = 0.8990$$

This value of $\gamma'_\pm$ is close to that in Table 13.2 for 0.01 m NaCl.

## Concentrated Electrolytes

Concentrated electrolytes deviate considerably from (13.23) and (13.24), and additional terms are necessary to express $\ln \gamma_\pm$ as a useful function of molality. For this purpose, we select the following equations:*

$$55.508 \ln a_1 = -\nu m + Bm^{1.5} + Cm^2 + Dm^{2.5} + Em^3 + \cdots \quad (13.27)$$

$$\nu \ln \gamma_\pm = -\frac{1.5}{0.5} Bm^{0.5} - \frac{2}{1} Cm - \frac{2.5}{1.5} Dm^{1.5} - \frac{3}{2} Em^2 - \cdots \quad (13.28)$$

The substitution of $m_\pm^\nu = m_+^{\nu_+} m_-^{\nu_-} = m^\nu (\nu_+)^{\nu_+} (\nu_-)^{\nu_-}$ in (13.28) and then the substitution of (13.27) and (13.28) in (13.8) would show that the Gibbs-Duhem relation is satisfied by these equations. Instead of $m$, the ionic strength $I$ may be used in (13.27) and (13.28). The coefficient $B$ is related to $\alpha$ by

$$\nu \ln \gamma_\pm = -\nu z_+ z_- \alpha I^{0.5} = -3Bm^{0.5} \quad (13.29)$$

Thus for $m$ molal $CaCl_2$ in water at 25°C, $\nu_+ = 1$, $\nu_- = 2$, $\nu = 3$, $z_+ = 2$, $z_- = -1$, $I = 0.5m (1 \times 2^2 + 2 \times 1^2) = 3m$, $\alpha = 1.171$, and from (13.29), $-3 \times 2 \times \alpha(3m)^{0.5} = -3Bm^{0.5}$; hence, $B = 4.0565$.

The values of $\gamma_\pm$ for multicomponent *dilute* solutions can be estimated by using (13.23) at $I < 0.05$, but for the higher range of concentration, an equation similar to that for multicomponent nonelectrolytic solutions is necessary as shown in Appendix IV.

The temperature dependence of (13.27) can best be expressed by using the available calorimetric data for $\overline{H}_1 - H_1^\circ$ for each composition in the following equation:

$$RT \ln a_1 = \overline{H}_1 - H_1^\circ - T(\overline{S}_1 - S_1^\circ) \quad (13.30)$$

Calorimetric data for a number of electrolytes have been compiled in Gen. Refs. (I)–(III). In the absence of calorimetric data, at least two sets of values for $a_1$ are necessary to write two equations for $55.508 \ln a_1$, one for each temperature. The values of $C, D, E, \ldots$ from each equation can then be used to express these parameters as linear functions of temperature as in the solutions of nonelectrolytes in Chapters X and XI.

**Example 3** We use the data for HCl at 25°C in Table 13.2 for 0.01, 0.1 and 0.5 molal solutions to represent the listed data at $m < 0.6$ with (13.28).

**Solution** For HCl, $\nu_+ = \nu_- = 1$, and $\nu = 2$, and $I = m$; hence,

$$\ln \gamma_\pm = -1.5Bm^{0.5} - Cm - 0.83333Dm^{1.5} - 0.75Em^2 \quad (13.31)$$

From (13.29), $1.5B = \alpha = 1.171$, as expected. The substitution of the three values of $\gamma_\pm$ at 0.01, 0.1 and 0.5 molal yields three simultaneous equations with

---

*N. A. Gokcen, "Determination and Estimation of Ionic Activities of Metal Salts in Water," U.S. Bureau of Mines RI 8372 (1979).

three unknown coefficients, $C$, $D$, and $E$. The solution of these equations gives the values of these coefficients, and their substitution in (13.31) yields the required equation as follows:

$$C = -1.8588, \qquad D = 1.9252, \qquad E = -1.0004$$

$$\ln \gamma_\pm = -1.171 m^{0.5} + 1.8588 m - 1.60433 m^{1.5} + 0.7503 m^2 \tag{13.31a}$$

This equation represents the listed data nearly perfectly at $m < 0.6$, and in excess of this composition, additional terms beyond $E$ are necessary. We substitute the foregoing values of $C$, $D$, and $E$ in (13.27) to express the activity of water:

$$55.508 \ln a_1 = -2m + 0.78067 m^{1.5} - 1.8588 m^2 + 1.9252 m^{2.5} - 1.0004 m^3$$

The activity of water is $a_1 = ZP/(Z°P°) = \gamma_1 x_1$, where the numerator is the compressibility factor $Z$ and the pressure $P$ for the solution, and the denominator is $Z°$ and $P°$ for pure water [Gen. Ref. (6) and (25)], with $Z \approx Z°$ for $x_1 > 0.9$.

## Determination of Activities

We shall now consider experimental determination of activities of electrolytes. The most precise methods are based on (1) determination of the activity of solvent and, (2) electrochemical measurements. While the first set of methods is, in principle, identical with that used for nonelectrolytes, the second is useful for electrolytes. An outline of these methods is presented in the following sections.

*Depression of Freezing Point* The depression of the freezing point of the solvent is convenient and very precise in determining the activities of electrolytes. The solvent in solid state is pure and its Gibbs energy $G_1°$ (solid) is equal to the partial Gibbs energy of the liquid solvent $\overline{G}_1$ (liq.), i.e., $G_1°(s) = \overline{G}_1(l) = G_1°(l) + RT \ln a_1$ as in solutions of nonelectrolytes; therefore,

$$G_1°(l) - G_1°(s) = \Delta_{fus} G_1° = -RT \ln a_1 \tag{13.32}$$

where $\Delta_{fus} G_1°$ is the standard Gibbs energy of fusion for the pure solvent. Very precise data are available for $\Delta_{fus} G_1°$ as a function of temperature. Thus, for water,* the standard enthalpy of fusion is given by $\Delta_{fus} H_1°/R = 722.79$ K, and the standard heat capacity of fusion, by $\Delta_{fus} C_p°/R = 4.484$; hence, $\Delta_{fus} H_1°$ and $\Delta_{fus} S_1°$ can be expressed as functions of temperature by (8.14) and (8.15) respectively, and then substituted in (13.32) to obtain

$$\ln a_1 = 722.79 \left( \frac{1}{273.15} - \frac{1}{273.15 - \Theta} \right) + \frac{4.484 \Theta}{273.15 - \Theta}$$

$$+ 4.484 \ln \left( \frac{273.15 - \Theta}{273.15} \right); \qquad (T = 273.15 - \Theta) \tag{13.33}$$

---

*See General Ref. (2).

where $\Theta$ is the depression of the freezing point, taken to be a positive quantity so that $T = 273.15 - \Theta$. For example, if the depression of the freezing point is 10°C, i.e., $\Theta = 10$, then from (13.33), $a_1 = 0.9072$.

**Vapor Pressure of Solvent** The activity of solvent can be determined from the vapor pressure measurements, from which $a_\pm$ can be calculated. For dilute aqueous solutions, the accuracy is not high but for concentrated solutions the results are satisfactory. The limitation of the old methods arises from the fact that the reproducibility of vapor pressure measurements is about 0.005 percent which is equivalent to the lowering of the freezing point by approximately 0.0014 molal solution of a 1:1 strong electrolyte. An improved method utilizes a differential pressure transducer,[*] one arm of which is connected to a flask containing pure water, and the other arm, to a flask containing a solution of electrolyte, all immersed in a thermostat controlled to ±0.005°C. The transducer is capable of reading a small pressure difference of $\Delta P = 300$ Pa with an accuracy of 0.01% so that $\Delta P = 300 \pm 0.03$. Let the pressure of pure water be $P° = 3{,}000$ Pa (at about 24.1°C); therefore, the pressure over the solution is $P = 3{,}000 - 300 + 0.03 = 2{,}700 + 0.03$, and $P$ can be determined with an accuracy of 0.0011%. Several solutions can be connected to the arm of the transducer in such a way that the data for several compositions can be obtained. It is essential that the electrolyte itself has no contribution to $\Delta P$, and that the system is leakproof. By this method, accurate values of $a_1$ can be obtained, from which $\gamma_\pm$ can be calculated by using (13.27) and (13.28).

**Vapor Pressures of Solute** The vapor pressures of most electrolytes are immeasurably small in one or two molal aqueous solutions even when the vapor pressures of pure electrolytes are fairly high as in the case of hydrogen halides. The pressure of HCl(g) over the solutions containing 4 to 10 m HCl, and of HBr(g) over 6 to 11 m, HBr and HI(g) over 6 to 11 m have been measured.[†] The results are only useful in extending the activity measurements to higher concentrations. As an example, consider HCl(g) = $H^+ + Cl^-$ for which $P_{HCl} = 2.43$ Pa $= 2.43 \times 10^{-5}$ bar at $m = 4.0$ and 25°C. For this reaction, the equilibrium constant is given by $K_p = m^2 \gamma_\pm^2 / P_{HCl}$. From Table 13.2, $\gamma_\pm$ is 1.762, and $K_p = (4.0 \times 1.762)^2/(2.43 \times 10^{-5}) = 2.04 \times 10^6$. From another experimentally measured value of $P_{HCl}$, (e.g., $P = 0.00112$ bar at 25°C, and $m = 8$), $\gamma_\pm^2 = 0.00112 \times 2.04 \times 10^6/64$; hence $\gamma_\pm = 5.97$.

**Osmotic Pressure** Consider a U-tube divided by a membrane permeable to water but impermeable to a solute as shown in Fig. 13.2. Such membranes are usually called semipermeable membranes. When the right arm is filled with pure water, and the left arm to the same level with a solution of a particular solute, and both sides are initially at 1 bar, it is soon observed that the level of solution at the left rises. This phenomenon is caused by the diffusion of water through the membrane

---
[*]See N. A. Gokcen, U.S. Bureau of Mines RI8372 (1979).
[†]Hala, et al. loc. cit.

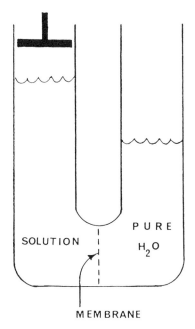

**Fig. 13.2** Osmotic pressure; left arm contains water solution and right arm, pure water.

because the activity of water is higher in the right side. The pressure over the left arm may be increased to $P$ in order to stop the diffusion of water through the membrane; the activities of solvent on both sides are then equal. The pressure $P$ increases the activity of solvent in the solution to unity and thus stops the diffusion of water. The activity of water for the left side in Fig. 13.2 is defined by

$$\overline{G}_1(P, T) = G_1^\circ(P = 1, T) + RT \ln a_1(P, T), \text{ (constant } m);$$

(left side) (13.34)

where $P$ and $T$ in parentheses are the independent variables and $G_1^\circ$ always refers to one bar. The effect of pressure $P$ is to increase $a_1$ so that $\overline{G}_1$ in the left arm becomes equal to $G_1^\circ$ for pure water in the right arm, and then $a_1(P, T)$ in (13.40) becomes unity. Since $\partial \overline{G}_1/\partial P = \overline{V}_1$, where $\overline{V}_1$ is the partial volume of water in the solution, integration gives

$$\int_1^P d\overline{G}_1 = \overline{G}_1(P, T) - \overline{G}_1(P = 1, T) = \int_1^P \overline{V}_1 dP, \quad \text{(left side)}$$

(13.35)

When diffusion stops at $P$, $\overline{G}_1(P, T)$ becomes equal $G^\circ(1, T)$ of the right arm, [cf. (13.34)], and the terms between the equal signs in (13.35) become $G_1^\circ - \overline{G}_1(1, T)$, and from the definition of activity,

$$G_1^\circ - \overline{G}_1(1, T) = -RT \ln a_1(1, T) \tag{13.36}$$

260    Thermodynamics

Substitution of this equation into (13.35) gives

$$\ln a_1(1, T) = -\frac{1}{RT}\int_1^P \overline{V}_1 dP; \quad \text{(exact)} \tag{13.37}$$

This equation is exact, and for accurate integration of the right side, $\overline{V}_1$ must be determined experimentally at each composition as a function of pressure. In dilute solutions, it is safe to take $\overline{V}_1 = V_1^\circ$ and to carry out the integration, although $\overline{V}_1$ should be used whenever possible. Further, when the solution is dilute $V_1^\circ$ for the solvent is independent of moderate changes in pressure, and with these assumptions (13.37) becomes

$$\ln a_1(1, T) = -(P - 1)V_1^\circ/RT \equiv -\Pi V_1^\circ/RT; \quad \left(\overline{V}_1 = V_1^\circ\right) \tag{13.38}$$

where $\Pi = P - 1$ is called the *osmotic pressure*. The osmotic coefficient $\phi$, also called the practical osmotic coefficient in Gen. Ref. (I), and the *molal osmotic coefficient* in Gen. Ref. (II), and fortunately denoted by the same symbol in most books, is defined by

$$\ln a_1 = -\phi \sum_i m_i M_1/1000 \tag{13.39}$$

where $m_i$ is the molality of ion $i$, and $M_1$, the molecular weight of the solvent. The rational osmotic coefficient, $g$, introduced by Bjerrum (1907), is defined by

$$\ln a_1 = g \ln x_1, \quad \text{or} \quad \ln \gamma_1 = (g - 1) \ln x_1 \tag{13.40}$$

where the second equation is obtained from the first by substituting $\gamma_1 x_1$ for $a_1$, and rearranging the result. For solvents obeying Raoult's law $\gamma_1 = 1$ and $g = 1$, therefore, $g$ is another measure of the deviation from Raoult's law.

*Example 4* The aqueous solution in the left arm of the U-tube in Fig. 13.2 contains 0.1 m of NaCl. The vapor pressure of pure water on the right side of the U-tube is 0.03167 bar. The activity of water, $a_1$ from the most reliable measurements, including the depression of the freezing point, is 0.996646. Calculate $\Pi, \phi, g, \beta, \gamma_\pm$ and the vapor pressure of solution. [Problem based on data in Ref. (II), p. 476].

*Solution* The molar volume $V_1^\circ$ of water is $18.071 \times 10^{-6}$ m$^3$, $R = 8.31451$ J mol$^{-1}$ K$^{-1}$, and $T = 298.15$; substitution of these values and $a_1 = 0.996646$ in (13.38) gives

$$\Pi = -8.31451 \times 298.15 \ln 0.996646/18.071 \times 10^{-6}$$

$$= 4.61 \times 10^5 \text{ Pa} = 4.61 \text{ bars}$$

This pressure would be about half as much for sucrose of the same concentration because sucrose does not ionize. Likewise, from (13.39), and from $\sum m_i = 0.2$, and $M_1 = 18.0152$,

$$\phi = -55.508(\ln 0.996646)/0.2 = 0.9324$$

The mole fraction of water is given by $55.508/(55.508 + 0.2) = 0.99641$; therefore, from the first equation in (13.40),

$$g = (\ln 0.996646)/\ln 0.99641 = 0.93415$$

The foregoing example shows that the magnitude of $\Pi$ is large, and therefore it should be possible to measure $\Pi$ accurately and to calculate the activity of a solvent by substituting $\Pi$ in (13.38). The experiments require that the membrane be insoluble and truly semipermeable. For solutes having large molecular weights, such as the high polymers, it is not difficult to obtain nearly semipermeable membranes, but small molecules often diffuse to equalize the concentration. With adequate precaution, however, precise values of $a_\pm$ have been obtained at low concentrations.

*Activity from Solubility Measurements* The activity $a_s$ of pure solid electrolyte is unity, and when the electrolyte is in contact with its saturated solution, the equilibrium constant is given by

$$K_p = a_\pm^\nu/a_s = (a_+)^{\nu_+} \cdot (a_-)^{\nu_-} = (\gamma_\pm)^\nu (m_+)^{\nu_+}(m_-)^{\nu_-} \quad (13.41)$$

where $K_p$ refers to $A_{\nu_+}B_{\nu_-} = \nu_+ A^{z+} + \nu_- A^{z-}$. The product $\overset{\blacktriangle}{S} = m_+^{\nu_+} \cdot m_-^{\nu_-}$ is called the solubility product. Equation (13.41) yields, for a 1:1 electrolyte,

$$\ln m_+ \cdot m_- = \ln K_p - 2\ln \gamma_\pm$$

Substitution of (13.24) for $\ln \gamma_\pm$ at 25°C into this equation yields

$$\ln m_+ \cdot m_- = \ln \overset{\blacktriangle}{S} = \ln K_p + 2 \times 1.171 I^{0.5}, \quad (25°C) \quad (13.42)$$

The ionic strength, $I$, may be varied by adding strong electrolytes and the changes in $I$ affect the experimentally determined solubility product as required by (13.42) since $K_p$ must remain constant. A plot of $\ln \overset{\blacktriangle}{S}$ versus $I^{0.5}$ must yield a linear plot for values of $I^{0.5}$ smaller than preferably 0.05 and extrapolation to $I^{0.5} = 0$ yields the intercept which is equal to $\ln K_p$. From $K_p$, and from (13.41), $\gamma_\pm$ can be calculated.

The activity coefficients can also be determined, (1) by galvanic cells* as will be seen in Chapter XIV, and (2) by measurements of diffusion of ions in dilute solutions, such as the Onsager-Fuoss method, which has been improved by isotopic techniques to obtain accurate results.* The activity coefficients of a selected number of electrolytes, obtained by various methods, are presented in Table 13.2.

Considerable amount of data on $a_\pm$ and $a_1$ has been generated during the past 75 years as shown in Ref. (III), but the task is far from being adequate, particularly

---

*See H. S. Harned and W. J. Hamer, Eds., "The Structure of Electrolytic Solutions," John Wiley & Sons, (1959); I. M. Klotz and R. M. Rosenberg, "Chemical Thermodynamics: Basic Theory and Methods," John Wiley & Sons (1994).

for multicomponent solutions. Therefore, successful methods of estimation have been devised by a number of investigators. A fairly simple method by Meissner and Kusik is presented in Appendix IV.

**Weak Electrolytes**

A weak electrolyte dissociates to a limited extent even in a dilute solution so that its undissociated molecules are in equilibrium with its ions. A typical example is acetic acid HOAc(=HOCH$_3$CO) which dissociates as follows:

$$\text{HOAc} = \text{H}^+ + \text{OAc}^- \qquad (13.43)$$
$$m(1-\delta), \quad (\delta m) \quad (\delta m) \qquad (\delta m = m_+ = m_- = I)$$

where $m$ is the number of moles of HOAc in one kg H$_2$O prior to dissociation and $m(1-\delta)$, the undissociated acid in equilibrium with $\delta m$ moles each of H$^+$ and OAc$^-$, $\delta$ being the degree of dissociation defined by the moles of dissociated acid $\delta m$, divided by the initial number of moles $m$ of the acid. The degree of dissociation can be determined by various methods but most accurately by conductance measurements. Let a solution containing $m$ moles of acetic acid in one kg water be confined in an open top quadrangular container. The opposite vertical and parallel walls of the container are conducting, the remaining vertical and parallel walls, as well as the bottom, are insulating so that the electrical conductance of the solution can be measured (not the conductivity). The conductance is due to the moles of ions present between the conducting plates, i.e., $2\delta m$, as indicated in (13.43). Let the initial measured conductance be designated by $\Lambda(m)$. Progressive addition of water increases the conductance in proportion to $2\delta m$ because the dilution of solution in this manner increases the number of moles of ions. At infinite dilution, all the acid is ionized so that the number of moles of ions becomes $2m$. The ratio of conductance at the initial molality, $\Lambda(m)$, to the conductance at infinite dilution, $\Lambda_\infty$, is then equal to the corresponding ratio of ions, $2\delta m/2m = \delta$; hence, the degree of dissociation is given by

$$\delta = \Lambda(m)/\Lambda_\infty \qquad (13.44)$$

The experimental measurements of MacInnes and Shedlovski and their values of $\delta$ are listed in Table 13.3 where the second column gives $\delta = \Lambda(m)/\Lambda_\infty$. The activity coefficient $\gamma_\pm$ and the ionization constant $K_i$ are computed from the following equations:

$$\ln \gamma_\pm = -\frac{1.171(\delta m)^{0.5}}{1+(\delta m)^{0.5}} \qquad K_i = \frac{a_+ \cdot a_-}{a_{ac}} = \frac{(\gamma_\pm \cdot \delta m)^2}{(1-\delta)m} \qquad (13.45)$$

where $\delta m$ is the ionic strength and $a_{ac} = (1-\delta)m$ is the activity of undissociated acid, which is equal to its molality on the Henrian scale at least up to its molality of about 0.01 in Table 13.3. For example, for the fifth row, $m = 0.0024211$, $\delta = 0.08290$, $I = \delta m = 0.00020071$, $I^{0.5} = 0.014167$, $\ln \gamma_\pm = -1.171 \times 0.014167/1.014167 = -0.016350$; hence, $\gamma = 0.98378$, and

**Table 13.3** Ionization constant of acetic acid at 25°C; adapted from data of MacInnes and Shedlovski;* see also Ref. (II).

| $10^4 m/mol^{-1}/kg^{-1}$ | $\delta = \Lambda(m)/\Lambda_\infty$ | $10^5 K_i$ |
|---|---|---|
| 0.2809 | 0.5393 | 1.757 |
| 1.1168 | 0.3277 | 1.759 |
| 2.1908 | 0.2477 | 1.756 |
| 10.313 | 0.12375 | 1.756 |
| 24.211 | 0.08290 | 1.756 |
| 59.288 | 0.05401 | 1.754 |
| 200.6 | 0.02987 | 1.744 |
| 501.5 | 0.01905 | 1.730 |
| 1002.9 | 0.013493 | 1.704 |
| 2005.9 | 0.009494 | 1.655 |

*D. A. MacInnes and T. Shedlovski, J. Am. Chem. Soc., 54, 1429 (1932).

$K_i = (0.98378 \times 0.00020071)^2/[(1 - 0.08290) \times 0.0024211] = 1.756 \times 10^{-5}$
which is listed in the table. In the most dilute range, the average for the set of five ionization constants yield nearly identical result:

$$K_i = \frac{(\delta \gamma_\pm)^2 m}{1 - \delta} = 1.757 \times 10^{-5} \tag{13.46}$$

This value will be confirmed by the results of measurements with galvanic cells in the next chapter. Equation (13.46), or its equivalent in (13.45), shows that as $m$ decreases, both $\gamma_\pm$ and $\delta$ approach unity, since $K_i$ is a finite constant, but $m$ has to be very small for $\delta \to 1$. For example, $m$ has to be $10^{-8}$ for $\delta = 0.9994$.

The results for $K_i$ at higher concentrations than $m = 0.005$ show a decreasing trend with increasing concentration. This is partly due to the neglect of the activity coefficient $\gamma_{ac}$ of undissociated acid.

The effect of temperature on $K_i$ is rather unusual since $\ln K_i$ versus $1/T$ gives a curve rather than a straight line because $\Delta C_p^\circ/R$ for ionization is rather high, i.e., $-18.37$.

The degree of dissociation $\delta$ becomes smaller with increasing molality, $m$, and for weak acids and bases, the ionic species usually become negligible in comparison to the undissociated molecules. Certain electrolytes such as $H_2CO_3$ dissociate in two stages and for each state of ionization, $H_2CO_3 = H^+ + HCO_3^-$ and $HCO_3^- = H^+ + CO_2^{--}$, the corresponding ionization constant has been determined. See Ref. (I). Water, which is considered to be a weak acid, dissociates slightly to yield $H^+$ and $OH^-$ ions, and the corresponding ionization constant can be accurately determined from the galvanic cell measurements as discussed in the next chapter.

## Temkin Rule

Molten salts, oxides, and their mixtures are ionized to various extents as indicated by their ionic conduction. In a fused salt solution, such as $K_2O(1)$-$YF_3(2)$ where (1) and (2) designate the components, we can measure the activities of components referred to their pure salts as the standard states, not the activities of individual ions. Temkin* assumed that at any instantaneous moment, the salt mixture may be regarded as having a lattice occupied randomly by the cations, and another interpenetrating lattice occupied randomly by the anions. The ionization of $x_1$ mole of $K_2O$ and $x_2$ mole of $YF_3$, with $x_1 + x_2 = 1$, yield the following moles of species:

Cations: $2x_1$ of $K^+$; and $x_2$ of $Y^{3+}$; total cations: $2x_1 + x_2 = 1 + x_1$
Anions: $x_1$ of $O^{2-}$; and $3x_2$ of $F^-$; total anions: $x_1 + 3x_2 = 1 + 2x_2$
Cationic fractions: $[K^+] = 2x_1/(1+x_1)$; $[Y^{3+}] = x_2/(1+x_1)$
Anionic fractions: $[O^{2-}] = x_1/(1+2x_2)$; $[F^-] = 3x_2/(1+2x_2)$

where [ ] indicate the ionic fractions. The sum of the cationic fractions, as well as the anionic fractions, is unity. The distribution of ions within each lattice is assumed to be random; hence, the activity of each ionic species is equal to its ionic fraction. With this in mind, (13.2) without water but with four ions is

$$2x_1 \, d \ln[K^-] + x_2 \, d \ln[Y^{3+}] + x_1 \, d \ln[O^{2-}] + 3x_2 \, d \ln[F^-]$$
$$= x_1 \, d \ln\{[K^+]^2 \cdot [O^{2-}]\} + x_2 \, d \ln\{[Y^{3+}] \cdot [F^-]^3\} = 0 \qquad (13.47)$$

Since we can measure only the activities of the component salts $a_1$ and $a_2$, then (13.2) is simply

$$x_1 \, d \ln a_1 + x_2 \, d \ln a_2 = 0 \qquad (13.48)$$

Comparison of (13.47) and (13.48) term by term shows that

$$\boxed{a_1 = [K^+]^2 \cdot [O^-]; \qquad a_2 = [Y^{3+}] \cdot [F^-]^3} \qquad (13.49)$$

where the stoichiometric coefficients have become exponents as in the ionic solutions. Equation (13.49) is a statement of the *Temkin rule*. A much longer derivation of (13.49) by using random entropies yield the same result. The Temkin equation does not consider attractive forces between cations and anions, and repulsive or attractive forces between cations-cations and anions-anions, which require the equations in Chapter XI. Nevertheless, it provides estimates on the activities in molten salts in the absence of experimental data.† (See Problem 13.13).

---

*M. Temkin, Acta Physicochim., U.R.S.S. (Russia), vol. 20, 411 (1945). For other models, see E. T. Turkdogan, "Physicochemical Properties of Molten Slags and Glasses, The Metals Society, London (1983), and Y. K. Rao, loc. cit.
†R. G. Reddy and S. G. Kumar, Met. Trans. vol. 24B, 1031 (1993), and vol. 25B, 91 (1994).

## PROBLEMS

13.1 Calculate $m_+$, $m_-$, $m_\pm$, $I$, and $\gamma_\pm$ for a solution containing 0.0001 molal $CaCl_2$ at 25°C.

*Partial ans.:* $I = 0.0003$, $\gamma_\pm = 0.9609$

13.2 The freezing point of 0.05 molal NaCl in water is 272.974 K. Calculate the activity and activity coefficient of water, and the mean ionic activity coefficient of NaCl at 273.150 K. Neglect the effect of temperature within 0.2°C.

13.3 Fit the data in Table 13.2 for $LaCl_3$ to (13.28) by using the data at 0.001, 0.005, and 0.01 m and compute the missing value of $\gamma_\pm$ at 0.002 m.   *Ans.:* $\gamma_\pm = 0.728$

13.4 From (13.21), written as $\ln \gamma_i = -z_i^2 \alpha I^{0.5}/(1+I^{0.5})$, derive (13.23) in detail. The important step is to show that $\nu z_+ z_- = -\nu_+ z_+^2 - \nu_- z_-^2$ from $\nu = \nu_+ + \nu_-$, and from $\nu_+ z_+ = -\nu_- z_-$.

13.5 The ionic strength of a solution containing $LaCl_3$ and NaCl is 0.92. Calculate (a) the molality of each electrolyte if the ionic concentration of the chloride ion is 0.56 m, and (b) the mean ionic molality of $LaCl_3$.

*Partial ans.:* (a) $m(LaCl_3) = 0.12$

13.6 Assume that $1.76 \times 10^{-4}$ mole of $AgIO_3$ is soluble in water at 25°C. Calculate the equilibrium constant $K_i$ and $\Delta G°$ for $AgIO_3(s) = Ag^+ + IO_3^-$.

*Partial ans.:* $K_i = 3.003 \times 10^{-8}$

13.7 Calculate $\gamma_\pm$(HCl), and $\gamma_\pm$(KCl) at 25°C for a solution containing 0.007, and 0.0008 molal of HCl, and KCl respectively.

13.8 Calculate the molality of HCl that depresses the freezing point of water 0.0100°C.

*Ans.:* 0.00269

13.9 Calculate the vapor pressure and activity of water containing 2 m solution of NaCl at 25°C. The vapor pressure of pure water at 25°C is 3167 Pa. Use the values of $\gamma_\pm$ for NaCl in Table 13.2.

*Partial ans.:* activity of $H_2O = 0.9316$

13.10 Calculate the osmotic pressure, the osmotic coefficient and the rational osmotic coefficient at 25°C for a solution containing (a) 0.0001 m of HCl and 0.1 m sucrose (b) 0.1 m of KCl by using Table 13.2.

*Partial ans.:* (b) $\phi = 0.9287$; $g = 0.9304$

13.11 Consider a solution containing $Li^+$(1 m), $K^+$(2 m), and $F^-$(1.5 m) and $Cl^-$(1.5 m) with their molalities shown in parentheses. The solution can be formed by using a set of three independent salt components out of four salts, LiCl, LiF, KCl, and KF; hence the system contains four components including the solvent. (a) How many different ways can the components be chosen. (b) Write $a_\pm^2$ and $\gamma_\pm^2$ for each electrolyte in terms of $a_+$, $a_-$, $\gamma_+$, and $\gamma_-$ and show that

$$\frac{a_\pm(\text{LiF})}{a_\pm(\text{LiCl})} = \frac{a_\pm(\text{KF})}{a_\pm(\text{KCl})}, \quad \text{and} \quad \frac{\gamma_\pm(\text{LiF})}{\gamma_\pm(\text{LiCl})} = \frac{\gamma_\pm(\text{KF})}{\gamma_\pm(\text{KCl})}$$

13.12 Express $\Delta G°$, $\Delta H°$, and $\Delta S°$ for $HOAc = H^+ + OAc^-$ from the following values of $10^5 K_i$: 1.652 (at 0°C) and 1.761 (30°C), and $\Delta C_p°/R = -18.37$.

*Partial ans.:* $\Delta G°/R = 5{,}459.7 - 112.033 T + 18.37 T \ln T$.

13.13 Calculate the activities of equimolar fused salt of $Li_2O + CaF_2$ by using Temkin's rule.

*Partial ans.:* $a_{Li_2O} = 0.148$

CHAPTER XIV

# REVERSIBLE GALVANIC CELLS

## Introduction

A reversible galvanic cell, often called a cell in this chapter, has two electrodes, capable of conducting electricity, and a reaction medium, a liquid solution or a solid solution of an electrolyte. Gases, liquids, and solids participating in the cell reaction may be in contact with the electrodes and the solution. When the electrodes are connected externally by a wire, electrons flow from one electrode to the other by electronic conduction through the wire; conduction within the electrolyte, however, is ionic, i.e., the current is carried by the negatively charged anions and positively charged cations. This chapter deals with the thermodynamics of electrochemical reactions in reversible galvanic cells.

The change in Gibbs energy accompanying an electric current at a constant temperature and pressure is

$$\Delta G_{P,T} = W'' \quad \text{(net work)} \tag{14.1}$$

where $W''$ is the net work other than the work of expansion, and it is the amount of electricity in Coulombs multiplied by the electromotive force, emf. If the cell reaction requires $z$ moles of electrons, and $F$, or 96,485.3 Coulombs per mole of electrons, with $E$ volts of reversible emf, then the electrical work is related to the change in Gibbs energy by

$$\Delta G_{P,T} = -zFE = -96{,}485.31zE, \text{ J} \tag{14.2}$$

The constant $F$ is called the Faraday constant. The symbol $z$ without subscripts should not be confused with ionic charges $z_+$ and $z_-$ although $z$ is always a multiple of $z_+$ and, quite often, equal to $z_+$. One mole of electrons, i.e., $z = 1$, is equivalent to the number of electrons given off when one mole of a monovalent element becomes a cation. When the reactants and the products for the cell reaction are in their standard states, then (14.2) becomes

$$\Delta G_T^\circ = -zFE^\circ \tag{14.3}$$

where $E°$ is the standard emf of the cell. The entropy change is obtained by differentiating $\Delta G$ in (14.2) and $\Delta G°$ in (14.3) with respect to temperature, i.e.,

$$\Delta S_p = zF(\partial E/\partial T)_p; \qquad \Delta S_p° = zF(\partial E°/\partial T)_p \qquad (14.4)$$

The determination of $\Delta S_p$ and $\Delta S_p°$ requires the measurement of the change in $E$ and $E°$ with temperature. The change in enthalpy can be obtained by substituting (14.2), (14.3), and (14.4) in $\Delta H = \Delta G + T\Delta S$:

$$\Delta H_{p,T} = -zF[E - T(\partial E/\partial T)_p];$$
$$\Delta H_{p,T}° = -zF[E° - T(\partial E°/\partial T)_p] \qquad (14.5)$$

The relationships in (14.5) are known as the Gibbs-Helmholtz equations.

## Properties of Reversible Cells

A thermodynamic investigation of a reaction in a galvanic cell requires carefully selected cell components to assure that the cell functions reversibly. Details of construction and operation of various cells depend on the reaction to be investigated. The choice of electrolytes, electrodes, gas, liquid and solid components, is dictated not only by the cell reaction, but also by experience with various types of cells. It is essential that side reactions are absent, and that the emf of the cell does not vary either with time or with externally applied potential differing slightly from the emf of the cell. In numerous cases, it is necessary to design different cells for the same reaction, or to invoke indirect data to ascertain that the reversible emf corresponds entirely to the desired cell reaction. In this chapter, it is assumed that each cell is constructed properly as described in numerous papers and books on electrochemistry,* and therefore, the term reversible will not always be repeated.

## Single Electrode Reactions

The single electrode reaction when a metal, or a gas such as hydrogen, is in contact with its ions, is written as

$$L = L^{z+} + z_+ e^- \qquad (14.6)$$

where $L$ is a metal or hydrogen in its neutral state, $L^{z+}$, the cation, $e^-$, the electron, $z_+$ the valence of the cation, and $z = z_+$ in this reaction. Since the electrons generated by the reaction cannot be dissipated by the solution, the electrode becomes overcharged with electrons.

If the electrode is a non-metal $A$ in contact with its anions $A^{z-}$, the reaction may be written as

$$A^{z-} = A + ze^- \qquad (14.7)$$

where $z = |z_-|$.

---

*See for example, P. Rieger, "Electrochemistry," Prentice-Hall (1987); J. Koryta, J. Dvorak, and L. Kavan, "Principle of Electrochemistry," John Wiley & Sons (1993).

When an *inert electrode* is in contact with a solution containing the ions of the same metal, e.g., $Fe^{++}, Fe^{+++}$ ions in contact with a *platinum electrode* the reaction is similar to (14.6)

$$Fe^{++} = Fe^{+++} + e^- \qquad (14.8)$$

and the function of the electrode is to make electrical contact with the solution. In all these cases an element or an ion loses electrons as the electrode reaction proceeds from the left to the right. The loss of electrons is called oxidation, and the gain of electrons, reduction. The state in which an electron has been lost is called the oxidized state, and the state prior to giving off an electron, the reduced state; hence the electrode reactions may be written as

$$\text{Reduced State} = \text{Oxidized State} + ze^- \qquad (14.9)$$

$$\text{Oxidized State} + ze^- = \text{Reduced State} \qquad (14.10)$$

An appropriate combination of electrode reactions yields the overall cell reaction with the requirement that the electrons cancel out in the final reaction.

## Convention in Notation

The notation used in writing a cell must be consistent and informative as to the sign of emf, the type of electrodes, the electrolyte, and the chemical reaction. When the left-hand side electrode reaction is oxidizing, i.e.,

$$\text{Reduced State} = \text{Oxidized State} + ze^-, \quad \text{(left side)} \qquad (14.11)$$

and the right-hand side reaction is reducing, i.e.,

$$\text{Oxidized State} + ze^- = \text{Reduced}, \quad \text{(right side)} \qquad (14.12)$$

then if the electrodes are connected *externally* with a wire, the electrons flow from the left to the right, and the resulting emf, $E$ is taken as positive in value. If the electrons flow in the opposite direction, the emf for the cell is negative, and the cell and the reactions should be reversed. The cell is generally written so that its constituents are in a proper sequence, and each phase is separated from the adjoining phases by a vertical bar; e.g.,

$$Pt|H_2(1 \text{ bar})|HCl(aq., 0.5m)|AgCl(s)|Ag$$

where $Pt|H_2(1 \text{ bar})$ is the negative electrode or anode, 0.5 molal solution of HCl in water is the electrolyte, and $AgCl(s)|Ag$ is the positive electrode or cathode, as shown in Fig. 14.1. (The cathode is where electrons enter in a cell). HCl is also written as $H^+, Cl^-$. The electrode reactions for this cell are

$$0.5H_2(1 \text{ bar}) = H^+(m') + e^-$$

$$AgCl(s) + e^- = Ag(s) + Cl^-(m'')$$

where $m'$ and $m''$ are the molalities of $H^+$ or $Cl^-$ ions respectively. When an electrode is in contact with a gas or a salt of limited solubility, the electrode is

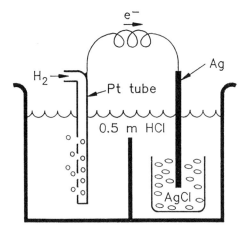

**Fig. 14.1** Galvanic Cell for Pt|H$_2$ (1 bar)|HCl(aq., 0.5$m$)|AgCl(s)|Ag. Ag is the cathode.

written first if it is on the left side as in Pt|H$_2$ above; if the electrode is on the right side the reverse convention is adopted, as in AgCl(s)|Ag(s). The cell reaction is therefore the sum of the preceding reactions, i.e.,

$$0.5\text{H}_2(\text{g}) + \text{AgCl}(\text{s}) = \text{Ag}(\text{s}) + \text{H}^+(m') + \text{Cl}^-(m'')$$

Since AgCl is in equilibrium with its ions in solution, it is possible to rewrite this reaction as $0.5\text{H}_2(\text{g}) + \text{Ag}^+ = \text{Ag}(\text{s}) + \text{H}^+$, a reaction in which H$_2$ is oxidized and Ag$^+$ is reduced. Such reactions are called *redox* reactions.

As an additional example, the following cell is considered:

$$\text{Ag}(\text{s})|\text{AgNO}_3(\text{aq., 0.1 }m) \vdots \text{AgNO}_3(\text{aq., 0.5}m))\text{Ag}(\text{s})$$

In this cell, there are two electrolytes containing different concentrations of AgNO$_3$ separated by a porous membrane as indicated by $\vdots$. Since $\overline{G}_i$ of silver ion in the 0.1$m$ solution is lower than that in the 0.5$m$ solution, the silver is dissolved on the left and plated out on the right, i.e.,

$$\text{Ag}(\text{s}) = \text{Ag}^+ + e^- \quad \text{(left side)}$$
$$\text{Ag}^+ + e^- = \text{Ag}(\text{s}) \quad \text{(right side)}$$

The overall reaction is Ag(left side) = Ag(right side). This cell is called a concentration cell and the junction separating the electrolytes, the liquid junction. The velocities of positive and negative ions are, in general, considerably different from each other so that the concentration profile at the junction changes with time. The ions moving at different rates at the junction create an emf which is called the liquid junction potential. The measured emf, therefore represents the sum of the potential from the reaction, and that from the liquid junction. The liquid junction potential must be minimized whenever possible in studying the emf of cell reactions. In some aqueous cells the liquid junction potential can be minimized by placing a salt bridge between the two electrolytes. The salt bridge must have anions and cations of similar mobilities, such as the ions of KCl.

An example of concentration cells at high temperatures is as follows:

Zn(pure liq.)|$ZnCl_2$ (in fused alkali chlorides)|Zn-Sb(liq. solution)

The electrolyte in this cell consists of fused alkali salts in which the small amount of $ZnCl_2$ is dissolved. The alkali salts are chosen so that Zn and $ZnCl_2$ cannot be involved in a side reaction such as $Zn(l) + 2Na^+ = Zn^{++} + 2Na(l)$. For example, if $CdCl_2$ were present in the electrolyte, the side reaction $Zn(l) + Cd^{++} = Zn^{++} + Cd(l)$ would occur and thus interfere with the emf measurement. In addition, it is important to avoid thermocouple effects in making external contacts from the electrodes to a potentiometer in measuring the emf, particularly with high temperature cells. The cell reaction in this case is

Zn(pure liq.) = Zn(in liq. Zn-Sb solution)

or simply the transfer of zinc from the pure zinc electrode to the zinc-antimony solution. The corresponding change in the Gibbs energy is related to the emf and the activity of zinc in Zn-Sb solution by

$$\overline{G}(\text{Zn in solution}) - G°(\text{pure Zn}) = -2FE = RT \ln a_{Zn}. \quad (14.13)$$

Direct determinations of $a_{Zn}$ from the vapor pressure of zinc over the Zn-Sb solutions agree with $a_{Zn}$ obtained from (14.13), and assure that there are no side reactions in this cell. Similar cells have been used in numerous investigations* to obtain precise data for metallic solutions and intermetallic compounds. See Gen. Ref. (9).

### Reaction Isotherm and emf

An overall cell reaction may be written as

$$\nu_1 A_1 + \nu_2 A_2 + \cdots = \nu_I A_I + \nu_{II} A_{II} + \cdots \quad [\text{cf. (12.1)}]$$

The corresponding change in the Gibbs energy is

$$\Delta G_{P,T} = \Delta G_T° + RT \ln \frac{a_I^{\nu_I} \cdot a_{II}^{\nu_{II}} \cdots}{a_1^{\nu_1} \cdot a_2^{\nu_2} \cdots} = \Delta G_T° + RT \ln J \quad (14.14)$$

where $J$ is related to the activities of substances as shown by the expression in the center. Substitution of (14.2), (14.3), $R = 8.31451$ J mol$^{-1}$ K$^{-1}$ and $F = 96485.31$ Coulombs per gram equivalent yields

$$E = E° - 8.6173 \times 10^{-5} (T/z) \ln J; \quad (T \text{ in K}, E \text{ in volts}) \quad (14.15)$$

$$E = E° - (0.0256925/z) \ln J; \quad (25°C) \quad (14.16)$$

---

*Z. Moser and R. Castanet, Met. Trans., *10B*, 483 (1979); M. Le Bouteiller, A.M. Martre, R. Farhi, and C. Petot, ibid., *8B*, 339 (1977); Z. Moser, E. Kawecka, F. Sommer, and B. Predel, ibid., *13B*, 71 (1982).

As an example consider the following hydrogen-calomel cell carefully investigated by Hills and Ives* at 25°C;

$$\text{Pt}|\text{H}_2(\text{g})|\text{aq. HCl}(m)|\text{Hg}_2\text{Cl}_2(\text{s})|\text{Hg(l)} \tag{14.17}$$

The electrode reactions are

$$0.5\text{H}_2(\text{g}) = \text{H}^+ + e^- \quad \text{(left side)}$$
$$0.5\text{Hg}_2\text{Cl}_2(\text{s}) + e^- = \text{Hg(l)} + \text{Cl}^- \quad \text{(right side)}$$

and the cell reaction is the sum of these reactions, i.e.,

$$0.5\text{H}_2(\text{g}) + 0.5\text{Hg}_2\text{Cl}_2(\text{s}) = \text{Hg(l)} + \text{H}^+ + \text{Cl}^-$$

For unit fugacity of hydrogen, $J$ for this reaction is given by $J = a_+ \cdot a_-$ where $a_+$ and $a_-$ are the activities of $\text{H}^+$ and $\text{Cl}^-$ respectively. From the definition of the mean ionic activity by (13.5), it is evident that $J = m^2 \gamma_\pm^2$, and the substitution of this expression in (14.16) gives

$$E = E^\circ - 0.0513850(\ln m + \ln \gamma_\pm); \quad (25°\text{C}, E \text{ in volts}) \tag{14.18}$$

The measurements of $E$ at various molalities and the substitution in the linear form of (13.31) yields

$$E + 0.0513850 \ln m - 0.06017 m^{0.5} = E^\circ + 0.051385 Cm \tag{14.19}$$

The left-hand side of this equation is a linear function of $m$ at $m < 0.005$ as shown by the right side. A plot of the left side of this equation versus $m$, extrapolated to $m \to 0$ yields $E^\circ = 0.26791$ volts. The experimental values of $E$ and $m$, and the calculated value of $E^\circ$ from the preceding step are then substituted in (14.18) to obtain $\gamma_\pm$.

## Standard emf of Half-cells

The standard emf of a half-cell, also called the single electrode potential, refers to the emf of a particular full-cell. Consider a cell in which the right-hand side electrode is a reversible hydrogen electrode under a gaseous hydrogen atmosphere of unit fugacity in contact with a solution containing a hydrogen ion activity of unity, $a_+(\text{H}^+) = 1$, through a platinum or similar electrode. The left-hand side electrode may be a pure metal $M$, in contact with its ion at unit activity, or any other basically similar type of electrode. The emf of such a cell is then called the standard emf of the half-cell for $M$, or in particular, the *oxidation potential* of $M$. By convention, a cell for this purpose is always represented as follows:

$$\text{Electrode } M|M^{z+} \text{ at } a_+(\text{H}^+) = 1, \quad \text{H}^+(a_+ = 1)|\text{H}_2(\text{g})(P = 1 \text{ bar})|\text{Pt} \tag{14.20}$$

(The ions are in water unless specified otherwise; henceforth, aq. will not be repeated for the solvent). This convention is adopted despite the fact that the

---

*G. J. Hills and D. J. G. Ives, J. Chem. Soc., (London), 318 (1951).

emf might have a negative value but this is necessary for a meaningful sequence of tabulating the emf of various half-cells. The cell reaction of (14.20) is the oxidation at the left electrode and the reduction at the right electrode, i.e., generation of hydrogen at the right; therefore, the cell reaction is simply

$$\text{Reduced State} + z\,H^+ = \text{Oxidized State} + 0.5z\,H_2(1\text{ bar}) \tag{14.21}$$

The resulting emf, with $a_+(H^+) = 1$, is expressed by

$$E = E° - \frac{RT}{zF}\ln\frac{a_{\text{oxidized state}}}{a_{\text{reduced state}}}; \qquad (E\text{ always in volts}) \tag{14.22}$$

where the term after the logarithmic sign is unity as shown in (14.20) and the measured $E$ is equal to $E°$. The left-hand side electrode may be a pure metal such as Cd in which case the reduced state is Cd and the oxidized state is $Cd^{++}$ and $z = z_+ = 2$ in (14.21). The standard emf of a half-cell is defined in terms of the standard hydrogen half-cell, (SHHC), (also called the standard hydrogen electrode),

$$H^+(a_+ = 1)\,|\,H_2(g)(1\text{ bar})\,|\,Pt; \qquad (\text{Reaction: } H^+ + e^- = 0.5H_2(g))$$

to which a standard half-cell potential of zero is arbitrarily assigned at all temperatures i.e., $E°(\text{SHHC}) = 0$. Likewise, the standard half-cell for any substance $M$ is

$$M(a = 1)\,|\,M^{z+}(a_+ = 1); \qquad (\text{Reaction: } M = M^{z+} + ze^-)$$

and the standard emf of this half-cell, $E°$, is the actually measured emf when the two preceding half-cells are coupled to form a full-cell.

*Pressure Correction*  The pressure of $H_2$ over SHHC in earlier publications is 1 atm = 1.01325 bar (Gen. Ref. 25). Conversion into 1 bar = 0.1 MPa requires writing the following reactions:

$$\begin{array}{ll}
M + H^+(a_+ = 1) = M^+(a_+ = 1) & \Delta G° = -FE°\,(\text{old}) \\
\qquad\qquad + 0.5\,H_2(1\text{ atm}); & \\
0.5H_2(1\text{ atm}) \quad = 0.5H_2(1\text{ bar}); & \Delta G = 0.5RT\ln\left(\dfrac{1}{1.01325}\right) \\
 & = -FE\,(\text{for }\Delta P) \qquad (6.24) \\
\hline
M + H^+(a_+ = 1) = M^+(a_+ = 1) & \Delta G° = -FE°\,(\text{new}) \\
\qquad\qquad + 0.5H_2(1\text{ bar}); & = -F[E°\,(\text{old}) + E\,(\text{for }\Delta P)]
\end{array}$$

where the last equation is the sum of the preceding two equations. The second equation yields $E(\text{for }\Delta P) = 0.00017$ at 25°C; hence, the sum of first two emfs yield the final emf, i.e.,

$$E°(\text{new}) = E°(\text{old}) + 0.00017; \qquad (25°C) \tag{14.23}$$

The correction is small, and often less than the experimental errors. When a cell reaction does not involve a gas, there is no correction so that the old set of data for standard emf can be used for such a reaction without error. Selected data for $E°$, corrected according to (14.23), are listed in Table 14.1.

**Table 14.1** Selected values of standard emf of half-cells, $E°$, at 25°C; $E°$ is also called oxidation potential. Selected values from a table by P. Vanysek in Gen. Ref. (25) for one atm, corrected below to one bar by using (14.23). All metals and salts are in solid state except as indicated, the solvent is water.

| Electrode Reaction | $E°$, Volts | Electrode Reaction | $E°$, Volts |
|---|---|---|---|
| $Li = Li^+ + e^-$ | 3.0403 | $Ag + Br^- = AgBr + e^-$ | −0.07116 |
| $K = K^+ + e^-$ | 2.931 | $Cu^+ = Cu^{++} + 2e^-$ | −0.153 |
| $Ca = Ca^{++} + 2e^-$ | 2.868 | $Ag + Cl^- = AgCl + e^-$ | −0.22216 |
| $Na = Na^+ + e^-$ | 2.71 | $2Hg(l) + 2Cl^- = Hg_2Cl_2 + 2e^-$ | −0.26791 |
| $La = La^{3+} + 3e^-$ | 2.379 | $Cu = Cu^{++} + 2e^-$ | −0.3417 |
| $Mg = Mg^{++} + 2e^-$ | 2.372 | $2OH^- = H_2O + 0.5O_2(g) + 2e^-$ | −0.401 |
| $Al = Al^{3+} + 3e^-$ | 1.662 | $2I^- = I_2 + 2e^-$ | −0.5353 |
| $0.5H_2(g) + OH^- = H_2O + e^-$ | 0.82877 | $2Hg(l) + SO_4^{--} = Hg_2SO_4 + 2e^-$ | −0.6123 |
| $Zn = Zn^{++} + 2e^-$ | 0.7620 | $Fe^{++} = Fe^{3+} + e^-$ | −0.771 |
| $Fe = Fe^{++} + 2e^-$ | 0.447 | $2Hg(l) = Hg_2^{++} + 2e^-$ | −0.7971 |
| $Cd = Cd^{++} + 2e^-$ | 0.4031 | $Ag = Ag^+ + e^-$ | −0.7994 |
| $Co = Co^{++} + 2e^-$ | 0.28 | $Hg_2^{++} = 2Hg^{++} + 2e^-$ | −0.920 |
| $Ni = Ni^{++} + 2e^-$ | 0.257 | $2Br^- = Br_2(l) + 2e^-$ | −1.066 |
| $Ag + I^- = AgI + e^-$ | 0.15241 | $2H_2O = O_2(g) + 4H^+ + 4e^-$ | −1.229 |
| $Sn(white) = Sn^{++} + 2e^-$ | 0.1377 | $2Cl^- = Cl_2(g) + 2e^-$ | −1.35810 |
| $Pb = Pb^{++} + 2e^-$ | 0.1260 | $2F^- = F_2(g) + 2e^-$ | −2.866 |
| $H_2(g) = 2H^+ + 2e^-$ | 0.000 | | |

*Example 1*  The Daniell cell consists of Zn electrode in contact with $ZnSO_4$ solution, separated from $CuSO_4$ solution by a liquid junction or a membrane, and $CuSO_4$ solution in contact with copper; i.e.,

$$Zn|ZnSO_4(m) \vdots\vdots CuSO_4(m)|Cu$$

The electrode reactions are $Zn = Zn^{++} + 2e^-$, and $Cu^{++} + 2e^- = Cu$ so that the cell reaction is

$$Zn + Cu^{++} = Zn^{++} + Cu$$

The emf of this cell is given by (14.15); i.e.,

$$E = E° - 4.3087 \times 10^{-5} T \ln \frac{a_+(Zn^{++})}{a_+(Cu^{++})}$$

If the molalities of $ZnSO_4$ and $CuSO_4$ are kept equal and decreased, the second term in this equation vanishes and the measured emf becomes identical with $E°$. It is necessary, however, to keep the liquid junction potential to a minimum or to make a small correction for it all low molalities. At 25°C, $E° = 1.1037$ and if the value of $E° = -0.3417$ for the copper electrode is known from other measurements, then for $Zn = Zn^{++} + 2e^-$, $E° = 1.1037 - 0.3417 = 0.7620$, as listed in Table 14.1. Similar cells for numerous metals, e.g., Cd, Fe, Co, Ni, Tl, etc., have been used successfully for this purpose.

**Example 2** Experimental measurement of $E°$ for $2Hg(l) = Hg_2^{++} + 2e^-$ requires first $E°$ for the hydrogen-calomel cell (14.17) for which the cell reaction is

(I) $\quad H_2(g) + Hg_2Cl_2(s) = 2Hg(\ell) + 2H^+ + 2Cl^-$;

$$E° = 0.26791; \qquad (25°C)$$

The next required reaction is

(II) $\quad Hg_2Cl_2(s) = Hg_2^{++} + 2Cl^-; \qquad K_i = 1.3 \times 10^{-18}; \qquad (25°C)$

The standard emf for this reaction cannot be determined electrochemically, but it can be calculated from $\Delta G° = -RT \ln K_i = -zFE°$; thus,

$$E° = (0.0256925/z) \ln K_i; \qquad E° = -0.5291; \qquad (25°C)$$

The final reaction is

(III) $\quad H_2(g) = 2H^+ + 2e^-; \qquad E° = 0$

where $E°$ is zero by definition. When Reactions (II) and (III) are added and (I) is subtracted simultaneously with their $-zFE°$, the following result is obtained

(IV) $\quad 2Hg(\ell) = Hg_2^{++} + 2e^-; \qquad E° = -0.7970; \qquad (25°C)$

This value is very close to that listed in Table 14.1. The procedure is simply the algebraic sum of $\Delta G°$ for three reactions and the calculation of $E°$ from $\Delta G° = -zFE°$. In this example, for each reaction, $z$ is 2, and therefore, $-0.7970 = -0.5291 - 0.26791$.

Reaction (I) for the hydrogen-calomel cell and its emf, subtracted from $H_2(g) = 2H^+ + 2e^-$, and $E° = 0$, yields

$$2Hg(l) + 2Cl^- = Hg_2Cl_2(s) + 2e^-; \qquad E° = -0.26791$$

This value is also listed in Table 14.1.

**Example 3** The standard electrode potentials of certain cells may be computed from the standard Gibbs energy of formation of the compound and its solubility in water. An interesting example is AgCl(s) for which $\Delta_f G°_{298} = -109{,}834$ J mol$^{-1}$. The ionic concentration $m_+$ of Ag$^+$, which is equal to $m_-$ of Cl$^-$, obtained by conductance method, is $m_+ = m_- = 1.30 \times 10^{-5}$ mol (kg water)$^{-1}$, when the solution is saturated with AgCl(s) at 25°C. The value of $\gamma_\pm$ from the Debye-Hückel equation is $\gamma_\pm = 0.99574$; hence the equilibrium constant for the ions in equilibrium with solid AgCl is $K_i = (1.30 \times 10^{-5})^2 (0.99574)^2$

$= 1.676 \times 10^{-10}$. The reactions and the corresponding standard emf, $E°$, calculated from $\Delta G° = -RT \ln K_i = -zFE°$ are as follows:

$$\text{Ag(s)} + 0.5\text{Cl}_2(\text{g}) = \text{AgCl(s)} \qquad E° = \frac{\Delta G°}{-F} = \frac{-109{,}834}{-96{,}485.3}$$
$$= 1.1384 \text{ volt, } (25°\text{C})$$

$$\text{AgCl(s)} = \text{Ag}^+ + \text{Cl}^- \qquad E° = \frac{RT \ln K_i}{F} = \frac{RT \ln(1.676 \times 10^{-10})}{96{,}485.3}$$
$$= -0.5783 \text{ volt, } (25°\text{C})$$

---

$$\text{Ag(s)} + 0.5\text{Cl}_2(\text{g}) = \text{Ag}^+ + \text{Cl}^- \qquad E° = 0.5601 \text{ volt, } (25°\text{C})$$

The addition of the electrode reaction and $E°$ for $\text{Cl}_2$ electrode, from Table 14.1, i.e.,

$$\text{Cl}^- = 0.5\text{Cl}_2(\text{g}) + e^-; \qquad E° = -1.3581 \text{ volt, } (25°\text{C})$$

to the immediately preceding reaction and its $E°$ gives

$$\text{Ag(s)} = \text{Ag}^+ + e^-; \qquad E° = -0.7980 \text{ volt, } (25°\text{C})$$

This value differs by 0.0013 volt from the listed value in Table 14.1, because, a small error in the solubility of AgCl causes a large error in $E°$.

To obtain the standard emf of a cell, an appropriate combination of the two half-cells and their balanced combined final reaction and emf, $E°$, is sufficient. For example, the following combination gives

$$\text{Ni(s)} = \text{Ni}^{++} + 2e^- \qquad E° = 0.257 \text{ volt, } (25°\text{C})$$
$$\text{Cu}^{++} + 2e^- = \text{Cu(s)} \qquad -(E° = -0.3417) \text{ volt, } (25°\text{C})$$
$$\overline{\text{Ni(s)} + \text{Cu}^{++} = \text{Cu(s)} + \text{Ni}^{++}} \qquad \overline{E° = 0.5987 \text{ volt, } (25°\text{C})}$$

The corresponding cell is

$$\text{Ni} | \text{NiCl}_2 \vdots \text{CuCl}_2 | \text{Cu}$$

where the two solutions in the center above are separated to prevent plating out copper on nickel. The cell could also be written out as two cells in series to avoid complications at the liquid junction, i.e.,

$$\text{Ni} | \text{NiCl}_2, \text{HCl} | \overset{\downarrow}{\text{H}_2} | \text{Pt} \} \{ \text{Pt} | \overset{\downarrow}{\text{H}_2} | \text{CuCl}_2, \text{HCl} | \text{Cu}$$

where $\text{H}_2$ generated by the left cell is consumed by the right cell, as shown by the arrow, and the two neighboring Pt electrodes are connected by an external wire.

### Variation of emf with Temperature and Pressure

The effect of temperature on the emf is obtained from the measurements of $E$ at various temperatures. A plot of $E$ versus $T$ for a given concentration, i.e., at fixed $m$, usually gives a line with a very small curvature. The slope of the line yields

$$(\partial E/\partial T)_{P,m} = \Delta S/zF \qquad (14.24)$$

from which $\Delta S$ for the cell reaction is calculated. When $E°$ is substituted for $E$, the corresponding change in the entropy is $\Delta S°$. For example, $\partial E/\partial T = 3.40 \times 10^{-4}$ volt $K^{-1}$ at 25°C for the following cell:

$$Ag(s)|AgCl(s)|NaCl(m)|Hg_2Cl_2(s)|Hg(l)$$

The cell reaction is

$$Ag(s) + 0.5Hg_2Cl_2(s) = AgCl(s) + Hg(l)$$

where the reactants and the products in this case are in their standard states; hence $E = E°$, and

$$\Delta S° = zF(\partial E°/\partial T) = 96{,}485.3 \times 3.40 \times 10^{-4} = 32.81 \text{ J mol}^{-1} \text{ K}^{-1}$$

The corresponding entropy change from the third law is close, i.e.,

$$\Delta S° = 75.90 + 96.23 - 96.27 - 42.55 = 33.31 \text{ J mol}^{-1} \text{ K}^{-1}$$
$$\text{(Hg)} \quad \text{(AgCl)} \quad \text{(0.5Hg}_2\text{Cl}_2\text{)} \quad \text{(Ag)}$$

From Table 14.1, the following result is obtained:

$$Ag(s) + Cl^- = AgCl(s) + e^- \qquad E° = -0.22216 \text{ volt, (25°C)}$$
$$-[Hg(l) + Cl^- = 0.5Hg_2Cl_2(s) + e^-] \quad -[E° = -0.26791] \text{ volt, (25°C)}$$

$$Ag(s) + 0.5Hg_2Cl_2(s) = AgCl(s) + Hg(l) \qquad E° = 0.04575 \text{ volt, (25°C)}$$

Substitution of $\partial E°/\partial T = 3.40 \times 10^{-4}$ and $E° = 0.04575$ in (14.5) yields $\Delta H°_{P,T}$ for (14.24); i.e.,

$$\Delta H°_{P,T} = -96485.3[0.04575 - 298.15 \times 3.40 \times 10^{-4}] = 5{,}367 \text{ J}$$

The value of $\Delta H°_{P,T}$ from Gen. Ref. (2) is $4.184[-30{,}370(\text{AgCl}) + 63{,}319 \times 0.5 \, (0.5\text{Hg}_2\text{Cl}_2)] = 5{,}395$.

The effect of pressure on $E$ is given by

$$\Delta V = -zF(\partial E/\partial P)_{T,m} \tag{14.25}$$

where $\Delta V$ refers to the cell reaction. In general $\Delta V$ is very small for moderate changes in pressure but large when a gas is either consumed or generated at one electrode. The effect of pressure on $E$ in this case is accounted for in the expression for $J$ (14.15). [See also (6.24).]

## Ionization Constant of Water

The ionization constant $K_{i,w}$ is defined by

$$K_{i,w} = a_+(\text{H}^+) \cdot a_-(\text{OH}^-)/a_{\text{H}_2\text{O}} = m_{\text{H}^+} \cdot m_{\text{OH}^-} \cdot \gamma_{\text{H}^+}\gamma_{\text{OH}^-}/a_{\text{H}_2\text{O}} \tag{14.26}$$

where the activity of water, $a_{\text{H}_2\text{O}}$ is on the mole fraction scale. For pure water, $a_{\text{H}_2\text{O}}$ is unity and $K_{i,w}$ is

$$K_{i,w} = m_{\text{H}^+} \cdot m_{\text{OH}^-}, \qquad \text{(pure water)} \tag{14.27}$$

where $m_+$ and $m_-$ are very small and the activity coefficients are unity as required by the Debye-Hückel equation. An acid is defined as a proton donor, and a base, a proton acceptor, e.g., HCl and $HNO_3$ are acids because they give $H^+$ in water and $Cl^-$ and $NO_3^-$ are bases because they accept protons. Accordingly, water is an acid, and indeed a very weak acid. Conductivity measurements with water shows that $K_{i,w} \approx 1.04 \times 10^{-14}$ mol$^2$ kg$^{-2}$ at 25°C. A more precise value can be obtained from the following cell containing $m_1$ molal KOH, and $m_2$ molal KCl:

$$Pt|H_2(g)(1 \text{ atm})|KOH(m_1), KCl(m_2)|AgCl|Ag \qquad (14.28)$$

Potassium ion, $K^+$, and $Cl^-$ may be replaced by any other alkali, and halide ions respectively with the corresponding salt of Ag, e.g., NaOH, NaI, and AgI. The electrode reactions for the cell are $0.5H_2(g) = H^+ + e^-$ and $AgCl + e^- = Ag + Cl^-$ so that the cell reaction is

$$0.5H_2(g) + AgCl = Ag + H^+ + Cl^- \qquad (14.29)$$

and the emf is given by

$$E = E° - (RT/F) \ln(m_{H^+} \cdot m_{Cl^-} \cdot \gamma_{H^+} \cdot \gamma_{Cl^-}) \qquad (14.30)$$

where $m_{Cl^-} = m_2$, and it is understood that the fugacity of hydrogen is unity. Solving for $\ln m_{H^+}$ from (14.26) and substituting in (14.30), with $m_{OH^-} = m_1$ and with the numerical value of $F/R = 11604.45$ K volt$^{-1}$, yields the following equation:

$$\frac{11604.45(E° - E)}{T} - \ln \frac{m_2}{m_1} = \ln \frac{\gamma_{Cl^-} \cdot a_{H_2O}}{\gamma_{OH^-}} + \ln K_{i,w} \qquad (14.31)$$

The procedure of eliminating $m_{H^+}$ by means of $K_i$ is common in obtaining any type of ionization constant as will also be seen in the next section. The ionic strength $I$, defined by (13.19), is given by $I = m = m_1 + m_2$, i.e., the sum of the molalities of KOH and KCl. The term $\ln(\gamma_{Cl^-}/\gamma_{OH^-})$ in the range of $I = m < 0.1$ is proportional to $I = m$ as is (13.27); hence,

$$\ln(\gamma_i/\gamma_j) = \beta_i(m_1 + m_2) - \beta_j(m_1 + m_2)$$

where $i$ and $j$ are for $Cl^-$ and $OH^-$ respectively. In addition $\ln a_{H_2O} \approx \ln x_{H_2O} = \ln[1 - (m_1 + m_2)/55.508] \approx -(m_1 + m_2)/55.508$; therefore the first term after the equal sign in (14.31) is linear in $m = m_1 + m_2$. A plot of the left-hand side of (14.31), (LHS), versus $m$ is therefore linear and the intercept at $m = 0$ is $\ln K_{i,w}$. The data for KCl, LiCl, NaCl, and $CaCl_2$ at 25°C with their respective hydroxides are shown in Fig. 14.2. The intercept yields

$$K_{i,w} = 1.002 \times 10^{-14} \text{ mol}^2 \text{ kg}^{-2}; \quad (25°C) \qquad (14.32)$$

in very close agreement with the conductivity measurements. At ordinary pressures, $K_{i,w}$, given by Harned and Owen in Ref. (I), in Chapter XIII is

$$\log K_{i,w} = (-6013.79/T) - 23.6521 \log T + 64.6968 \qquad (14.33)$$

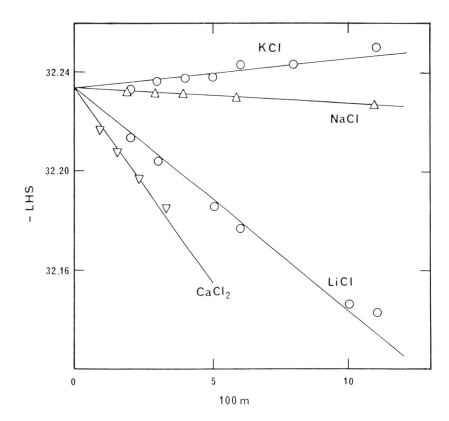

**Fig. 14.2** Ionization constant $K_{i,w}$ of water; see (14.31). [Adapted from Ref. (I) cited in Chapter XIII].

The last term in their equation has been corrected slightly in this text to 64.6968 to obtain agreement between (14.32) and (14.33). As the pressure increases, $K_{i,w}$ also increases, e.g., at 0.8 kilobar and 25°C, $K_{i,w}$ is twice as large as at 1 bar, and at 10 kilobars, water becomes significantly ionized.*

The ionization of weak acids and bases can be investigated by the same technique. For example, the ionization constant of acetic acid, determined by using an appropriate cell, is $K_i = 1.757 \times 10^{-5}$, as discussed in Refs. (I) and (II) in Chapter XIII.

The ionization constant of pure water shows that $\log m_+$ (for H$^+$) = $-7$, and also $\log m_-$ (for OH$^-$) = $-7$, and the addition of a small amount of acid would drastically change the concentration of OH$^-$, and the addition of a small amount of alkali would have the reverse effect. For example, the addition of approximately 1

---

*For effects of pressure on electrolytes see H. Köster and E. U. Franck, Ber. Bunsenges. Phys. Chem., 73, 716 (1969).

mole of HCl in 1 kg of water would increase $m_+$ to approximately 1, and decrease $m_-$ to $10^{-14}$. This behavior has practical ramifications and lead to the definition of pH by

$$\text{pH} \equiv -\log a_+ \approx -\log m_+(\text{H})^+ \tag{14.34}$$

Since the single ion activity cannot be determined, an "operational definition" of pH by using appropriate cells and buffers is necessary as described by Covington [Gen. Ref. (25)], but for our purposes, the approximate form of (14.34) is sufficient. Thus, pH is approximately 0 for 1 molal HCl, and 14 for about 1 molal KOH. In (14.34) molarity (mol L$^{-1}$) is used in preference to molality, $m_+$, but the difference in pH defined by molarity or molality is usually negligible.

A similar phenomenon occurs in semiconductors in which the electrons, $e^-$, in the conduction band, and holes, $e^+$, in the valence band correspond to OH$^-$ and H$^+$ respectively in water. Therefore, the product of concentrations of electrons and holes, $e^-$(electrons) $\times$ $e^+$(holes), is nearly constant for each semiconductor at a given temperature, e.g., $e^- \times e^+ = 2.1 \times 10^{19}$ cm$^{-6}$ for silicon at 298 K, when the concentrations are expressed in particles per cm$^3$. The concentrations of $e^-$ or $e^+$ can be varied drastically by doping, i.e., by alloying with very small amounts of appropriate elements that contribute $e^-$, e.g. P and As; or $e^+$, e.g. B(boron). [See Gen. Ref. (15).]

## Cells with Solid Electrolytes

All the solutions considered in this chapter, except fused salts mentioned earlier, are aqueous solutions. Fused electrolytes are similar to both aqueous solutions and solid electrolytes. It is therefore appropriate to consider briefly some of the solid electrolytes useful not only in thermodynamic measurements but also in chemical analysis of liquid metals, alloys, salts, and slags.* These electrolytes, alone or in solution, are basically the same as the aqueous solutions from a thermodynamic point of view. They present experimental difficulties at high temperatures often not encountered in aqueous solutions at ambient temperatures. For example, it is important that the conduction must be ionic, not metallic or semi-conductive; diffusion of foreign components must not interfere with the emf measurements; ionic concentration gradients should be minimal; and temperature nonuniformity

---

*K. Kiukkola and C. Wagner, J. Electrochem. Soc., *104*, 308 and 379 (1957); J. B. Wagner and C. Wagner, ibid., p. 509; R. A Rapp and D. A. Shores, "Techniques in Metals Research," R. F. Bunshah, Editor, IV, Pt. 2, 123, John Wiley & Sons (1970); T. H. Etsell and S. N. Flengas, Chem. Rev., *70*, 339 (1970); G. J. Janz, editor, "Molten Salts Handbook," Academic Press (1967); J. N. Pratt, Met. Trans., *21A*, 1223 (1990), (a comprehensive review); M. Cima and L. Brewer, ibid., *19B*, 893 (1994); Q. Han, X. Feng, S. Liu, H. Niu, and Z. Tang, ibid., *21B*, 295 (1990); H. Majima and Y. Awakura, ibid., *12B*, 141 (1981); M. Iwase and A. McLean, ibid., *14B*, 765 (1983); R. J. Brisley and D. J. Fray, ibid., p. 435. S. Otsuka, Y. Kurose, and Z. Kozuka, ibid., *15B*, 141 (1984); Y. Feutelais, B. Legendre, S. Misra, and T. J. Anderson, J. Phase Equil., *15*, 171 (1994).

must be absent.* The most important problem for any type of cell, particularly for the cells with solid electrolytes, is to identify the measured emf with the electrochemical reaction under investigation. It is sufficient here to give a few examples because the experimental difficulties are best understood by reading the original papers.

The technique for the galvanic cells with solid electrolytes has been perfected by C. Wagner, et al.,[†] in a series of careful investigations. Solid halides of silver and lead, oxides of zirconium, tantalum and yttrium have been found to be suitable electrolytes because they become ionic conductors at sufficiently high temperatures. For example, AgI becomes an ionic conductor above 420 K and it has been used by Kiukkola and Wagner in the following cell in the range of 150 to 425°C.

$$\text{Ag(s)}|\text{AgI(s)}|\text{Ag}_2\text{S}|\text{S(l)}|\text{C(graphite)} \tag{14.35}$$

The cell reaction is $2\text{Ag(s)} + \text{S(l)} = \text{Ag}_2\text{S(s)}$ hence the measured emf is the standard emf $E°$. From $E° = 244 \pm 1$ millivolts and $\partial E°/\partial T = \Delta E°/\Delta T = 0.16$ mV K$^{-1}$ at 300°C, $\Delta_f G° = -47,085$ J mol$^{-1}$, $\Delta_f S° = 30.88$ J mol$^{-1}$ K$^{-1}$, and $\Delta_f H° = -29,386$ J mol$^{-1}$ from (14.3) and (14.4). $\Delta_f G°$ obtained by this technique agrees within 0.8 kJ mol$^{-1}$ of that obtained from the gas-condensed-phase equilibrium in $\text{Ag}_2\text{S(s)} + \text{H}_2\text{(g)} = 2\text{Ag(s)} + \text{H}_2\text{S(g)}$ by Rosenqvist[‡] and thus confirms that the cell is suitable for this type of investigation. Likewise, their data for $\Delta_f G°$ of cobaltous oxide, CoO, with ZrO$_2$ containing 0.5 mole fraction of CaO as the solid electrolyte, and for $\Delta_f G°$ of PbS, with PbCl$_2$ containing 0.5 weight percent KCl as the solid electrolyte, agree with the corresponding values from the gas-condensed-phase equilibria. [See also Gen. Ref. (4).]

Galvanic cells with solid electrolytes are necessary to obtain thermodynamic data difficult or impossible to obtain by other existing techniques. For example, a solid solution of thoria, ThO$_2$ containing 6 weight percent yttrium oxide, Y$_2$O$_3$, has been used by Singhal and Worrell[§] in the following cell to obtain the activities and related thermodynamic properties of tantalum-molybdenum, Ta-Mo, solid solutions:

$$\text{Ta(s)(pure Ta} + \text{Ta}_2\text{O}_5 \text{ compact)} |\text{ThO}_2(6\% \text{ Y}_2\text{O}_3)|$$
$$\text{Ta-Mo(s)(alloy} + \text{Ta}_2\text{O}_5 \text{ compact)} \tag{14.36}$$

The electrode reactions are $\text{Ta} = \text{Ta}^{5+} + 5e^-$ and $\text{Ta}^{5+} + 5e^- = \text{Ta}$ (in Ta-Mo); (see Problem 14.8). A similar cell with solid ZrO$_2$ containing 0.15 mole fraction of either CaO or Y$_2$O$_3$ has been used for the determination of the activities in

---

*C. Wagner, Advances in Electrochem. and Electrochem. Eng., P. Delahay, Ed., IV, 2, Interscience Publ. (1966).
[†] See footnote on previous page.
[‡] T. Rosenqvist, Trans. AIME, *185*, 451 (1949).
[§] S. C. Singhal and W. L. Worrel, in "Metallurgical Thermochemistry," edited by O. Kubaschewski, Nat. Phys. Lab., Her Majesty's Stationery Office, London (1972).

solid Fe-Ni system by Henriet, Gatellier and Olette.[*] These measurements are very useful in constructing the phase diagrams from the Gibbs energy of mixing, $\Delta_{mix}G$, as a function of composition and temperature. In general, it is possible to obtain fairly good phase diagrams involving condensed phases even from poor sets of values for $\Delta_{mix}G$. However, exceptionally accurate phase diagrams yield only fair values of $\Delta_{mix}G$ at temperatures above 500 K.

Another interesting and convenient solid electrolyte is $CaF_2$. It has been used by a number of investigators, e.g., by Schaefer[†] in the following cell:

$$\text{Pt},(Fe+FeF_2)|CaF_2|(NiF_2+Ni),\text{Pt} \tag{14.37}$$

All the phases are solid in their standard states, and overall cell reaction is

$$Fe(s)+NiF_2(s)=FeF_2(s)+Ni(s) \tag{14.38}$$

The electrode reactions can be written as in (14.11) and (14.14). The essential point is that the pressure of $F_2$ over $(Ni+NiF_2)$ is higher than that over $(Fe+FeF_2)$; hence, $F^-$ ions travel from the right to the left through $CaF_2$ and react to form $FeF_2$. Schaefer measured the emf of this cell from 897.5 to 1098.7 K, from which the values of $\Delta_r G°$ and $\Delta_r H°$ were obtained. For example, $E° = 0.3939$ volt at 1019.7 K, from which

$$\Delta_r G° = -zFE° = -2 \times 96{,}485.3 \times 0.3939 = -76{,}011 \text{ J}$$

For $NiF_2$, $\Delta_f G° = -499{,}298$ J mol$^{-1}$ is known fairly accurately [Gen Ref. (2)]; therefore, $\Delta_f G°$ for $FeF_2$ is

$$\Delta_f G°(FeF_2) = -76{,}011 - 499{,}298 = -575{,}309 \text{ J mol}^{-1} \tag{14.39}$$

Solid fluoride electrolytes are used to determine fluoride concentration in melts, and oxide electrolytes are used for measuring dissolved oxygen in slags and metals, e.g., in liquid iron, cobalt, nickel, copper, etc.; therefore, they are technologically important.

## PROBLEMS

14.1 Calculate the emf of the following cell at 25°C:

$$\text{Pt}|H_2(1\text{ bar})|\text{HCl(aq.)}|H_2(0.1\text{ bar with }0.9\text{ bar of argon})$$

*Ans.: 0.0296 volt*

14.2 Calculate the pressures of hydrogen capable of plating out lead at 25°C from the solutions in which (a) the ionic activities of $Pb^{++}$ and $H^+$ are unity, and (b) the concentrations of $Pb^{++}$ and $H^+$ are 0.002 and 0.001 molal respectively. Assume that hydrogen is an ideal gas.

*Ans.: (a) 18,184 bars*

---

[*] D. Henriet, C. Gatellier, and M. Olette, ibid., last footnote after (14.36), p. 97. For interesting applications see G. R. Fitterer, ibid., p. 589; A. J. Rollino and S. Aronson, J. Chem. Ed., *49*, 825 (1972).
[†] S. C. Schaefer and N. A. Gokcen, High Temp. Science, *14*, 153 (1981).

14.3 Construct the reversible cells in which the following reactions occur and determine their standard emf at 25°C from Table 14.1.
(I)  $Fe + Cu^{++} = Fe^{++} + Cu$      (III) $Fe + 2HCl(aq.) = FeCl_2(aq.) + H_2$
(II) $H_2 + 0.5O_2 = H_2O$              (IV) $2Fe^{++} + 2Hg^{++} = Hg_2^{++} + 2Fe^{3+}$

14.4 (a) The standard Gibbs energy of formation of $Hg_2Cl_2(s)$ is $-210,330$ J mol$^{-1}$ at 25°C. Construct a cell in which $2Hg(\ell) + Cl_2(g) = Hg_2Cl_2(s)$ takes place and compute the emf of the cell. (b) Repeat the problem for AgI(s), for which $\Delta_f G°/R = -7,976$.
Ans.: (a) $E° = 1.090$ volts; (b) 0.6873 volt.

14.5 Construct a cell for $Ag + Fe^{3+} = Ag^+ + Fe^{++}$ at 25°C using the nitrate salts and calculate (a) $E°$ from Table 14.1, and (b) $K_i = (m_+)(m_{++})/(m_{3+})$ at infinite dilution.
Ans.: (a) $E° = -0.0284$ volt; (b) $K_i = 0.331$.

14.6 Express $\Delta G°$, $\Delta H°$, $\Delta S°$, and $\Delta C_p°$ for the ionization equilibrium, $H_2O = H^+ + OH^-$ for pure liquid water. Calculate also $E°$ and $\partial E°/\partial T$ for ionization at 25°C.

14.7 The following cell has been investigated by K. Schwerdtfeger and H. J. Engell, Arch. Eisenhüttenwesen, 35, 533 (1964):

$Pt|Si(l)|liquid(CaO \cdot xSiO_2)$ in $SiO_2$ crucible$|O_2(g)|Pt$

The liquid electrolyte is saturated with $SiO_2(s)$ and the cell reaction is $Si(l) + O_2(g) = SiO_2(s)$. Calculate $E°$ and $\partial E°/\partial T$ at 1800 K from $\Delta G°/R = -71,030$ and $\Delta S°/R = -23.80$.
Ans.: $E° = 1.5302$ volts; $\partial E°/\partial T = -5.127 \times 10^{-4}$ volt K$^{-1}$.

14.8 The values $E$ in volts and $(\partial E/\partial T)_{x_2}$ at 1200 K for various mole fractions $x_2$ of tantalum, from S. C. Singhal and W. L. Worrell, loc. cit., are as follows:

| $x_2$: | 0.113 | 0.201 | 0.328 | 0.400 | 0.499 | 0.615 | 0.696 | 0.801 | 0.904 |
|---|---|---|---|---|---|---|---|---|---|
| $10^3 E/V$: | 103.3 | 85.7 | 64.1 | 52.2 | 37.9 | 24.9 | 17.0 | 8.9 | 3.2 |
| $10^6 \partial E/\partial T$: | 34.3 | 24.7 | 15.6 | 13.4 | 8.7 | 6.0 | 4.6 | 2.8 | 1.4 |

Calculate $\overline{G}_2 - G_2°$, $\overline{H}_2 - H_2°$, $\overline{S}_2 - S_2°$ and the related excess thermodynamic properties and express $G^e$ as a function of $x_2$ in terms of three parameters with each parameter as a linear function of temperature.
Partial ans.: at $x_2 = 0.400$: $\overline{G}_2 - G_2° = -25,183$ J atom-mol$^{-1}$

14.9 The equilibrium constant for $e^-$ and $e^+$ in silicon was given as $2.1 \times 10^{19}$ cm$^{-6}$ at 298 K in the text. Calculate (a) $\Delta G°$ for the electron-hole equilibrium in units of eV used by physicists and (b) the change in concentration of $e^+$ in silicon when $10^{13}$ atoms of $P$ is alloyed with one cm$^3$ of Si, assuming that each $P$ atom contributes one $e^-$. [cf. Gen. Ref. (15), Chapter 7].
Ans.: (a) $-1.14$ eV; (b) loss of $4.58 \times 10^9$ holes cm$^{-3}$; $e^+$(final) $= 2.099 \times 10^6$.

CHAPTER XV

# PHASE DIAGRAMS

## Introduction*

A phase diagram represents a map of coexisting phase boundaries as affected by the variables of state. A number of phase diagrams have been presented in Chapters VIII and X for developing the thermodynamics of phase equilibria. In this chapter, the relationships between the phase diagrams and the molar Gibbs energy diagrams are presented in detail. We shall limit our discussion to condensed phase diagrams in which pressure is not a variable of state. The removal of the restriction on pressure does not require a special treatment. We shall consider the temperature-composition phase diagrams involving solids and liquids.†

## Binary Phase Diagrams

A condensed binary phase diagram, represented on temperature-composition coordinates, consists of the single-phase regions separated by the two-phase regions. The curves separating the single phases from two phases represent the compositions of single phases. When a homogeneous liquid phase freezes to form a homogeneous solid, the phase diagram may be (1) as in the lower portion of Fig. 10.4, (2) as in Fig. 10.8 with a common minimum point for the phase boundary curves, or (3) as in Fig. 10.9 with a common maximum point. It can be shown that the minimum and maximum points for the phase boundaries in the condensed phase diagrams must also obey the Gibbs-Konovalow theorems, i.e., a pair of phase boundary curves for a two-phase region must be tangent to each other at their common extremum point. When the liquidus and solidus curves are depressed toward the center of a phase diagram and the system freezes into two immiscible solid phases, a eutectic may be formed as shown in Fig. 10.5. At the eutectic temperature, three phases coexist, but above and below this temperature

---

*The sections "$\Delta G$ Diagrams for Other Phases" up to "Ternary Phase Diagrams" are for advanced readers.
†For extensive presentations, see Gen. Refs. (10–13), (18), (20) and (26); see also P. Gordon, "Principles of Phase Diagrams in Materials Systems," Krieger Publ. Co. (1983); T. B. Massalski Met. Trans., 20A, 445; 1295 (1989); I. Ansara, C. Chatillon, H. L. Lukas, T. Nishizawa, H. Ohtani, K. Ishida, M. Hillert, B. Sundman, B. B. Argent, A. Watson, T. G. Chart, and T. Anderson, CALPHAD, 18(2), 177 (1994).

at least one phase must disappear as required by the phase rule. The temperature at which three phases coexist, as in Fig. 10.5, is called the *eutectic temperature* and the corresponding phase transition, the *eutectic reaction*. There are other similar reactions involving three phases with various possible combinations. Each reaction is named according to the states of aggregation of phases and the number of reactant phases. The suffix "-tectic" is used for reactions involving one or two liquid phases, and "-tectoid" for three solid phases. Various types of three-phase reactions are presented in Figs. 15.1 and 15.2, and summarized as follows with the coexisting phases marked as in these figures:

| NAME | DIAGRAM | REACTION UPON COOLING |
|---|---|---|
| *Transitions involving liquid phases:* | | |
| Eutectic | $\alpha \vdash\!\!\!\stackrel{L_1}{\vee}\!\!\!\dashv \beta$ | Liquid($L_1$) → Solid($\alpha$) + Solid($\beta$) |
| Monotectic | $L_2 \vdash\!\!\!\stackrel{L_3}{\vee}\!\!\!\dashv \epsilon$ | Liquid($L_3$) → Liquid($L_2$) + Solid($\epsilon$) |
| Peritectic | $\beta \vdash\!\!\!\stackrel{\wedge}{\lambda}\!\!\!\dashv L_2$ | Solid($\beta$) + Liquid($L_2$) → Solid($\lambda$) |
| Syntectic | $L_1 \vdash\!\!\!\stackrel{\wedge}{\beta}\!\!\!\dashv L_2$ | Liquid($L_1$) + Liquid($L_2$) → Solid($\beta$) |
| Catatectic* (metatectic) | $\alpha \vdash\!\!\!\stackrel{\beta}{\vee}\!\!\!\dashv L$ | Solid($\beta$) → Solid($\alpha$) + Liquid($L$) |
| *Transitions involving only solid phases:* | | |
| Eutectoid | $\alpha \vdash\!\!\!\stackrel{\beta}{\vee}\!\!\!\dashv \lambda$ | Solid($\beta$) → Solid($\alpha$) + Solid($\lambda$) |
| Peritectoid | $\alpha \vdash\!\!\!\stackrel{\wedge}{\eta}\!\!\!\dashv \lambda$ | Solid($\alpha$) + Solid($\lambda$) → Solid($\eta$) |

These transformations consist of (1) eutectic-type, in which the reactant is a single phase, and (2) the peritectic-type, in which the reactants consist of two phases. Thus there are four eutectic-type and three peritectic-type transitions in the preceding list.

A critical point exists when one phase dissociates into two phases at a point where the phase boundary has a horizontal inflection point, i.e., when $\partial T/\partial x_2$ and $\partial^2 T/\partial x_2^2$ are both zero as shown in Fig. 15.1 for $L \to L_1 + L_2$, $L \to L_2 + L_3$ and $\delta \to \delta_1 + \delta_2$. It should be noted that the maximum point in the $\eta$-region does not have an inflection point and the phase boundaries for $\eta$ and $\sigma$ are tangent to each other at their common maximum points. The maximum point in the center of Fig. 15.2 is for a congruently melting compound $AB$, capable of

---

*"Catatectic" has been suggested by S. Wagner and D. A. Rigney, Metall. Trans., 5, 2155 (1974); "metatectic" is often but not always used for this transformation.

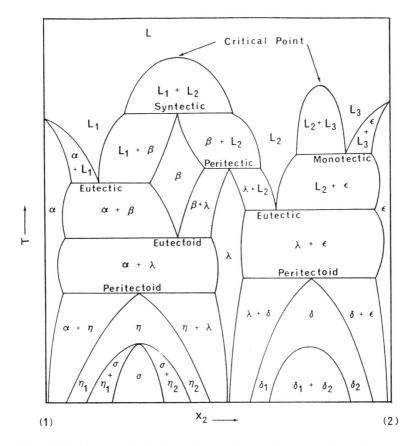

**Fig. 15.1** Hypothetical phase diagram for various transformations.

dissolving its component elements to limited extents as indicated by the phase-region $AB_x$. An unusual phase diagram of the nicotine-water or ethylpyridine-water type, illustrated by Fig. 15.3, consists of a sac of two-phase region with an upper critical point $D$, and a lower critical point $E$. The foregoing diagrams contain all the possible phase equilibria encountered in the condensed phase binary systems. **Numerous such diagrams are presented in Gen. Refs. (9, 18–20), and illustrated by Figs. 15.27–15.39 at the end of this chapter.**\*

---

\*For determination of phase diagrams by thermal analysis see, e.g., K. Ishida, T. Shumiya, T. Nomura, H. Ohtani, and T. Nishizawa, J. Less-Common Met., *142*, 135 (1988); H. Ohtani and K. Ishida, J. Electronic Mat., *23*, 747 (1994); Y. T. Zhu, J. H. Devletian, and A. Manthiram, J. Phase Equi., *15*, 37 (1994); S.-W. Chen, H. W. Beumler, and Y. A. Chang, Met. Trans., *22A*, 203 (1991); K. W. Richter and H. Ipser, J. Phase Equil., *15*, 165 (1994). Other methods are presented in Gen. Ref. (10–13); see also thermodynamic methods in Chapters XII and XIV; J. E. Morral, R. S. Schiffman, and S. M. Merchant, editors, "Experimental Methods of Phase Diagram Determination," ASM Intl. (1993).

288    Thermodynamics

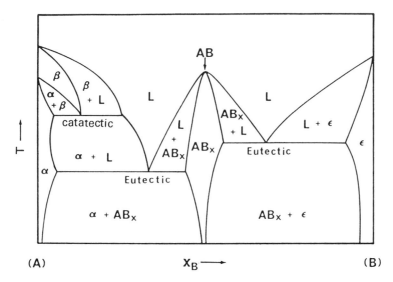

**Fig. 15.2** Hypothetical phase diagram for catatectic (metatectic) and eutectic transformations, and for compound AB capable of dissolving its components to limited extents.

### Erroneous Diagrams

A number of important aspects of the phase equilibria summarized in Figs. 15.1 and 15.2 must be observed to avoid errors in drawing the phase boundaries. All such errors violate (a) the phase rule at constant pressure, i.e., $\Upsilon = c - \phi + 1$, cf. (8.44), (b) the Gibbs-Konovalow theorems, and (c) the requirement that the extended portions of the phase boundaries terminate in the two-phase regions. The Gibbs energy requirement for (c) will be presented later. Examples of these violations are illustrated in Fig. 15.4, and summarized as follows:[*]

(a) Along $BC$, $\alpha$, $L_1$, $L_2$, and $\beta$ coexist and thus violate the phase rule since the degree of freedom $\Upsilon$ with four coexisting phases, is $-1$ and $\Upsilon$ cannot assume a negative value. This error can be corrected by joining $L_1$ and $L_2$ at one point on $BC$. On $D$, $\alpha$, $\epsilon$, $\lambda$ and $\beta$ coexist because at $E$ there are more than two curves and one straight line intersecting one another; the phase boundary curve $ME$ must therefore not terminate at $E$. Accordingly, there must be no more than one straight line and two curves at a point of intersection, or coincidence. On $F$, a eutectoid and a peritectoid on the same horizontal line signifies that $\alpha$, $\epsilon$, $\delta$ and $\lambda$ coexist and violate the phase rule. There are three phases on the straight line $J$, $\Upsilon = 0$, and since $J$ is not horizontal these phases coexist over a temperature range and violate the phase rule since $\Upsilon = 1$. At $K$, there

---

[*]See also Gen. Ref. (20), and H. Okamoto and T. B. Massalski, J. Phase Equil., *12*, 148 (1991).

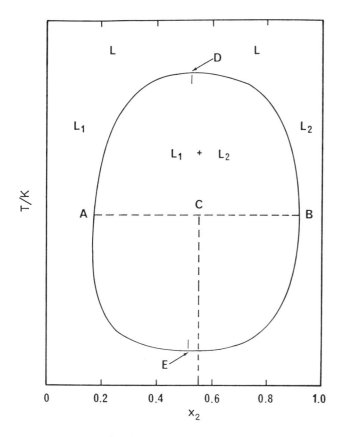

**Fig. 15.3** Solubility gap in a liquid solution. Upper critical point $D$, lower critical point $E$; between these points two liquids exist, above $D$, and below $E$ one liquid exists. Straight line $AB$ joining two coexisting phases is called "tieline."

are two phases $\lambda$ and $L$ over a temperature range for pure component 2, and $\Upsilon = -1$; the phase boundaries must intersect the vertical line for the pure component at the same point.

(b) At $I$, $M$ and $N$, the Gibbs-Konovalow theorems are violated, i.e., each pair of phase boundaries does not meet at an extremum point. At $M$, the phase boundaries should not meet at all, and an entirely different construction must be made to eliminate the error, particularly because of the erroneous separation of $\beta$ and $\lambda$ along a curve $EM$ instead of a two-phase field.

(c) The extended portion of the phase boundary at $B$, shown by the broken extended curve, and the unstable liquid on the extended portion of $AC$ are in equilibrium. Since the liquid below $C$ is unstable, the dotted curve below $B$

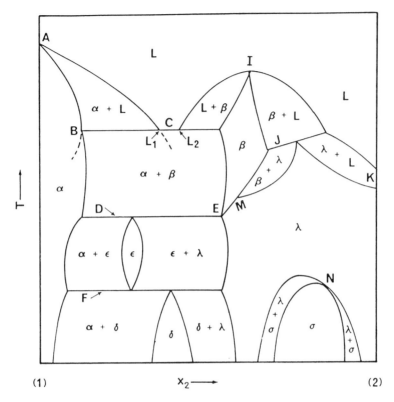

**Fig. 15.4** Errors in a hypothetical phase diagram violating phase rule or other thermodynamic principles.

must also be in an unstable region, not in a one-phase region which is stable. Hence the extended portion below $B$ must terminate in a two-phase region, or within the area $BCDE$. This is possible when $\alpha$ phase boundary curves and the horizontal line $BC$ intersect one another at an angle smaller than 180 degrees. All other such intersections of two curves and one horizontal line are therefore properly constructed in this respect.

## Lever Rule

Consider Fig. 15.3 at the temperature corresponding to the tieline $AB$. As point $C$ moves from $A$ to $B$ the compositions of $L_1$ and $L_2$ remain at $x_A$ and $x_B$ respectively but the relative proportions of $L_1$ and $L_2$ vary according to the bulk composition $x_C$ corresponding to the location of $C$. The compositions $x_A$, $x_B$, etc., refer to the mole fraction of the second component. Let $\eta(A)$ and $\eta(B)$ represent the amounts of $L_1$ (at $A$) and $L_2$ (at $B$) in moles with the restriction that $\eta(A) + \eta(B) = 1$ because in phase equilibria the basis of calculations and representations is one mole of the mixture. The mass balance for the system gives $x_A\eta(A) + x_B\eta(B) = x_C = x_C[\eta(A) + \eta(B)]$ from which, it is evident that

$$\frac{\eta(A)}{\eta(B)} = \frac{x_B - x_C}{x_C - x_A} = \frac{\overline{CB}}{\overline{AC}}; \qquad \{[\eta(A) + \eta(B)] = 1\}$$

Therefore, the relative amounts of $L_1$ (at $A$) and $L_2$ (at $B$) expressed by the left side of this equation is equal to the ratio of the line segments opposite to the coexisting phases. This relationship is known as the *lever rule*. Care must be exercised to distinguish the single phase composition and the bulk composition of two coexisting phases by observing the phase diagram since $x_i$ is used in numerous publications for the mole fraction of the system without distinction as to whether the system consists of one or more phases or the phases are solid or liquid. However, $x_i$ used for representing various Gibbs energies always refers to a single phase and we shall always denote the phase to which it refers as $x_i(l)$, $x_i(\alpha)$, $x_i(\beta)$ etc, when distinction is necessary. The preceding equation may be transformed into the following useful forms:

$$\eta(A) = (x_B - x_C)/(x_B - x_A); \qquad \eta(B) = (x_C - x_A)/(x_B - x_A) \qquad (15.1)$$

It is now appropriate to discuss the molar Gibbs energy-composition diagrams after the foregoing background on phase diagrams.

## Molar Gibbs Energy of Mixing—Composition Diagrams

The molar Gibbs energy of mixing-composition diagrams for simple ideal and regular solutions have been presented in Figs. 10.1 and 11.2 respectively. The diagrams for other solutions are generally complicated in shape and require various types of representation largely dictated by the coexisting phases. The simplest representation is for a liquid phase, $L$, dissociating into two liquid phases $L_1$ and $L_2$ as in Figs. 15.1 and 15.3, or a solid phase, $\delta$, dissociating into two solid phases, $\delta_1$ and $\delta_2$, as in Fig. 15.1. The molar Gibbs energy of mixing $\Delta_{\text{mix}}G$ versus the mole fraction of second component, $x_2$, for the temperature corresponding to the horizontal line $AB$ in Fig. 15.3, is represented in Fig. 15.5. The curve in Fig. 15.5 is drawn to scale by assuming that $\Delta_{\text{mix}}G$ is represented by a combination of (10.4) and (11.21) as follows:

$$\Delta_{\text{mix}}G \equiv G(\text{soln}) - x_1 G_1^\circ - x_2 G_2^\circ = x_1 x_2 (5{,}000 x_1 + 8{,}000 x_2)$$
$$+ 5{,}500 x_1^2 x_2^2 + 3{,}000(x_1 \ln x_1 + x_2 \ln x_2) \qquad (15.2)$$

where the last term with 3,000 represents $\Delta_{\text{mix}}G(\text{ideal})$, i.e., (10.4) for an ideal solution with $RT$ taken to be 3,000 J mol$^{-1}$ for convenience ($T = 360.82$ K). The coefficients in (15.2) are obtained by using one maximum and two minimum points in Fig. 15.5 to construct an interesting equation. The compositions at $A$ and $B$ in Fig. 15.3 correspond to $A$ and $B$ in Fig. 15.5. The intercepts of the straight line $AB$ to the curve for $\Delta_{\text{mix}}G$ represent $\Delta \overline{G}_1 = \overline{G}_1(l) - G_1^\circ(l)$ and $\Delta \overline{G}_2 = \overline{G}_2(l) - G_2^\circ(l)$ as shown in the figure. The points of tangency, $A$ and $B$ are

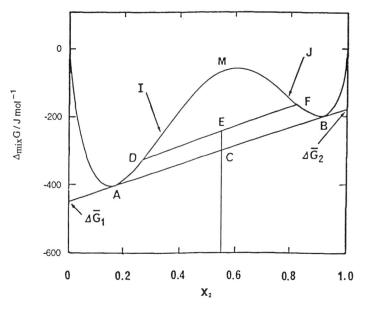

**Fig. 15.5** $\Delta_{mix}G$ versus $x_2$ for temperature corresponding to ACB in Fig. 15.3 Compositions at A, C and B are identical in both figures. Equation (15.2) is represented exactly by the curve.

not at the minimum points of the curve for $\Delta_{mix}G$. The phase equilibria require that, at a given temperature and pressure, two phases coexist when the partial Gibbs energy, $\overline{G}_i$, ($i = 1$ or 2), is the same in both phases; this is the criterion of equilibrium for the co-existing phases at the same pressure and temperature.

The portion of the curve on the left of A in Fig. 15.5 for the single phase $L_1$. At the origin, (0, 0), $\partial \Delta_{mix}G/\partial x_2$ is given by

$$\partial \Delta_{mix}G/\partial x_2 (x_2 \to 0) = 5{,}000 + 3{,}000(\ln x_2 - \ln x_1) = -\infty \qquad (15.3)$$

Therefore, the curve approaches $x_2 \to 0$ and $x_2 \to 1$ with a |slope| $= \infty$.

Consider now the section of the curve for $\Delta_{mix}G(L_1, L_2)$ for the mixtures of $L_1$ and $L_2$ between A and B. A mixture of two phases having a bulk composition of $x_C$ at C and consisting of the phases at A and B have the molar Gibbs energy of mixing expressed by

$$\Delta_{mix}G(L_1, L_2) = \frac{x_B - x_C}{x_B - x_A} \Delta_{mix}G(\text{at } A) + \frac{x_C - x_A}{x_B - x_A} \Delta_{mix}G(\text{at } B) \qquad (15.4)$$

as required by the lever rule because the Gibbs energy of a mixture of two phases is the sum of the Gibbs energies of its constituent phases. Equation (15.4) is linear in $x_C$, and the straight line between A and B represents this equation. Likewise,

$\Delta_{mix}G(L_D, L_F)$ of combined phases at $D$ and $F$ is represented by the ordinate of $E$ located on the straight line $DF$. For a selected composition $x_C$ it is evident that

$$\Delta_{mix}G(L_D, L_F; \text{ at } E) > \Delta_{mix}G(L_1, L_2; \text{ at } C) \tag{15.5}$$

Thus, at $x_C$, a combination of any pair of two immiscible phases above $AB$ has a higher value of $\Delta_{mix}G$ (for two phases) than $\Delta_{mix}G$ (for two phases at $A$ and $B$). Such mixtures as those formed by the intersection of the straight lines above $AB$ with the curve $AMB$ are therefore unstable. It will be seen later that the sections of the curve in the vicinity of $A$ and $B$ for two phases are represented by two different equations when the phases differ in their crystal structures or states of aggregation. In anticipation of such representations, we may write $\Delta_{mix}G(L_1) = \Delta_{mix}G(x_A)$, and $\Delta_{mix}G(L_2) = \Delta_{mix}G(x_B)$, i.e., $\Delta_{mix}G$ for each liquid is a function of its mole fraction in the vicinity of their stable composition, with $x_A$ and $x_B$ being regarded as their respective composition variables. The changes in the bulk composition of the mixtures of two phases, $x_C$, does not affect equilibrium; instead, the changes in $x_C$ cause the changes in the relative amounts of $L_1$ and $L_2$. Therefore, $\Delta_{mix}G(L_1, L_2)$ is a function of two variables, $x_A$ and $x_B$ for a given value of $x_C$. The condition of equilibrium requires that $\Delta_{mix}G(L_1, L_2)$ be a minimum at equilibrium for a given value of $x_C$, i.e.,

$$d\Delta_{mix}G(L_1, L_2) = 0, \quad \text{(closed system; } P \text{ and } T \text{ constant)} \tag{15.6}$$

It was shown in Chapter II, (2.60) and (2.61), that the total differential of a function $u = f(x_A, x_B)$ is zero when the partial derivatives $\partial u/\partial x_A$ and $\partial u/\partial x_B$ are both zero. Equation (15.6) therefore requires that at a constant value of $x_C$, the following partial derivatives be zero:

$$\partial \Delta_{mix}G(L_1, L_2)\partial x_A = 0; \quad \partial \Delta_{mix}G(L_1, L_2)/\partial x_B = 0 \tag{15.7}$$

These equations are readily verified by substituting (15.2) in (15.4), differentiating, and then setting $x_A = 0.170$, $x_B = 0.919$ and $x_C = 0.500$. We can verify the first partial derivative in (15.7) as follows:

$$\partial \Delta_{mix}G(L_1, L_2)/\partial x_A = \{\Delta_{mix}G(\text{at } A) + (x_B - x_A)[\partial \Delta_{mix}G(\text{at } A)/\partial x_A]$$
$$- \Delta_{mix}G(\text{at } B)\}(x_B - x_C)/(x_B - x_A)^2$$
$$= \{-480.7 + 0.749 \times 327.4 + 235.5\}0.747 = 0.0$$

Figure 15.5 shows that $\partial \Delta_{mix}G(\text{at } A)/\partial x_A = \partial \Delta_{mix}G(\text{at } B)/\partial x_B$, i.e., the slopes at $A$ and $B$ are identical, and further, $x_B - x_A \equiv (1 - x_A) - (1 - x_B)$; therefore, the terms inside the braces in the preceding equation yield $\Delta\overline{G}_1(\text{at } A) = \Delta\overline{G}_1(\text{at } B)$, cf. (11.8)]. The second equation in (15.7) can be similarly verified.

The curve for $\Delta_{mix}G$ in Fig. 15.5 has two inflection points at $I$, ($x_2 = 0.331$) and at $J$, ($x_2 = 0.791$) as can be shown by substituting these values of $x_2$ in the second derivatives of (15.2). These points are called the <u>spinodes</u>, which are important in kinetics of nucleation and growth of new phases from the supersaturated single phases. The equations for the activities of components can be derived from (15.2) and the results can be plotted versus $x_2$. It can be shown that the maximum and the minimum points in such activity versus composition diagrams for both components coincide with the spinodes (see Problem 15.2).

The shape of the curve in Fig. 15.5 changes with increasing temperature as required by the phase diagram. Thus, $A$ and $B$ in Fig. 15.5 approach each other and finally, the minimum, maximum and inflection points coincide at the horizontal inflection point $D$ in Fig. 15.3 as required by the phase diagram. A similar statement is also valid for the lower horizontal inflection point $E$ in the phase diagram. In general, the mutual solubilities for the overwhelming majority of binary systems increase with increasing temperature, and lower critical points such as that in Fig. 15.3 are rare.

**Hereafter in this chapter, we often use $\Delta G$ for $\Delta_{\mathrm{mix}}G$ for brevity when no confusion is possible.**

## $\Delta G$ Diagrams for Other Phases

The diagram shown in Fig. 15.5 is for a single liquid phase decomposing into two liquid phases, and it may also be for a solid phase $\delta$ decomposing into two closely related solid phases, as in Fig. 15.1. For all other types of transformation, it is customary to represent $\Delta G$ versus $x_2$ for each phase at each selected temperature on the same diagram. If the phases in equilibrium are solid and liquid, the convention for writing $\Delta G$ for each phase through the entire range of composition at a selected temperature $T$ is as follows:

$$\Delta G(\mathrm{l}) \equiv G(\mathrm{l}) - x_1(\mathrm{l})G_1^\circ(\text{stable phase at } T)$$
$$- x_2(\mathrm{l})G_2^\circ(\text{stable phase at } T) \tag{15.8}$$

$$\Delta G(\mathrm{s}) \equiv G(\mathrm{s}) - x_1(\mathrm{s})G_1^\circ(\text{stable phase at } T)$$
$$- x_2(\mathrm{s})G_2^\circ(\text{stable phase at } T) \tag{15.9}$$

where $G(\mathrm{l})$ and $G(\mathrm{s})$ are the molar Gibbs energies of the liquid and solid solutions. Equation (15.8) at $x_2 = 1$ becomes $\Delta G(\mathrm{l}) = G^\circ(\mathrm{l}) - G_2^\circ(\mathrm{s})$ when the stable pure phase for component 2 is solid at $T$, because for $x_2 = 1$, $G(\mathrm{l})$ is the same as $G_2^\circ(\mathrm{l})$ for the pure liquid 2. If, however, the stable phase for component 2 is liquid, (15.8) is then zero at $x_2(\mathrm{l}) = 1$. The convention for $\Delta G$ adopted in the preceding chapters referred to the pure components and the solutions in the same state of aggregation, and therefore should not be confused with $\Delta G$ used in the molar Gibbs energy diagrams.

We reconsider (15.8) and (15.9) at a temperature $T$ greater than the melting points of both components so that

$$\Delta G(\mathrm{l}) = G(\mathrm{l}) - x_1(\mathrm{l})G_1^\circ(\mathrm{l}) - x_2(\mathrm{l})G_2^\circ(\mathrm{l}) \tag{15.10}$$

$$\Delta G(\mathrm{s}) = G(\mathrm{s}) - x_1(\mathrm{s})G_1^\circ(\mathrm{l}) - x_2(\mathrm{s})G_2^\circ(\mathrm{l}) \tag{15.11}$$

We add and subtract $x_1(s)G_1^\circ(s) + x_2(s)G_2^\circ(s)$ to (15.11) and rearrange the terms to obtain

$$\Delta G(s) = G(s) - x_1(s)G_1^\circ(s) - x_2(s)G_2^\circ(s)$$
$$- x_1(s)\Delta_{fus}G_1^\circ - x_2(s)\Delta_{fus}G_2^\circ \quad (15.12)$$

where $\Delta_{fus}G_i^\circ$ is the standard molar Gibbs energy of fusion of pure component $i$. The first three terms after the equal sign can be transformed into $RTx_1(s)\ln a_1(s) + RTx_2(s)\ln a_2(s)$ by using the definition of activity; i.e., $\overline{G}_i(s) = G_i^\circ(s) + RT\ln a_i(s)$ and observing that $G(s) = x_1(s)\overline{G}_1 + x_2(s)\overline{G}_2$. Likewise, we can transform the three terms on the right side of (15.10) and rewrite (15.10) and (15.11) as follows:

$$\Delta G(l) = RTx_1(l)\ln a_1(l) + RTx_2(l)\ln a_2(l) \quad (15.13)$$

$$\Delta G(s) = RTx_1(s)\ln a_1(s) + RTx_2(s)\ln a_2(s)$$
$$- x_1(s)\Delta_{fus}G_1^\circ - x_2(s)\Delta_{fus}G_2^\circ \quad (15.14)$$

We now present a simple example for these equations, with the following arbitrarily selected values

$T_{fus1} = 900\,K \qquad \Delta_{fus}H_1^\circ = 900R \qquad \Delta_{fus}S_1^\circ = R$

$T_{fus2} = 300\,K \qquad \Delta_{fus}H_2^\circ = 300R \qquad \Delta_{fus}S_2^\circ = R$

where $T_{fus\,i}$ is the melting point of $i$. From the definition of $\Delta G^\circ$ as $\Delta G^\circ = \Delta H^\circ - T\Delta S^\circ$, it is evident that

$$\Delta_{fus}G_1^\circ = 900R - RT; \qquad \Delta_{fus}G_2^\circ = 300R - RT \quad (15.15)$$

If, in addition, we set the activities equal to the mole fractions, i.e., the liquid and solid phases are ideal, (15.13) and (15.14) become

$$\Delta G(l) = RTx_1(l)\ln x_1(l) + RTx_2(l)\ln x_2(l) \quad (15.16)$$

$$\Delta G(s) = RTx_1(s)\ln x_1(s) + RTx_2(s)\ln x_2(s)$$
$$- x_1(s)\Delta_{fus}G_1^\circ - x_2(s)\Delta_{fus}G_2^\circ \quad (15.17)$$

Equations (15.16) and (15.17) are plotted in Fig. 15.6 for $T = 1000\,K$. The phase diagram for this system, calculated from (10.17) and (10.18) is shown in the lower portion of Fig. 15.7. We note that $\Delta G(s)$ for the solid in Fig. 15.6 is higher than $\Delta G(l)$ for the liquid; hence the liquid phase is stable relative to the solid phase at 1000 K. The difference between the vertical intercepts (at $x_1 = 1$ and $x_2 = 1$) for these curves represent $\Delta_{fus}G_1^\circ = -100R$ and $\Delta_{fus}G_2^\circ = -700R$. The positions of the curves below 300 K are reversed, i.e., the curve for $\Delta G(l)$ is above that for $\Delta G(s)$ because the solid phase is stable relative to the liquid phase below 300 K as can be shown with a similar figure.

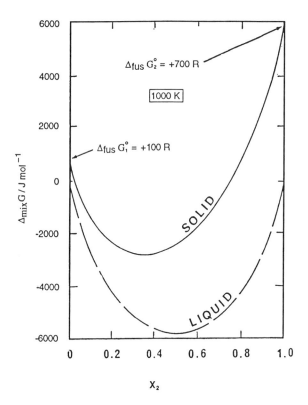

**Fig. 15.6** Diagram for $\Delta G(l)$ and $\Delta G(s)$ at 1000 K from (15.16) and (15.17) respectively. $\Delta G(s) > \Delta G(l)$ throughout; therefore solid phase is unstable relative to liquid phase. Phase diagram for this system is in lower portion of Fig. 15.7.

We discuss next a set of curves at 550 K where the liquid and solid phases coexist at appropriate concentrations. These curves are given by

$$\Delta G(l) = G(l) - x_1(l)G_1^\circ(l) - x_2(l)G_2^\circ(l) + x_1(l)\Delta_{fus}G_1^\circ \quad (15.18)$$

$$\Delta G(s) = G(s) - x_1(s)G_1^\circ(s) - x_2(s)G_2^\circ(s) - x_2(s)\Delta_{fus}G_2^\circ \quad (15.19)$$

Substitution of $a_i = x_i$ and $\overline{G}_i = G_i^\circ + RT \ln x_i$ in these equations gives

$$\Delta G(l) = RTx_1(l) \ln x_1(l) + RTx_2(l) \ln x_2(l) + x_1(l)\Delta_{fus}G_1^\circ \quad (15.20)$$

$$\Delta G(s) = RTx_1(s) \ln x_1(s) + RTx_2(s) \ln x_2(s) - x_2(s)\Delta_{fus}G_2^\circ \quad (15.21)$$

Substitution of (15.15) in these equations gives

$$\Delta G(l) = 550Rx_1(l) \ln x_1(l) + 550Rx_2(l) \ln x_2(l) + x_1(l)350R \quad (15.22)$$

$$\Delta G(s) = 550Rx_1(s) \ln x_1(s) + 550Rx_2(s) \ln x_2(s) + x_2(s)250R \quad (15.23)$$

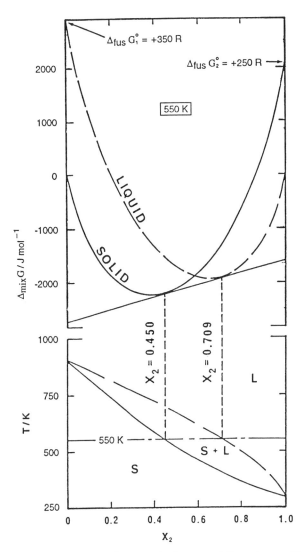

**Fig. 15.7** Upper diagram is for $\Delta_{mix}G(l)$ and $\Delta_{mix}G(s)$ at 550 K from (15.22) and (15.23) respectively. Phase diagram in lower part can be readily calculated from (10.17) and (10.18) in Chapter X.

Equations (15.22) and (15.23) are represented in Fig. 15.7 as $\Delta_{mix}G = \Delta G$. The straight line tangent to both curves gives the compositions of solid and liquid at $x_2 = x_2(s) = 0.450$ and $x_2 = x_2(l) = 0.709$ respectively. The liquid phase is stable from $x_2 = 0.709$ to $x_2 = 1$, and the solid phase is stable from $x_2 = 0.0$ to $x_2 = 0.450$, as required by the relatively lower values of $\Delta G$ represented by the lower sections of the curves. At the point of intersection of the curves $\Delta G(l)$ and $\Delta G(s)$ are equal but $\overline{G}_1(l)$ and $\overline{G}_1(s)$, as well as $\overline{G}_2(l)$ and $\overline{G}_2(s)$, are not equal; therefore there is no equilibrium at this point.

It is interesting to apply (15.6) and (15.7) to (15.20) and (15.21) at $x_2(s) = 0.450$, $x_2(l) = 0.709$ and $x_2(s,l) = x_C = 0.6$. The result for the first partial derivative in (15.7), with $\Delta G(s, l)$ written according to (15.4), is as follows:

$$\partial \Delta G(s, l)/\partial x_2(s) = \{\Delta G(s) + (0.709 - 0.450)[\partial \Delta G(s)/\partial x_2(s)] - \Delta G(l)\}$$
$$\times 0.109/0.0671 = \{-528.54 + 71.85 + 456.7\}1.624 = 0$$

The vertical intercepts of the curves in Fig. 15.7 correspond to $\Delta_{\text{fus}} G_1^\circ = 350R$ and $\Delta_{\text{fus}} G_2^\circ = -250R$, as indicated in the figure. The absolute values of the slopes at the intercepts are infinity as can be shown by substituting $x_i = 1$ in the derivatives of $\Delta G$ with respect to $x_i$. Deviations from ideality modify the curves for $\Delta G$ represented in the preceding figures; however, the principles involved in the representation are basically the same. The curves, similar to those in Figs. 15.5–15.7, for *nonideal solutions* require (15.13) and (15.14) with $a_i \neq x_i$. In these equations, substitution of $a_1 = \gamma_1 x_1$ and $a_2 = \gamma_2 x_2$, and expansion yields $G^E = RT(x_1 \ln \gamma_1 + x_2 \ln \gamma_2)$. An appropriate analytical equation for $G^E$ must be substituted in each of these equations for representing them as functions of composition and temperature.

## $\Delta G$ Diagrams for Complex Systems

The diagrams of $\Delta G$ versus $x_2$ for complex phase diagrams may be illustrated schematically by the eutectic system shown in Fig. 15.8. The $\Delta G$ diagram for the liquid and $\alpha$ phases at $T_I$ between the melting points of components 1, and 2 is, in principle, the same as the diagram in Fig. 15.7 and therefore not presented (see Problem 4). At $T_{II}$, below the melting points of components and above the eutectic temperature $T_e$, the curves are shown in the upper portion of Fig. 15.8. The straight lines tangent to each pair of curves give the compositions of each pair of phases. Single phases are stable outside the two-phase regions where $\Delta G$ is lower for the stable phases than the unstable phases. Fig. 15.9 shows the diagrams at $T_e$ and $T_{III}$ indicated on the phase diagram in Fig. 15.8. At $T_e$, three phases coexist as shown by the single tangent line to the three curves. As the temperature decreases the curve for the liquid, $\Delta G(l)$, moves up and the remaining curves move down so that at $T_{III}$ the $\alpha$-phase at $e$ is in equilibrium with the $\beta$-phase at $h$. Let us assume that we can supercool the liquid and $\alpha$-phases as shown by the extended dotted portions of the phase boundaries in Fig. 15.8. The liquid at $g$, and the $\alpha$-phase at $f$ must then coexist as shown in Fig. 15.8, and in Fig. 15.9, by the tangent line at $f$ and $g$. The curve for $\Delta G(l)$ at $T_{III}$ is, and must always be higher than the curves for $\Delta G(\alpha)$ and $\Delta G(\beta)$ and must be located between these curves; therefore the tangent at $f$ and $g$ must be in the two-phase region and since the values of $\Delta G$ for the two phases at $f$ and $g$ are higher than those at $e$ and $h$, the phases at $f$

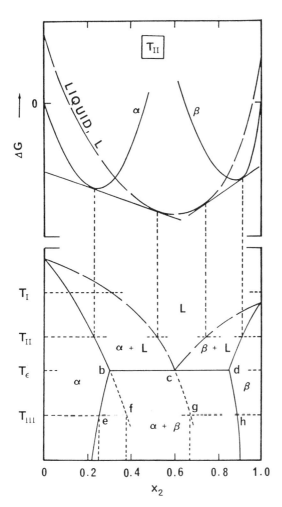

**Fig. 15.8** Upper figure schematically shows $\Delta G$ for liquid $\alpha$ and $\beta$ phases at $T_{II}$. Lower figure is corresponding phase diagram.

and $g$ are unstable. The extended portions of the phase boundaries in Fig. 15.8, terminating in the two-phase regions, are therefore properly constructed. This is possible when the angles between the adjoining curves and the eutectic line at $b$ and at $d$ are less than 180 degrees.

## Calculation of Phase Diagrams from Thermodynamic Data

The criteria for equilibrium is expressed by (1) by equality of $\overline{G}_1$ and $\overline{G}_2$ among the coexisting phases, or (2) the minimization of $\Delta G$(two phases), i.e., by setting the

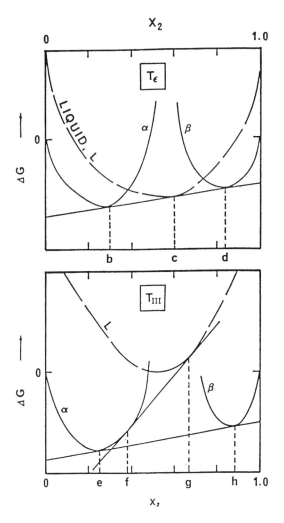

**Fig. 15.9** Upper figure schematically shows $\Delta_{mix}G = \Delta G$ for liquid, $\alpha$ and $\beta$ phases at $T_\epsilon$, and lower figure, at $T_{III}$. For $T_\epsilon$, $T_{III}$, and compositions, see phase diagram of Fig. 15.8.

partial derivatives of $\Delta G$ (two phases) with respect to the independent composition variables of coexisting phases equal to zero as shown by (15.7). Note that the number of independent composition variables for a closed system at equilibrium is the same as the sum of the independent composition variables of all phases. The first method is convenient for binary systems and the second method may have advantages for ternary and multicomponent systems.* We illustrate the basic principles involved in the first method and refer to the reader to the original paper on the second method.*

The calculations by the first method for a binary system require writing two equations based on $\overline{G}_1(\alpha) - \overline{G}_1(\beta) = 0$ and $\overline{G}_2(\alpha) - \overline{G}_2(\beta) = 0$ for two coexisting phases $\alpha$ and $\beta$. The result for component 1 is

$$G_1^\circ(\beta) - G_1^\circ(\alpha) = \Delta G_1^\circ(\alpha \to \beta) = \Delta H^\circ(\alpha \to \beta) - T\Delta S^\circ(\alpha \to \beta)$$
$$= RT[\ln a_1(\alpha) - \ln a_1(\beta)] \qquad (15.24)$$

Substitution of $\overline{G}_i^E = g_i = RT \ln \gamma_i$ and rearrangement of result give

$$\Delta G_1^\circ(\alpha \to \beta) - \overline{G}_1^E(\alpha) + \overline{G}_1^E(\beta)$$
$$+ RT \ln\{[1 - x_2(\beta)]/[1 - x_2(\alpha)]\} = 0 \qquad (15.25)$$

Likewise, for component 2,

$$\Delta G_2^\circ(\alpha \to \beta) - \overline{G}_2^E(\alpha) + \overline{G}_2^E(\beta) + RT \ln[x_2(\beta)/x_2(\alpha)] = 0 \qquad (15.26)$$

These equations can be solved to obtain the values of $x_2(\alpha)$ and $x_2(\beta)$ at various temperatures, representing the phase boundaries. There are various types of computer programs for this purpose,* depending on the complexity of functions chosen for $g_i$. The method of Gaye and Lupis based on stepwise calculations is simple and convenient. Calculation of $x_2(\alpha)$ and $x_2(\beta)$ at a selected temperature from (15.25) and (15.26) necessitates a computer because of the logarithmic terms, as well as the additional terms involving $x_2^n$ with $n \geq 2$, depending on the analytical form selected to represent $g_i$. We present a brief description of their method. The computation is started with a known set of values of $x_2(\alpha, T)$ and $x_2(\beta, T)$ at a known temperature $T$. The value of $x_2(\alpha, T + \Delta T)$ at $T + \Delta T$ is then obtained from the Taylor expansion as follows:

$$x_2(\alpha, T + \Delta T) = x_2(\alpha, T) + \frac{dx_2(\alpha, T)}{dT}\Delta T$$
$$+ \frac{d^2 x_2(\alpha, T)}{dT^2}(\Delta T)^2 \qquad (15.27)$$

---

*H. Gaye and C. H. P. Lupis, Scripta Met., *4*, 685 (1970); see also L. Kaufman and H. Bernstein, "Computer Calculation of Phase Diagrams," Academic Press, New York (1970); H. L. Lukas and S. G. Fries, J. Phase Equil., *13*, 532 (1992), (BINGSS Program); J. Charles, M. Notin, M. Rahmane, and J. Hertz, ibid., p. 497 (NANCYUN program for binary alloys); J. A. D. Connolly and D. M. Kerrick, CALPHAD, *11*, 1 (1987); P. Dorner, E. T. Henig, H. Krieg, H. L. Lukas, and G. Petzov, ibid., *4*, 241 (1980); G. Eriksson and K. Hack, ibid., *8*, 15 (1984); (Solgaskmix program and multicomponent systems); A. Schultz, Y.-Y., Chuang and Y. A. Chang, Bull. Alloy Phase Diag., *6*, 304 (1985), (computer representation); D. D. Lee, J. H. Choy, and J. K. Lee, J. Phase Equil., *13*, 365 (1992), (computer generation of diagrams); P. J. van Ekeren, A. C. G. van Genderen, M. H. G. Jacobs, and H. A. J. Oonk, CALPHAD, *12(3)*, 303 (1988); M. E. Schlesinger and J. W. Newkisk, J. Phase Equil., *14*, 54 (1993).

$$x_2(\beta, T + \Delta T) = x_2(\beta, T) + \frac{dx_2(\beta, T)}{dT}\Delta T$$
$$+ \frac{d^2 x_2(\beta, T)}{dT^2}(\Delta T)^2 \qquad (15.28)$$

where the derivatives are obtained at the known initial values of $x_2(\alpha, T), x_2(\beta, T)$, and $T$ from the derivatives of (15.25) and (15.26). After $x_2(\alpha, T + \Delta T)$ and $x_2(\beta, T + \Delta T)$ are computed in this manner, the succeeding values at $T + 2\Delta T$ are obtained by repeating the procedure described by (15.27) and (15.28). The first terms after the equal signs in the new set of equations are the left sides of (15.27) and (15.28) at $T + \Delta T$. Calculations by using a few sets of adequate thermodynamic data for a limited number of phase diagrams show that if $\Delta T$ is taken to be 1 K, the accuracy of the calculated values of $x_2$ is about $10^{-6}$ mole fraction as shown by Gaye and Lupis. One of their examples, the system Au-Ni with Au as component 1 and Ni as component 2, is presented as follows. ($x_2$ is the mole fraction of Ni). The necessary data are

Au: $\quad T_{\text{fus }1} = 1336$ K;
$\Delta G_1^\circ(\text{s} \to \text{l}) \text{ cal mol}^{-1} = 2271 - 1.34T \ln T + 7.116T + 0.00062T^2$

Ni: $\quad T_{\text{fus }2} = 1725$ K;
$\Delta G_2^\circ(\text{s} \to \text{l}) \text{ cal mol}^{-1} = 1368 - 3.20 T \ln T + 21.504T + 0.0009T^2$

$G^E(\text{s}) \text{ cal mol}^{-1} = x_1(\text{s})x_2(\text{s})[5770x_1(\text{s}) + 9150x_2(\text{s})$
$\qquad - 3400x_1(\text{s})x_2(\text{s})][1 - (T/2660)]$

$G^E(\text{l}) \text{ cal mol}^{-1} = x_1(\text{l})x_2(\text{l})[1040x_1(\text{l}) + 460x_2(\text{l})].$

The results of their calculations are plotted in Fig. 15.10.

The foregoing methods can be reversed to obtain thermodynamic properties of mixtures, in particular $G^E$, from the phase diagrams. In general, the accuracy of the experimental methods for the determination of the phase boundaries, particularly at high temperatures, is not sufficient for reliable values of $G^E$. The phase boundaries of ambient temperatures are more accurate than those at high temperatures and the results for $G^E$ are often fair in accuracy. The computational procedure in this process is much simpler because it does not involve solving for the unknowns with the logarithmic terms.

### Ternary Phase Diagrams

We discuss the outline of geometrical representation of ternary systems with emphasis on thermodynamic aspects. Consideration of all the possible diagrams and their intricacies is beyond our scope, and all such diagrams are discussed in an excellent classical monograph by Masing (cf. bibliography at the end of this chapter). We consider the ternary phase diagrams in which the external pressure is constant and sufficiently high to suppress the vapor phase. Therefore, such a ternary diagram has three independent variables of state, i.e., two composition variables, and

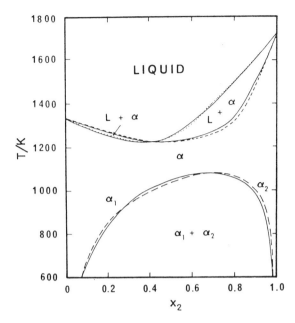

**Fig. 15.10** Phase diagram for Au(component 1) and Ni(component 2). Solid curves are calculated, and broken curves are experimental. (From Gaye and Lupis, loc. cit.).

one temperature variable. A ternary phase diagram can then be represented by a prism with an equilateral triangle for its base. The triangle represents the compositions of components, and the height, the temperature. Figure 15.11 shows the triangular base, which is a two-dimensional cartesian coordinate, bent from 90° to 60° to form a triangle; it is called the Gibbs triangle. On each parallel line facing the pure Component $A$, the mole fraction of $A$ is constant, and similarly for $B$ and $C$, as in cartesian coordinates. For example, at point $P$, $x_A = 0.3$, $x_B = 0.2$ and $x_C = 0.5$, as indicated by the broken lines. Thus, the Gibbs triangle conveniently gives all three mole fractions.

Figure 15.12 shows a simple ternary diagram in which all three binary systems, as well as the ternary system, form one homogeneous liquid and one homogeneous solid above and below their two-phase regions. Such a system is called an isomorphous system. The vertical axis is the temperature, and the lateral vertical surfaces are the three binary isomorphous phase diagrams. The top curved surface is the liquidus surface, and the lower curved surface in the solidus surface, and neither surface has a maximum or a minimum point. A selected temperature, $T'$, is a plane surface that intersects the phase boundaries as shown in the figure, and represented in two-dimension in Fig. 15.13, which is called an isothermal section. The dotted lines in both figures represent the tielines, i.e., the conjugate phases.

304    *Thermodynamics*

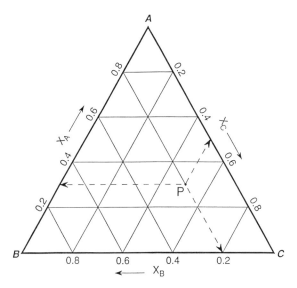

**Fig. 15.11** Gibbs triangle.

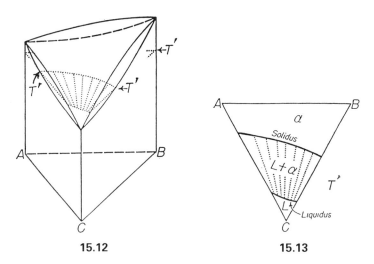

**Fig. 15.12 & 13** Isomorphous phase diagram and isothermic (or isothermal) section. Adapted from Rhines (cf. bibliography).

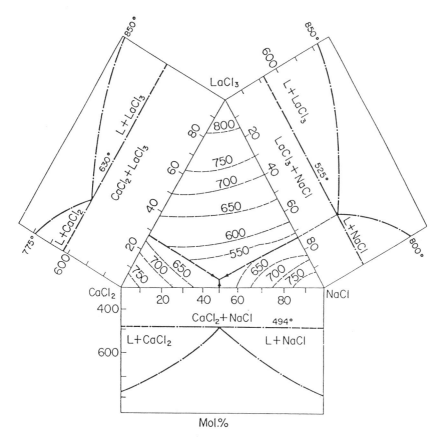

**Fig. 15.14** System NaCl-CaCl$_2$-LaCl$_3$. Adapted from I. S. Morozov, Z. N. Shevtsova, and L. V. Klyukina, Zhur. Neorg. Khim., 2, 1640 (1957).

As the temperature increases, the liquid area in the isothermal sections increases, and finally, it dominates the entire area of the triangle. The reverse phenomenon occurs as the temperature decreases.

The phase rule for constant-pressure ternary systems is

$$\Upsilon = c - p + 1 = 4 - p \tag{15.29}$$

If the temperature is held constant, as in Fig. 15.13, then $\Upsilon = 3 - p$. In the two-phase region in the figure, $\Upsilon = 1$, and at a selected tieline, the composition of one phase is fixed, $\Upsilon = 0$, hence, the composition of the other phase is also fixed.

A ternary system in which there is complete miscibility in liquid state and nearly complete mutual insolubility in solid state is shown in Fig. 15.14 for NaCl-CaCl$_2$-LaCl$_3$. The triangular portion is a contour mapping of the liquidus surfaces

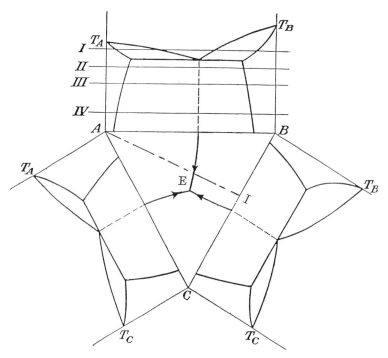

**Fig. 15.15** Ternary eutectic system. Adapted from Marsh (cf. bibliography).

with light broken curves. The heavy broken curves represent the intersections of pairs of planes (valleys) that lead to the ternary eutectic point with is lower than the three binary eutectic points in this case. Each surface represents a surface of one-fold saturation; hence, the valleys represent the lines of two-fold saturation.

A more interesting hypothetical diagram is shown in Fig. 15.15 where significant solid solubilities exist. The melting points of pure components are indicated by $T_A$, $T_B$, and $T_C$. Again the intersections of the planes leading to the ternary eutectic $E$ are indicated inside the triangle. For simplicity, the temperature contours are not indicated. The isothermal sections at temperatures I, II, III, and IV in Fig. 15.15, are shown in Figs. 15.16–15.19. The vertical section at $AI$ in Fig. 15.15 is shown in Fig. 15.20. This section is called an isopleth in which $X_B/X_C$ is a constant. Isopleths can also be drawn at a constant mole fraction of one of the components. It is an easy task to draw an isopleth when there are sufficient numbers of isothermal sections for a ternary system. Generally, isopleths represent the regions of existence of various phase fields without the tielines. When there is a strong binary compound in one of the binary systems, the isopleth passing through the compound and the opposing pure component may be regarded as a pseudo-binary system, and as such, the isopleth may contain pseudo-binary tielines. The

Phase Diagrams 307

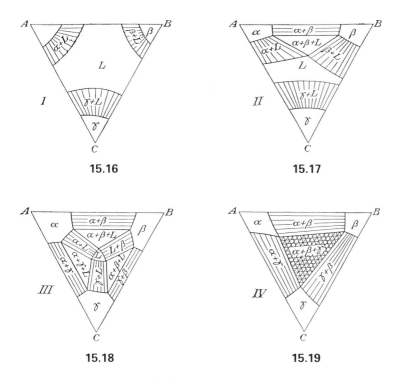

**Fig. 15.16–19** Ternary isothermal sections in Fig. 15.15. Adapted from Marsh.

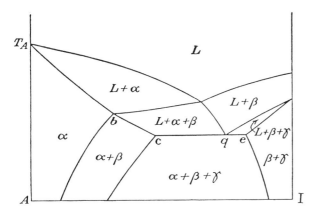

**Fig. 15.20** Isopleth at Al in Fig. 15.15. Adapted from Marsh.

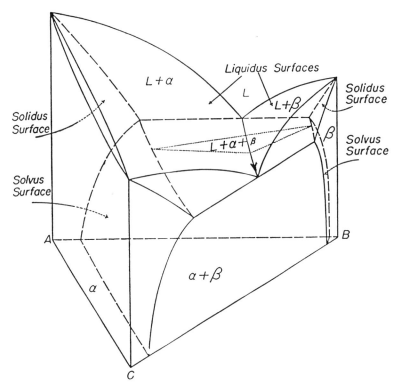

**Fig. 15.21** Ternary system formed by two binary eutectic and one binary isomorphous systems. Reproduced from Rhines.

isopleths are much less frequently encountered in the literature than the isothermal sections and the contour plots (with and without the component binary systems). The latter two types are more informative than the former type.

A ternary system formed by two binary eutectic systems and one isomorphous binary system is shown in Fig. 15.21. The three phase equilibrium isotherm between the two binary eutectic temperatures is shown as a dotted triangle, $L + \alpha + \beta$. The apex closest to the binary eutectic point of B-C intersects the line of two-fold saturation. The isothermal section containing this triangle is shown in Fig. 15.22.

The oxide ternary systems are usually more complex than the metallic systems, in part, due to the formation of numerous compounds and compound-like phases and many invariant points. We illustrate the complexity of a ternary oxide system by replicating the $CaO$-$SiO_2$-$Al_2O_3$ system in Fig. 15.23 from "Phase Diagrams for Ceramists" Gen. Ref. (20). A complete description of this system is beyond the scope of this book; Fig. 15.23 is simply for illustration.

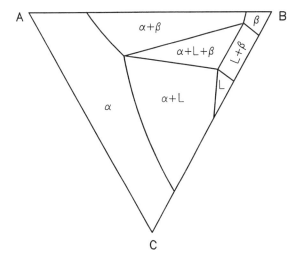

**Fig. 15.22** Isothermal section at $L + \alpha + \beta$ triangle in Fig. 15.21.

A quaternary system can be represented by a tetrahedron, and only for each temperature, and a geometrical representation of a pentanary and higher order systems is not possible. The "flow diagrams," tabular forms of multiphase equilibria, usually in decreasing sequence of temperature, have been used to simplify the presentation of complex ternary and multicomponent diagrams, e.g., as described by Connell.* The advent of modern computerized differential thermal analyzers are specially useful as an experimental tool for this purpose.

### Tielines

The tielines in a two-phase region, as in Fig. 15.12, can be used to determine the fraction of each phase in a bulk mass of two phases, as in a binary system. The bulk mass is taken as one mole, but it can also be taken as one kg. When there are three phases in equilibrium, they form a tie-triangle, as the inner triangle in Figs. 15.19, 15.25, and 15.26. Such a triangle generally has unequal edges. The lever rule for a point $w$ in Fig. 15.26 is expressed by

$$\text{Fraction of } \alpha \text{ phase (at } w) = wy/ay \qquad (15.30)$$

The fractions of the remaining phases are determined similarly, and all the fractions add up to unity representing the bulk of a mass of three phases.

### Thermodynamic Consideration

The intersections of the phase boundaries must obey an extension of the requirements in binary diagrams, i.e., the extended portions of the phase boundaries must

---

*R. G. Connell, Jr., J. Phase Equil., *15*, 6 (1994).

310  Thermodynamics

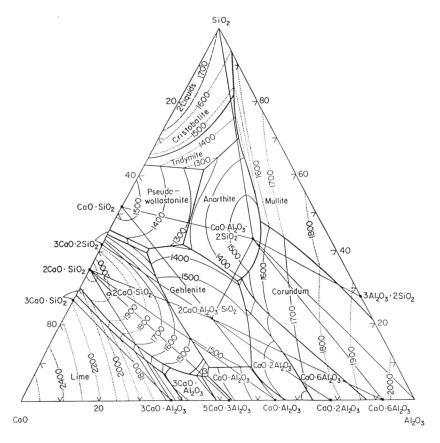

**Fig. 15.23** CaO-SiO$_2$-Al$_2$O$_3$ phase diagram with temperature contours. From Levin, Robbins and McMurdie Gen. Ref. (20). Reproduced by permission.

have higher Gibbs energies than the stable fields in which they terminate. To clarify an important point in that respect, we present a trivial property at the eutectoid point in a binary system in Fig. 15.24. It has to be emphasized at first that the eutectoid line expands into two dimensions as the third component is added to the binary system, illustrated schematically with a dotted triangle in the figure, see also Fig. 15.21. Therefore, we consider Fig. 15.24 with this point in view. The extensions of the $\alpha$ phase boundaries terminate correctly in the two-phase fields in Fig. 15.24, and this is also the case in the $\alpha$ boundaries in Fig. 15.25. The $\beta$ boundaries terminate within the $\beta$ phase field, and they are incorrect in both figures. The $\gamma$ phase boundaries extend correctly by crossing the three-phase line (three-phase field in ternaries), as indicated by $d$ and $e$, again in both figures. The Gibbs energies would not permit the extension of the $\gamma$ boundary on the right

Phase Diagrams 311

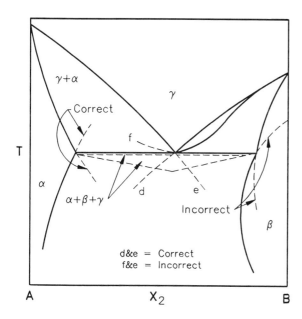

**Fig. 15.24** Correct and incorrect phase boundaries. Broken triangle indicates ternary phase field when third component is added.

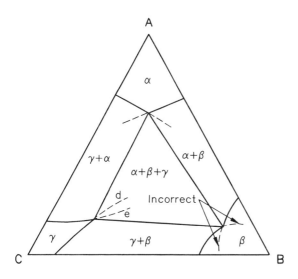

**Fig. 15.25** Incorrect beta phase boundaries.

312    *Thermodynamics*

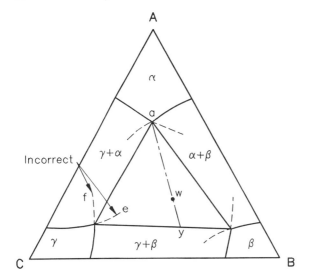

**Fig. 15.26**   Incorrect gamma phase boundaries.

side in Fig. 15.24 to extend into the $\alpha + \gamma$ field without crossing the three-phase eutectic line (field in a ternary) as indicated by $f$, while the left side correctly terminates by crossing the three-phase eutectic line (or field in a ternary) as shown by $e$; hence $f$ and $e$ as a pair are incorrect in both Figs. 15.24 and 15.26. It should be noted that the $\beta$ phase field in Fig. 15.26 is now correct. (The line $awy$ in Fig. 15.26 is the lever for calculating the relative mass of $\alpha$ phase in a mixture of three phases at $w$).

The Gibbs-Konovalow theorems require that the two phase boundary surfaces, both exhibiting maximum points (or both exhibiting minimum points) must be horizontally tangent to each other at the same point. Thus, one may proceed from one single-phase region to another single-phase region only at a point in ternary systems, as well as in binary systems.

The phase rule is a very important guideline in constructing ternary diagrams or sections without error. We describe the ternary systems by considering Figs. 15.16–15.19 and 15.22:

(1) One may proceed in any direction from a single phase-field to a two-phase field, or from a two-phase field to one-phase field along a curve as in Fig. 15.2.
(2) The single phase field touches the three phase field at one point so that a three-phase field is bounded by single phase points in Figs. 15.16–15.19 and 15.22.
(3) The boundaries of a three-phase field are the two-phase boundary tielines, i.e., one can proceed from a three-phase field to a two-phase field along a tieline.
(4) One can proceed from a two-phase field to another two-phase field at a single point on an isothermal ternary section or isopleth.

In general, the curves and the lines in Fig. 15.22 become surfaces and the points become lines when a ternary diagram is represented in three dimensions. In addition, it can be deduced from Fig. 15.14 or 15.15 that at the ternary eutectic point, where the degree of freedom is zero, the liquid composition must be located within a three-phase field for $\alpha + \beta + \gamma$ in such a way that the liquid composition falls within that triangle to form a four-phase plane. Phase diagrams can be obtained by thermal analysis, micrography, and physical property measurements.*

## Second Order Transitions

Thus far, we have dealt with the phase equilibria in alloys involving only the first-order phase transitions. The Gibbs energy of a given multicomponent system is a function of its variables of state $P, T, n_1, n_2, \ldots, n_c$, i.e., $\mathcal{G} = \mathcal{G}(P, T, n_1, n_2 \ldots, n_c)$. This function is continuous for the first-order phase transitions but its derivative, with respect to one of its variables, becomes discontinuous upon a first-order transition. The variable of the greatest importance is the temperature; therefore, we limit our discussion to the derivatives of $\mathcal{G}$ with respect to $T$. A first-order transition is accompanied with a discontinuity in the first derivatives of $\mathcal{G}$; thus,

$$\frac{\partial \mathcal{G}}{\partial T} = -\mathcal{S}; \quad \text{or} \quad \frac{\partial(\mathcal{G}/T)}{\partial(1/T)} = \mathcal{H} \tag{15.31}$$

would show a discontinuity in the entropy or enthalpy in the vicinity of a transition or transformation temperature. At a second-order transition, the second derivatives of $\mathcal{G}$ exhibit a discontinuity; i.e.,

$$\frac{\partial^2 \mathcal{G}}{\partial T^2} = -\frac{\mathcal{C}_p}{T}; \quad \text{or} \quad \frac{\partial^2(\mathcal{G}/T)}{\partial(1/T)\partial T} = \mathcal{C}_p \tag{15.32}$$

In summary, $\mathcal{S}$ and $\mathcal{H}$, or $S$ and $H$, are discontinuous for the first-order phase transitions and $\mathcal{C}_p$ or $C_p$ is discontinuous for the second-order phase transitions.

The order-disorder phenomena are second-order phase transitions in which discontinuities are observed in $C_p$ of metals and alloys. In Fig. 15.32 for C-Fe, the ordering below 760–770°C is due to the magnetic domain ordering where a distinct change occurs in $C_p$. In Fig. 15.33 for Cu-Zn, below 760–770°C and close to the equiatomic composition, Cu atoms occupy one interpenetrating lattice, and Zn atoms occupy another interpenetrating lattice and retain the same crystal structure as above these temperatures. This phenomenon is another form of ordering, and causes a sharp change in $C_p$ of the alloy [see Chapter 5 in Gen. Ref. (15)]. In phase diagrams, the second order transitions are indicated with dotted or broken lines, while solid lines are retained for the first order transitions.

---

*See numerous articles in J. Phase Equil., and CALPHAD; e.g., R. C. Sharma and I. Mukerjee, J. Phase Equil., *13*, 5 (1992). See also the footnotes after Fig. 15.1.

# BIBLIOGRAPHY

1. G. Masing, "Ternary Systems," translated by B. A. Rogers, Reinhold Publishing Co. (1944).
2. F. N. Rhines, "Phase Diagrams in Metallurgy," McGraw-Hill Book Co. (1956).
3. J. S. Marsh, "Principles of Phase Diagrams," McGraw-Hill Book Co. (1935).
4. "Phase Diagrams for Ceramists," in 5 volumes by Am. Cer. Soc.-NIST, in 10 volumes, 1964–1994.
5. "Handbook of Ternary Alloy Phase Diagrams," P. Villars, A. Prince, and H. Okamoto, editors, ASM International (1994–1995).

# PROBLEMS

15.1 (a) Calculate the values of $x_2$ for the maximum, minimum and inflection points for the curve in Fig. 15.5, and $\Delta \overline{G}_1$ and $\Delta \overline{G}_2$ at these points. (b) Calculate $\Delta \overline{G}_1$ for the straight line $AB$ in Fig. 15.5. Use (11.9) and (15.2).

*Partial ans.*: (b) $-448$ J mol$^{-1}$

15.2 Plot $a_2$ versus $x_2$ for (15.2) and show analytically that the maximum and minimum points coincide with the spinodes.

15.3 Calculate (a) the ratio of the slopes of the liquidus and solidus curves in Fig. 15.7 at $T = 900$ K, (b) the segregation coefficient at 890 K, and (c) $\Delta \overline{G}_1$ and $\Delta \overline{G}_2$ at 550 K. (d) Replot the phase diagram in Fig. 15.7 after adding $100 R x_1(l) x_2(l)$ and $100 R x_1(s) x_2(s)$ to (15.16) and (15.17) respectively; i.e., assuming that the liquid and solid solutions follow a regular behavior.

15.4 Give schematic representations of $\Delta G$ for $\alpha$, $\beta$ and liquid in Fig. 15.8 (a) at $T = T_{\text{fus},1}$ and (b) $T = T_1$.

15.5 Express (15.27) and (15.28) for the liquidus and solidus curves for Au–Ni system from the data and the equations given in the text.

15.6 The phase diagram for Ag-Cu is in Fig. 15.27. The melting points are 1234.93 K for Ag, and 1357.77 K for Cu. The standard Gibbs energies of fusion are as follows: $\Delta_{\text{fus}} G° = 11{,}297 - 9.148 T$ for Ag, and $\Delta_{\text{fus}} G° = 13{,}055 - 9.614 T$ for Cu, both in J mol$^{-1}$. (a) Calculate the liquidus by assuming ideality and using the solidus compositions from the diagram. (b) Recalculate the liquidus by using the solidus compositions from the figure with ideality in solids and $G^E = 14{,}100 x_1 x_2$ J mol$^{-1}$ for the liquid phase. Scale off the solidus as necessary.

15.7 CaF$_2$ and LiF form a simple eutectic with negligible mutual solid solubilities. Construct their binary phase diagram by using the following data from Gen. Ref. (2):

Melting, $T_{\text{fus}}/\text{K}$: 1690 for CaF$_2$    1121.3 for LiF
$\Delta_{\text{fus}} H°/\text{J mol}^{-1}$: 29,790 for CaF$_2$    27,050 for LiF

The entropy of fusion is $\Delta_{\text{fus}} S° = \Delta_{\text{fus}} H° / T_{\text{fus}}$; therefore, the standard Gibbs energies of fusion are as follows:

$\Delta_{\text{fus}} G° = 29{,}790 - 17.6272 T$ for CaF$_2$;
$\Delta_{\text{fus}} G° = 27{,}050 - 24.1238 T$ for LiF

*Partial ans.*: Calculated eutectic point is $x(\text{LiF}) = 0.76$ at 750 K; the experimental diagram is in Gen. Ref. (20) as Diagram No. 1470.

15.8   Calculate the binary and ternary eutectic temperatures and compositions for the system shown in Fig. 15.14 by assuming ideality in liquids and negligible mutual solid solubilities. Use the following data from Gen. Ref. (2):

|  | NaCl | CaCl$_2$ | LaCl$_3$ |
|---|---|---|---|
| $T_{fus}/K$: | 1073.8 | 1045 | 1131 |
| $\Delta_{fus}H°/J\,mol^{-1}$: | 28,284 | 28,158 | 54,392 |

15.9   Replot $\Delta_{mix}G$ in Fig. 15.7 versus $x_2$ for 500 K and show, within the accuracy of the plot, that the common tangent is identical with the equality of partial Gibbs energies in liquid and solid. See (15.16) and (15.17) and the text.

15.10  Derive $\Delta \overline{G}_2 = 500R \ln x_2(l)$ and $\Delta \overline{G}_2(s) = 500R \ln x_2(s) + 200R$ for Problem 15.9, and verify numerically that $x_2(l)/x_2(s) = 1.492$.

## SELECTED BINARY PHASE DIAGRAMS

The following illustrative phase diagrams are reproduced by permission from Gen. Ref. (20) unless cited otherwise.

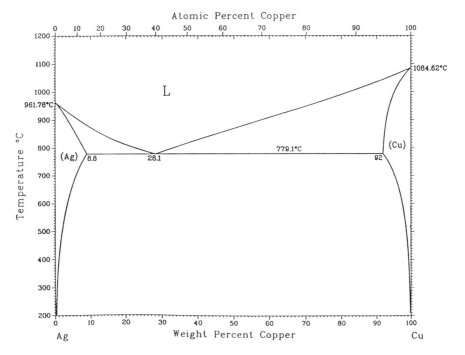

**Fig. 15.27** Ag-Cu phase diagram assessed by P. R. Subramanian and J. H. Perepezko, J. Phase Equil., *14*, 62 (1994).

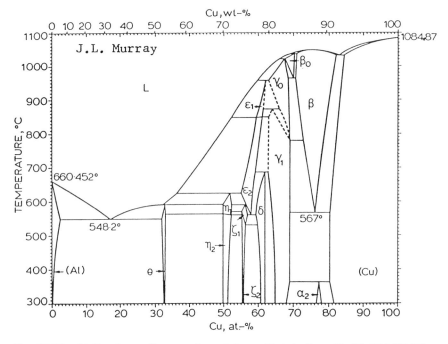

**Fig. 15.28** Al-Cu phase diagram from Intern. Metals Rev. *30*, (5), 211 (1985); also in Gen. Ref. (18).

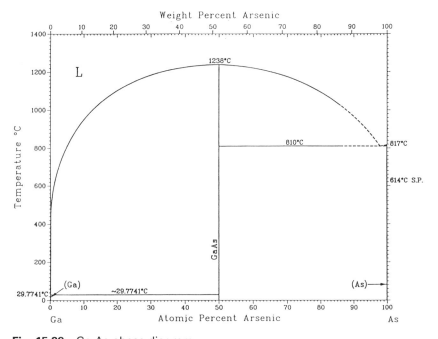

**Fig. 15.29** Ga-As phase diagram.

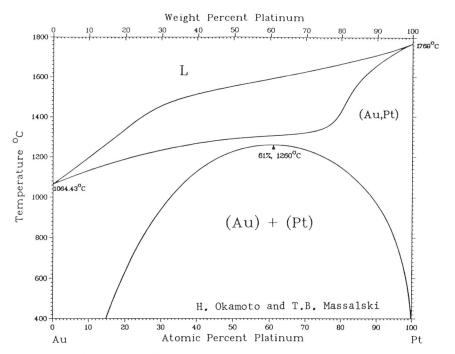

**Fig. 15.30** Au-Pt phase diagram.

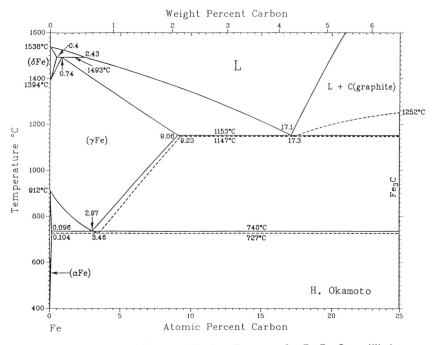

**Fig. 15.31** C-Fe phase diagram. Broken lines are for Fe-Fe$_3$C equilibria.

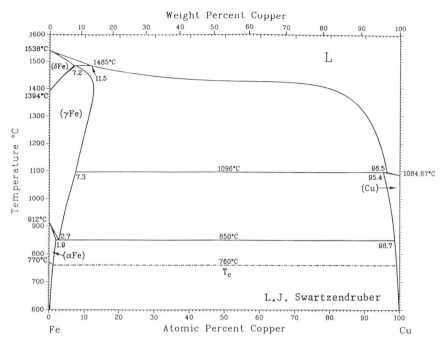

**Fig. 15.32** Cu-Fe phase diagram. The line with dots is for second order phase transition (magnetic tr.).

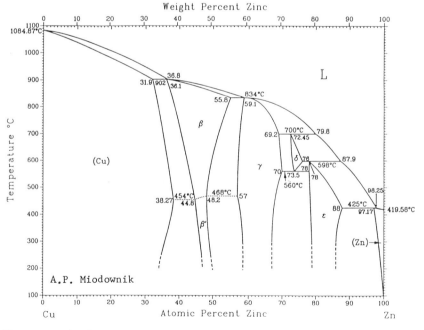

**Fig. 15.33** Cu-Zn phase diagram. The line with dots is for second order transition (ordering).

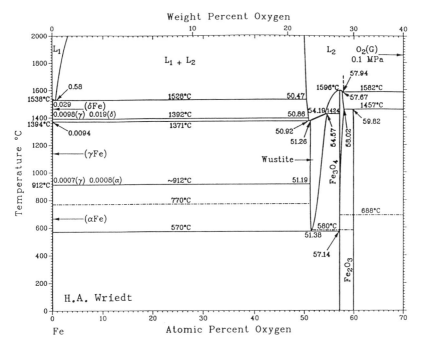

**Fig. 15.34** Fe-O phase diagram.

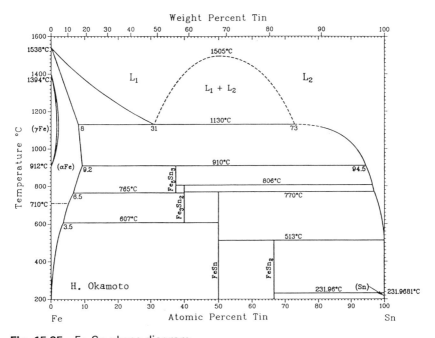

**Fig. 15.35** Fe-Sn phase diagram.

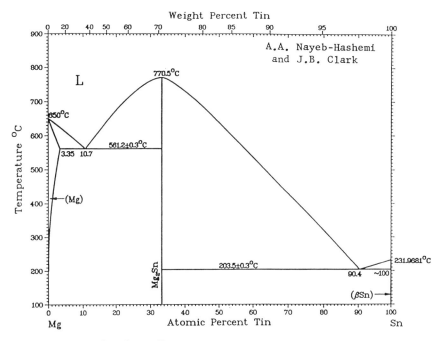

**Fig. 15.36** Mg-Sn phase diagram.

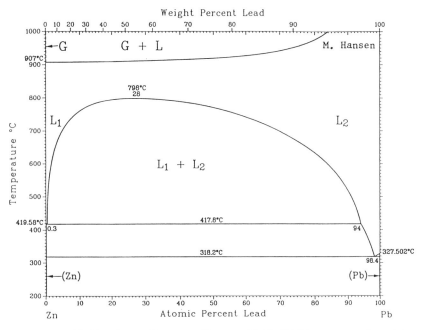

**Fig. 15.37** Pb-Zn phase diagram. Figures 15.28–15.37 are reproduced from Gen. Ref. (18) by permission.

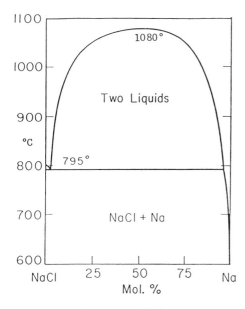

**Fig. 15.38** System Na-NaCl. M. A. Bredig and H. R. Bronstein, J. Phys. Chem., *64*, 65 (1960).

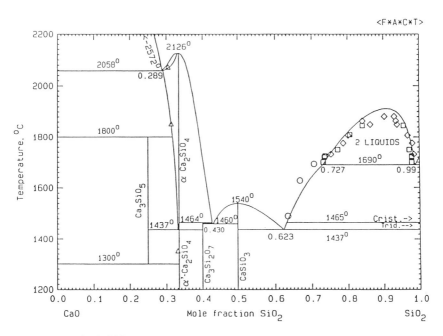

**Fig. 15.39** CaO-SiO$_2$ phase diagram; courtesy, A. Romero-Serrano and A. D. Pelton from a paper to be published.

CHAPTER XVI

# SPECIAL TOPICS

## PART I  SURFACE TENSION

### Properties of Surfaces

The effects of surfaces are generally small when a condensed phase consists of a single mass or a few large pieces so that the ratio of the surface area to the volume is small. The effects may become large as the surface to volume ratio becomes significantly large for a given amount of a single or multicomponent condensed phases. For simplicity, we consider a binary system of two phases since the results can readily be extended to any number of components and phases without difficulty. An interfacial surface exists when there are two phases in contact with each other. The transition from one phase to the other across the surface separating the phases is not considered to be geometrically sharp, but rather gradual on a molecular scale from a chemical point of view. For condensed phases not too close to their critical temperatures of any type, various observations indicate that the transition layer between two phases is only a few molecules in thickness. The properties of this region can be investigated by postulating that a two-dimensional phase, called the *surface phase*, exists *between two phases*. The term "surface phase" is a misnomer since it does not refer to a phase in ordinary sense but it has been established by tradition. We shall first assume the existence of a hypothetical surface between two phases on which the surface phase is located. This procedure is necessary for deriving the appropriate equations first in order to define mathematically the appropriate location of the surface phase.

The concentration of a component in the surface phase is expressed in terms of molecules per unit area and therefore the surface phase may be regarded as a sheet of molecules. The surface excess, positive or negative in value, for a constituent of the system is defined as follows. If $n_i$ is the total number of molecules of constituent $i$ in the system of two phases, $c_i^I$ and $c_i^{II}$ are the bulk concentrations of $i$ in phases I and II, and $V_i^I$ and $V_i^{II}$ are their volumes, the surface excess $n_i^s$ is defined by

$$n_i^s = n_i - c_i^I V_i^I - c_i^{II} V_i^{II} = c_i^s A \qquad (16.1)$$

where $c_i^s$ is the excess surface concentration of $i$ in moles per m² and $A$ in m² is the area of interface between two phases. A similar equation may also be written with the concentrations expressed in molalities or mole fractions.

## Criteria for Equilibrium

The Gibbs energy $\mathcal{G}$ of a system of two phases, in which the surface forces are important, is the sum of the Gibbs energies of the phases, $\mathcal{G}^{\text{I}}$ and $\mathcal{G}^{\text{II}}$, and the Gibbs energy of the surface phase $\mathcal{G}^s$, i.e.,

$$\mathcal{G} = \mathcal{G}^{\text{I}} + \mathcal{G}^{\text{II}} + \mathcal{G}^s \tag{16.2}$$

where $\mathcal{G}^{\text{I}}$ and $\mathcal{G}^{\text{II}}$ refer to the bulk phases. $\mathcal{G}^{\text{I}}$ and $\mathcal{G}^{\text{II}}$ are functions of $P, T, n_1$ and $n_2$ of each phase for a binary system, and $\mathcal{G}^s$ is a function of $T, A, n_1^s$ and $n_2^s$; therefore,

$$d\mathcal{G}^{\text{I}} = \mathcal{V}^{\text{I}} dP - \mathcal{S}^{\text{I}} dT + \overline{G}_1^{\text{I}} dn_1^{\text{I}} + \overline{G}_2^{\text{I}} dn_2^{\text{I}} \tag{16.3}$$

$$d\mathcal{G}^{\text{II}} = \mathcal{V}^{\text{II}} dP - \mathcal{S}^{\text{II}} dT + \overline{G}_1^{\text{II}} dn_1^{\text{II}} + \overline{G}_2^{\text{II}} dn_2^{\text{II}} \tag{16.4}$$

$$d\mathcal{G}^s = \frac{\partial \mathcal{G}^s}{\partial T} dT + \frac{\partial \mathcal{G}^s}{\partial A} dA + \overline{G}_1^s dn_1^s + \overline{G}_2^s dn_2^s \tag{16.5}$$

where the partial Gibbs surface energy $\overline{G}_i^s$ is defined by $\partial G^s/\partial n_i^s$. The first law of thermodynamics for the entire closed system is

$$d\mathcal{U} = T d\mathcal{S} - P d\mathcal{V} + \sigma dA \tag{16.6}$$

where $\sigma$ is called the surface tension or interfacial tension, $A$, the surface area and $\sigma\, dA$ is the surface work imparted to the system by generating a surface area of $dA$. We add the differential of $(P\mathcal{V} - T\mathcal{S})$ on both sides of this equation to obtain an alternative equation for $\mathcal{G}$ in (16.2); thus

$$d\mathcal{G} = \mathcal{V} dP - \mathcal{S} dT + \sigma dA \tag{16.7}$$

This equation shows that $\partial \mathcal{G}/\partial A = \sigma$, and since $\mathcal{G}^{\text{I}}$ and $\mathcal{G}^{\text{II}}$ are independent of $A$, it is evident that

$$\partial \mathcal{G}/\partial A = \partial \mathcal{G}^s/\partial A = \sigma \tag{16.8}$$

In parallel with $\partial \mathcal{G}^{\text{I}}/\partial T = -\mathcal{S}^{\text{I}}$, it is appropriate to define the surface entropy $\mathcal{S}^s$ by

$$\partial \mathcal{G}^s/\partial T = -\mathcal{S}^s \tag{16.9}$$

Substitution of (16.8) and (16.9) into (16.5) gives

$$d\mathcal{G}^s = -\mathcal{S}^s dT + \sigma dA + \overline{G}_1^s dn_1^s + \overline{G}_2^s dn_2^s \tag{16.10}$$

The total differential of (16.2) may now be written as

$$d\mathcal{G} = \mathcal{V} dP - \mathcal{S} dT + \sigma dA + \overline{G}_1^{\text{I}} dn_1^{\text{I}} + \overline{G}_2^{\text{I}} dn_2^{\text{I}} + \overline{G}_1^{\text{II}} dn_1^{\text{II}} + \overline{G}_2^{\text{II}} dn_2^{\text{II}}$$
$$+ \overline{G}_1^s dn_1^s + \overline{G}_2^s dn_2^s; \quad (\mathcal{V} = \mathcal{V}^{\text{I}} + \mathcal{V}^{\text{II}}; \mathcal{S} = \mathcal{S}^{\text{I}} + \mathcal{S}^{\text{II}} + \mathcal{S}^s) \tag{16.11}$$

For a system under surface forces, the system must be closed and $d\mathcal{G}$ must be zero at constant $P$, $T$ and $A$, so that according to (16.7), $d\mathcal{G}_{P,T,n,A} = 0$. At equilibrium, therefore, $d\mathcal{G}$ in (16.11) is zero, and from $dn_1 = 0 = dn_1^{\text{I}} + dn_1^{\text{II}} + dn_1^s$, and $dn_2 = 0 = dn_2^{\text{I}} + dn_2^{\text{II}} + dn_2^s$, (16.11) at constant $P$, $T$, and $A$ becomes

$$\left(\overline{G}_1^{\text{I}} \, dn_1^{\text{I}} + \overline{G}_1^{\text{II}} \, dn_1^{\text{II}} + \overline{G}_1^s \, dn_1^s\right) + \left(\overline{G}_2^{\text{I}} \, dn_2^{\text{I}} + \overline{G}_2^{\text{II}} \, dn_2^{\text{II}} + \overline{G}_2^s \, dn_2^s\right) = 0$$

Lagrange's method of undetermined multipliers, as applied in Chapter VIII requires that this equation can be satisfied if and only if

$$\overline{G}_1^{\text{I}} = \overline{G}_1^{\text{II}} = \overline{G}_1^s = \overline{G}_1; \quad \text{and} \quad \overline{G}_2^{\text{I}} = \overline{G}_2^{\text{II}} = \overline{G}_2^s = \overline{G}_2 \tag{16.12}$$

Therefore the partial Gibbs surface energy of a component is equal to its partial Gibbs energy $\overline{G}_i$ in the bulk under equilibrium conditions. Equations (16.12) represent the criteria for equilibrium when the surface forces play an important role.

Equation (16.9) for $\mathcal{S}^s$ and $\mathcal{G}^s$ define $\mathcal{H}^s$ but we shall not use this property because of limitation in space.

## Gibbs Adsorption Equation

The Gibbs surface energy $\mathcal{G}^s$ due to the surface tension is $\sigma A$, and the additional terms are $n_i \overline{G}_i^s$; thus for a binary system

$$\mathcal{G}^s = \sigma A + \overline{G}_1^s n_1^s + \overline{G}_2^s n_2^s \tag{16.13}$$

From the total differential of (16.13), we subtract (16.10) at constant temperature, to obtain

$$A \, d\sigma + n_1^s \, d\overline{G}_1^s + n_2^s \, d\overline{G}_2^s = 0; \quad \text{(constant } T \text{ and } P\text{)} \tag{16.14}$$

We recall from (16.1) that $n_1^s = c_i^s A$, and substitute $c_i^s$ in (16.14), with $\overline{G}_i^s = \overline{G}_i$ from (16.12) so that

$$d\sigma + c_1^s \, d\overline{G}_1 + c_2^s \, d\overline{G}_2 = 0; \quad \text{(constant } T \text{ and } P\text{)} \tag{16.15}$$

This equation permits us to define the location of the surface phase as dictated by convenience in the study of solutions in which component 1 is taken to be the solvent. *The location of the surface phase is chosen* to make $c_1^s$ the surface excess of component 1, the solvent, zero. This choice simplifies (16.15) into

$$d\sigma + c_2^s \, d\overline{G}_2 = 0 \tag{16.16}$$

The activity of solute 2, based on molarity is $a_2 = \gamma_2 c_2$ where $c_2$ is molarity, and $\overline{G}_2 = G_2^\infty + RT \ln \gamma_2 c_2$. Substitution of this equation in (16.15) leads to

$$\boxed{c_2^s = -\frac{1}{RT}\left(\frac{\partial \sigma}{\partial \ln a_2}\right)_{T,P}}; \quad \boxed{c_2^s(\text{dilute}) \approx -\frac{c}{RT}\left(\frac{\partial \sigma}{\partial c_2}\right)_{T,P}} \tag{16.17}$$

where the last equation is for a dilute solution in which $a_2 \approx c_2$ or $\gamma_2 \approx 1$. Equations (16.17) are called the *Gibbs adsorption equations*, which state that if a component is added in a solvent i.e., if $c_2$ is increased, and if this increase causes a decrease in the surface tension, $\sigma$, then $\partial\sigma/\partial c_2$ is negative and the surface concentration $c_2^s$ of solute is positive i.e., the adsorption of the solute occurs. Such substances are called surfactants. Conversely when $\partial\sigma/\partial c_2$ is positive $c_2^s$ is negative and the negative adsorption of the solute occurs, i.e., the surface concentration of the solute becomes less than that in the bulk.

## Vapor Pressure of Droplets

The surface tension plays an important role in growth and vaporization of solid and liquid droplets depending on the prevailing vapor pressure. An interesting example is the nucleation, growth and vaporization of clouds consisting of water or ice particles. We assume that the droplets or particles under examination are spherical to simplify our treatment. We consider a droplet of radius $r$ in equilibrium with its vapor at a fugacity $f$. The fugacity of a flat surface of the same pure component at the same temperature is $f°$. The Gibbs energy change in transferring a very small amount of the vapor ($dn$ moles) at $f°$ onto the droplet is

$$d\mathcal{G} = (dn)RT \ln(f/f°); \qquad \text{(constant } P \text{ and } T) \tag{16.18}$$

The change in $\mathcal{G}$, $d\mathcal{G}$, is caused by the increase in the surface area of the droplet at a given pressure and temperature, and if the system consists of a single component, $d\mathcal{G}$ from (16.13) is given by $d\mathcal{G} = \sigma\,dA$, and (16.18) becomes

$$(dn)RT \ln(f/f°) = \sigma\,dA \tag{16.19}$$

If the molar volume of the substance is $V$ and the volume of the spherical droplet is $4\pi r^3/3$ then the number of moles $n$ in the droplet is $n = 4\pi r^3/3V$, and then $dn$ is given by $dn = 4\pi r^2\,dr/V$. The surface area $A$ of the droplet is $A = 4\pi r^2$; therefore $dA = 8\pi r\,dr$. Substitution of $dn$ and $dA$ in (16.19) gives

$$\boxed{RT \ln(f/f°) = 2V\sigma/r; \qquad \text{(constant } P \text{ and } T)} \tag{16.20}$$

This equation states that as the droplet radius $r$ decreases the fugacity and the vapor pressure of the particle increase and the droplet disappears by distilling onto the larger particles or on the flat surface of the substance. For example the vapor pressure of a flat surface of water at 298.15 K is $P = 3167$ Pa $\approx f°$, its surface tension is 0.072 N m$^{-1}$ (72 dynes cm$^{-1}$) and its molar volume is $18 \times 10^{-6}$ m$^3$ mol$^{-1}$. Substitution of these values and $R = 8.3145$ J mol$^{-1}$ K$^{-1}$ from Table 1.1 into (16.20), for a droplet size of $r = 10^{-8}$ meter, gives $\ln(f/f°) = 2 \times 18 \times 10^{-6} \times 0.072/(10^{-8} \times 8.3145 \times 298.15) = 0.10456$. Hence, $f \approx P = 3516$ Pa, i.e., $f$ is 11% higher than $f°$.

Equation (16.20) explains that a nucleus must be formed for the formation of droplets even when $f$ is very much larger than $f°$ and this explains the reason for the cloud nucleation by fine particles such as atomized silver halides. In addition if the nucleating agent also adsorbs the vapor and decreases its surface tension $\sigma$, then the fugacity $f$ need not be much larger than $f°$. The absence of nucleating particles has led to remarkable experiments in which unstable phases have been drastically supercooled; e.g., liquid gold has been supercooled several hundred degrees below its melting point. Numerous liquid elements and solutions have also been quenched rapidly to retain liquid-like (amorphous) properties at room temperature.*

Equation (16.20) is not only valid for the external surface of a droplet, but also for a bubble formed inside a liquid. If we substitute $P°V$ for $RT$, $P°$ for $f°$, and $P$ for $f$ in (16.20), we obtain

$$P - P° = 2\sigma/r; \quad \text{(bubble formation in liquids)} \tag{16.21}$$

where the Stirling approximation has been used for $\ln(P/P_o)$. Equation (16.21), known as the Laplace equation (1806), is valid when $P$ and $P°$ are not greatly different. At $P° = 1$ bar, the additional pressure to form a bubble of $r = 1$ mm in water at 25°C with helium is $2\sigma/0.001 = 144$ Pa, if $\sigma$ is not affected by He.

## PART II  GRAVITATIONAL ELECTRIC AND MAGNETIC FIELDS

Gravitational, electric and magnetic fields exert their effects on the system externally, unlike temperature surface tension, etc. For this reason they are also called *remote forces*. We discuss the gravitational field first in sufficient detail for our purposes and briefly describe electric and magnetic fields.

### Gravitational Field

A gravitational field affects the properties of a system by exerting external forces on the component molecules. The properties may vary according to the intensity of the field at various points in the system. From the first law of thermodynamics it was seen that for one mole of a substance $i$ of molecular weight $M$

$$dU = T\,dS - P\,dV + dW''$$

where $dW''$ is the gravitational work. It can be shown as for the surface work that

$$dW'' = dG_{i,P,T} \tag{16.22}$$

Therefore, the gravitational work is a form of available work. When $M$ is lifted up a vertical distance of $h$ higher than a reference point selected to be $h = 0$ under a uniform gravitational force of the earth, the gravitational work gained by the substance at constant $P$ and $T$ is

$$dW'' = dG_{i,P,T} = Mg\,dh \tag{16.23}$$

where $g$ is the acceleration. We emphasize that $h$ is positive when it is vertically upward.

We now consider the effect of hydrostatic pressure on the properties of a gas or liquid column consisting of a pure substance $i$ confined in a container at

---

*S. R. Elliott, "Physics of Amorphous Materials," Halsted Printing, div. John Wiley & Sons (1990).

a fixed position. In this case $h = 0$ is the bottom of the container, and again $h > 0$ in upward direction. The molar Gibbs energy $G$ is a function of $P$, $T$, and $h$; therefore at a constant temperature

$$dG = \left(\frac{\partial G}{\partial P}\right)_{T,h} dP + \left(\frac{\partial G}{\partial h}\right)_{P,T} dh = V\, dP + Mg\, dh \tag{16.24}$$

At equilibrium when the entire system is closed and the temperature and the external pressure are fixed, the left side of (16.24) is zero. However, there is no restriction on the internal pressure which may vary according to the height $h$ of the gas or liquid column. The value of $dP$ in (16.24) when the external pressure is fixed is then equal to the change in the internal pressure due to gravity so that $dP = dP$ (internal). Therefore, (16.24) may be rewritten as

$$V\, dP(\text{int}) = -Mg\, dh; \quad \text{(at equilibrium)} \tag{16.25}$$

It should be observed that $P(\text{int})$ is given by $P(\text{int}) = P(h) = P(\text{at } h = 0) + P(\text{hydrostatic})$. Substitution of $V = RT/P(h)$ in (16.25) and integration at constant $T$ yields

$$\ln[P(h)/P(h=0)] = -Mgh/RT \tag{16.26}$$

As a numerical example, we consider the air at $h = 100$ m and $T = 273$ K, with $M \approx 0.029$ kg mol$^{-1}$. With these values, the right side of (16.26) becomes $-0.029 \times 9.80 \times 100/(273 \times 8.3145 \times 10^7) = -0.01252$ and the corresponding change in the pressure is about 1,261 Pa. At higher altitudes many natural phenomena including changes in $T$ and $g$, invalidate (16.26). The most useful application of (16.25) is therefore for liquids for which $V$ is virtually constant, and $g$ is the artificially created acceleration by a centrifuge capable of attaining several thousand revolutions per second.

## Solutions

The partial Gibbs energy $\overline{G}_i$ of a component is a function of $P$, $T$, $x_i$, and $h$; therefore, for a binary system and for component 2, as the solute,

$$d\overline{G}_2 = \overline{V}_2\, dP - \overline{S}_2\, dT + \left(\frac{\partial \overline{G}_2}{\partial h}\right)_{P,T,x_2} dh + \left(\frac{\partial \overline{G}_2}{\partial x_2}\right)_{P,T,h} dx_2 \tag{16.27}$$

At constant $P(\text{ext})$ and $T$ for a closed system, $\overline{G}_2$ is uniform in the system or $d\overline{G}_2$ is zero. Equation (16.25) for the entire solution may be written as

$$dP(\text{int}) = -\rho g\, dh; \quad \text{(at equilibrium; } \rho = M/V)$$

where $\rho = M/V$ has been substituted. For component 2, $\partial \overline{G}_2/\partial h$ in (16.27) is equal to $M_2 g$ as can be seen from (16.23). Substitution of these equations in (16.27) gives

$$0 = M_2 g\, dh - \overline{V}_2 \rho g\, dh + (\partial \overline{G}_2/\partial x_2)_{P,T,h}\, dx_2$$

and upon rearrangement

$$\frac{dh}{dx_2} = \frac{1}{g(\overline{V}_2\rho - M_2)} \left(\frac{\partial \overline{G}_2}{\partial x_2}\right)_{P,T,h} \tag{16.28}$$

Variation of $x_2$ with $h$ can be obtained if $\overline{G}_2$ is known as a function of $x_2$ in a solution, i.e., if $\overline{G}_2 = G_2^\circ + RT \ln a_2$ is known from the measurements of the activity. If the solution is ideal, $\partial \overline{G}_2/\partial x_2 = RT \partial \ln x_2/\partial x_2 = RT/x_2$, and (16.28) simplifies into

$$\frac{dh}{dx_2} = \frac{RT}{x_2 g\left(V_2^\circ \rho - M_2\right)} \tag{16.29}$$

where $\overline{V}_2 = V_2^\circ$ has been substituted since the solution is assumed to be ideal. For a nonideal solution $\partial \overline{G}_2/\partial x_2 = RT \partial \ln a_2/\partial x_2$ must be substituted in (16.28) to obtain the corresponding equation.

## Centrifugal Force

When a solution in a tube is placed in a centrifuge, it is subjected to very large values of acceleration under rapid rotations, and the bottom of the tube is away from the axis of rotation. The centrifugal acceleration $g$, replaces the gravitational acceleration because the solution is held horizontally. Let $r$ be the distance of a portion of solution from the axis of rotation, and $\omega$, the angular velocity in radians per second so that $g$ is given by $g = \omega^2 r$. Since $h = h_\circ - r$, where $h_\circ$ is the location of the axis measured from the bottom of the sample, it is obvious that $dh = -dr$. Substitution of $g = \omega^2 r$ and $dh = -dr$ in (16.29) gives

$$\frac{dr}{dx_2} = \frac{-RT}{x_2 \omega^2 r \left(V_2^\circ \rho - M_2\right)}; \qquad (r = \text{distance from axis of rotation})$$

Separation of variables and integration from $r_1$ to $r_2$, $(r_1 < r_2)$, as the distances from the center of rotation gives

$$\ln \left[\frac{x_2(\text{at } r_2)}{x_2(\text{at } r_1)}\right] = \left(\frac{M_2 - V_2^\circ \rho}{2RT}\right)(r_2^2 - r_1^2)\omega^2 \tag{16.30}$$

For dilute liquid solutions, $\rho \approx M_1/V_1^\circ$, and if the solute is heavy and dense, i.e., $M_2 > M_1$, $V_2^\circ \approx V_1^\circ$; hence, the right side of (16.30) is positive, and the concentration of $x_2$ increases toward the bottom of the containing tube and under favorable conditions component 2 may precipitate out.

Equation (16.28) is also applicable to ideal gas mixtures. Further simplification of (16.28) is possible prior to integration since the partial molar volumes of the components are equal. (See Problem 16.3.)

## Electric and Magnetic Fields

We describe the effects of electric and magnetic fields very briefly because these forces are in the domains of physics and chemical physics. Detailed treatments are given in other books.* Our purpose is to show here the similarities of these fields with the gravitational field.

An electric field exists between two parallel plates when an electric potential $E$ is applied to the plates. When a dielectric substance is placed between the plates a change in the charge $q$ in Coulombs occurs. The differential of the energy of charging is then equal to $E\,dq$. The charging process does work on the dielectric system so that in the absence of other types of work, but including the work of expansion, the first law of thermodynamics can be written as

$$dU = T\,dS - P\,dV + E\,dq \tag{16.31}$$

When we add the total differential of $(PV - TS)$ this equation becomes

$$dG = V\,dP - S\,dT + E\,dq \tag{16.32}$$

From this equation, it is evident that

$$(\partial G/\partial q)_{P,T} = E \tag{16.33}$$

Again $G = G(P, T, q, x_1, x_2 \ldots)$, and similarly for $\overline{G}_i$. All the related thermodynamic properties, as affected by an electric field, can be obtained from (16.32). Likewise, if a substance has a magnetic moment per mole $\mathcal{M}$ in a magnetic field $\mathcal{H}$ and it is subjected to changes in $\mathcal{M}$ by changes in $\mathcal{H}$, the magnetic work is expressed by $\mathcal{H}d\mathcal{M}$ and the first law of thermodynamics is again written as

$$dU = T\,dS - P\,dV + \mathcal{H}\,d\mathcal{M} \tag{16.34}$$

Again $(\partial G/\partial \mathcal{M})_{P,T} = \mathcal{H}$, and all the related thermodynamic properties can be obtained as in the electric field.

## PART III[†] LONG-RANGE ORDER

The Gibbs energy of a given multicomponent system is a function of its variables of state so that $G = G(P, T, x_1, x_2, \ldots)$. This function is continuous for the first-order phase transitions but its derivative, with respect to one of its variables, becomes discontinuous upon a first-order transition. A first-order transition is accompanied with a discontinuity in the first derivatives of $G$; thus, with respect to $T$,

$$\frac{\partial G}{\partial T} = -S; \quad \text{or} \quad \frac{\partial(G/T)}{\partial(1/T)} = H \tag{16.35}$$

would show a discontinuity in the entropy or enthalpy. At a second-order transition, the second derivatives of $G$ exhibit discontinuities; i.e.,

$$\frac{\partial^2 G}{\partial T^2} = -\frac{C_p}{T}; \quad \text{or} \quad \frac{\partial^2(G/T)}{\partial(1/T)\partial T} = C_p \tag{16.36}$$

---

*See for example Gen. Ref. (22), and H. Eyring, D. Henderson, B. J. Stover, and E. M. Eyring, "Statistical Mechanics and Dynamics," John Wiley & Sons (1982).
[†]Part III has been adapted from Gen. Ref. (15), with permission from the Plenum Press.

In summary, $S$, or $H$, is discontinuous for the first-order phase transitions and $C_p$ is discontinuous for the second-order phase transitions.

The order-disorder phenomena are second-order phase transitions in which discontinuities are observed in $C_p$. In pure elements, such as iron, order-disorder in the vicinity of 1043 K produces magnetic order-disorder, but in alloys such as Cu-Zn at 742 K, the arrangement in the crystal lattice causes order-disorder. Superconductivity is another example for second-order transition.

## Ordering and Clustering

The number of AB bonds in a substitutional binary solution is more than the random number $N_A \bar{z} x_A x_B$ [see Chapter XI, remarks prior to (11.51)] when the exchange energy $W = (\bar{z}/2)(2e_{AB} - e_{AA} - e_{BB})$ is negative, and less when $W$ is positive. Very large negative values of $W/kT$ may cause formation of one or more intermetallic compounds in the same system. For positive values of $W/kT$, clustering, or the association of like atoms, occurs, and in fact, when $W/kT$ is sufficiently large, the clustering causes separation of a phase into two phases. On the other hand, if $W/kT$ is negative to some optimal degree that cannot always be quantified, A atoms may form nearly entirely A-B bonds with virtually no A-A bonds if $x_A \leq x_B$. This is possible when certain sites in the crystal lattice are occupied by A atoms and others by B atoms. In such substitutional solid solutions, we shall see that the A atoms may be regarded as having formed their own lattice, interpenetrating the lattice formed by the B atoms. This phenomenon is known as the *long-range* order. Since $W$ itself may not often be sufficiently negative, the long-range order is expected to occur at sufficiently low temperatures; however, the kinetic energy of atoms may not often be sufficient to move the atoms into their ordered state below ambient temperatures. The clustering may occur in liquid and solid alloys but the long-range order may occur only in solid solutions; therefore, this section is largely concerned with solid solutions.

## Order-Disorder in Binary Alloys

We consider order-disorder phenomena in binary substitutional alloys wherein the lattice structure of solid solution is retained upon transition from the ordered state to the disordered state. It will be assumed that minor changes in crystal structure upon transformation are not of primary importance. This type of transition is therefore a second-order transition, which may be called the *Curie-type transition*, or the *coherent transition*, but we retain the term order-disorder transition as the more descriptive term for our purposes. For a body-centered cubic (bcc) structure, shown in Fig. 16.1(a), the random distribution of A and B atoms for an equimolar or equiatomic solution exhibits no particular preference in the locations of A and B atoms. Each lattice site in such a solid solution may therefore be occupied by either A or B whether the solution is ideal or nonideal. However, if all the A atoms occupy only the corners of the cube as shown in Fig. 16.1(b), and all the B atoms, the body-centered sites, then we have a perfectly ordered structure as in beta brass, Cu-Zn.

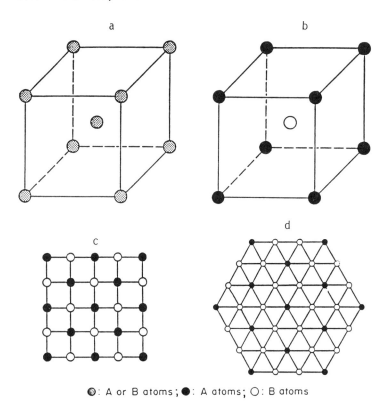

◑: A or B atoms; ●: A atoms; ○: B atoms

**Fig. 16.1** (a) Disordered structure in body-centered cubic crystals; each site is occupied randomly by either type of atom. (b) Ordered AB type of alloy; corner atoms occupied by A atoms form cubes interpenetrating cubes formed by B atoms occupying body-centered positions. (c) ZnS-type tetrahedral ordered structure projected on two-dimensional coordinates; A = Zn, B = S. (d) Two-dimensional $AB_2$-type hexagonal close-packaed structure. Each A atom is entirely surrounded by B atoms, and every third atom in any direction is an A atom.

We designate the ordered sites by a for A atoms, and by b for B atoms when the ordering is perfect, but when the ordering is partial, the sites still retain their identities, even if they are occupied by a number of wrong atoms. We note that 1/8 of each corner atom belongs to each cube, and that the nearest neighbors lie along the diagonals of the cube by 0.866 of the cube edge. All the A atoms may be regarded as forming their cubic lattice by interpenetrating a similar cubic lattice formed by the B atoms, and for this reason an ordered structure is said to form a superlattice. Tetrahedral crystals of the type exemplified by ZnS, may be projected on two dimensions as in Fig. 16.1(c) to illustrate this type of order in which each atom has four unlike nearest neighbors. The picture is somewhat different for a close-packed two-dimensional ordered structure for $AB_2$-type phase as shown in Fig. 16.1(d).

The existence of order can be determined semiquantitatively, or in favorable cases quantitatively, by X-ray diffraction. New diffraction lines are observed when ordering initiates in a random alloy. The intensities of the new lines are then a measure of the degree of ordering. Electron and neutron diffraction methods, and some of the physical properties also determine the extent of order with various degrees of accuracy.

The atomic ratio of A to B may not always be a ratio of whole numbers, and the numbers of a- and b-lattice sites may or may not be determined by the crystal structure. As the composition of a perfectly ordered phase is changed, the degree of ordering also changes because the excess number of atoms of one of the components must be accommodated in the wrong sites. Atomic vibrations increase with increasing temperature to such an extent that, finally at a critical temperature, the degree of order disappears and the solution becomes disordered in the sense that no measurable degree of long-range order can be observed.

## Long-Range Order Parameter

Thermodynamic treatment of ordering in alloys requires a convenient definition of the *long-range order parameter*, $r$, as follows:

$$\text{A atoms on a-sites} = N_A^a = 0.5 \times \text{all A atoms} \times (1+r) \tag{16.37}$$

For equal numbers of a- and b-sites, each designated by $L$, and for equal numbers of A and B atoms, or $x_A = x_B$, this equation is written as

$$N_A^a = x_A(1+r)L = 0.5(1+r)L \tag{16.38}$$

Thus, if all A atoms are distributed randomly, $r$ is zero and the probability of finding A atoms on a-sites is equal to its mole fraction. The parameter $r$ is unity for a perfectly ordered alloy and all the A atoms are on the a-sites, i.e., $N_A^a = N_A = L$. The sets of a- and b-sites are sometimes called the a- and b-sublattice respectively. Each site has $\bar{z}$ neighbors, $\bar{z}$ being the coordination number.

The parameter $r$ for the preceding simple case with (16.38) and $x_A = 0.5$ leads to

$$\left.\begin{array}{l}\text{A atoms on a-sites} = N_A^a = 0.5(1+r)L \\ \text{A atoms on b-sites} = N_A^b = 0.5(1-r)L \\ \text{B atoms on b-sites} = N_B^b = 0.5(1+r)L \\ \text{B atoms on a-sites} = N_B^a = 0.5(1-r)L\end{array}\right\} \tag{16.39}$$

The ordering parameter $r$ varies between zero and unity, i.e.,

$$0 \leq r \leq 1 \tag{16.40}$$

When $r$ is zero, the solution is disordered in the following discussions.

## Gorsky and Bragg-Williams (GBW) Approximation

The simplest treatment of order-disorder in alloys is the zeroth approximation due to GBW [see Gen. Ref. (15)]. This model is similar to the zeroth approximation to the regular solutions. For simplicity, we consider the body-centered cubic (bcc) lattice in our discussions with $N_A = N_B = L$ as in the Cu-Zn system. The nearest neighbors to an atom on the body-centered site are the corner sites, and since $\bar{z} = 8$, we take into account the eight bonds emanating from an a-site to the neighboring b-sites. The fraction of $L$ sites occupied by A atoms in $N_A^a/L$ and this is also equal to the probability of finding an A atom on the a-sites. The probability of finding an A atom on the b-sites is $N_A^b/L$ so that the probability of finding an AA pair from (16.39) is

$$\text{Probability of AA} = \frac{N_A^a}{L}\frac{N_A^b}{L} = (1-r^2)/4 \tag{16.41}$$

Likewise,

$$\text{Probability of BB} = (1-r^2)/4 \tag{16.42}$$

The probability of AB pairs is the sum of two probabilities, i.e., the probability of A on a-sites times B on b-sites, i.e., $0.25(1+r)^2$, plus the probability of A on b-sites times B on a-sites, i.e., $0.25(1-r)^2$; hence,

$$\text{Probability of AB} = (1+r)^2/4 + (1-r)^2/4 = (1+r^2)/2 \tag{16.43}$$

Multiplication of each probability term by the total number of bonds, $\bar{z}L$, gives the total number of each type of bond, and multiplication of each set of bonds by its bond energy* $e_{ij}$ gives the total energy of the alloy as in the zeroth approximation to the regular solutions, i.e.,

$$U \approx H(r) \approx (\bar{z}L/4)(1-r^2)e_{AA} + (\bar{z}L/4)(1-r^2)e_{BB}$$
$$+ (\bar{z}L/2)(1+r^2)e_{AB}$$
$$= (\bar{z}L/4)(e_{AA} + e_{BB} + 2e_{AB}) + (\bar{z}Lr^2/4)(2e_{AB} - e_{AA} - e_{BB}) \tag{16.44}$$

For the disordered solution, $r$ is zero and (16.44) becomes

$$H(r=0) = (\bar{z}L/4)(e_{AA} + e_{BB} + 2e_{AB}) \tag{16.45}$$

For convenience, we rewrite the exchange energy of (16.44) as

$$W = (\bar{z}/2)(2e_{AB} - e_{AA} - e_{BB}) \tag{16.46}$$

Equations (16.44–16.46) give

$$H(r) - H(r=0) = Lr^2W/2 \tag{16.47}$$

It is assumed that the atoms in each group as given in each of the equations (16.39) can be rearranged randomly and this is the reason that the model is

---

*We use $e_{ij}$ here, instead of $E_{ij}$ of XI, in conformity with the notation of Gen. Ref. (15).

called the zeroth approximation. The resulting distribution for all the atoms on the a-sites is

$$D_a = \frac{L!}{[(1+r)L/2]![(1-r)L/2]!} \tag{16.48}$$

The distribution $D_b$ for the atoms on the b-sites is identical, i.e., $D_a = D_b$; hence, the overall distribution $D_r$ is

$$D_r = D_a D_b = \left[\frac{L!}{[(1+r)L/2]![(1-r)L/2]!}\right]^2 \tag{16.49}$$

The entropy of solution is $S = k \ln D_r$; hence, by using the Stirling approximation, we obtain

$$S(r) = 2k \ln D_r = Lk[2\ln 2 - (1+r)\ln(1+r) - (1-r)\ln(1-r)] \tag{16.50}$$

Here, $S(r)$ is the entropy of the solution. For $r = 0$, $S(r = 0) = 2Lk \ln 2$ as expected since the distribution of atoms is then random; therefore,

$$S(r) - S(r = 0) = -Lk[(1+r)\ln(1+r) + (1-r)\ln(1-r)] \tag{16.51}$$

Substitution of (16.47) and (16.51) into the Gibbs energy yields

$$G(r) - G(r = 0) = \frac{Lr^2 W}{2} + LkT[(1+r)\ln(1+r) + (1-r)\ln(1-r)] \tag{16.52}$$

The equilibrium state of the alloy corresponds to the value of $r$ which makes $\partial G/\partial r$ zero:

$$\frac{\partial G}{\partial r} = LrW + LkT[\ln(1+r) - \ln(1-r)] = 0 \tag{16.53}$$

At the critical temperature, $T_c$, the second derivative of $G(r)$ with respect to $r$ is zero as $r$ approaches zero, i.e., the order disappears; thus,

$$\frac{\partial^2 G(r)}{\partial r^2} = LW + LkT\left(\frac{1}{1+r} + \frac{1}{1-r}\right) = 0; \quad \text{for } r = 0 \text{ and } T = T_c \tag{16.54}$$

Substitution of $r \to 0$ yields

$$T_c = -W/2k; \quad (W < 0) \tag{16.55}$$

The critical temperature must be positive; therefore, $W$ must be negative. Substitution of $W = -2kT_c$ [obtained from (16.55)], in (16.53) gives

$$\frac{1+r}{1-r} = \exp\left(\frac{-rW}{kT}\right) = \exp\left(\frac{2rT_c}{T}\right) \tag{16.56}$$

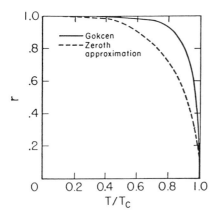

**Fig. 16.2** Variation of long-range order parameter with temperature. $T_c$ is 742 K. The solid curve shows the first approximation, which is much closer to experimental results, cf. Gen. Ref. (15).

A relationship equivalent to (16.56) was first derived by Gorsky (1928)* and later by Bragg and Williams. This equation can be rearranged to obtain

$$r = \frac{\exp\left(\frac{-rW}{kT}\right) - 1}{\exp\left(\frac{-rW}{kT}\right) + 1} \equiv \tanh\left(\frac{-rW}{2kT}\right) = \tanh\left(\frac{rT_c}{T}\right) \qquad (16.57)$$

where the identity sign defines tanh. The numerical computation of $T_c/T$ from this equation is simple. For this purpose, it is necessary to select an arbitrary value of $\alpha = rT_c/T$ and read the value of tanh $\alpha$ from an appropriate calculator or obtain it from the ratio immediately after the first equal sign in (16.57). Then, tanh $\alpha$ is the value of $r$, and this value substituted in $\alpha = rT_c/T$ yields the value of $T$ from the experimentally known value of the critical temperature. Thus, for the equiatomic Cu-Zn alloy having $T_c = 742$ K with $\alpha = 0.7$, tanh $0.7 = 0.6044 = r$, and then $T = rT_c/\alpha = 640.66$ K. The equilibrium values of $r$ and the corresponding temperature are plotted in Fig. 16.2, as the reduced temperature $T/T_c$ versus $r$, with the broken curve.

## Heat Capacity

The heat capacity $C (=C_p)$ can be obtained by differentiation of (16.47) with respect to temperature:

$$\frac{\partial H(r)}{\partial T} - \frac{\partial H(r=0)}{\partial T} = C(r) - C(r=0) = C^E = LrW\frac{\partial r}{\partial T} \qquad (16.58)$$

*Original references are cited in Gen. Ref. (15).

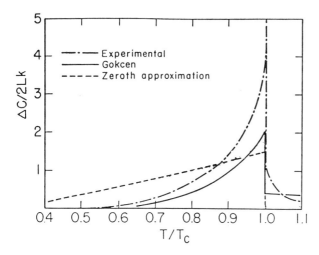

**Fig. 16.3** Variation of heat capacity with temperature for equiatomic Cu-Zn alloy. $T_c$ is 742 K, and measured heat capacities of pure components were subtraced from that of alloy to obtain experimental curve. $\Delta C$ is the same as the excess heat capacity $C^E$ as shown in text. The solid curve is for the first approximation dicussed in Gen. Ref. (15).

where $LW = -2LkT_c = -RT_c$ from (16.57). The expression for $\partial r/\partial T = dr/dT$ can be obtained by taking the logarithm of equation (16.57) and then differentiating it; the result, substituted in (16.58), is

$$\Delta C = C(r) - C(r = 0) = \frac{Rr^2 T_c^2 (1 - r^2)}{T^2 - T_c T(1 - r^2)} \quad (16.59)$$

where $C(r = 0)$ is the heat capacity of disordered alloy. The heat capacity $C(r = 0)$ of a random disordered solution is the same as the heat capacities of its component elements. The values of $r$ and $T$ are related by (16.57) and shown in Fig. 16.2; therefore, for each value of $r$, $T$ must be computed from (16.57) and then the result must be substituted in (16.59) to obtain $\Delta C$. The result for the previous example, for which $T = 640.66$ K, $T_c = 742$ K, or $T/T_c = 0.8634$, and $r = 0.6044$, is simply $\Delta C/R = 1.174$. Above the critical temperature, $T > T_c$, $r$ is zero and $\Delta C$ rapidly approaches zero because of $r^2$ in the numerator of (16.59). The values of $\Delta C$ calculated from (16.59) are plotted in Fig. 16.3. We have thus far considered the equal numbers of A and B atoms distributed on equal numbers of a- and b-sites. The unequal numbers of A and B atoms on equal numbers of a- and b-sites are considered in Gen. Ref. (15) together with the first approximation represented by the solid lines in Fig. 16.3.

## PROBLEMS

16.1 The surface tension of Hg(l) is 0.470 N m$^{-1}$ at 100°C [Gen. Ref. (25)]. Calculate the radius of Hg-droplets having 20% higher vapor pressure than the flat-surfaced Hg.

*Ans.: $r = 2.45 \times 10^{-8}$ m.*

16.2 Give the composition of water containing sucrose at 25.4°C, whose vapor is in equilibrium with pure water droplets at 25°C and 100 nanometers in radius. The molar volume of pure water is 18 cm$^3$, and its vapor pressure, 3,167 Pa at 25°C, and 3,243 Pa at 25.4°C.

*Ans.: Mole fraction of water is 0.987.*

16.3 Derive an equation similar to (16.30) for an ideal gas mixture.

16.4 An ideal binary solution is formed from $M_2 = 100 = 5M_1$, with $V_1^\circ = V_2^\circ = V$ (solution at 300 K), and $x_2 = 0.40$. Find the concentration of component 2 at the bottom of the test tube containing a sample of this solution, 10 cm high, when it is centrifuged at 150,000 revolutions per minute. The top of the solution is 15 cm from the center of the centrifuge.

*Ans.: $x_2 \approx 1$ at the bottom.*

16.5 Derive the following relationships in detail:

$$\left(\frac{\partial \mathcal{S}}{\partial A}\right)_{P,T} = -\left(\frac{\partial \sigma}{\partial T}\right)_{P,A}, \quad \left(\frac{\partial S}{\partial \mathcal{M}}\right)_{T,P} = -\left(\frac{\partial \mathcal{H}}{\partial T}\right)_{P,\mathcal{M}}$$

$$\left(\frac{\partial S}{\partial q}\right)_{P,T} = -\left(\frac{\partial E}{\partial T}\right)_{P,q}, \quad \left(\frac{\partial S}{\partial \mathcal{M}}\right)_{U,V} = -\mathcal{H}/T$$

$$\left(\frac{\partial \mathcal{V}}{\partial A}\right)_{S,V} = \sigma$$

16.6 Calculate $r$ and $\Delta C_p$ at $T/T_c = 0.8, 0.9, 0.95$ and $0.98$ for Cu-Zn alloy in Figs. 16.2 and 16.3.

*Answers for $T/T_c = 0.8$ are $r = 0.7104$ and $\Delta C/R = 1.026$.*

APPENDIX I

# GENERAL REFERENCES

1. *JANAF Thermochemical Tables*, 3rd edition, by M. W. Chase, Jr., C. A. Davies, J. R. Downey, Jr., D. J. Frurip, R. A. McDonald, and A. N. Syverud, published by Am. Chem. Soc. and Am. Inst. Phys. for NBS (now NIST), Washington, DC (1986).
2. *Thermodynamic Properties of Elements and Oxides*, U. S. Bur. Mines Bull. 672 (1982); *Thermodynamic Properties of Halides*, Bull. 674 (1984); *Thermodynamic Properties of Carbides, Nitrides, and Other Selected Substances*, Bull. 696 (1995), all by L. B. Pankratz. *Thermodynamic Properties of Sulfides*, Bull. 689, by L. B. Pankratz, A. D. Mah, and S. W. Watson (1987). *Thermodynamic Data for Mineral Technology*, Bull. 677, by L. B. Pankratz, J. M. Stuve, and N. A. Gokcen (1984). (U. S. Government Printing Office, Washington, DC).
3. *Thermochemical Data of Pure Substances*, by I. Barin, in collaboration with F. Sauert, E. Schultze-Rhonhof, and W. S. Sheng, VCH Publishers (1992).
4. *Materials Thermochemistry*, sixth edition, by O. Kubaschewski and C. B. Alcock, Elsevier Publ. Co. (1993).
5. *Thermodynamic Properties of Minerals and Related Substances at* 298.15 $K$ *and* 1 *Bar* ($10^5$ *Pascals*)*Pressure and at Higher Temperatures*, by R. A. Robie and B. S. Hemingway, Geological Survey Bull. (to be published in 1995), previously, Bull. 1452, U. S. Government Printing Office, Washington, (1978).
6. *Table of Thermodynamic and Transport Properties*, by J. Hilsenrath, C. W. Beckett, W. S. Benedict, L. Fano, H. J. Hoge, J. F. Masi, R. L. Nuttall, Y. S. Touloukian, and H. W. Wooley, Pergamon Press (now Elsevier) (1960).
7. *The Chemical Thermodynamics of Organic Compounds*, by D. R. Stull, E. F. Westrum, Jr., and G. C. Sinke, R. E. Krieger Publishing Co. (1987).
8. Data books published by the *Center for Information and Numerical Data Analysis and Synthesis* (*CINDAS*), Purdue University, West Lafayette, IN, in 13 volumes; ongoing effort. Earlier publications in 6 volumes are *Thermophysical Properties of Matter*, by Y. S. Touloukian et al., Macmillan Co. (1967).
9. *Selected Values of the Thermodynamic Properties of Binary Alloys*, by R. Hultgren, P. D. Desai, D. T. Hawkins, M. Gleiser, and K. K. Kelley, ASM, Materials Park, OH (1973).
10. *Bulletin of Alloy Phase Diagrams*, which became *Journal of Phase Equilibria in 1991*; published by ASM International, Materials Park, OH.
11. *Landolt-Bornstein Tables*, Springer-Verlag; ongoing publication.

12. *Journal of Physical and Chemical Reference Data*, Am. Chem. Soc.
13. *CALPHAD*, a journal published by Pergamon Press (now Elsevier).
14. *International Data Series, Selected Data on Mixtures*, in 22 volumes, ongoing effort, Thermodynamic Research Center, Texas A&M University, College Station, TX.
15. *Statistical Thermodynamics of Alloys*, by N. A. Gokcen, Plenum Press (1986).
16. *Solubilities*, W. F. Linke, 4th edition, Books on Demand, Ann Arbor, MI, vol. 1 (1958), vol. 2 (1966).
17. *Solubilities of Inorganic and Organic Compounds*, by H. Stephen and T. Stephen, editors, (1979); see also *Gas Solubilities*, by W. Garrard, Franklin Book Co. (1980).
18. *Binary Alloy Phase Diagrams*, 2nd edition, by T. B. Massalski, O. Okamoto, P. R. Subramanian, and L. Kacprzak, editors, in 3 volumes, ASM International OH, ongoing effort, (1990).
19. *Properties of Gases and Liquids*, R. C. Reid, J. M. Prausnitz and B. E. Poling, McGraw-Hill (1987).
20. *Phase Diagrams for Ceramists*, in 10 volumes, assessed and edited by the staff of NIST, and Am. Cer. Soc., ongoing effort, published by Am. Cer. Soc.; vol. I (1964) vol. X (1994).
21. *Modern High Temperature Science*, by J. L. Margrave, Humana Press (1984).
22. *Thermodynamics*, by E. A. Guggenheim, Elsevier Publ. Co. (1985).
23. *Vapour Pressure of the Elements*, translated and edited by J. I. Carasso, Academic Press (1963).
24. *Cohesion in Metals*, by F. R. de Boer, R. Boom, W. C. M. Mattens, A. R. Miedema, and A. K. Niessen, North Holland Publ. Co. (1988).
25. *Handbook of Chemistry and Physics*, 74th edition, edited by D. R. Lide, CRC Press (1993–94); see also *CRC Handbook of Solubility Parameters and Other Cohesion Parameters*, by A. F. Barton, CRC Press (1991).
26. *Smithells Metals Reference Book*, edited by E. A. Brandes and G. B. Brooks, Butterworth-Heinemann (1992).
27. *Proceedings of the Workshop on Techniques for Measurement of Thermodynamic Properties*, edited by N. A. Gokcen, R. V. Mrazek, and L. B. Pankratz, U. S. Bur. Mines IC-8853 (1981).
28. *Chemical Metallurgy—A Tribute to Carl Wagner*, edited by N. A. Gokcen, Conference Proc., TMS-AIME (1981).
29. *Computerized Metallurgical Databases*, edited by J. R. Cuthill, N. A. Gokcen, and J. E. Morral, TMS (1988); "Computer Software in Chemical and Extractive Metallurgy," edited by W. T. Thompson, F. Ajersch, and G. Ericsson, Pergamon Press (now Elsevier) (1989).

APPENDIX II

# TABLES OF THERMODYNAMIC DATA FOR EXAMPLES AND PROBLEMS IN TEXT

**[Adapted from Gen. Ref. (1) by permission]**

Aluminum Oxide, Alpha ($Al_2O_3$)  $\qquad$ $Al_2O_3$(cr)

| | ($J\,K^{-1}\,mol^{-1}$)* | | | ($kJ\,mol^{-1}$)* | | | |
|---|---|---|---|---|---|---|---|
| $T/K$ | $C_p°$ | $S°$ | $-[G° - H°(T_r)]/T$ | $H° - H°(T_r)$ | $\Delta_f H°$ | $\Delta_f G°$ | $Log\,K_f$ |
| 0 | 0. | 0. | Infinite | −10.020 | −1663.608 | −1663.608 | Infinite |
| 100 | 12.855 | 4.295 | 101.230 | −9.693 | −1668.606 | −1641.642 | 857.506 |
| 200 | 51.120 | 24.880 | 57.381 | −6.500 | −1673.383 | −1612.656 | 421.183 |
| 298.15 | 79.015 | 50.950 | 50.950 | 0. | −1675.692 | −1582.275 | 277.208 |
| 300 | 79.416 | 51.440 | 50.951 | 0.147 | −1675.717 | −1581.696 | 275.398 |
| 400 | 96.086 | 76.779 | 54.293 | 8.995 | −1676.342 | −1550.226 | 202.439 |
| 500 | 106.131 | 99.388 | 61.098 | 19.145 | −1676.045 | −1518.718 | 158.659 |
| 600 | 112.545 | 119.345 | 69.177 | 30.101 | −1675.300 | −1487.319 | 129.483 |
| 700 | 116.926 | 137.041 | 77.632 | 41.586 | −1674.391 | −1456.059 | 108.652 |
| 800 | 120.135 | 152.873 | 86.065 | 53.447 | −1673.498 | −1424.931 | 93.038 |
| 900 | 122.662 | 167.174 | 94.296 | 65.591 | −1672.744 | −1393.908 | 80.900 |
| 1000 | 124.771 | 180.210 | 102.245 | 77.965 | −1693.394 | −1361.437 | 71.114 |
| 1100 | 126.608 | 192.189 | 109.884 | 90.535 | −1692.437 | −1328.286 | 63.075 |
| 1200 | 128.252 | 203.277 | 117.211 | 103.280 | −1691.366 | −1295.228 | 56.380 |
| 1300 | 129.737 | 213.602 | 124.233 | 116.180 | −1690.190 | −1262.264 | 50.718 |
| 1400 | 131.081 | 223.267 | 130.965 | 129.222 | −1688.918 | −1229.393 | 45.869 |
| 1500 | 132.290 | 232.353 | 137.425 | 142.392 | −1687.561 | −1196.617 | 41.670 |
| 1600 | 133.361 | 240.925 | 143.628 | 155.675 | −1686.128 | −1163.934 | 37.999 |
| 1700 | 134.306 | 249.039 | 149.592 | 169.060 | −1684.632 | −1131.342 | 34.762 |
| 1800 | 135.143 | 256.740 | 155.333 | 182.533 | −1683.082 | −1098.841 | 31.888 |
| 1900 | 135.896 | 264.067 | 160.864 | 196.085 | −1681.489 | −1066.426 | 29.318 |
| 2000 | 136.608 | 271.056 | 166.201 | 209.710 | −1679.858 | −1034.096 | 27.008 |
| 2100 | 137.319 | 277.738 | 171.354 | 223.407 | −1678.190 | −1001.849 | 24.920 |
| 2200 | 138.030 | 284.143 | 176.336 | 237.174 | −1676.485 | −969.681 | 23.023 |
| 2300 | 138.741 | 290.294 | 181.158 | 251.013 | −1674.743 | −937.593 | 21.293 |
| 2327.000 | 138.934 | 291.914 | 182.434 | 254.761 | Alpha ↔ Liquid | | |
| 2400 | 139.453 | 296.214 | 185.829 | 264.922 | −1672.963 | −905.582 | 19.709 |
| 2500 | 140.206 | 301.922 | 190.360 | 278.905 | −1671.142 | −873.645 | 18.254 |
| 2600 | 140.959 | 307.435 | 194.757 | 292.963 | −1669.279 | −841.781 | 16.912 |
| 2700 | 141.754 | 312.770 | 199.030 | 307.098 | −1667.369 | −809.990 | 15.670 |
| 2800 | 142.591 | 317.940 | 203.185 | 321.315 | −2253.212 | −776.335 | 14.483 |
| 2900 | 143.511 | 322.960 | 207.229 | 335.620 | −2249.002 | −723.665 | 13.035 |
| 3000 | 144.474 | 327.841 | 211.168 | 350.019 | −2244.729 | −671.139 | 11.686 |

*Enthalpy Reference Temperature = $T_r$ = 298.15 K; Standard State Pressure = $p°$ = 0.1 MPa.

## Carbon (C)

C₁(ref)

| T/K | $C_p^\circ$ (J K⁻¹ mol⁻¹)* | $S^\circ$ | $-[G^\circ - H^\circ(T_r)]/T$ | $H^\circ - H^\circ(T_r)$ (kJ mol⁻¹)* | $\Delta_f H^\circ$ | $\Delta_f G^\circ$ | Log $K_f$ |
|---|---|---|---|---|---|---|---|
| 0 | 0. | 0. | Infinite | −1.051 | 0. | 0. | 0. |
| 100 | 1.674 | 0.952 | 10.867 | −0.991 | 0. | 0. | 0. |
| 200 | 5.006 | 3.082 | 6.407 | −0.665 | 0. | 0. | 0. |
| 250 | 6.816 | 4.394 | 5.871 | −0.369 | 0. | 0. | 0. |
| 298.15 | 8.517 | 5.740 | 5.740 | 0. | 0. | 0. | 0. |
| 300 | 8.581 | 5.793 | 5.741 | 0.016 | 0. | 0. | 0. |
| 350 | 10.241 | 7.242 | 5.851 | 0.487 | 0. | 0. | 0. |
| 400 | 11.817 | 8.713 | 6.117 | 1.039 | 0. | 0. | 0. |
| 450 | 13.289 | 10.191 | 6.487 | 1.667 | 0. | 0. | 0. |
| 500 | 14.623 | 11.662 | 6.932 | 2.365 | 0. | 0. | 0. |
| 600 | 16.844 | 14.533 | 7.961 | 3.943 | 0. | 0. | 0. |
| 700 | 18.537 | 17.263 | 9.097 | 5.716 | 0. | 0. | 0. |
| 800 | 19.827 | 19.826 | 10.279 | 7.637 | 0. | 0. | 0. |
| 900 | 20.824 | 22.221 | 11.475 | 9.672 | 0. | 0. | 0. |
| 1000 | 21.610 | 24.457 | 12.662 | 11.795 | 0. | 0. | 0. |
| 1100 | 22.244 | 26.548 | 13.831 | 13.989 | 0. | 0. | 0. |
| 1200 | 22.766 | 28.506 | 14.973 | 16.240 | 0. | 0. | 0. |
| 1300 | 23.204 | 30.346 | 16.085 | 18.539 | 0. | 0. | 0. |
| 1400 | 23.578 | 32.080 | 17.167 | 20.879 | 0. | 0. | 0. |
| 1500 | 23.904 | 33.718 | 18.216 | 23.253 | 0. | 0. | 0. |
| 1600 | 24.191 | 35.270 | 19.234 | 25.658 | 0. | 0. | 0. |
| 1700 | 24.448 | 36.744 | 20.221 | 28.090 | 0. | 0. | 0. |
| 1800 | 24.681 | 38.149 | 21.178 | 30.547 | 0. | 0. | 0. |
| 1900 | 24.895 | 39.489 | 22.107 | 33.026 | 0. | 0. | 0. |
| 2000 | 25.094 | 40.771 | 23.008 | 35.525 | 0. | 0. | 0. |
| 2100 | 25.278 | 42.000 | 23.883 | 38.044 | 0. | 0. | 0. |
| 2200 | 25.453 | 43.180 | 24.734 | 40.581 | 0. | 0. | 0. |
| 2300 | 25.618 | 44.315 | 25.561 | 43.134 | 0. | 0. | 0. |
| 2400 | 25.775 | 45.408 | 26.365 | 45.704 | 0. | 0. | 0. |
| 2500 | 25.926 | 46.464 | 27.148 | 48.289 | 0. | 0. | 0. |
| 2600 | 26.071 | 47.483 | 27.911 | 50.889 | 0. | 0. | 0. |
| 2700 | 26.212 | 48.470 | 28.654 | 53.503 | 0. | 0. | 0. |
| 2800 | 26.348 | 49.426 | 29.379 | 56.131 | 0. | 0. | 0. |
| 2900 | 26.481 | 50.353 | 30.086 | 58.773 | 0. | 0. | 0. |
| 3000 | 26.611 | 51.253 | 30.777 | 61.427 | 0. | 0. | 0. |
| 3100 | 26.738 | 52.127 | 31.451 | 64.095 | 0. | 0. | 0. |
| 3200 | 26.863 | 52.978 | 32.111 | 66.775 | 0. | 0. | 0. |
| 3300 | 26.986 | 53.807 | 32.756 | 69.467 | 0. | 0. | 0. |
| 3400 | 27.106 | 54.614 | 33.387 | 72.172 | 0. | 0. | 0. |
| 3500 | 27.225 | 55.401 | 34.005 | 74.889 | 0. | 0. | 0. |
| 3600 | 27.342 | 56.170 | 34.610 | 77.617 | 0. | 0. | 0. |
| 3700 | 27.459 | 56.921 | 35.203 | 80.357 | 0. | 0. | 0. |
| 3800 | 27.574 | 57.655 | 35.784 | 83.109 | 0. | 0. | 0. |
| 3900 | 27.688 | 58.372 | 36.354 | 85.872 | 0. | 0. | 0. |
| 4000 | 27.801 | 59.075 | 36.913 | 88.646 | 0. | 0. | 0. |

*Enthalpy Reference Temperature = $T_r$ = 298.15 K; Standard State Pressure = $p^\circ$ = 0.1 MPa.

## Carbon (C)

$C_1(g)$

| T/K | $C_p^\circ$ (J K$^{-1}$ mol$^{-1}$)* | $S^\circ$ | $-[G^\circ - H^\circ(T_r)]/T$ | $H^\circ - H^\circ(T_r)$ (kJ mol$^{-1}$)* | $\Delta_f H^\circ$ | $\Delta_f G^\circ$ | Log $K_f$ |
|---|---|---|---|---|---|---|---|
| 0 | 0. | 0. | Infinite | −6.536 | 711.185 | 711.185 | Infinite |
| 100 | 21.271 | 135.180 | 176.684 | −4.150 | 713.511 | 700.088 | −365.689 |
| 200 | 20.904 | 149.768 | 160.007 | −2.048 | 715.287 | 685.950 | −179.152 |
| 250 | 20.861 | 154.427 | 158.443 | −1.004 | 716.035 | 678.527 | −141.770 |
| 298.15 | 20.838 | 158.100 | 158.100 | 0. | 716.670 | 671.244 | −117.599 |
| 300 | 20.838 | 158.228 | 158.100 | 0.039 | 716.693 | 670.962 | −116.825 |
| 350 | 20.824 | 161.439 | 158.354 | 1.080 | 717.263 | 663.294 | −98.991 |
| 400 | 20.815 | 164.219 | 158.917 | 2.121 | 717.752 | 655.550 | −85.606 |
| 450 | 20.809 | 166.671 | 159.645 | 3.162 | 718.165 | 647.749 | −75.189 |
| 500 | 20.804 | 168.863 | 160.459 | 4.202 | 718.507 | 639.906 | −66.851 |
| 600 | 20.799 | 172.655 | 162.185 | 6.282 | 719.009 | 624.135 | −54.336 |
| 700 | 20.795 | 175.861 | 163.916 | 8.362 | 719.315 | 608.296 | −45.392 |
| 800 | 20.793 | 178.638 | 165.587 | 10.441 | 719.474 | 592.424 | −38.681 |
| 900 | 20.792 | 181.087 | 167.175 | 12.520 | 719.519 | 576.539 | −33.461 |
| 1000 | 20.791 | 183.278 | 168.678 | 14.600 | 719.475 | 560.654 | −29.286 |
| 1100 | 20.791 | 185.259 | 170.097 | 16.679 | 719.360 | 544.777 | −25.869 |
| 1200 | 20.793 | 187.068 | 171.437 | 18.758 | 719.188 | 528.913 | −23.023 |
| 1300 | 20.796 | 188.733 | 172.704 | 20.837 | 718.968 | 513.066 | −20.615 |
| 1400 | 20.803 | 190.274 | 173.905 | 22.917 | 718.709 | 497.237 | −18.552 |
| 1500 | 20.814 | 191.710 | 175.044 | 24.998 | 718.415 | 481.427 | −16.765 |
| 1600 | 20.829 | 193.053 | 176.128 | 27.080 | 718.092 | 465.639 | −15.202 |
| 1700 | 20.850 | 194.317 | 177.162 | 29.164 | 717.744 | 449.871 | −13.823 |
| 1800 | 20.878 | 195.509 | 178.148 | 31.250 | 717.373 | 434.124 | −12.598 |
| 1900 | 20.912 | 196.639 | 179.092 | 33.340 | 716.984 | 418.399 | −11.503 |
| 2000 | 20.952 | 197.713 | 179.996 | 35.433 | 716.577 | 402.694 | −10.517 |
| 2100 | 20.999 | 198.736 | 180.864 | 37.530 | 716.156 | 387.010 | −9.626 |
| 2200 | 21.052 | 199.714 | 181.699 | 39.633 | 715.722 | 371.347 | −8.817 |
| 2300 | 21.110 | 200.651 | 182.503 | 41.741 | 715.277 | 355.703 | −8.078 |
| 2400 | 21.174 | 201.551 | 183.278 | 43.855 | 714.821 | 340.079 | −7.402 |
| 2500 | 21.241 | 202.417 | 184.026 | 45.976 | 714.357 | 324.474 | −6.780 |
| 2600 | 21.313 | 203.251 | 184.750 | 48.103 | 713.884 | 308.888 | −6.206 |
| 2700 | 21.387 | 204.057 | 185.450 | 50.238 | 713.405 | 293.321 | −5.675 |
| 2800 | 21.464 | 204.836 | 186.129 | 52.381 | 712.920 | 277.771 | −5.182 |
| 2900 | 21.542 | 205.591 | 186.787 | 54.531 | 712.429 | 262.239 | −4.723 |
| 3000 | 21.621 | 206.322 | 187.426 | 56.689 | 711.932 | 246.723 | −4.296 |

*Enthalpy Reference Temperature = $T_r$ = 298.15 K; Standard State Pressure = $p^\circ$ = 0.1 MPa.

## Methane (CH₄)   $C_1H_4(g)$

| T/K | $C_p^\circ$ (J K$^{-1}$ mol$^{-1}$)* | $S^\circ$ | $-[G^\circ - H^\circ(T_r)]/T$ | $H^\circ - H^\circ(T_r)$ (kJ mol$^{-1}$)* | $\Delta_f H^\circ$ | $\Delta_f G^\circ$ | Log $K_f$ |
|---|---|---|---|---|---|---|---|
| 0      | 0.      | 0.      | Infinite | −10.024 | −66.911 | −66.911 | Infinite |
| 100    | 33.258  | 149.500 | 216.485  | −6.698  | −69.644 | −64.353 | 33.615 |
| 200    | 33.473  | 172.577 | 189.418  | −3.368  | −72.027 | −58.161 | 15.190 |
| 250    | 34.216  | 180.113 | 186.829  | −1.679  | −73.426 | −54.536 | 11.395 |
| 298.15 | 35.639  | 186.251 | 186.251  | 0.      | −74.873 | −50.768 | 8.894 |
| 300    | 35.708  | 186.472 | 186.252  | 0.066   | −74.929 | −50.618 | 8.813 |
| 350    | 37.874  | 192.131 | 186.694  | 1.903   | −76.461 | −46.445 | 6.932 |
| 400    | 40.500  | 197.356 | 187.704  | 3.861   | −77.969 | −42.054 | 5.492 |
| 450    | 43.374  | 202.291 | 189.053  | 5.957   | −79.422 | −37.476 | 4.350 |
| 500    | 46.342  | 207.014 | 190.614  | 8.200   | −80.802 | −32.741 | 3.420 |
| 600    | 52.227  | 215.987 | 194.103  | 13.130  | −83.308 | −22.887 | 1.993 |
| 700    | 57.794  | 224.461 | 197.840  | 18.635  | −85.452 | −12.643 | 0.943 |
| 800    | 62.932  | 232.518 | 201.675  | 24.675  | −87.238 | −2.115  | 0.138 |
| 900    | 67.601  | 240.205 | 205.532  | 31.205  | −88.692 | 8.616   | −0.500 |
| 1000   | 71.795  | 247.549 | 209.370  | 38.179  | −89.849 | 19.492  | −1.018 |
| 1100   | 75.529  | 254.570 | 213.162  | 45.549  | −90.750 | 30.472  | −1.447 |
| 1200   | 78.833  | 261.287 | 216.895  | 53.270  | −91.437 | 41.524  | −1.807 |
| 1300   | 81.744  | 267.714 | 220.558  | 61.302  | −91.945 | 52.626  | −2.115 |
| 1400   | 84.305  | 273.868 | 224.148  | 69.608  | −92.308 | 63.761  | −2.379 |
| 1500   | 86.556  | 279.763 | 227.660  | 78.153  | −92.553 | 74.918  | −2.609 |
| 1600   | 88.537  | 285.413 | 231.095  | 86.910  | −92.703 | 86.088  | −2.810 |
| 1700   | 90.283  | 290.834 | 234.450  | 95.853  | −92.780 | 97.265  | −2.989 |
| 1800   | 91.824  | 296.039 | 237.728  | 104.960 | −92.797 | 108.445 | −3.147 |
| 1900   | 93.188  | 301.041 | 240.930  | 114.212 | −92.770 | 119.624 | −3.289 |
| 2000   | 94.399  | 305.853 | 244.057  | 123.592 | −92.709 | 130.802 | −3.416 |
| 2100   | 95.477  | 310.485 | 247.110  | 133.087 | −92.624 | 141.975 | −3.531 |
| 2200   | 96.439  | 314.949 | 250.093  | 142.684 | −92.521 | 153.144 | −3.636 |
| 2300   | 97.301  | 319.255 | 253.007  | 152.371 | −92.409 | 164.308 | −3.732 |
| 2400   | 98.075  | 323.413 | 255.854  | 162.141 | −92.291 | 175.467 | −3.819 |
| 2500   | 98.772  | 327.431 | 258.638  | 171.984 | −92.174 | 186.622 | −3.899 |

*Enthalpy Reference Temperature $= T_r = 298.15$ K; Standard State Pressure $= p^\circ = 0.1$ MPa.

## Ethyne (C$_2$H$_2$)  C$_2$H$_2$(g)

| T/K | $C_p^\circ$ (J K$^{-1}$ mol$^{-1}$)* | $S^\circ$ | $-[G^\circ - H^\circ(T_r)]/T$ | $H^\circ - H^\circ(T_r)$ (kJ mol$^{-1}$)* | $\Delta_f H^\circ$ | $\Delta_f G^\circ$ | Log $K_f$ |
|---|---|---|---|---|---|---|---|
| 0 | 0. | 0. | Infinite | −10.012 | 235.755 | 235.755 | Infinite |
| 100 | 29.347 | 163.294 | 234.338 | −7.104 | 232.546 | 236.552 | −123.562 |
| 200 | 35.585 | 185.097 | 204.720 | −3.925 | 229.685 | 241.663 | −63.116 |
| 298.15 | 44.095 | 200.958 | 200.958 | 0. | 226.731 | 248.163 | −43.477 |
| 300 | 44.229 | 201.231 | 200.959 | 0.082 | 226.674 | 248.296 | −43.232 |
| 400 | 50.480 | 214.856 | 202.774 | 4.833 | 223.568 | 255.969 | −33.426 |
| 500 | 54.869 | 226.610 | 206.393 | 10.108 | 220.345 | 264.439 | −27.626 |
| 600 | 58.287 | 236.924 | 210.640 | 15.771 | 216.993 | 273.571 | −23.816 |
| 700 | 61.149 | 246.127 | 215.064 | 21.745 | 213.545 | 283.272 | −21.138 |
| 800 | 63.760 | 254.466 | 219.476 | 27.992 | 210.046 | 293.471 | −19.162 |
| 900 | 66.111 | 262.113 | 223.794 | 34.487 | 206.522 | 304.111 | −17.650 |
| 1000 | 68.275 | 269.192 | 227.984 | 41.208 | 202.989 | 315.144 | −16.461 |
| 1100 | 70.245 | 275.793 | 232.034 | 48.136 | 199.451 | 326.530 | −15.506 |
| 1200 | 72.053 | 281.984 | 235.941 | 55.252 | 195.908 | 338.239 | −14.723 |
| 1300 | 73.693 | 287.817 | 239.709 | 62.540 | 192.357 | 350.244 | −14.073 |
| 1400 | 75.178 | 293.334 | 243.344 | 69.985 | 188.795 | 362.523 | −13.526 |
| 1500 | 76.530 | 298.567 | 246.853 | 77.572 | 185.216 | 375.057 | −13.061 |
| 1600 | 77.747 | 303.546 | 250.242 | 85.286 | 181.619 | 387.830 | −12.661 |
| 1700 | 78.847 | 308.293 | 253.518 | 93.117 | 177.998 | 400.829 | −12.316 |
| 1800 | 79.852 | 312.829 | 256.688 | 101.053 | 174.353 | 414.041 | −12.015 |
| 1900 | 80.760 | 317.171 | 259.758 | 109.084 | 170.680 | 427.457 | −11.752 |
| 2000 | 81.605 | 321.335 | 262.733 | 117.203 | 166.980 | 441.068 | −11.520 |
| 2100 | 82.362 | 325.335 | 265.620 | 125.401 | 163.250 | 454.864 | −11.314 |
| 2200 | 83.065 | 329.183 | 268.422 | 133.673 | 159.491 | 468.838 | −11.132 |
| 2300 | 83.712 | 332.890 | 271.145 | 142.012 | 155.701 | 482.984 | −10.969 |
| 2400 | 84.312 | 336.465 | 273.793 | 150.414 | 151.881 | 497.295 | −10.823 |
| 2500 | 84.858 | 339.918 | 276.369 | 158.873 | 148.029 | 511.767 | −10.693 |

* Enthalpy Reference Temperature = $T_r$ = 298.15 K; Standard State Pressure = $p^\circ$ = 0.1 MPa.

## Carbon Monoxide (CO) $C_1O_1(g)$

| T/K | $C_p^\circ$ (J K$^{-1}$ mol$^{-1}$)* | $S^\circ$ | $-[G^\circ - H^\circ(T_r)]/T$ | $H^\circ - H^\circ(T_r)$ (kJ mol$^{-1}$)* | $\Delta_f H^\circ$ | $\Delta_f G^\circ$ | Log $K_f$ |
|---|---|---|---|---|---|---|---|
| 0 | 0. | 0. | Infinite | −8.671 | −113.805 | −113.805 | Infinite |
| 100 | 29.104 | 165.850 | 223.539 | −5.769 | −112.415 | −120.239 | 62.807 |
| 200 | 29.108 | 186.025 | 200.317 | −2.858 | −111.286 | −128.526 | 33.568 |
| 298.15 | 29.142 | 197.653 | 197.653 | 0. | −110.527 | −137.163 | 24.030 |
| 300 | 29.142 | 197.833 | 197.653 | 0.054 | −110.516 | −137.328 | 23.911 |
| 400 | 29.342 | 206.238 | 198.798 | 2.976 | −110.102 | −146.338 | 19.110 |
| 500 | 29.794 | 212.831 | 200.968 | 5.931 | −110.003 | −155.414 | 16.236 |
| 600 | 30.443 | 218.319 | 203.415 | 8.942 | −110.150 | −164.486 | 14.320 |
| 700 | 31.171 | 223.066 | 205.890 | 12.023 | −110.469 | −173.518 | 12.948 |
| 800 | 31.899 | 227.277 | 208.305 | 15.177 | −110.905 | −182.497 | 11.916 |
| 900 | 32.577 | 231.074 | 210.628 | 18.401 | −111.418 | −191.416 | 11.109 |
| 1000 | 33.183 | 234.538 | 212.848 | 21.690 | −111.983 | −200.275 | 10.461 |
| 1100 | 33.710 | 237.726 | 214.967 | 25.035 | −112.586 | −209.075 | 9.928 |
| 1200 | 34.175 | 240.679 | 216.988 | 28.430 | −113.217 | −217.819 | 9.481 |
| 1300 | 34.572 | 243.431 | 218.917 | 31.868 | −113.870 | −226.509 | 9.101 |
| 1400 | 34.920 | 246.006 | 220.761 | 35.343 | −114.541 | −235.149 | 8.774 |
| 1500 | 35.217 | 248.426 | 222.526 | 38.850 | −115.229 | −243.740 | 8.488 |
| 1600 | 35.480 | 250.707 | 224.216 | 42.385 | −115.933 | −252.284 | 8.236 |
| 1700 | 35.710 | 252.865 | 225.839 | 45.945 | −116.651 | −260.784 | 8.013 |
| 1800 | 35.911 | 254.912 | 227.398 | 49.526 | −117.384 | −269.242 | 7.813 |
| 1900 | 36.091 | 256.859 | 228.897 | 53.126 | −118.133 | −277.658 | 7.633 |
| 2000 | 36.250 | 258.714 | 230.342 | 56.744 | −118.896 | −286.034 | 7.470 |
| 2100 | 36.392 | 260.486 | 231.736 | 60.376 | −119.675 | −294.372 | 7.322 |
| 2200 | 36.518 | 262.182 | 233.081 | 64.021 | −120.470 | −302.672 | 7.186 |
| 2300 | 36.635 | 263.809 | 234.382 | 67.683 | −121.278 | −310.936 | 7.062 |
| 2400 | 36.321 | 265.359 | 235.641 | 71.324 | −122.133 | −319.164 | 6.946 |
| 2500 | 36.836 | 266.854 | 236.860 | 74.985 | −122.994 | −327.356 | 6.840 |
| 2600 | 36.924 | 268.300 | 238.041 | 78.673 | −123.854 | −335.514 | 6.741 |
| 2700 | 37.003 | 269.695 | 239.188 | 82.369 | −124.731 | −343.638 | 6.648 |
| 2800 | 37.083 | 271.042 | 240.302 | 86.074 | −125.623 | −351.729 | 6.562 |
| 2900 | 37.150 | 272.345 | 241.384 | 89.786 | −126.532 | −359.789 | 6.480 |
| 3000 | 37.217 | 273.605 | 242.437 | 93.504 | −127.457 | −367.816 | 6.404 |
| 3100 | 37.279 | 274.827 | 243.463 | 97.229 | −128.397 | −375.812 | 6.332 |
| 3200 | 37.338 | 276.011 | 244.461 | 100.960 | −129.353 | −383.778 | 6.265 |
| 3300 | 37.392 | 277.161 | 245.435 | 104.696 | −130.325 | −391.714 | 6.200 |
| 3400 | 37.443 | 278.278 | 246.385 | 108.438 | −131.312 | −399.620 | 6.139 |
| 3500 | 37.493 | 279.364 | 247.311 | 112.185 | −132.313 | −407.497 | 6.082 |
| 3600 | 37.543 | 280.421 | 248.216 | 115.937 | −133.329 | −415.345 | 6.027 |
| 3700 | 37.589 | 281.450 | 249.101 | 119.693 | −134.360 | −423.165 | 5.974 |
| 3800 | 37.631 | 282.453 | 249.965 | 123.454 | −135.405 | −430.956 | 5.924 |
| 3900 | 37.673 | 283.431 | 250.811 | 127.219 | −136.464 | −438.720 | 5.876 |
| 4000 | 37.715 | 284.386 | 251.638 | 130.989 | −137.537 | −446.457 | 5.830 |

*Enthalpy Reference Temperature = $T_r$ = 298.15 K; Standard State Pressure = $p^\circ$ = 0.1 MPa.

## Carbon Dioxide (CO$_2$)   C$_1$O$_2$(g)

| T/K | $C_p^\circ$ | $S^\circ$ | $-[G^\circ - H^\circ(T_r)]/T$ | $H^\circ - H^\circ(T_r)$ | $\Delta_f H^\circ$ | $\Delta_f G^\circ$ | Log $K_f$ |
|---|---|---|---|---|---|---|---|
| | (J K$^{-1}$ mol$^{-1}$)* | | | (kJ mol$^{-1}$)* | | | |
| 0 | 0. | 0. | Infinite | −9.364 | −393.151 | −393.151 | Infinite |
| 100 | 29.208 | 179.009 | 243.568 | −6.456 | −393.208 | −393.683 | 205.639 |
| 200 | 32.359 | 199.975 | 217.046 | −3.414 | −393.404 | −394.085 | 102.924 |
| 298.15 | 37.129 | 213.795 | 213.795 | 0. | −393.522 | −394.389 | 69.095 |
| 300 | 37.221 | 214.025 | 213.795 | 0.069 | −393.523 | −394.394 | 68.670 |
| 400 | 41.325 | 225.314 | 215.307 | 4.003 | −393.583 | −394.675 | 51.539 |
| 500 | 44.627 | 234.901 | 218.290 | 8.305 | −393.666 | −394.939 | 41.259 |
| 600 | 47.321 | 243.283 | 221.772 | 12.907 | −393.803 | −395.182 | 34.404 |
| 700 | 49.564 | 250.750 | 225.388 | 17.754 | −393.983 | −395.398 | 29.505 |
| 800 | 51.434 | 257.494 | 228.986 | 22.806 | −394.188 | −395.586 | 25.829 |
| 900 | 52.999 | 263.645 | 232.500 | 28.030 | −394.405 | −395.748 | 22.969 |
| 1000 | 54.308 | 269.299 | 235.901 | 33.397 | −394.623 | −395.886 | 20.679 |
| 1100 | 55.409 | 274.528 | 239.178 | 38.884 | −394.838 | −396.001 | 18.805 |
| 1200 | 56.342 | 279.390 | 242.329 | 44.473 | −395.050 | −396.098 | 17.242 |
| 1300 | 57.137 | 283.932 | 245.356 | 50.148 | −395.257 | −396.177 | 15.919 |
| 1400 | 57.802 | 288.191 | 248.265 | 55.896 | −395.462 | −396.240 | 14.784 |
| 1500 | 58.379 | 292.199 | 251.062 | 61.705 | −395.668 | −396.288 | 13.800 |
| 1600 | 58.886 | 295.983 | 253.753 | 67.569 | −395.876 | −396.323 | 12.939 |
| 1700 | 59.317 | 299.566 | 256.343 | 73.480 | −396.090 | −396.344 | 12.178 |
| 1800 | 59.701 | 302.968 | 258.840 | 79.431 | −396.311 | −396.353 | 11.502 |
| 1900 | 60.049 | 306.205 | 261.248 | 85.419 | −396.542 | −396.349 | 10.896 |
| 2000 | 60.350 | 309.293 | 263.574 | 91.439 | −396.784 | −396.333 | 10.351 |
| 2100 | 60.622 | 312.244 | 265.822 | 97.488 | −397.039 | −396.304 | 9.858 |
| 2200 | 60.865 | 315.070 | 267.996 | 103.562 | −397.309 | −396.262 | 9.408 |
| 2300 | 61.086 | 317.781 | 270.102 | 109.660 | −397.596 | −396.209 | 8.998 |
| 2400 | 61.287 | 320.385 | 272.144 | 115.779 | −397.900 | −396.142 | 8.622 |
| 2500 | 61.471 | 322.890 | 274.124 | 121.917 | −398.222 | −396.062 | 8.275 |
| 2600 | 61.647 | 325.305 | 276.046 | 128.073 | −398.562 | −395.969 | 7.955 |
| 2700 | 61.802 | 327.634 | 277.914 | 134.246 | −398.921 | −395.862 | 7.658 |
| 2800 | 61.952 | 329.885 | 279.730 | 140.433 | −399.299 | −395.742 | 7.383 |
| 2900 | 62.095 | 332.061 | 281.497 | 146.636 | −399.695 | −395.609 | 7.126 |
| 3000 | 62.229 | 334.169 | 283.218 | 152.852 | −400.111 | −395.461 | 6.886 |
| 3100 | 62.347 | 366.211 | 284.895 | 159.081 | −400.545 | −395.298 | 6.661 |
| 3200 | 62.462 | 338.192 | 286.529 | 165.321 | −400.998 | −395.122 | 6.450 |
| 3300 | 62.573 | 340.116 | 288.124 | 171.573 | −401.470 | −394.932 | 6.251 |
| 3400 | 62.681 | 341.986 | 289.681 | 177.836 | −401.960 | −394.726 | 6.064 |
| 3500 | 62.785 | 343.804 | 291.202 | 184.109 | −402.467 | −394.506 | 5.888 |
| 3600 | 62.884 | 345.574 | 292.687 | 190.393 | −402.991 | −394.271 | 5.721 |
| 3700 | 62.980 | 347.299 | 294.140 | 196.686 | −403.532 | −394.022 | 5.563 |
| 3800 | 63.074 | 348.979 | 295.561 | 202.989 | −404.089 | −393.756 | 5.413 |
| 3900 | 63.166 | 350.619 | 296.952 | 209.301 | −404.662 | −393.477 | 5.270 |
| 4000 | 63.254 | 352.219 | 298.314 | 215.622 | −405.251 | −393.183 | 5.134 |
| 4100 | 63.341 | 353.782 | 299.648 | 221.951 | −405.856 | −392.874 | 5.005 |
| 4200 | 63.426 | 355.310 | 300.955 | 228.290 | −406.475 | −392.550 | 4.882 |
| 4300 | 63.509 | 356.803 | 302.236 | 234.637 | −407.110 | −392.210 | 4.764 |
| 4400 | 63.588 | 358.264 | 303.493 | 240.991 | −407.760 | −391.857 | 4.652 |
| 4500 | 63.667 | 359.694 | 304.726 | 247.354 | −408.426 | −391.488 | 4.544 |

*Continued.*

## Carbon Dioxide Continued.

| T/K | $C_p^\circ$ | $S^\circ$ | $-[G^\circ - H^\circ(T_r)]/T$ | $H^\circ - H^\circ(T_r)$ | $\Delta_f H^\circ$ | $\Delta_f G^\circ$ | Log $K_f$ |
|---|---|---|---|---|---|---|---|
| | ($J\,K^{-1}\,mol^{-1}$)* | | | ($kJ\,mol^{-1}$)* | | | |
| 4600 | 63.745 | 361.094 | 305.937 | 253.725 | −409.106 | −391.105 | 4.441 |
| 4700 | 63.823 | 362.466 | 307.125 | 260.103 | −409.802 | −390.706 | 4.342 |
| 4800 | 63.893 | 363.810 | 308.292 | 266.489 | −410.514 | −390.292 | 4.247 |
| 4900 | 63.968 | 365.128 | 309.438 | 272.882 | −411.242 | −389.862 | 4.156 |
| 5000 | 64.046 | 366.422 | 310.565 | 279.283 | −411.986 | −389.419 | 4.068 |
| 5100 | 64.128 | 367.691 | 311.673 | 285.691 | −412.746 | −388.959 | 3.984 |
| 5200 | 64.220 | 368.937 | 312.762 | 292.109 | −413.522 | −388.486 | 3.902 |

*Enthalpy Reference Temperature = $T_r$ = 298.15 K; Standard State Pressure = $p^\circ$ = 0.1 MPa.

Appendix II    349

Copper (Cu)                                                                        Cu₁(ref)

| T/K | $C_p^\circ$ (J K⁻¹ mol⁻¹)* | $S^\circ$ | $-[G^\circ - H^\circ(T_r)]/T$ | $H^\circ - H^\circ(T_r)$ (kJ mol⁻¹)* | $\Delta_f H^\circ$ | $\Delta_f G^\circ$ | Log $K_f$ |
|---|---|---|---|---|---|---|---|
| 0 | 0. | 0. | Infinite | −5.007 | 0. | 0. | 0. |
| 100 | 16.010 | 10.034 | 53.414 | −4.338 | 0. | 0. | 0. |
| 200 | 22.631 | 23.730 | 35.354 | −2.325 | 0. | 0. | 0. |
| 298.15 | 24.442 | 33.164 | 33.164 | 0. | 0. | 0. | 0. |
| 300 | 24.462 | 33.315 | 33.164 | 0.045 | 0. | 0. | 0. |
| 400 | 25.318 | 40.484 | 34.136 | 2.539 | 0. | 0. | 0. |
| 500 | 25.912 | 46.206 | 35.997 | 5.105 | 0. | 0. | 0. |
| 600 | 26.481 | 50.982 | 38.107 | 7.725 | 0. | 0. | 0. |
| 700 | 26.996 | 55.103 | 40.247 | 10.399 | 0. | 0. | 0. |
| 800 | 27.494 | 58.739 | 42.336 | 13.123 | 0. | 0. | 0. |
| 900 | 28.049 | 62.009 | 44.343 | 15.899 | 0. | 0. | 0. |
| 1000 | 28.662 | 64.994 | 46.261 | 18.733 | 0. | 0. | 0. |
| 1100 | 29.479 | 67.763 | 48.091 | 21.638 | 0. | 0. | 0 |
| 1200 | 30.519 | 70.368 | 49.840 | 24.633 | 0. | 0. | 0. |
| 1300 | 32.143 | 72.871 | 51.516 | 27.762 | 0. | 0. | 0. |
| 1358.000 | 33.353 | 74.300 | 52.459 | 29.660 | Crystal ↔ Liquid | | |
| 1358.000 | 32.844 | 83.974 | 52.459 | 42.798 | Transition | | |
| 1400 | 32.844 | 84.974 | 53.419 | 44.177 | 0. | 0. | 0. |
| 1500 | 32.844 | 87.240 | 55.599 | 47.462 | 0. | 0. | 0. |
| 1600 | 32.844 | 89.360 | 57.644 | 50.746 | 0. | 0. | 0. |
| 1700 | 32.844 | 91.351 | 59.569 | 54.031 | 0. | 0. | 0. |
| 1800 | 32.844 | 93.229 | 61.387 | 57.315 | 0. | 0. | 0. |
| 1900 | 32.844 | 95.004 | 63.110 | 60.600 | 0. | 0. | 0. |
| 2000 | 32.844 | 96.689 | 64.747 | 63.884 | 0. | 0. | 0. |
| 2100 | 32.844 | 98.292 | 66.307 | 67.168 | 0. | 0. | 0. |
| 2200 | 32.844 | 99.819 | 67.795 | 70.453 | 0. | 0. | 0. |
| 2300 | 32.844 | 101.279 | 69.220 | 73.737 | 0 | 0. | 0. |
| 2400 | 32.844 | 102.677 | 70.585 | 77.022 | 0. | 0. | 0. |
| 2500 | 32.844 | 104.018 | 71.896 | 80.306 | 0. | 0. | 0. |
| 2600 | 32.844 | 105.306 | 73.156 | 83.591 | 0. | 0. | 0. |
| 2700 | 32.844 | 106.546 | 74.370 | 86.875 | 0. | 0. | 0. |
| 2800 | 32.844 | 107.740 | 75.540 | 90.159 | 0. | 0. | 0. |
| 2843.261 | 32.844 | 108.244 | 76.034 | 91.580 | Liquid ↔ Ideal Gas | | |
| 2843.261 | 24.379 | 213.995 | 76.034 | 392.257 | Fugacity = 1 bar | | |
| 2900 | 24.653 | 214.479 | 78.738 | 393.648 | 0. | 0. | 0. |
| 3000 | 25.147 | 215.323 | 83.277 | 396.138 | 0. | 0. | 0. |
| 3100 | 25.652 | 216.156 | 87.550 | 398.678 | 0. | 0. | 0. |
| 3200 | 26.162 | 216.978 | 91.582 | 401.268 | 0. | 0. | 0. |
| 3300 | 26.673 | 217.791 | 95.394 | 403.910 | 0. | 0. | 0. |
| 3400 | 27.180 | 218.595 | 99.006 | 406.603 | 0. | 0. | 0. |
| 3500 | 27.680 | 219.390 | 102.434 | 409.346 | 0. | 0. | 0. |

*Enthalpy Reference Temperature = $T_r$ = 298.15 K; Standard State Pressure = $p^\circ$ = 0.1 MPa.

## Copper Oxide (CuO)                                                    Cu$_1$O$_1$(cr)

| | ($J\,K^{-1}\,mol^{-1}$)* | | | ($kJ\,mol^{-1}$)* | | | |
|---|---|---|---|---|---|---|---|
| T/K | $C_p^\circ$ | $S^\circ$ | $-[G^\circ - H^\circ(T_r)]/T$ | $H^\circ - H^\circ(T_r)$ | $\Delta_f H^\circ$ | $\Delta_f G^\circ$ | Log $K_f$ |
| 0 | 0. | 0. | Infinite | −7.092 | −153.807 | −153.807 | Infinite |
| 100 | 16.499 | 9.768 | 74.176 | −6.441 | −155.277 | −146.585 | 76.568 |
| 200 | 34.795 | 27.018 | 46.256 | −3.848 | −156.152 | −137.461 | 35.901 |
| 298.15 | 42.246 | 42.594 | 42.594 | 0. | −156.063 | −128.292 | 22.476 |
| 300 | 42.364 | 42.856 | 42.595 | 0.078 | −156.057 | −128.120 | 22.308 |
| 400 | 46.806 | 55.723 | 44.319 | 4.562 | −155.553 | −118.875 | 15.523 |
| 500 | 49.261 | 66.451 | 47.703 | 9.374 | −154.836 | −109.785 | 11.469 |
| 600 | 50.935 | 75.588 | 51.608 | 14.388 | −154.022 | −100.851 | 8.780 |
| 700 | 52.240 | 83.541 | 55.614 | 19.549 | −153.162 | −92.056 | 6.869 |
| 800 | 53.349 | 90.591 | 59.554 | 24.829 | −152.274 | −83.387 | 5.445 |
| 900 | 54.340 | 96.932 | 63.361 | 30.215 | −151.368 | −74.830 | 4.343 |
| 1000 | 55.262 | 102.706 | 67.011 | 35.695 | −150.453 | −66.375 | 3.467 |
| 1100 | 56.135 | 108.014 | 70.500 | 41.265 | −149.542 | −58.012 | 2.755 |
| 1200 | 56.978 | 112.935 | 73.834 | 46.921 | −148.656 | −49.730 | 2.165 |
| 1300 | 57.800 | 117.528 | 77.020 | 52.660 | −147.837 | −41.520 | 1.668 |
| 1400 | 58.601 | 121.841 | 80.069 | 58.480 | −160.239 | −32.963 | 1.230 |
| 1500 | 59.395 | 125.911 | 82.991 | 64.380 | −159.444 | −23.899 | 0.832 |
| 1600 | 60.181 | 129.769 | 85.795 | 70.359 | −158.583 | −14.891 | 0.486 |
| 1700 | 60.958 | 133.441 | 88.491 | 76.416 | −157.657 | −5.938 | 0.182 |
| 1800 | 61.726 | 136.947 | 91.086 | 82.550 | −156.665 | 2.958 | −0.086 |
| 1900 | 62.494 | 140.305 | 93.589 | 88.761 | −155.608 | 11.798 | −0.324 |
| 2000 | 63.259 | 143.530 | 96.006 | 95.049 | −154.486 | 20.580 | −0.537 |

*Enthalpy Reference Temperature = $T_r$ = 298.15 K; Standard State Pressure = $p^\circ$ = 0.1 MPa.

## Copper Oxide (Cu$_2$O)

Cu$_2$O$_1$(cr, l)

| T/K | ($J\ K^{-1}\ mol^{-1}$)* | | | ($kJ\ mol^{-1}$)* | | | |
|---|---|---|---|---|---|---|---|
| | $C_p^\circ$ | $S^\circ$ | $-[G^\circ - H^\circ(T_r)]/T$ | $H^\circ - H^\circ(T_r)$ | $\Delta_f H^\circ$ | $\Delta_f G^\circ$ | Log $K_f$ |
| 0 | 0. | 0. | Infinite | −12.600 | −168.952 | −168.952 | Infinite |
| 100 | 39.499 | 37.092 | 141.510 | −10.442 | −169.584 | −162.621 | 84.944 |
| 200 | 53.619 | 69.143 | 97.826 | −5.737 | −170.360 | −155.348 | 40.573 |
| 298.15 | 62.543 | 92.360 | 92.360 | 0. | −170.707 | −147.886 | 25.909 |
| 300 | 62.664 | 92.747 | 92.361 | 0.116 | −170.709 | −147.745 | 25.725 |
| 400 | 67.688 | 111.531 | 94.888 | 6.657 | −170.641 | −140.092 | 18.294 |
| 500 | 70.922 | 127.000 | 99.809 | 13.596 | −170.363 | −132.484 | 13.840 |
| 600 | 73.448 | 140.161 | 105.465 | 20.818 | −169.961 | −124.944 | 10.877 |
| 700 | 75.645 | 151.651 | 111.259 | 28.274 | −169.479 | −117.478 | 8.766 |
| 800 | 77.666 | 161.886 | 116.959 | 35.941 | −168.929 | −110.087 | 7.188 |
| 900 | 79.493 | 171.141 | 122.473 | 43.801 | −168.325 | −102.767 | 5.964 |
| 1000 | 81.245 | 179.606 | 127.769 | 51.837 | −167.688 | −95.517 | 4.989 |
| 1100 | 83.385 | 187.443 | 132.842 | 60.061 | −167.029 | −88.331 | 4.194 |
| 1200 | 86.110 | 194.812 | 137.702 | 68.532 | −166.322 | −81.208 | 3.535 |
| 1300 | 89.173 | 201.824 | 142.367 | 77.294 | −165.608 | −74.144 | 2.979 |
| 1400 | 92.489 | 208.552 | 146.855 | 86.375 | −191.165 | −66.321 | 2.474 |
| 1500 | 95.987 | 215.052 | 151.186 | 95.798 | −190.132 | −57.438 | 2.000 |
| 1516.700 | 96.587 | 216.118 | 151.895 | 97.406 | Crystal ↔ Liquid | | |
| 1516.700 | 99.914 | 258.821 | 151.895 | 162.174 | Transition | | |
| 1600 | 99.914 | 264.163 | 157.603 | 170.497 | −123.836 | −52.197 | 1.704 |
| 1700 | 99.914 | 270.220 | 164.051 | 180.488 | −122.259 | −47.768 | 1.468 |
| 1800 | 99.914 | 275.931 | 170.109 | 190.480 | −120.694 | −43.431 | 1.260 |
| 1900 | 99.914 | 281.333 | 175.822 | 200.471 | −119.142 | −39.181 | 1.077 |
| 2000 | 99.914 | 286.458 | 181.227 | 210.463 | −117.600 | −35.013 | 0.914 |

*Enthalpy Reference Temperature = $T_r$ = 298.15 K; Standard State Pressure = $p^\circ$ = 0.1 MPa.

## Iron (Fe)

Fe₁(ref)

| T/K | $C_p^\circ$ (J K⁻¹ mol⁻¹)* | $S^\circ$ | $-[G^\circ - H^\circ(T_r)]/T$ | $H^\circ - H^\circ(T_r)$ (kJ mol⁻¹)* | $\Delta_f H^\circ$ | $\Delta_f G^\circ$ | Log $K_f$ |
|---|---|---|---|---|---|---|---|
| 0 | 0. | 0. | Infinite | −4.507 | 0. | 0. | 0. |
| 100 | 12.101 | 6.065 | 46.905 | −4.084 | 0. | 0. | 0. |
| 200 | 21.588 | 17.949 | 29.524 | −2.315 | 0. | 0. | 0. |
| 298.15 | 25.094 | 27.321 | 27.321 | 0. | 0. | 0. | 0. |
| 300 | 25.140 | 27.476 | 27.321 | 0.046 | 0. | 0. | 0. |
| 400 | 27.386 | 35.021 | 28.335 | 2.674 | 0. | 0. | 0. |
| 500 | 29.702 | 41.377 | 30.323 | 5.527 | 0. | 0. | 0. |
| 600 | 32.049 | 46.997 | 32.642 | 8.613 | 0. | 0. | 0. |
| 700 | 34.602 | 52.119 | 35.063 | 11.939 | 0. | 0. | 0. |
| 800 | 37.949 | 56.940 | 37.498 | 15.553 | 0. | 0. | 0. |
| 900 | 43.095 | 61.679 | 39.922 | 19.582 | 0. | 0. | 0. |
| 1000 | 54.434 | 66.672 | 42.342 | 24.329 | 0. | 0. | 0. |
| 1042.000 | 83.680 | 69.245 | 43.372 | 26.960 | $C_p$ Lambda Maximum | | |
| 1042.000 | 83.681 | 69.245 | 43.372 | 26.960 | Transition | | |
| 1100 | 46.401 | 72.062 | 44.815 | 29.972 | 0. | 0. | 0. |
| 1184.000 | 41.422 | 75.260 | 46.865 | 33.619 | Alpha ↔ Gamma | | |
| 1184.000 | 33.882 | 76.020 | 46.865 | 34.519 | Transition | | |
| 1200 | 34.016 | 76.476 | 47.257 | 35.062 | 0. | 0. | 0. |
| 1300 | 34.853 | 79.231 | 49.612 | 38.505 | 0. | 0. | 0. |
| 1400 | 35.690 | 81.845 | 51.822 | 42.033 | 0. | 0. | 0. |
| 1500 | 36.526 | 84.336 | 53.907 | 45.643 | 0. | 0. | 0. |
| 1600 | 37.363 | 86.720 | 55.884 | 49.338 | 0. | 0. | 0. |
| 1665.000 | 37.907 | 88.218 | 57.117 | 51.784 | Gamma ↔ Delta | | |
| 1665.000 | 41.112 | 88.721 | 57.117 | 52.621 | Transition | | |
| 1700 | 41.463 | 89.580 | 57.776 | 54.066 | 0. | 0. | 0. |
| 1800 | 42.468 | 91.978 | 59.610 | 58.263 | 0. | 0. | 0. |
| 1809.000 | 42.558 | 92.190 | 59.772 | 58.645 | Delta ↔ Liquid | | |
| 1809.000 | 46.024 | 99.823 | 59.772 | 72.452 | Transition | | |
| 1900 | 46.024 | 102.082 | 61.744 | 76.641 | 0. | 0. | 0. |
| 2000 | 46.024 | 104.442 | 63.821 | 81.243 | 0. | 0. | 0. |
| 2100 | 46.024 | 106.688 | 65.809 | 85.845 | 0. | 0. | 0. |
| 2200 | 46.024 | 108.829 | 67.716 | 90.448 | 0. | 0. | 0. |
| 2300 | 46.024 | 110.875 | 69.549 | 95.050 | 0. | 0. | 0. |
| 2400 | 46.024 | 112.833 | 71.312 | 99.653 | 0. | 0. | 0. |
| 2500 | 46.024 | 114.712 | 73.010 | 104.255 | 0. | 0. | 0. |
| 2600 | 46.024 | 116.517 | 74.649 | 108.857 | 0. | 0. | 0. |
| 2700 | 46.024 | 118.254 | 76.232 | 113.460 | 0. | 0. | 0. |
| 2800 | 46.024 | 119.928 | 77.763 | 118.062 | 0. | 0. | 0. |
| 2900 | 46.024 | 121.543 | 79.245 | 122.665 | 0. | 0. | 0. |
| 3000 | 46.024 | 123.103 | 80.681 | 127.267 | 0. | 0. | 0. |
| 3133.345 | 46.024 | 125.105 | 82.529 | 133.404 | Liquid ↔ Ideal Gas | | |
| 3133.345 | 26.642 | 236.674 | 82.530 | 482.989 | Fugacity =1 bar | | |
| 3200 | 26.856 | 237.237 | 85.746 | 484.772 | 0. | 0. | 0. |
| 3400 | 27.511 | 238.885 | 94.706 | 490.208 | 0. | 0. | 0. |
| 3600 | 28.192 | 240.477 | 102.761 | 495.778 | 0. | 0. | 0. |
| 3800 | 28.905 | 242.020 | 110.050 | 501.487 | 0. | 0. | 0. |
| 4000 | 29.654 | 243.521 | 116.686 | 507.342 | 0. | 0. | 0. |

*Enthalpy Reference Temperature = $T_r$ = 298.15 K; Standard State Pressure = $p^\circ$ = 0.1 MPa.

## Iron Oxide (FeO)

$Fe_1O_1(cr, 1)$

| T/K | $C_p^\circ$ ($J\,K^{-1}\,mol^{-1}$)* | $S^\circ$ ($J\,K^{-1}\,mol^{-1}$)* | $-[G^\circ - H^\circ(T_r)]/T$ ($J\,K^{-1}\,mol^{-1}$)* | $H^\circ - H^\circ(T_r)$ ($kJ\,mol^{-1}$)* | $\Delta_f H^\circ$ ($kJ\,mol^{-1}$)* | $\Delta_f G^\circ$ ($kJ\,mol^{-1}$)* | Log $K_f$ |
|---|---|---|---|---|---|---|---|
| 0 | | | | | | | |
| 100 | | | | | | | |
| 200 | | | | | | | |
| 298.15 | 49.915 | 60.752 | 60.752 | 0. | −272.044 | −251.429 | 44.049 |
| 300 | 49.999 | 61.061 | 60.753 | 0.092 | −272.025 | −251.301 | 43.755 |
| 400 | 51.840 | 75.704 | 62.737 | 5.187 | −271.044 | −244.543 | 31.934 |
| 500 | 53.388 | 87.438 | 66.541 | 10.449 | −270.164 | −238.022 | 24.866 |
| 600 | 54.894 | 97.309 | 70.868 | 15.865 | −269.414 | −231.666 | 20.168 |
| 700 | 56.149 | 105.866 | 75.270 | 21.418 | −268.814 | −225.425 | 16.821 |
| 800 | 57.321 | 113.443 | 79.577 | 27.093 | −268.422 | −219.256 | 14.316 |
| 900 | 58.283 | 120.249 | 83.724 | 32.872 | −268.374 | −213.118 | 12.369 |
| 1000 | 59.371 | 126.447 | 87.691 | 38.756 | −268.969 | −206.955 | 10.810 |
| 1100 | 60.234 | 132.146 | 91.477 | 44.736 | −270.385 | −200.670 | 9.529 |
| 1200 | 61.086 | 137.424 | 95.089 | 50.802 | −271.184 | −194.316 | 8.458 |
| 1300 | 61.975 | 142.349 | 98.537 | 56.956 | −270.265 | −187.947 | 7.552 |
| 1400 | 62.760 | 146.971 | 101.833 | 63.194 | −269.361 | −181.649 | 6.777 |
| 1500 | 63.396 | 151.323 | 104.989 | 69.502 | −268.484 | −175.415 | 6.108 |
| 1600 | 64.015 | 155.435 | 108.014 | 75.873 | −267.642 | −169.238 | 5.525 |
| 1650.000 | 64.319 | 157.409 | 109.481 | 79.081 | Crystal ↔ Liquid | | |
| 1650.000 | 68.199 | 171.990 | 109.481 | 103.139 | Transition | | |
| 1700 | 68.199 | 174.026 | 111.350 | 106.549 | −243.540 | −163.826 | 5.034 |
| 1800 | 68.199 | 177.924 | 114.941 | 113.369 | −242.774 | −159.160 | 4.619 |
| 1900 | 68.199 | 181.611 | 118.354 | 120.189 | −256.202 | −153.831 | 4.229 |
| 2000 | 68.199 | 185.109 | 121.605 | 127.009 | −255.866 | −148.452 | 3.877 |

*Enthalpy Reference Temperature = $T_r$ = 298.15 K, Standard State Pressure = $p^\circ$ = 0.1 MPa.

## Iron Oxide, Hematite (Fe$_2$O$_3$)

Fe$_2$O$_3$(cr)

| T/K | (J K$^{-1}$ mol$^{-1}$)* | | | (kJ mol$^{-1}$)* | | | |
|---|---|---|---|---|---|---|---|
| | $C_p^\circ$ | $S^\circ$ | $-[G^\circ - H^\circ(T_r)]/T$ | $H^\circ - H^\circ(T_r)$ | $\Delta_f H^\circ$ | $\Delta_f G^\circ$ | Log $K_f$ |
| 0 | 0. | 0. | Infinite | −15.560 | −819.025 | −819.025 | Infinite |
| 100 | 31.497 | 14.611 | 159.795 | −14.518 | −823.186 | −797.438 | 416.539 |
| 200 | 76.567 | 51.279 | 96.132 | −8.970 | −825.542 | −770.573 | 201.253 |
| 298.15 | 103.763 | 87.400 | 87.400 | 0. | −825.503 | −743.523 | 130.262 |
| 300 | 104.182 | 88.043 | 87.402 | 0.192 | −825.485 | −743.014 | 129.370 |
| 400 | 120.123 | 120.299 | 91.691 | 11.443 | −823.947 | −715.727 | 93.464 |
| 500 | 131.796 | 148.400 | 100.284 | 24.058 | −821.626 | −688.929 | 71.972 |
| 600 | 141.168 | 173.270 | 110.415 | 37.713 | −818.882 | −662.642 | 57.688 |
| 700 | 149.729 | 195.675 | 121.020 | 52.259 | −815.870 | −636.837 | 47.521 |
| 800 | 158.218 | 216.225 | 131.653 | 67.658 | −812.705 | −611.477 | 39.925 |
| 900 | 166.490 | 235.340 | 142.124 | 83.894 | −809.634 | −586.510 | 34.040 |
| 950.000 | 170.570 | 244.451 | 147.271 | 92.321 | I ↔ II | | |
| 950.000 | 150.624 | 245.156 | 147.271 | 92.990 | Transition | | |
| 1000 | 150.624 | 252.882 | 152.360 | 100.522 | −807.695 | −561.866 | 29.349 |
| 1050.000 | 150.624 | 260.231 | 157.323 | 108.053 | II ↔ III | | |
| 1050.000 | 140.407 | 260.231 | 157.323 | 108.053 | Transition | | |
| 1100 | 140.775 | 266.771 | 162.150 | 115.083 | −809.682 | −537.171 | 25.508 |
| 1200 | 141.511 | 279.052 | 171.388 | 129.197 | −811.071 | −512.374 | 22.303 |
| 1300 | 142.248 | 290.408 | 180.112 | 143.385 | −809.145 | −487.562 | 19.590 |
| 1400 | 142.984 | 300.976 | 188.372 | 157.647 | −807.358 | −462.893 | 17.271 |
| 1500 | 143.720 | 310.866 | 196.212 | 171.982 | −805.706 | −438.347 | 15.265 |
| 1600 | 144.566 | 320.170 | 203.671 | 186.397 | −804.180 | −413.906 | 13.513 |
| 1700 | 145.266 | 328.956 | 210.785 | 200.890 | −804.681 | −389.520 | 11.968 |
| 1800 | 145.780 | 337.275 | 217.583 | 215.445 | −804.094 | −365.117 | 10.595 |
| 1900 | 146.231 | 345.169 | 224.092 | 230.045 | −831.859 | −339.338 | 9.329 |
| 2000 | 146.789 | 352.684 | 230.335 | 244.696 | −832.056 | −313.410 | 8.185 |
| 2100 | 147.277 | 359.858 | 236.334 | 259.400 | −832.235 | −287.474 | 7.151 |
| 2200 | 147.695 | 366.719 | 242.105 | 274.149 | −832.403 | −261.528 | 6.209 |
| 2300 | 148.044 | 373.292 | 247.667 | 288.937 | −832.566 | −235.576 | 5.350 |
| 2400 | 148.323 | 379.599 | 253.034 | 303.756 | −832.731 | −209.617 | 4.562 |
| 2500 | 148.532 | 385.658 | 258.219 | 318.599 | −832.905 | −183.650 | 3.837 |

*Enthalpy Reference Temperature $= T_r = 298.15$ K; Standard State Pressure $= p^\circ = 0.1$ MPa.

## Iron Oxide, Magnetite (Fe$_3$O$_4$)   Fe$_3$O$_4$(cr)

| T/K | $C_p^\circ$ | $S^\circ$ | $-[G^\circ - H^\circ(T_r)]/T$ | $H^\circ - H^\circ(T_r)$ | $\Delta_f H^\circ$ | $\Delta_f G^\circ$ | Log $K_f$ |
|---|---|---|---|---|---|---|---|
| | (J K$^{-1}$ mol$^{-1}$)* | | | (kJ mol$^{-1}$)* | | | |
| 0 | 0. | 0. | Infinite | −24.571 | −1114.579 | −1114.579 | Infinite |
| 100 | 56.335 | 27.611 | 254.203 | −22.659 | −1119.743 | −1086.024 | 567.281 |
| 114.150 | 155.535 | 37.694 | 226.660 | −21.570 | $C_p$ Lambda Maximum | | |
| 114.150 | 155.279 | 37.698 | 226.660 | −21.570 | Transition | | |
| 200 | 116.945 | 92.367 | 157.872 | −13.101 | −1121.314 | −1051.624 | 274.656 |
| 298.15 | 147.235 | 145.266 | 145.266 | 0. | −1120.894 | −1017.438 | 178.251 |
| 300 | 147.695 | 146.179 | 145.269 | 0.273 | −1120.869 | −1016.797 | 177.040 |
| 400 | 171.126 | 191.952 | 151.352 | 16.240 | −1118.728 | −982.386 | 128.286 |
| 500 | 192.380 | 232.441 | 163.593 | 34.424 | −1115.219 | −948.681 | 99.108 |
| 600 | 212.547 | 269.299 | 178.181 | 54.671 | −1110.550 | −915.794 | 79.727 |
| 700 | 232.714 | 303.578 | 193.672 | 76.934 | −1104.774 | −883.776 | 65.948 |
| 800 | 252.881 | 335.969 | 209.452 | 101.214 | −1098.011 | −852.657 | 55.673 |
| 900 | 273.050 | 366.919 | 225.241 | 127.510 | −1090.611 | −822.428 | 47.732 |
| 900.000 | 273.050 | 366.919 | 225.241 | 127.511 | $C_p$ Lambda Maximum | | |
| 900.000 | 200.832 | 366.919 | 225.241 | 127.511 | Transition | | |
| 1000 | 200.832 | 388.079 | 240.485 | 147.594 | −1091.694 | −792.602 | 41.401 |
| 1100 | 200.832 | 407.220 | 254.786 | 167.677 | −1095.556 | −762.463 | 36.206 |
| 1200 | 200.832 | 424.695 | 268.228 | 187.760 | −1097.841 | −732.139 | 31.869 |
| 1300 | 200.832 | 440.770 | 280.890 | 207.843 | −1095.254 | −701.771 | 28.197 |
| 1400 | 200.832 | 455.653 | 292.848 | 227.927 | −1092.980 | −671.590 | 25.057 |
| 1500 | 200.832 | 469.509 | 304.169 | 248.010 | −1091.011 | −641.562 | 22.341 |
| 1600 | 200.832 | 482.471 | 314.912 | 268.093 | −1089.346 | −611.654 | 19.968 |
| 1700 | 200.832 | 494.646 | 325.131 | 288.176 | −1090.831 | −581.785 | 17.876 |
| 1800 | 200.832 | 506.125 | 334.870 | 308.259 | −1090.769 | −551.846 | 16.014 |
| 1900 | 200.832 | 516.984 | 344.172 | 328.343 | −1133.299 | −519.794 | 14.290 |
| 2000 | 200.832 | 527.285 | 353.072 | 348.426 | −1134.548 | −487.472 | 12.731 |

*Enthalpy Reference Temperature = $T_r$ = 298.15 K; Standard State Pressure = $p^\circ$ = 0.1 MPa.

## Hydrogen ($H_2$)

$H_2$(ref)

| T/K | $C_p^\circ$ | $S^\circ$ | $-[G^\circ - H^\circ(T_r)]/T$ | $H^\circ - H^\circ(T_r)$ | $\Delta_f H^\circ$ | $\Delta_f G^\circ$ | Log $K_f$ |
|---|---|---|---|---|---|---|---|
| | ($J K^{-1} mol^{-1}$)* | | | ($kJ mol^{-1}$)* | | | |
| 0 | 0. | 0. | Infinite | −8.467 | 0. | 0. | 0. |
| 100 | 28.154 | 100.727 | 155.408 | −5.468 | 0. | 0. | 0. |
| 200 | 27.447 | 119.412 | 133.284 | −2.774 | 0. | 0. | 0. |
| 250 | 28.344 | 125.640 | 131.152 | −1.378 | 0. | 0. | 0. |
| 298.15 | 28.836 | 130.680 | 130.680 | 0. | 0. | 0. | 0. |
| 300 | 28.849 | 130.858 | 130.680 | 0.053 | 0. | 0. | 0. |
| 350 | 29.081 | 135.325 | 131.032 | 1.502 | 0. | 0. | 0. |
| 400 | 29.181 | 139.216 | 131.817 | 2.959 | 0. | 0. | 0. |
| 450 | 29.229 | 142.656 | 132.834 | 4.420 | 0. | 0. | 0. |
| 500 | 29.260 | 145.737 | 133.973 | 5.882 | 0. | 0. | 0. |
| 600 | 29.327 | 151.077 | 136.392 | 8.811 | 0. | 0. | 0. |
| 700 | 29.441 | 155.606 | 138.822 | 11.749 | 0. | 0. | 0. |
| 800 | 29.624 | 159.548 | 141.171 | 14.702 | 0. | 0. | 0. |
| 900 | 29.881 | 163.051 | 143.411 | 17.676 | 0. | 0. | 0. |
| 1000 | 30.205 | 166.216 | 145.536 | 20.680 | 0. | 0. | 0. |
| 1100 | 30.581 | 169.112 | 147.549 | 23.719 | 0. | 0. | 0. |
| 1200 | 30.992 | 171.790 | 149.459 | 26.797 | 0. | 0. | 0. |
| 1300 | 31.423 | 174.288 | 151.274 | 29.918 | 0. | 0. | 0. |
| 1400 | 31.861 | 176.633 | 153.003 | 33.082 | 0. | 0. | 0. |
| 1500 | 32.298 | 178.846 | 154.652 | 36.290 | 0. | 0. | 0. |
| 1600 | 32.725 | 180.944 | 156.231 | 39.541 | 0. | 0. | 0. |
| 1700 | 33.139 | 182.940 | 157.743 | 42.835 | 0. | 0. | 0. |
| 1800 | 33.537 | 184.846 | 159.197 | 46.169 | 0. | 0. | 0. |
| 1900 | 33.917 | 186.669 | 160.595 | 49.541 | 0. | 0. | 0. |
| 2000 | 34.280 | 188.418 | 161.943 | 52.951 | 0. | 0. | 0. |
| 2100 | 34.624 | 190.099 | 163.244 | 56.397 | 0. | 0. | 0. |
| 2200 | 34.952 | 191.718 | 164.501 | 59.876 | 0. | 0. | 0. |
| 2300 | 35.263 | 193.278 | 165.719 | 63.387 | 0. | 0. | 0. |
| 2400 | 35.559 | 194.785 | 166.899 | 66.928 | 0. | 0. | 0. |
| 2500 | 35.842 | 196.243 | 168.044 | 70.498 | 0. | 0. | 0. |
| 2600 | 36.111 | 197.654 | 169.155 | 74.096 | 0. | 0. | 0. |
| 2700 | 36.370 | 199.021 | 170.236 | 77.720 | 0. | 0. | 0. |
| 2800 | 36.618 | 200.349 | 171.288 | 81.369 | 0. | 0. | 0. |
| 2900 | 36.856 | 201.638 | 172.313 | 85.043 | 0. | 0. | 0. |
| 3000 | 37.087 | 202.891 | 173.311 | 88.740 | 0. | 0. | 0. |
| 3100 | 37.311 | 204.111 | 174.285 | 92.460 | 0. | 0. | 0. |
| 3200 | 37.528 | 205.299 | 175.236 | 96.202 | 0. | 0. | 0. |
| 3300 | 37.740 | 206.457 | 176.164 | 99.966 | 0. | 0. | 0. |
| 3400 | 37.946 | 207.587 | 177.072 | 103.750 | 0. | 0. | 0. |
| 3500 | 38.149 | 208.690 | 177.960 | 107.555 | 0. | 0. | 0. |

*Enthalpy Reference Temperature $= T_r = 298.15$ K; Standard State Pressure $= p^\circ = 0.1$ MPa.

## Water ($H_2O$)      $H_2O_1(g)$

| | ($J\,K^{-1}\,mol^{-1}$)* | | | ($kJ\,mol^{-1}$)* | | | |
|---|---|---|---|---|---|---|---|
| $T/K$ | $C_p^\circ$ | $S^\circ$ | $-[G^\circ - H^\circ(T_r)]/T$ | $H^\circ - H^\circ(T_r)$ | $\Delta_f H^\circ$ | $\Delta_f G^\circ$ | Log $K_f$ |
| 0      | 0.      | 0.      | Infinite | −9.904 | −238.921 | −238.921 | Infinite |
| 100    | 33.299  | 152.388 | 218.534  | −6.615 | −240.083 | −236.584 | 123.579 |
| 200    | 33.349  | 175.485 | 191.896  | −3.282 | −240.900 | −232.766 | 60.792 |
| 298.15 | 33.590  | 188.834 | 188.834  | 0.     | −241.826 | −228.582 | 40.047 |
| 300    | 33.596  | 189.042 | 188.835  | 0.062  | −241.844 | −228.500 | 39.785 |
| 400    | 34.262  | 198.788 | 190.159  | 3.452  | −242.846 | −223.901 | 29.238 |
| 500    | 35.226  | 206.534 | 192.685  | 6.925  | −243.826 | −219.051 | 22.884 |
| 600    | 36.325  | 213.052 | 195.550  | 10.501 | −244.758 | −214.007 | 18.631 |
| 700    | 37.495  | 218.739 | 198.465  | 14.192 | −245.632 | −208.812 | 15.582 |
| 800    | 38.721  | 223.825 | 201.322  | 18.002 | −246.443 | −203.496 | 13.287 |
| 900    | 39.987  | 228.459 | 204.084  | 21.938 | −247.185 | −198.083 | 11.496 |
| 1000   | 41.268  | 232.738 | 206.738  | 26.000 | −247.857 | −192.590 | 10.060 |
| 1100   | 42.536  | 236.731 | 209.285  | 30.191 | −248.460 | −187.033 | 8.881 |
| 1200   | 43.768  | 240.485 | 211.730  | 34.506 | −248.997 | −181.425 | 7.897 |
| 1300   | 44.945  | 244.035 | 214.080  | 38.942 | −249.473 | −175.774 | 7.063 |
| 1400   | 46.054  | 247.407 | 216.341  | 43.493 | −249.894 | −170.089 | 6.346 |
| 1500   | 47.090  | 250.620 | 218.520  | 48.151 | −250.265 | −164.376 | 5.724 |
| 1600   | 48.050  | 253.690 | 220.623  | 52.908 | −250.592 | −158.639 | 5.179 |
| 1700   | 48.935  | 256.630 | 222.655  | 57.758 | −250.881 | −152.883 | 4.698 |
| 1800   | 49.749  | 259.451 | 224.621  | 62.693 | −251.138 | −147.111 | 4.269 |
| 1900   | 50.496  | 262.161 | 226.526  | 67.706 | −251.368 | −141.325 | 3.885 |
| 2000   | 51.180  | 264.769 | 228.374  | 72.790 | −251.575 | −135.528 | 3.540 |
| 2100   | 51.823  | 267.282 | 230.167  | 77.941 | −251.762 | −129.721 | 3.227 |
| 2200   | 52.408  | 269.706 | 231.909  | 83.153 | −251.934 | −123.905 | 2.942 |
| 2300   | 52.947  | 272.048 | 233.604  | 88.421 | −252.092 | −118.082 | 2.682 |
| 2400   | 53.444  | 274.312 | 235.253  | 93.741 | −252.239 | −112.252 | 2.443 |
| 2500   | 53.904  | 276.503 | 236.860  | 99.108 | −252.379 | −106.416 | 2.223 |
| 2600   | 54.329  | 278.625 | 238.425  | 104.520 | −252.513 | −100.575 | 2.021 |
| 2700   | 54.723  | 280.683 | 239.952  | 109.973 | −252.643 | −94.729 | 1.833 |
| 2800   | 55.089  | 282.680 | 241.443  | 115.464 | −252.771 | −88.878 | 1.658 |
| 2900   | 55.430  | 284.619 | 242.899  | 120.990 | −252.897 | −83.023 | 1.495 |
| 3000   | 55.748  | 286.504 | 244.321  | 126.549 | −253.024 | −77.163 | 1.344 |
| 3100   | 56.044  | 288.337 | 245.711  | 132.139 | −253.152 | −71.298 | 1.201 |
| 3200   | 56.323  | 290.120 | 247.071  | 137.757 | −253.282 | −65.430 | 1.068 |
| 3300   | 56.583  | 291.858 | 248.402  | 143.403 | −253.416 | −59.558 | 0.943 |
| 3400   | 56.828  | 293.550 | 249.705  | 149.073 | −253.553 | −53.681 | 0.825 |
| 3500   | 57.058  | 295.201 | 250.982  | 154.768 | −253.696 | −47.801 | 0.713 |
| 3600   | 57.276  | 296.812 | 252.233  | 160.485 | −253.844 | −41.916 | 0.608 |
| 3700   | 57.480  | 298.384 | 253.459  | 166.222 | −253.997 | −36.027 | 0.509 |
| 3800   | 57.675  | 299.919 | 254.661  | 171.980 | −254.158 | −30.133 | 0.414 |
| 3900   | 57.859  | 301.420 | 255.841  | 177.757 | −254.326 | −24.236 | 0.325 |
| 4000   | 58.033  | 302.887 | 256.999  | 183.552 | −254.501 | −18.334 | 0.239 |
| 4100   | 58.199  | 304.322 | 258.136  | 189.363 | −254.684 | −12.427 | 0.158 |
| 4200   | 58.357  | 305.726 | 259.252  | 195.191 | −254.876 | −6.516 | 0.081 |
| 4300   | 58.507  | 307.101 | 260.349  | 201.034 | −255.078 | −0.600 | 0.007 |
| 4400   | 58.650  | 308.448 | 261.427  | 206.892 | −255.288 | 5.320 | −0.063 |
| 4500   | 58.787  | 309.767 | 262.486  | 212.764 | −255.508 | 11.245 | −0.131 |

*Continued.*

## Water Continued.

| T/K | $C_p^\circ$ | $S^\circ$ | $-[G^\circ - H^\circ(T_r)]/T$ | $H^\circ - H^\circ(T_r)$ | $\Delta_f H^\circ$ | $\Delta_f G^\circ$ | Log $K_f$ |
|---|---|---|---|---|---|---|---|
| | (J K$^{-1}$ mol$^{-1}$)* | | | (kJ mol$^{-1}$)* | | | |
| 4600 | 58.918 | 311.061 | 263.528 | 218.650 | −255.738 | 17.175 | −0.195 |
| 4700 | 59.044 | 312.329 | 264.553 | 224.548 | −255.978 | 23.111 | −0.257 |
| 4800 | 59.164 | 313.574 | 265.562 | 230.458 | −256.229 | 29.052 | −0.316 |
| 4900 | 59.275 | 314.795 | 266.554 | 236.380 | −256.491 | 34.998 | −0.373 |
| 5000 | 59.390 | 315.993 | 267.531 | 242.313 | −256.763 | 40.949 | −0.428 |
| 5100 | 59.509 | 317.171 | 268.493 | 248.258 | −257.046 | 46.906 | −0.480 |
| 5200 | 59.628 | 318.327 | 269.440 | 254.215 | −257.338 | 52.869 | −0.531 |

*Enthalpy Reference Temperature = $T_r$ = 298.15 K; Standard State Pressure = $p^\circ$ = 0.1 MPa.
For H$_2$O(l): $S_{298}^\circ$ = 69.95; $H_T^\circ - H_{298}^\circ$ = 75.48$T$ − 22,506; $\Delta_{fus}H_{273}^\circ$ = 6,009.5; $\Delta_{vap}H_{298}^\circ$ = 44,007.

## Hydrogen Sulfide (H$_2$S)    H$_2$S$_1$(g)

| T/K | $C_p^\circ$ | $S^\circ$ | $-[G^\circ - H^\circ(T_r)]/T$ | $H^\circ - H^\circ(T_r)$ | $\Delta_f H^\circ$ | $\Delta_f G^\circ$ | Log $K_f$ |
|---|---|---|---|---|---|---|---|
| | (J K$^{-1}$ mol$^{-1}$)* | | | (kJ mol$^{-1}$)* | | | |
| 0 | 0. | 0. | Infinite | −9.962 | −17.584 | −17.584 | Infinite |
| 100 | 33.259 | 168.972 | 235.330 | −6.636 | −17.947 | −23.519 | 12.285 |
| 200 | 33.380 | 192.038 | 208.586 | −3.310 | −18.957 | −28.754 | 7.510 |
| 298.15 | 34.192 | 205.757 | 205.757 | 0. | −20.502 | −33.329 | 5.839 |
| 300 | 34.208 | 205.969 | 205.758 | 0.063 | −20.534 | −33.408 | 5.817 |
| 400 | 35.581 | 215.992 | 207.115 | 3.551 | −24.549 | −37.343 | 4.876 |
| 500 | 37.192 | 224.102 | 209.726 | 7.188 | −27.762 | −40.179 | 4.197 |
| 600 | 38.936 | 231.037 | 212.713 | 10.994 | −30.470 | −42.399 | 3.691 |
| 700 | 40.740 | 237.174 | 215.777 | 14.978 | −32.771 | −44.201 | 3.298 |
| 800 | 42.518 | 242.731 | 218.804 | 19.142 | −34.771 | −45.694 | 2.984 |
| 900 | 44.212 | 247.838 | 221.750 | 23.479 | −89.665 | −45.870 | 2.662 |
| 1000 | 45.786 | 252.579 | 224.599 | 27.980 | −90.024 | −40.984 | 2.141 |
| 1100 | 47.200 | 257.011 | 227.346 | 32.631 | −90.284 | −36.066 | 1.713 |
| 1200 | 48.467 | 261.173 | 229.993 | 37.415 | −90.465 | −31.129 | 1.355 |
| 1300 | 49.593 | 265.098 | 232.544 | 42.320 | −90.584 | −26.179 | 1.052 |
| 1400 | 50.593 | 268.810 | 235.003 | 47.330 | −90.656 | −21.222 | 0.792 |
| 1500 | 51.476 | 272.331 | 237.375 | 52.434 | −90.693 | −16.261 | 0.566 |
| 1600 | 52.262 | 275.679 | 239.666 | 57.622 | −90.705 | −11.298 | 0.369 |
| 1700 | 52.961 | 278.869 | 241.879 | 62.883 | −90.699 | −6.336 | 0.195 |
| 1800 | 53.589 | 281.914 | 244.019 | 68.212 | −90.681 | −1.374 | 0.040 |
| 1900 | 54.145 | 284.827 | 246.091 | 73.599 | −90.656 | 3.587 | −0.099 |
| 2000 | 54.656 | 287.617 | 248.098 | 79.039 | −90.627 | 8.547 | −0.223 |
| 2100 | 55.107 | 290.295 | 250.044 | 84.528 | −90.597 | 13.505 | −0.336 |
| 2200 | 55.522 | 292.868 | 251.932 | 90.059 | −90.566 | 18.461 | −0.438 |
| 2300 | 55.898 | 295.345 | 253.766 | 95.631 | −90.538 | 23.416 | −0.532 |
| 2400 | 56.250 | 297.731 | 255.549 | 101.238 | −90.512 | 28.370 | −0.617 |
| 2500 | 56.568 | 300.034 | 257.282 | 106.879 | −90.490 | 33.323 | −0.696 |

*Enthalpy Reference Temperature = $T_r$ = 298.15 K; Standard State Pressure = $p^\circ$ = 0.1 MPa.

## Nitrogen (N₂)　　　　　　　　　　　　　　　　　　　　　　　　　　　　　　　　　　　　N₂(ref)

| T/K | $C_p^\circ$ | $S^\circ$ | $-[G^\circ - H^\circ(T_r)]/T$ | $H^\circ - H^\circ(T_r)$ | $\Delta_f H^\circ$ | $\Delta_f G^\circ$ | Log $K_f$ |
|---|---|---|---|---|---|---|---|
| | ($J\,K^{-1}\,mol^{-1}$)* | | | ($kJ\,mol^{-1}$)* | | | |
| 0      | 0.      | 0.      | Infinite | −8.670 | 0. | 0. | 0. |
| 100    | 29.104  | 159.811 | 217.490  | −5.768 | 0. | 0. | 0. |
| 200    | 29.107  | 179.985 | 194.272  | −2.857 | 0. | 0. | 0. |
| 250    | 29.111  | 186.481 | 192.088  | −1.402 | 0  | 0. | 0. |
| 298.15 | 29.124  | 191.609 | 191.609  | 0.     | 0. | 0. | 0. |
| 300    | 29.125  | 191.789 | 191.610  | 0.054  | 0. | 0. | 0. |
| 350    | 29.165  | 196.281 | 191.964  | 1.511  | 0. | 0. | 0. |
| 400    | 29.249  | 200.181 | 192.753  | 2.971  | 0. | 0. | 0. |
| 450    | 29.387  | 203.633 | 193.774  | 4.437  | 0. | 0. | 0. |
| 500    | 29.580  | 206.739 | 194.917  | 5.911  | 0. | 0. | 0. |
| 600    | 30.110  | 212.176 | 197.353  | 8.894  | 0. | 0. | 0. |
| 700    | 30.754  | 216.866 | 199.813  | 11.937 | 0. | 0. | 0. |
| 800    | 31.433  | 221.017 | 202.209  | 15.046 | 0. | 0. | 0. |
| 900    | 32.090  | 224.757 | 204.510  | 18.223 | 0. | 0. | 0. |
| 1000   | 32.697  | 228.170 | 206.708  | 21.463 | 0. | 0. | 0. |
| 1100   | 33.241  | 231.313 | 208.804  | 24.760 | 0. | 0. | 0. |
| 1200   | 33.723  | 234.226 | 210.802  | 28.109 | 0. | 0. | 0. |
| 1300   | 34.147  | 236.943 | 212.710  | 31.503 | 0. | 0. | 0. |
| 1400   | 34.518  | 239.487 | 214.533  | 34.936 | 0. | 0. | 0. |
| 1500   | 34.843  | 241.880 | 216.277  | 38.405 | 0. | 0. | 0. |
| 1600   | 35.128  | 244.138 | 217.948  | 41.904 | 0. | 0. | 0. |
| 1700   | 35.378  | 246.275 | 219.552  | 45.429 | 0. | 0. | 0. |
| 1800   | 35.600  | 248.304 | 221.094  | 48.978 | 0. | 0. | 0. |
| 1900   | 35.796  | 250.234 | 222.577  | 52.548 | 0. | 0. | 0. |
| 2000   | 35.971  | 252.074 | 224.006  | 56.137 | 0. | 0. | 0. |
| 2100   | 36.126  | 253.833 | 225.385  | 59.742 | 0. | 0. | 0. |
| 2200   | 36.268  | 255.517 | 226.717  | 63.361 | 0. | 0. | 0. |
| 2300   | 36.395  | 257.132 | 228.004  | 66.995 | 0. | 0. | 0. |
| 2400   | 36.511  | 258.684 | 229.250  | 70.640 | 0. | 0. | 0. |
| 2500   | 36.616  | 260.176 | 230.458  | 74.296 | 0. | 0. | 0. |
| 2600   | 36.713  | 261.614 | 231.629  | 77.963 | 0. | 0. | 0. |
| 2700   | 36.801  | 263.001 | 232.765  | 81.639 | 0. | 0. | 0. |
| 2800   | 36.883  | 264.341 | 233.869  | 85.323 | 0. | 0. | 0. |
| 2900   | 36.959  | 265.637 | 234.942  | 89.015 | 0. | 0. | 0. |
| 3000   | 37.030  | 266.891 | 235.986  | 92.715 | 0. | 0. | 0. |
| 3100   | 37.096  | 268.106 | 237.003  | 96.421 | 0. | 0. | 0. |
| 3200   | 37.158  | 269.285 | 237.993  | 100.134| 0. | 0. | 0. |
| 3300   | 37.216  | 270.429 | 238.959  | 103.852| 0. | 0. | 0. |
| 3400   | 37.271  | 271.541 | 239.901  | 107.577| 0. | 0. | 0. |
| 3500   | 37.323  | 272.622 | 240.821  | 111.306| 0. | 0. | 0. |
| 3600   | 37.373  | 273.675 | 241.719  | 115.041| 0. | 0. | 0. |
| 3700   | 37.420  | 274.699 | 242.596  | 118.781| 0. | 0. | 0. |
| 3800   | 37.465  | 275.698 | 243.454  | 122.525| 0. | 0. | 0. |
| 3900   | 37.508  | 276.671 | 244.294  | 126.274| 0. | 0. | 0. |
| 4000   | 37.550  | 277.622 | 245.115  | 130.027| 0. | 0. | 0. |

*Continued.*

## Nitrogen Continued.

| T/K | $C_p^\circ$ | $S^\circ$ | $-[G^\circ - H^\circ(T_r)]/T$ | $H^\circ - H^\circ(T_r)$ | $\Delta_f H^\circ$ | $\Delta_f G^\circ$ | Log $K_f$ |
|---|---|---|---|---|---|---|---|
| | ($J\,K^{-1}\,mol^{-1}$)* | | | ($kJ\,mol^{-1}$)* | | | |
| 4100 | 37.590 | 278.549 | 245.919 | 133.784 | 0. | 0. | 0. |
| 4200 | 37.629 | 279.456 | 246.707 | 137.545 | 0. | 0. | 0. |
| 4300 | 37.666 | 280.341 | 247.479 | 141.309 | 0. | 0. | 0. |
| 4400 | 37.702 | 281.208 | 248.236 | 145.078 | 0. | 0. | 0. |
| 4500 | 37.738 | 282.056 | 248.978 | 148.850 | 0. | 0. | 0. |
| 4600 | 37.773 | 282.885 | 249.706 | 152.625 | 0. | 0. | 0. |
| 4700 | 37.808 | 283.698 | 250.420 | 156.405 | 0. | 0. | 0. |
| 4800 | 37.843 | 284.494 | 251.122 | 160.187 | 0. | 0. | 0. |
| 4900 | 37.878 | 285.275 | 251.811 | 163.973 | 0. | 0. | 0. |
| 5000 | 37.912 | 286.041 | 252.488 | 167.763 | 0. | 0. | 0. |
| 5100 | 37.947 | 286.792 | 253.153 | 171.556 | 0. | 0. | 0. |
| 5200 | 37.981 | 287.529 | 253.807 | 175.352 | 0. | 0. | 0. |

*Enthalpy Reference Temperature = $T_r$ = 298.15 K; Standard State Pressure = $p^\circ$ = 0.1 MPa.

## Oxygen (O$_2$)    O$_2$(ref)

| T/K | $C_p^\circ$ | $S^\circ$ | $-[G^\circ - H^\circ(T_r)]/T$ | $H^\circ - H^\circ(T_r)$ | $\Delta_f H^\circ$ | $\Delta_f G^\circ$ | Log $K_f$ |
|---|---|---|---|---|---|---|---|
| | (J K$^{-1}$ mol$^{-1}$)* | | | (kJ mol$^{-1}$)* | | | |
| 0 | 0. | 0. | Infinite | −8.683 | 0. | 0. | 0. |
| 100 | 29.106 | 173.307 | 231.094 | −5.779 | 0. | 0. | 0. |
| 200 | 29.126 | 193.485 | 207.823 | −2.868 | 0. | 0. | 0. |
| 250 | 29.201 | 199.990 | 205.630 | −1.410 | 0. | 0. | 0. |
| 298.15 | 29.376 | 205.147 | 205.147 | 0. | 0. | 0. | 0. |
| 300 | 29.385 | 205.329 | 205.148 | 0.054 | 0. | 0. | 0. |
| 350 | 29.694 | 209.880 | 205.506 | 1.531 | 0. | 0. | 0. |
| 400 | 30.106 | 213.871 | 206.308 | 3.025 | 0. | 0. | 0. |
| 450 | 30.584 | 217.445 | 207.350 | 4.543 | 0. | 0. | 0. |
| 500 | 31.091 | 220.693 | 208.524 | 6.084 | 0. | 0. | 0. |
| 600 | 32.090 | 226.451 | 211.044 | 9.244 | 0. | 0. | 0. |
| 700 | 32.981 | 231.466 | 213.611 | 12.499 | 0. | 0. | 0. |
| 800 | 33.733 | 235.921 | 216.126 | 15.835 | 0. | 0. | 0. |
| 900 | 34.355 | 239.931 | 218.552 | 19.241 | 0. | 0. | 0. |
| 1000 | 34.870 | 243.578 | 220.875 | 22.703 | 0. | 0. | 0. |
| 1100 | 35.300 | 246.922 | 223.093 | 26.212 | 0. | 0. | 0. |
| 1200 | 35.667 | 250.010 | 225.209 | 29.761 | 0. | 0. | 0. |
| 1300 | 35.988 | 252.878 | 227.229 | 33.344 | 0. | 0. | 0. |
| 1400 | 36.277 | 255.556 | 229.158 | 36.957 | 0. | 0. | 0. |
| 1500 | 36.544 | 258.068 | 231.002 | 40.599 | 0. | 0. | 0. |
| 1600 | 36.796 | 260.434 | 232.768 | 44.266 | 0. | 0. | 0. |
| 1700 | 37.040 | 262.672 | 234.462 | 47.958 | 0. | 0. | 0. |
| 1800 | 37.277 | 264.796 | 236.089 | 51.673 | 0. | 0. | 0. |
| 1900 | 37.510 | 266.818 | 237.653 | 55.413 | 0. | 0. | 0. |
| 2000 | 37.741 | 268.748 | 239.160 | 59.175 | 0. | 0. | 0. |
| 2100 | 37.969 | 270.595 | 240.613 | 62.961 | 0. | 0. | 0. |
| 2200 | 38.195 | 272.366 | 242.017 | 66.769 | 0. | 0. | 0. |
| 2300 | 38.419 | 274.069 | 243.374 | 70.600 | 0. | 0. | 0. |
| 2400 | 38.639 | 275.709 | 244.687 | 74.453 | 0. | 0. | 0. |
| 2500 | 38.856 | 277.290 | 245.959 | 78.328 | 0. | 0. | 0. |
| 2600 | 39.068 | 278.819 | 247.194 | 82.224 | 0. | 0. | 0. |
| 2700 | 39.276 | 280.297 | 248.393 | 86.141 | 0. | 0. | 0. |
| 2800 | 39.478 | 281.729 | 249.558 | 90.079 | 0. | 0. | 0. |
| 2900 | 39.674 | 283.118 | 250.691 | 94.036 | 0. | 0. | 0. |
| 3000 | 39.864 | 284.466 | 251.795 | 98.013 | 0. | 0. | 0. |
| 3100 | 40.048 | 285.776 | 252.870 | 102.009 | 0. | 0. | 0. |
| 3200 | 40.225 | 287.050 | 253.918 | 106.023 | 0. | 0. | 0. |
| 3300 | 40.395 | 288.291 | 254.941 | 110.054 | 0. | 0. | 0. |
| 3400 | 40.559 | 289.499 | 255.940 | 114.102 | 0. | 0. | 0. |
| 3500 | 40.716 | 290.677 | 256.916 | 118.165 | 0. | 0. | 0. |
| 3600 | 40.868 | 291.826 | 257.870 | 122.245 | 0. | 0. | 0. |
| 3700 | 41.013 | 292.948 | 258.802 | 126.339 | 0. | 0. | 0. |
| 3800 | 41.154 | 294.044 | 259.716 | 130.447 | 0. | 0. | 0. |
| 3900 | 41.289 | 295.115 | 260.610 | 134.569 | 0. | 0. | 0. |
| 4000 | 41.421 | 296.162 | 261.485 | 138.705 | 0. | 0. | 0. |
| 4100 | 41.549 | 297.186 | 262.344 | 142.854 | 0. | 0. | 0. |
| 4200 | 41.674 | 298.189 | 263.185 | 147.015 | 0. | 0. | 0. |
| 4300 | 41.798 | 299.171 | 264.011 | 151.188 | 0. | 0. | 0. |
| 4400 | 41.920 | 300.133 | 264.821 | 155.374 | 0. | 0. | 0. |
| 4500 | 42.042 | 301.076 | 265.616 | 159.572 | 0. | 0. | 0. |

*Continued.*

## Oxygen Continued.

| T/K | $C_p^\circ$ (J K$^{-1}$ mol$^{-1}$)* | $S^\circ$ | $-[G^\circ - H^\circ(T_r)]/T$ | $H^\circ - H^\circ(T_r)$ (kJ mol$^{-1}$)* | $\Delta_f H^\circ$ | $\Delta_f G^\circ$ | Log $K_f$ |
|------|--------|---------|---------|---------|------|------|------|
| 4600 | 42.164 | 302.002 | 266.397 | 163.783 | 0. | 0. | 0. |
| 4700 | 42.287 | 302.910 | 267.164 | 168.005 | 0. | 0. | 0. |
| 4800 | 42.413 | 303.801 | 267.918 | 172.240 | 0. | 0. | 0. |
| 4900 | 42.542 | 304.677 | 268.660 | 176.488 | 0. | 0. | 0. |
| 5000 | 42.675 | 305.538 | 269.389 | 180.749 | 0. | 0. | 0. |
| 5100 | 42.813 | 306.385 | 270.106 | 185.023 | 0. | 0. | 0. |
| 5200 | 42.956 | 307.217 | 270.811 | 189.311 | 0. | 0. | 0. |

*Enthalpy Reference Temperature = $T_r$ = 298.15 K; Standard State Pressure = $p^\circ$ = 0.1 MPa.

## Sulfur (S)   S₁(ref)

| T/K | $C_p^\circ$ | $S^\circ$ | $-[G^\circ - H^\circ(T_r)]/T$ | $H^\circ - H^\circ(T_r)$ | $\Delta_f H^\circ$ | $\Delta_f G^\circ$ | Log $K_f$ |
|---|---|---|---|---|---|---|---|
| | (J K⁻¹ mol⁻¹)* | | | (kJ mol⁻¹)* | | | |
| 0 | 0. | 0. | Infinite | −4.412 | 0. | 0. | 0. |
| 100 | 12.770 | 12.522 | 49.744 | −3.722 | 0. | 0. | 0. |
| 200 | 19.368 | 23.637 | 34.038 | −2.080 | 0. | 0. | 0. |
| 298.15 | 22.698 | 32.056 | 32.056 | 0. | 0. | 0. | 0. |
| 300 | 22.744 | 32.196 | 32.056 | 0.042 | 0. | 0. | 0. |
| 368.300 | 24.246 | 37.015 | 32.540 | 1.648 | Alpha ↔ Beta | | |
| 368.300 | 24.773 | 38.103 | 32.540 | 2.049 | Transition | | |
| 388.360 | 25.167 | 39.427 | 32.861 | 2.550 | Beta ↔ Liquid | | |
| 388.360 | 31.058 | 43.859 | 32.861 | 4.271 | Transition | | |
| 400 | 32.162 | 44.793 | 33.195 | 4.639 | 0. | 0. | 0. |
| 432.020 | 53.808 | 47.431 | 34.151 | 5.737 | $C_p$ Lambda Maximum | | |
| 432.020 | 53.806 | 47.431 | 34.151 | 5.737 | Transition | | |
| 500 | 37.986 | 53.532 | 36.398 | 8.567 | 0. | 0. | 0. |
| 600 | 34.308 | 60.078 | 39.825 | 12.152 | 0. | 0. | 0. |
| 700 | 32.681 | 65.241 | 43.099 | 15.499 | 0. | 0. | 0. |
| 800 | 31.699 | 69.530 | 46.143 | 18.710 | 0. | 0. | 0. |
| 882.117 | 31.665 | 72.624 | 48.467 | 21.310 | Liquid ↔ Ideal Gas | | |
| 882.117 | 18.454 | 133.077 | 48.467 | 74.636 | Fugacity = 1 bar | | |
| 900 | 18.483 | 133.448 | 50.152 | 74.967 | 0. | 0. | 0. |
| 1000 | 18.638 | 135.403 | 58.581 | 76.823 | 0. | 0. | 0. |
| 1100 | 18.792 | 137.187 | 65.647 | 78.694 | 0. | 0. | 0. |
| 1200 | 18.947 | 138.829 | 71.678 | 80.581 | 0. | 0. | 0. |
| 1300 | 19.103 | 140.352 | 76.903 | 82.484 | 0. | 0. | 0. |
| 1400 | 19.257 | 141.773 | 81.486 | 84.402 | 0. | 0. | 0. |
| 1500 | 19.409 | 143.107 | 85.550 | 86.335 | 0. | 0. | 0. |
| 1600 | 19.556 | 144.364 | 89.187 | 88.283 | 0. | 0. | 0. |
| 1700 | 19.697 | 145.554 | 92.468 | 90.246 | 0. | 0. | 0. |
| 1800 | 19.830 | 146.684 | 95.449 | 92.223 | 0. | 0. | 0. |
| 1900 | 19.956 | 147.759 | 98.174 | 94.212 | 0. | 0. | 0. |
| 2000 | 20.072 | 148.786 | 100.679 | 96.213 | 0. | 0. | 0. |

*Enthalpy Reference Temperature = $T_r$ = 298.15 K; Standard State Pressure = $p^\circ$ = 0.1 MPa.

## Sulfur ($S_2$) — $S_2(g)$

| T/K | $C_p^\circ$ | $S^\circ$ | $-[G^\circ - H^\circ(T_r)]/T$ | $H^\circ - H^\circ(T_r)$ | $\Delta_f H^\circ$ | $\Delta_f G^\circ$ | Log $K_f$ |
|---|---|---|---|---|---|---|---|
| | ($J\,K^{-1}\,mol^{-1}$)* | | | ($kJ\,mol^{-1}$)* | | | |
| 0 | 0. | 0. | Infinite | −9.124 | 128.300 | 128.300 | Infinite |
| 100 | 29.367 | 195.067 | 255.684 | −6.062 | 129.983 | 112.980 | −59.015 |
| 200 | 30.452 | 215.621 | 231.072 | −3.090 | 129.670 | 96.001 | −25.073 |
| 250 | 31.513 | 222.529 | 228.695 | −1.541 | 129.180 | 87.637 | −18.311 |
| 298.15 | 32.490 | 228.165 | 228.165 | 0. | 128.600 | 79.687 | −13.961 |
| 300 | 32.525 | 228.366 | 228.166 | 0.060 | 128.576 | 79.384 | −13.822 |
| 350 | 33.387 | 233.447 | 228.565 | 1.709 | 127.892 | 71.238 | −10.632 |
| 400 | 34.090 | 237.953 | 229.462 | 3.396 | 122.718 | 63.371 | −8.275 |
| 450 | 34.656 | 242.002 | 230.634 | 5.115 | 120.586 | 56.063 | −6.508 |
| 500 | 35.111 | 245.678 | 231.958 | 6.860 | 118.326 | 49.019 | −5.121 |
| 600 | 35.781 | 252.142 | 234.798 | 10.407 | 114.703 | 35.511 | −3.092 |
| 700 | 36.260 | 257.695 | 237.681 | 14.009 | 111.612 | 22.562 | −1.684 |
| 800 | 36.637 | 262.562 | 240.493 | 17.655 | 108.835 | 10.033 | −0.655 |
| 882.117 | 36.908 | 266.155 | 242.717 | 20.674 | Fugacity = 1 bar | | |
| 900 | 36.966 | 266.896 | 243.191 | 21.335 | 0. | 0. | 0. |
| 1000 | 37.277 | 270.807 | 245.760 | 25.047 | 0. | 0. | 0. |
| 1100 | 37.584 | 274.374 | 248.201 | 28.790 | 0. | 0. | 0. |
| 1200 | 37.894 | 277.658 | 250.521 | 32.564 | 0. | 0. | 0. |
| 1300 | 38.205 | 280.703 | 252.727 | 36.369 | 0. | 0. | 0. |
| 1400 | 38.514 | 283.546 | 254.828 | 40.205 | 0. | 0. | 0. |
| 1500 | 38.818 | 286.213 | 256.832 | 44.072 | 0. | 0. | 0. |
| 1600 | 39.112 | 288.728 | 258.748 | 47.968 | 0. | 0. | 0. |
| 1700 | 39.394 | 291.108 | 260.582 | 51.894 | 0. | 0. | 0. |
| 1800 | 39.661 | 293.367 | 262.341 | 55.847 | 0. | 0. | 0. |
| 1900 | 39.911 | 295.518 | 264.031 | 59.825 | 0. | 0. | 0. |
| 2000 | 40.144 | 297.571 | 265.657 | 63.828 | 0. | 0. | 0. |
| 2100 | 40.352 | 299.535 | 267.224 | 67.853 | 0. | 0. | 0. |
| 2200 | 40.547 | 301.417 | 268.736 | 71.898 | 0. | 0. | 0. |
| 2300 | 40.729 | 303.223 | 270.196 | 75.962 | 0. | 0. | 0. |
| 2400 | 40.897 | 304.960 | 271.609 | 80.044 | 0. | 0. | 0. |
| 2500 | 41.052 | 306.633 | 272.977 | 84.141 | 0. | 0. | 0. |
| 2600 | 41.179 | 308.246 | 274.302 | 88.253 | 0. | 0. | 0. |
| 2700 | 41.299 | 309.802 | 275.588 | 92.377 | 0. | 0. | 0. |
| 2800 | 41.414 | 311.306 | 276.837 | 96.512 | 0. | 0. | 0. |
| 2900 | 41.523 | 312.761 | 278.051 | 100.659 | 0. | 0. | 0. |
| 3000 | 41.626 | 314.171 | 279.232 | 104.817 | 0. | 0. | 0. |

*Enthalpy Reference Temperature = $T_r$ = 298.15 K; Standard State Pressure = $p^\circ$ = 0.1 MPa.

## Sulfur Oxide (SO)                                                    $O_1S_1(g)$

| T/K | $C_p°$ | $S°$ | $-[G° - H°(T_r)]/T$ | $H° - H°(T_r)$ | $\Delta_f H°$ | $\Delta_f G°$ | Log $K_f$ |
|---|---|---|---|---|---|---|---|
| | ($J\,K^{-1}\,mol^{-1}$)* | | | ($kJ\,mol^{-1}$)* | | | |
| 0 | 0. | 0. | Infinite | −8.733 | 5.028 | 5.028 | Infinite |
| 100 | 29.106 | 189.916 | 248.172 | −5.826 | 5.793 | −3.281 | 1.714 |
| 200 | 29.269 | 210.114 | 224.668 | −2.911 | 5.610 | −12.337 | 3.222 |
| 250 | 29.634 | 216.680 | 222.437 | −1.439 | 5.333 | −16.793 | 3.509 |
| 298.15 | 30.173 | 221.944 | 221.944 | 0. | 5.007 | −21.026 | 3.684 |
| 300 | 30.197 | 222.130 | 221.944 | 0.056 | 4.994 | −21.187 | 3.689 |
| 350 | 30.867 | 226.835 | 222.314 | 1.582 | 4.616 | −25.521 | 3.809 |
| 400 | 31.560 | 231.002 | 223.144 | 3.143 | 1.998 | −29.711 | 3.880 |
| 450 | 32.221 | 234.758 | 224.230 | 4.738 | 0.909 | −33.619 | 3.902 |
| 500 | 32.826 | 238.184 | 225.456 | 6.364 | −0.238 | −37.391 | 3.906 |
| 600 | 33.838 | 244.263 | 228.097 | 9.699 | −2.067 | −44.643 | 3.887 |
| 700 | 34.612 | 249.540 | 230.791 | 13.124 | −3.617 | −51.614 | 3.851 |
| 800 | 35.206 | 254.202 | 233.432 | 16.616 | −5.005 | −58.374 | 3.811 |
| 900 | 35.672 | 258.377 | 235.976 | 20.161 | −59.419 | −63.886 | 3.708 |
| 1000 | 36.053 | 262.155 | 238.408 | 23.748 | −59.420 | −64.382 | 3.363 |
| 1100 | 36.379 | 265.607 | 240.726 | 27.369 | −59.424 | −64.878 | 3.081 |
| 1200 | 36.672 | 268.785 | 242.933 | 31.022 | −59.432 | −65.374 | 2.846 |
| 1300 | 36.946 | 271.731 | 245.037 | 34.703 | −59.446 | −65.869 | 2.647 |
| 1400 | 37.210 | 274.479 | 247.043 | 38.411 | −59.462 | −66.362 | 2.476 |
| 1500 | 37.469 | 277.055 | 248.959 | 42.145 | −59.482 | −66.854 | 2.328 |
| 1600 | 37.725 | 279.482 | 250.791 | 45.905 | −59.505 | −67.345 | 2.199 |
| 1700 | 37.980 | 281.776 | 252.547 | 49.690 | −59.528 | −67.834 | 2.084 |
| 1800 | 38.232 | 283.954 | 254.232 | 53.501 | −59.552 | −68.323 | 1.983 |
| 1900 | 38.482 | 286.028 | 255.851 | 57.336 | −59.575 | −68.809 | 1.892 |
| 2000 | 38.727 | 288.008 | 257.410 | 61.197 | −59.597 | −69.294 | 1.810 |
| 2100 | 38.967 | 289.904 | 258.912 | 65.082 | −59.618 | −69.779 | 1.736 |
| 2200 | 39.200 | 291.722 | 260.363 | 68.990 | −59.636 | −70.262 | 1.668 |
| 2300 | 39.425 | 293.469 | 261.764 | 72.921 | −59.652 | −70.745 | 1.607 |
| 2400 | 39.641 | 295.152 | 263.121 | 76.875 | −59.666 | −71.227 | 1.550 |
| 2500 | 39.847 | 296.774 | 264.434 | 80.849 | −59.678 | −71.708 | 1.498 |
| 2600 | 40.043 | 298.341 | 265.709 | 84.844 | −59.687 | −72.189 | 1.450 |
| 2700 | 40.229 | 299.856 | 266.945 | 88.857 | −59.694 | −72.670 | 1.406 |
| 2800 | 40.404 | 301.322 | 268.147 | 92.889 | −59.699 | −73.150 | 1.365 |
| 2900 | 40.568 | 302.742 | 269.316 | 96.938 | −59.702 | −73.631 | 1.326 |
| 3000 | 40.721 | 304.120 | 270.453 | 101.002 | −59.705 | −74.111 | 1.290 |

*Enthalpy Reference Temperature = $T_r$ = 298.15 K; Standard State Pressure = $p°$ = 0.1 MPa.

## Sulfur Dioxide (SO$_2$)                                                                 O$_2$S$_1$(g)

| T/K | $C_p^\circ$ | $S^\circ$ | $-[G^\circ - H^\circ(T_r)]/T$ | $H^\circ - H^\circ(T_r)$ | $\Delta_f H^\circ$ | $\Delta_f G^\circ$ | Log $K_f$ |
|---|---|---|---|---|---|---|---|
| | J K$^{-1}$ mol$^{-1}$ | | | kJ mol$^{-1}$ | | | |
| 0 | 0.000 | 0.000 | Infinite | −10.552 | −294.299 | −294.299 | Infinite |
| 100 | 33.526 | 209.025 | 281.199 | −7.217 | −294.559 | −296.878 | 155.073 |
| 200 | 36.372 | 233.033 | 251.714 | −3.736 | −295.631 | −298.813 | 78.042 |
| 298.15 | 39.878 | 248.212 | 248.212 | 0.000 | −296.842 | −300.125 | 52.581 |
| 300 | 39.945 | 248.459 | 248.213 | 0.074 | −296.865 | −300.145 | 52.260 |
| 400 | 43.493 | 260.448 | 249.824 | 4.250 | −300.257 | −300.971 | 39.303 |
| 500 | 46.576 | 270.495 | 252.979 | 8.758 | −302.736 | −300.871 | 31.432 |
| 600 | 49.049 | 279.214 | 256.641 | 13.544 | −304.694 | −300.305 | 26.144 |
| 700 | 50.961 | 286.924 | 260.427 | 18.548 | −306.291 | −299.444 | 22.345 |
| 800 | 52.434 | 293.829 | 264.178 | 23.721 | −307.667 | −298.370 | 19.482 |
| 900 | 53.580 | 300.073 | 267.825 | 29.023 | −362.026 | −296.051 | 17.182 |
| 1000 | 54.484 | 305.767 | 271.339 | 34.428 | −361.940 | −288.725 | 15.081 |
| 1100 | 55.204 | 310.995 | 274.710 | 39.914 | −361.835 | −281.409 | 13.363 |
| 1200 | 55.794 | 315.824 | 277.937 | 45.464 | −361.720 | −274.102 | 11.931 |
| 1300 | 56.279 | 320.310 | 281.026 | 51.069 | −361.601 | −266.806 | 10.720 |
| 1400 | 56.689 | 324.496 | 283.983 | 56.718 | −361.484 | −259.518 | 9.683 |
| 1500 | 57.036 | 328.419 | 286.816 | 62.404 | −361.372 | −252.239 | 8.784 |
| 1600 | 57.338 | 332.110 | 289.533 | 68.123 | −361.268 | −244.967 | 7.997 |
| 1700 | 57.601 | 335.594 | 292.141 | 73.870 | −361.176 | −237.701 | 7.304 |
| 1800 | 57.831 | 338.893 | 294.647 | 79.642 | −361.096 | −230.440 | 6.687 |
| 1900 | 58.040 | 342.026 | 297.059 | 85.436 | −361.031 | −223.183 | 6.136 |
| 2000 | 58.229 | 345.007 | 299.383 | 91.250 | −360.981 | −215.929 | 5.639 |
| 2100 | 58.400 | 347.853 | 301.624 | 97.081 | −360.948 | −208.678 | 5.191 |
| 2200 | 58.555 | 350.573 | 303.787 | 102.929 | −360.931 | −201.427 | 4.782 |
| 2300 | 58.702 | 353.179 | 305.878 | 108.792 | −360.930 | −194.177 | 4.410 |
| 2400 | 58.840 | 355.680 | 307.902 | 114.669 | −360.947 | −186.927 | 4.068 |
| 2500 | 58.965 | 358.085 | 309.861 | 120.559 | −360.980 | −179.675 | 3.754 |
| 2600 | 59.086 | 360.400 | 311.761 | 126.462 | −361.030 | −172.422 | 3.464 |
| 2700 | 59.199 | 362.632 | 313.604 | 132.376 | −361.095 | −165.166 | 3.195 |
| 2800 | 59.308 | 364.787 | 315.394 | 138.302 | −361.175 | −157.908 | 2.946 |
| 2900 | 59.413 | 366.870 | 317.133 | 144.238 | −361.270 | −150.648 | 2.713 |
| 3000 | 59.513 | 368.886 | 318.825 | 150.184 | −361.379 | −143.383 | 2.497 |

*Enthalpy Reference Temperature = $T_r$ = 298.15 K; Standard State Pressure = $p^\circ$ = 0.1 MPa.

## Sulfur Trioxide (SO₃)   O₃S₁(g)

| T/K | $C_p^\circ$ (J K⁻¹ mol⁻¹)* | $S^\circ$ | $-[G^\circ - H^\circ(T_r)]/T$ | $H^\circ - H^\circ(T_r)$ (kJ mol⁻¹)* | $\Delta_f H^\circ$ | $\Delta_f G^\circ$ | Log $K_f$ |
|---|---|---|---|---|---|---|---|
| 0 | 0. | 0. | Infinite | −11.697 | −390.025 | −390.025 | Infinite |
| 100 | 34.076 | 212.371 | 295.976 | −8.361 | −391.735 | −385.724 | 201.481 |
| 200 | 42.336 | 238.259 | 261.145 | −4.577 | −393.960 | −378.839 | 98.943 |
| 250 | 46.784 | 248.192 | 257.582 | −2.348 | −394.937 | −374.943 | 78.340 |
| 298.15 | 50.661 | 256.769 | 256.769 | 0. | −395.765 | −371.016 | 65.000 |
| 300 | 50.802 | 257.083 | 256.770 | 0.094 | −395.794 | −370.862 | 64.573 |
| 350 | 54.423 | 265.191 | 257.402 | 2.726 | −396.543 | −366.646 | 54.719 |
| 400 | 57.672 | 272.674 | 258.849 | 5.530 | −399.412 | −362.242 | 47.304 |
| 450 | 60.559 | 279.637 | 260.777 | 8.487 | −400.656 | −357.529 | 41.501 |
| 500 | 63.100 | 286.152 | 262.992 | 11.580 | −401.878 | −352.668 | 36.843 |
| 600 | 67.255 | 298.041 | 267.862 | 18.107 | −403.675 | −342.647 | 29.830 |
| 700 | 70.390 | 308.655 | 272.945 | 24.997 | −405.014 | −332.365 | 24.801 |
| 800 | 72.761 | 318.217 | 278.017 | 32.160 | −406.068 | −321.912 | 21.019 |
| 900 | 74.570 | 326.896 | 282.973 | 39.531 | −460.062 | −310.258 | 18.007 |
| 1000 | 75.968 | 334.828 | 287.768 | 47.060 | −459.581 | −293.639 | 15.338 |
| 1100 | 77.065 | 342.122 | 292.382 | 54.714 | −459.063 | −277.069 | 13.157 |
| 1200 | 77.937 | 348.866 | 296.811 | 62.466 | −458.521 | −260.548 | 11.341 |
| 1300 | 78.639 | 355.133 | 301.060 | 70.296 | −457.968 | −244.073 | 9.807 |
| 1400 | 79.212 | 360.983 | 305.133 | 78.189 | −457.413 | −227.640 | 8.493 |
| 1500 | 79.685 | 366.465 | 309.041 | 86.135 | −456.863 | −211.247 | 7.356 |
| 1600 | 80.079 | 371.620 | 312.793 | 94.124 | −456.323 | −194.890 | 6.363 |
| 1700 | 80.410 | 376.485 | 316.398 | 102.149 | −455.798 | −178.567 | 5.487 |
| 1800 | 80.692 | 381.090 | 319.865 | 110.204 | −455.293 | −162.274 | 4.709 |
| 1900 | 80.932 | 385.459 | 323.203 | 118.286 | −454.810 | −146.009 | 4.014 |
| 2000 | 81.140 | 389.616 | 326.421 | 126.390 | −454.351 | −129.768 | 3.389 |
| 2100 | 81.319 | 393.579 | 329.525 | 134.513 | −453.919 | −113.549 | 2.824 |
| 2200 | 81.476 | 397.366 | 332.523 | 142.653 | −453.514 | −97.350 | 2.311 |
| 2300 | 81.614 | 400.990 | 335.422 | 150.807 | −453.137 | −81.170 | 1.843 |
| 2400 | 81.735 | 404.466 | 338.227 | 158.975 | −452.790 | −65.006 | 1.415 |
| 2500 | 81.843 | 407.805 | 340.944 | 167.154 | −452.472 | −48.855 | 1.021 |
| 2600 | 81.939 | 411.017 | 343.578 | 175.343 | −452.183 | −32.716 | 0.657 |
| 2700 | 82.025 | 414.111 | 346.133 | 183.541 | −451.922 | −16.587 | 0.321 |
| 2800 | 82.102 | 417.096 | 348.614 | 191.748 | −451.690 | −0.467 | 0.009 |
| 2900 | 82.171 | 419.978 | 351.026 | 199.961 | −451.487 | 15.643 | −0.282 |
| 3000 | 82.234 | 422.765 | 353.371 | 208.182 | −451.311 | 31.748 | −0.553 |

*Enthalpy Reference Temperature = $T_r$ = 298.15 K; Standard State Pressure = $p^\circ$ = 0.1 MPa.

## Titanium (Ti)                                                Ti₁(ref)

| T/K | $C_p°$ | $S°$ | $-[G° - H°(T_r)]/T$ | $H° - H°(T_r)$ | $\Delta_f H°$ | $\Delta_f G°$ | Log $K_f$ |
|---|---|---|---|---|---|---|---|
|  | ($J K^{-1} mol^{-1}$)* | | | ($kJ mol^{-1}$)* | | | |
| 0 | 0. | 0. | Infinite | −4.830 | 0. | 0. | 0. |
| 100 | 14.334 | 8.261 | 50.955 | −4.269 | 0. | 0. | 0. |
| 200 | 22.367 | 21.227 | 32.989 | −2.352 | 0. | 0. | 0. |
| 298.15 | 25.238 | 30.759 | 30.759 | 0. | 0. | 0. | 0. |
| 300 | 25.276 | 30.915 | 30.760 | 0.047 | 0. | 0. | 0. |
| 400 | 26.862 | 38.423 | 31.772 | 2.660 | 0. | 0. | 0. |
| 500 | 27.877 | 44.534 | 33.733 | 5.401 | 0. | 0. | 0. |
| 600 | 28.596 | 49.683 | 35.973 | 8.226 | 0. | 0. | 0. |
| 700 | 29.135 | 54.134 | 38.257 | 11.114 | 0. | 0. | 0. |
| 800 | 29.472 | 58.039 | 40.490 | 14.039 | 0. | 0. | 0. |
| 900 | 30.454 | 61.561 | 42.639 | 17.030 | 0. | 0. | 0. |
| 1000 | 32.074 | 64.848 | 44.697 | 20.151 | 0. | 0. | 0. |
| 1100 | 34.334 | 68.006 | 46.673 | 23.466 | 0. | 0. | 0. |
| 1166.000 | 36.175 | 70.058 | 47.938 | 25.791 | Alpha ↔ Beta | | |
| 1166.000 | 29.245 | 73.636 | 47.938 | 29.963 | Transition | | |
| 1200 | 29.459 | 74.479 | 48.679 | 30.961 | 0. | 0. | 0. |
| 1300 | 30.175 | 76.864 | 50.756 | 33.941 | 0. | 0. | 0. |
| 1400 | 31.023 | 79.131 | 52.702 | 37.000 | 0. | 0. | 0. |
| 1500 | 32.003 | 81.304 | 54.537 | 40.150 | 0. | 0. | 0. |
| 1600 | 33.115 | 83.404 | 56.276 | 43.405 | 0. | 0. | 0. |
| 1700 | 34.359 | 85.448 | 57.932 | 46.778 | 0. | 0. | 0. |
| 1800 | 35.736 | 87.451 | 59.517 | 50.281 | 0. | 0. | 0. |
| 1900 | 37.244 | 89.422 | 61.039 | 53.929 | 0. | 0. | 0. |
| 1939.000 | 37.868 | 90.186 | 61.617 | 55.394 | Beta ↔ Liquid | | |
| 1939.000 | 47.237 | 97.481 | 61.617 | 69.540 | Transition | | |
| 2000 | 47.237 | 98.944 | 62.734 | 72.421 | 0. | 0. | 0. |
| 2100 | 47.237 | 101.249 | 64.513 | 77.145 | 0. | 0. | 0. |
| 2200 | 47.237 | 103.446 | 66.233 | 81.869 | 0. | 0. | 0. |
| 2300 | 47.237 | 105.546 | 67.897 | 86.592 | 0. | 0. | 0. |
| 2400 | 47.237 | 107.557 | 69.508 | 91.316 | 0. | 0. | 0. |
| 2500 | 47.237 | 109.485 | 71.069 | 96.040 | 0. | 0. | 0. |
| 2600 | 47.237 | 111.338 | 72.582 | 100.764 | 0. | 0. | 0. |
| 2700 | 47.237 | 113.120 | 74.051 | 105.487 | 0. | 0. | 0. |
| 2800 | 47.237 | 114.838 | 75.477 | 110.211 | 0. | 0. | 0. |
| 2900 | 47.237 | 116.496 | 76.863 | 114.935 | 0. | 0. | 0. |
| 3000 | 47.237 | 118.097 | 78.211 | 119.659 | 0. | 0. | 0. |
| 3100 | 47.237 | 119.646 | 79.523 | 124.382 | 0. | 0. | 0. |
| 3200 | 47.237 | 121.146 | 80.800 | 129.106 | 0. | 0. | 0. |
| 3300 | 47.237 | 122.600 | 82.045 | 133.830 | 0. | 0. | 0. |
| 3400 | 47.237 | 124.010 | 83.259 | 138.554 | 0. | 0. | 0. |
| 3500 | 47.237 | 125.379 | 84.443 | 143.227 | 0. | 0. | 0. |
| 3600 | 47.237 | 126.710 | 85.598 | 148.001 | 0. | 0. | 0. |
| 3630.956 | 47.237 | 127.114 | 85.951 | 149.463 | Liquid ↔ Ideal Gas | | |
| 3630.956 | 34.219 | 240.028 | 85.951 | 559.447 | Fugacity = 1 bar | | |
| 3700 | 34.613 | 240.676 | 88.832 | 561.823 | 0. | 0. | 0. |
| 3800 | 35.173 | 241.607 | 92.840 | 565.313 | 0. | 0. | 0. |
| 3900 | 35.721 | 242.527 | 96.667 | 568.857 | 0. | 0. | 0. |
| 4000 | 36.254 | 243.439 | 100.324 | 572.456 | 0. | 0. | 0. |

*Enthalpy Reference Temperature = $T_r$ = 298.15 K; Standard State Pressure = $p°$ = 0.1 MPa.

Appendix II

## Zinc (Zn)                                                                                         Zn₁(ref)

| T/K | $C_p^\circ$ (J K⁻¹ mol⁻¹)* | $S^\circ$ | $-[G^\circ - H^\circ(T_r)]/T$ | $H^\circ - H^\circ(T_r)$ (kJ mol⁻¹)* | $\Delta_f H^\circ$ | $\Delta_f G^\circ$ | Log $K_f$ |
|---|---|---|---|---|---|---|---|
| 0 | 0. | 0. | Infinite | −5.669 | 0. | 0. | 0. |
| 100 | 19.455 | 16.523 | 63.224 | −4.670 | 0. | 0. | 0. |
| 200 | 24.050 | 31.820 | 44.005 | −2.437 | 0. | 0. | 0. |
| 298.15 | 25.387 | 41.717 | 41.717 | 0. | 0. | 0. | 0. |
| 300 | 25.406 | 41.874 | 41.717 | 0.047 | 0. | 0. | 0. |
| 400 | 26.346 | 49.314 | 42.725 | 2.636 | 0. | 0. | 0. |
| 500 | 27.386 | 55.301 | 44.660 | 5.321 | 0. | 0. | 0. |
| 600 | 28.588 | 60.399 | 46.868 | 8.118 | 0. | 0. | 0. |
| 692.730 | 29.802 | 64.591 | 48.965 | 10.825 | Crystal ↔ Liquid | | |
| 692.730 | 31.380 | 75.161 | 48.965 | 18.147 | Transition | | |
| 700 | 31.380 | 75.489 | 49.239 | 18.375 | 0. | 0. | 0. |
| 800 | 31.380 | 79.679 | 52.788 | 21.513 | 0. | 0. | 0. |
| 900 | 31.380 | 83.375 | 55.985 | 24.651 | 0. | 0. | 0. |
| 1000 | 31.380 | 86.681 | 58.892 | 27.789 | 0. | 0. | 0. |
| 1100 | 31.380 | 89.672 | 61.557 | 30.927 | 0. | 0. | 0. |
| 1180.173 | 31.380 | 91.880 | 63.542 | 33.443 | Liquid ↔ Ideal Gas | | |
| 1180.173 | 20.786 | 189.586 | 63.542 | 148.754 | Fugacity = 1 bar | | |
| 1200 | 20.786 | 189.933 | 65.628 | 149.166 | 0. | 0. | 0. |
| 1300 | 20.786 | 191.597 | 75.255 | 151.244 | 0. | 0. | 0. |
| 1400 | 20.786 | 193.137 | 83.620 | 153.323 | 0. | 0. | 0. |
| 1500 | 20.786 | 194.571 | 90.970 | 155.402 | 0. | 0. | 0. |
| 1600 | 20.786 | 195.913 | 97.487 | 157.480 | 0. | 0. | 0. |
| 1700 | 20.786 | 197.173 | 103.315 | 159.559 | 0. | 0. | 0. |
| 1800 | 20.786 | 198.361 | 108.562 | 161.637 | 0. | 0. | 0. |
| 1900 | 20.786 | 199.485 | 113.318 | 163.716 | 0. | 0. | 0. |
| 2000 | 20.786 | 200.551 | 117.653 | 165.795 | 0. | 0. | 0. |
| 2100 | 20.786 | 201.565 | 121.625 | 167.873 | 0. | 0. | 0. |
| 2200 | 20.786 | 202.532 | 125.281 | 169.952 | 0. | 0. | 0. |
| 2300 | 20.786 | 203.456 | 128.660 | 172.030 | 0. | 0. | 0. |
| 2400 | 20.786 | 204.341 | 131.795 | 174.109 | 0. | 0. | 0. |
| 2500 | 20.786 | 205.189 | 134.714 | 176.188 | 0. | 0. | 0. |
| 2600 | 20.786 | 206.004 | 137.440 | 178.266 | 0. | 0. | 0. |
| 2700 | 20.787 | 206.789 | 139.994 | 180.345 | 0. | 0. | 0. |
| 2800 | 20.787 | 207.545 | 142.393 | 182.424 | 0. | 0. | 0. |
| 2900 | 20.788 | 208.274 | 144.653 | 184.502 | 0. | 0. | 0. |
| 3000 | 20.789 | 208.979 | 146.785 | 186.581 | 0. | 0. | 0. |

*Enthalpy Reference Temperature = $T_r$ = 298.15 K; Standard State Pressure = $p^\circ$ = 0.1 MPa.

APPENDIX III

# THERMODYNAMIC SIMULATOR (TSIM) FOR THERMODYNAMIC CALCULATIONS

TSIM is a user-friendly *Microsoft Windows* based Thermodynamic SIMulator developed for this book. We have tried to provide an intuitive and user-friendly interface with simple point and click access to a variety of available options and functions. Please read the *readme.txt* file on the diskette for instructions on how to view and print the user guide.

The diskette contains a database of thermodynamic information for a large set of chemical species in the form of equations. Using this information, the current version of TSIM can display the following:

1. $C_p^\circ$ or Heat Capacity and $H_T^\circ - H_{298}^\circ$ calculations.
2. Enthalpy calculations.
3. Gibbs Energy calculations for chemical reactions.

TSIM allows you to search for a species and displays the equations as well as the tabular data obtained from these equations, and plots the resulting graphs. The display can be tabular or graphical and a user can specify the units, parameter range, step size for the range, and a number of other options (see the user guide in the diskette).

Thermodynamic equations given in general reference (2) have been used in all TSIM calculations and the data is included on the diskette.

Interesting computer programs on disks are also available from Professor A. E. Morris, Fulton Hall, University of Missouri, Rolla, MO 65401-0249.

APPENDIX IV

# ESTIMATION OF ACTIVITIES IN MULTICOMPONENT IONIC SOLUTIONS

The equations for the activities of multicomponent electrolyte solutions contain numerous terms as justified by the accuracy of data. For two electrolytes in water, i.e., a ternary solution, the following equation has been proposed* for the activity of water, $a_w$:

$$55.508 \ln a_w = -\nu_{12}m + Bm^{1.5} + Cm^2 + Dm^{2.5} + \cdots - \nu_{34}n + B'n^{1.5}$$
$$+ Jn^2 + Kn^{2.5} + Qmn + Smn^{1.5} + Umn^2 + \cdots Vm^{1.5}n$$
$$+ Wm^{1.5}n^{1.5} + \cdots + Xm^2n + Ym^2n^{1.5} + \cdots \quad \text{(IV.1)}$$

Here, $m$ and $n$ are the molalities of electrolytes, $\nu_{12}$ and $\nu_{34}$ are the sums of stoichiometric coefficients for electrolytes 12 and 34 respectively, e.g., as in (13.27) where $\nu$ corresponds to $\nu_{12}$ in (IV.1). The odd subscripts are used for cations ($+$) and the even subscripts are used for anions ($-$), and their combinations as 12 and 34 designate the compounds. The equations for $\gamma_{12}$ and $\gamma_{34}$ are

$$\ln \gamma_{12}^{\nu_{12}} = -\frac{1.5}{0.5}Bm^{0.5} - \frac{2}{1}Cm - \frac{2.5}{1.5}Dm^{1.5} - \cdots - \frac{1}{1}Qn - \frac{1}{1.5}Sn^{1.5}$$
$$-\frac{1}{2}Un^2 - \cdots - \frac{1.5}{1.5}Vm^{0.5}n - \frac{1.5}{2}Wm^{0.5}n^{1.5} - \cdots$$
$$-\frac{2}{2}Xmn - \frac{2}{2.5}Ymn^{1.5} - \cdots \quad \text{(IV.2)}$$

$$\ln \gamma_{34}^{\nu_{34}} = -\frac{1.5}{0.5}B'n^{0.5} - \frac{2}{1}Jn - \frac{2.5}{1.5}Kn^{1.5} - \cdots - \frac{1}{1}Qm - \frac{1.5}{1.5}Smn^{0.5}$$
$$-\frac{2}{2}Umn - \cdots - \frac{1}{1.5}Vm^{1.5} - \frac{1.5}{2}Wm^{1.5}n^{0.5} - \cdots$$
$$-\frac{1}{2}Xm^2 - \frac{1.5}{2.5}Ym^2n^{0.5} - \cdots \quad \text{(IV.3)}$$

The preceding equations require extensive data for the determination of their coefficients. Such data for ternary and multicomponent systems are scarce; therefore,

*N. A. Gokcen, U.S. Bur. of Mines, RI 8372 (1979).

374    Thermodynamics

a method of their estimation is technologically important. Several methods have been devised for this purpose, e.g., those by Kusik and Meissner, Bromley, and Pitzer.[†] In this Appendix, we present the method developed by Kusik and Meissner.

The notation used by Kusik and Meissner (KM) is slightly different from that in Chapter XIII, and convenient in designating several electrolytes in one solution as noted in conjunction with (IV.1) and (IV.2). For example, the notation for $Ba_3(PO_4)_2$ in Chapter XIII and in here is as follows:

$$\nu_+ = \nu_1 = 3, \qquad \nu_- = \nu_2 = 2, \qquad z_+ = z_1 = 2, \qquad -z_- = z_2 = +3,$$
$$m = m_{12}, \qquad m_+ = m_1 = 3m_{12}, \qquad m_- = 2m_{12}, \qquad \gamma_\pm = \gamma_{12},$$
$$a_\pm = a_{12}, \qquad m_\pm = m_{12\pm}. \text{ (for an additional electrolyte } m' = m_{34})$$

### Activity Coefficients in Binary Solutions

The Kusik-Meissner method is based on the observation that when $\log \mathring{\gamma}_{12}$ of only one electrolyte in each solution is plotted versus ionic strength $I$, the curves for various electrolytes cross one another, but when $\log \Gamma^\circ_{12} = (1/z_1 z_2) \log \mathring{\gamma}_{12}$ is plotted versus $I$, very few curves cross one another as indicated in Fig. IV.1. The property $\Gamma^\circ_{12}$ is called the reduced activity coefficient of one electrolyte in water and defined by

$$\Gamma^\circ_{12} = (\mathring{\gamma}_{12})^{1/z_1 z_2} \tag{IV.4}$$

For more than one electrolyte in solution, the reduced activity coefficient is similarly defined, but without the overscript; as follows:

$$\Gamma_{12} = (\gamma_{12})^{1/z_1 z_2}, \tag{IV.5}$$

where $\gamma_{12}$ is the activity coefficient in a ternary or multicomponent solution. In principle, a single value of $\Gamma^\circ_{12}$ at a sufficiently high value of $I$ determines the location of the entire curve for a given temperature. The value of the empirical parameter $q$ in Fig. IV.1 depends on the temperature and on the electrolyte, and it is obtained from

$$\Gamma^\circ_{12} = [1 + (0.75 - 0.065q)(1 + 0.1I)^q - (0.75 - 0.065q)]\Gamma^*, \tag{IV.6}$$

where $\Gamma^*$ is given by

$$\log \Gamma^* = -\frac{0.5107 I^{0.5}}{1 + cI^{0.5}}; \qquad \left(c = 1 + 0.55q \cdot e^{-0.023 I^3}\right). \tag{IV.7}$$

The terms in (IV.6) prior to $\Gamma^*$ constitute empirical corrections to the Debye-Hückel equation given by (IV.7). These equations have been obtained by trial and error on a computer to calculate the best coefficients capable of representing the activity coefficients of as many electrolytes as possible. Actually, each $q$ value is an average of various values based on $\mathring{\gamma}_{12}$ from experimental data points in the

---

[†]N. A. Gokcen and B. R. Staples, "Corrosion Science and Engineering," Edited by R. A. Rapp, N. A. Gokcen, and A. Pourbaix, CEBELCOR, Brussels, p. 211 (1989) contains references to the original investigations. See also N. A. Gokcen, loc. cit.

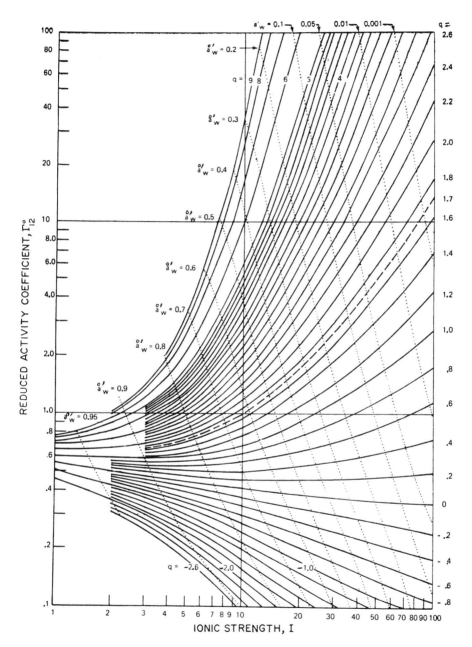

**Fig. IV.1** Kusik-Meissner diagram for isotherms of $\Gamma_{12}^\circ$ versus I with constant water activities for 1:1 electrolytes represented by dotted curves with $\mathring{a}' = (\mathring{a})'$ in text.

literature. If the value of $q$ has been determined for one temperature, such as $q_{25}$ for 25°C, the value at any other temperature, $q_t$, can be calculated from

$$\frac{q_t - q_{25}}{t - 25} = aq_{25} + b^*, \qquad (IV.8)$$

where $t$ is the temperature in °C, and $a = -0.0079$, $b^* = 0.0029$ for sulfates, and $a = -0.005$ and $b^* = +0.0085$ for all other electrolytes. Sulfuric acid, some sulfates of multivalent elements, thorium nitrate, and halides of zinc and cadmium do not obey the foregoing equations with sufficient degrees of accuracy. Table IV.1 shows the selected values of $q$. It is evident that except for high ionic concentrations of $HClO_4$, LiCl, LiBr, $LiNO_3$, and NaOH, the values of $\Gamma_{12}^{\circ}$ for 1:1 electrolytes are represented with standard deviations less than 10%. This is also true for higher valence electrolytes, with the exception of $CaCl_2$, $MgCl_2$, $UO_2SO_4$, and $ZnSO_4$. However, for multivalent electrolytes a small error in $\Gamma_{12}^{\circ}$ causes a larger error in $\mathring{\gamma}_{12}$, for example, an error of 5.2% in $\Gamma_{12}^{\circ}$ for $MnSO_4$ is equivalent to 22.5% error in $\mathring{\gamma}_{12}$ since $z_1 = z_2 = 2$ for this electrolyte. Therefore, the success of this method decreases with increasing values of the product $z_1 \cdot z_2$ for the same error in $\Gamma_{12}^{\circ}$.

## Activity of Water in Binary Solutions

The activity of water for 1:1 electrolytes, designated as $(\mathring{a}_w)'$, is represented in the Kusik-Meissner diagram by the dotted lines as shown in Fig. IV.1. The activity $(\mathring{a}_w)$ for electrolytes higher than 1:1 is computed from the following equation:

$$\log(\mathring{a}_w) = 0.0156 I \left(1 - \frac{1}{z_1 z_2}\right) + \log(\mathring{a}_w)'. \qquad (IV.9)$$

For example, at 25°C, $\Gamma_{12}^{\circ} = 1.7$ for $NiCl_2$ when $m_{12} = 4$ and $I = 12$, and from $\Gamma_{12}^{\circ}$ and $I$, Fig. IV.1 gives $(\mathring{a}_w)' = 0.52$. Substitution of $I$ and $(\mathring{a}_w)'$ in (IV.9) gives $(\mathring{a}_w) = 0.645$ in good agreement with the experimental value of 0.635.

## Activity Coefficients in Multicomponent Solutions

The activity coefficients in mixtures of electrolytes can be estimated from the following equation:

$$\log \Gamma_{ij} = \frac{z_i}{z_i + z_j} \left(V_{i2} I_2 \log \Gamma_{i2}^{\circ} + V_{i4} I_4 \log \Gamma_{i4}^{\circ} + \cdots\right) I^{-1}$$

$$+ \frac{z_j}{z_i + z_j} \left(V_{j1} I_1 \log \Gamma_{j1}^{\circ} + V_{j3} I_3 \log \Gamma_{j3}^{\circ} + \cdots\right) I^{-1}. \qquad (IV.10)$$

The values of $\Gamma_{ij}^{\circ}$ for binary solutions are determined at the temperature and the total ionic strength $I$ of the multicomponent mixture (total ionic strength is always without subscript). The terms $V$ and $I$ with various subscripts are defined by

$$V_{ij} = 0.5 \frac{(z_i + z_j)^2}{z_i z_j}; \qquad I_i = 0.5 m_i z_i^2 = \text{individual ionic strength.} \quad (IV.11)$$

**Table IV.1** Average $q$ values for electrolytes.

| | $q^*$ | $I(max)^\dagger$ | $S^\ddagger$ | | $q^*$ | $I(max)^\dagger$ | $S^\ddagger$ |
|---|---|---|---|---|---|---|---|
| *Electrolytes* | | | | *Higher electrolytes (continued)* | | | |
| $AgNO_3$ | −2.550 | 6.0 | 0.012 | $Cd(NO_3)_2$ | 1.530 | 7.5 | 0.043 |
| HCl | 6.690 | 6.0 | .050 | $CdSO_4$ | .016 | 14.0 | .049 |
| HCl | 6.100 | 16.0 | .092 | $CoCl_2$ | 2.250 | 12.0 | .039 |
| $HClO_4$ | 8.200 | 6.0 | .044 | $Co(NO_3)_2$ | 2.080 | 15.0 | .023 |
| $HClO_4$ | 9.300 | 16.0 | .189 | $CrCl_3$ | 1.720 | 7.2 | .022 |
| $HNO_3$ | 3.660 | 3.0 | .001 | $Cr(NO_3)_3$ | 1.510 | 8.4 | .028 |
| KBr | 1.150 | 5.5 | .003 | $Cr_2(SO_4)_3$ | .430 | 18.0 | .033 |
| KCl | .920 | 4.5 | .006 | $CuCl_2$ | 1.400 | 6.0 | .014 |
| $KClO_3$ | −1.700 | .7 | .012 | $Cu(NO_3)_2$ | 1.830 | 18.0 | .022 |
| KF | 2.130 | 4.0 | .025 | $CuSO_4$ | .000 | 5.6 | .008 |
| KI | 1.620 | 4.5 | .011 | $FeCl_2$ | 2.160 | 6.0 | .008 |
| $KNO_3$ | −2.330 | 3.5 | .008 | $K_2CrO_4$ | .163 | 10.5 | .018 |
| KOH | 4.770 | 6.0 | .029 | $K_2SO_4$ | −.250 | 2.1 | .001 |
| LiBr | 7.270 | 6.0 | .018 | $LaCl_3$ | 1.410 | 12.0 | .022 |
| LiBr | 7.800 | 20.0 | .144 | $Li_2SO_4$ | .570 | 9.0 | .014 |
| LiCl | 5.620 | 6.0 | .025 | $MgBr_2$ | 3.500 | 15.0 | .096 |
| LiCl | 5.650 | 20.0 | .148 | $MgCl_2$ | 2.900 | 15.0 | .103 |
| LiOH | −.080 | 4.0 | .018 | $MgI_2$ | 4.040 | 6.0 | .045 |
| $LiNO_3$ | 3.800 | 6.0 | .040 | $Mg(NO_3)_2$ | 2.320 | 15.0 | .026 |
| $LiNO_3$ | 3.400 | 12.0 | .111 | $MgSO_4$ | .150 | 12.0 | .040 |
| NaBr | 2.980 | 4.0 | .017 | $MnCl_2$ | 1.600 | 18.0 | .075 |
| NaCl | 2.230 | 6.0 | .041 | $MnSO_4$ | .140 | 16.0 | .052 |
| $NaClO_3$ | .410 | 3.5 | .007 | $Na_2CrO_4$ | .410 | 12.0 | .043 |
| $NaClO_4$ | 1.300 | 6.0 | .005 | $Na_2S_2O_3$ | .180 | 10.5 | .023 |
| $NaH_2PO_4$ | −1.590 | 6.0 | .062 | $Na_2SO_4$ | −.190 | 12.0 | .029 |
| NaI | 4.060 | 3.5 | .007 | $(NH_4)_2SO_4$ | −.250 | 12.0 | .037 |
| $NaNO_3$ | −.390 | 6.0 | .014 | $NiCl_2$ | 2.330 | 15.0 | .007 |
| NaOH | 3.000 | 6.0 | .066 | $NiSO_4$ | .025 | 10.0 | .029 |
| NaOH | 3.950 | 29.0 | .226 | $Pb(ClO_4)_2$ | 2.250 | 18.0 | .024 |
| $NH_4Cl$ | .820 | 6.0 | .013 | $Pb(NO_3)_2$ | −.970 | 6.0 | .010 |
| *Higher electrolytes* | | | | $UO_2Cl_2$ | 2.400 | 9.0 | .056 |
| $AlCl_3$ | 1.920 | 10.8 | .028 | $UO_2(ClO_4)_2$ | 5.640 | 16.5 | .064 |
| $Al_2(SO_4)_3$ | .360 | 15.0 | .016 | $UO_2(NO_3)_2$ | 2.900 | .0 | .034 |
| $BaBr_2$ | 1.920 | 6.0 | .021 | $UO_2SO_4$ | .066 | 8.0 | .257 |
| $BaCl_2$ | 1.480 | 5.4 | .027 | $ZnCl_2$ | .800 | 18.0 | .096 |
| $CaCl_2$ | 2.400 | 18.0 | .013 | $Zn(ClO_4)_2$ | 4.300 | 12.0 | .063 |
| $Ca(NO_3)_2$ | .930 | 18.0 | .037 | $Zn(NO_3)_2$ | 2.280 | 18.0 | .027 |
| $Ca(NO_3)_2$ | .900 | 60.0 | .049 | $ZnSO_4$ | .050 | 8.0 | .257 |

*Value of exponent $q$ in (IV.6).
†Maximum value of ionic strength at which (IV.6) is compared against experimental data.
‡Estimated standard deviation in $[\Gamma(\text{in Eq. IV.6})/\Gamma(\text{experimental})] - 1$.

378    Thermodynamics

As an example, consider $m_{12} = 5.30$ for $NaNO_3$, and $m_{32} = 5.27$ for $Ca(NO_3)_2$ so that $I = 21.11$; from (IV.11), $V_{12} = 2.0$ and $V_{32} = 2.25$; from Table IV.1 and Fig. IV.1, $\Gamma_{12}^\circ = 0.28$ and $\Gamma_{32}^\circ = 0.87$; substitution in (IV.10) gives $\Gamma_{12} = 0.49$ in approximate agreement with 0.38, based on experiments.

## Activity of Water in Multicomponent Solutions

The activity of water $(a_w)_{mix}$ for the multicomponent mixtures can be calculated from

$$\log(a_w)_{mix} = W_{12} \log(\mathring{a}_w)_{12} + W_{14} \log(\mathring{a}_w)_{14} + \cdots$$
$$+ W_{32} \log(\mathring{a}_w)_{32} + W_{34} \log(\mathring{a}_w)_{34} + \cdots$$
$$+ 0.0156 I \left( \frac{W_{12}}{z_1 z_2} + \frac{W_{23}}{z_2 z_3} + \cdots \right)$$
$$- 0.0156 \left( \frac{I_1}{z_1^2} + \frac{I_2}{z_2^2} + \frac{I_3}{z_3^2} + \cdots \right). \qquad (IV.12)$$

The activity of water in the binary solution $(\mathring{a}_w)_{ij}$ in this equation must be calculated at the total ionic strength of the multicomponent solution. It is important to remember that $(\mathring{a}_w)_{ij}$ for multivalent electrolytes must be computed from $(\mathring{a}_w)'$ for 1:1 electrolytes by (IV.9). The terms $W_{ij}$, $X_i$, and $Y_j$ are defined by

$$W_{ij} = \frac{X_i Y_j (z_i + z_j)^2}{z_i z_j} \left( \frac{I_c I_a}{I^2} \right); \qquad \left( X_i = \frac{I_i}{I_c} = \text{cationic fraction} \right)$$

$$Y_j = \frac{I_j}{I_a} = \text{anionic fraction};$$

$$(I_c = I_1 + I_3 \cdots +; \qquad I_a = I_2 + I_4 + \cdots). \qquad (IV.13)$$

For a solution in which all $z_i$ are equal, and all $z_j$ are equal, but $z_i$ is not necessarily equal to $z_j$, (IV.12) simplifies into the following form:

$$\log(a_w)_{mix} = X_1 Y_2 \log(\mathring{a}_w)_{12} + X_1 Y_4 \log(\mathring{a}_w)_{14} + \cdots$$
$$+ X_3 Y_2 \log(\mathring{a}_w)_{32} + X_3 Y_4 \log(\mathring{a}_w)_{34} + \cdots \qquad (IV.14)$$

## Estimations of $\Gamma_{ij}^\circ$ and $q$ for Binary Solutions

The foregoing methods of calculation are based on the existence of at least one experimental result for $\Gamma_{ij}^\circ$ of an electrolyte in a binary solution, so that $q$ can be calculated. If such a result is not available for an electrolyte, it can be estimated from the following empirical equation:

$$\Gamma_{ij}^\circ = A_j \Gamma_{iCl}^\circ + B_j, \qquad (IV.15)$$

where $i$ is the common cation and $j$ is the anion for which $\Gamma_{ij}^\circ$ is unknown, and $A_j$ and $B_j$ are constants for $j$, listed in Table IV.2. An example would illustrate

**Table IV.2** Values of $\Gamma^\circ_{iCl}$, $A_j$ and $B_j$ at $I = 2$ and 25°C

| iCl | $\Gamma^\circ_{iCl}$ | iCl | $\Gamma^\circ_{iCl}$ | j | $A_j$ | $B_j$ |
|---|---|---|---|---|---|---|
| AgCl | 0.550 | LiCl | 0.921 | $F^-$ | −1.482 | 1.521 |
| AlCl$_3$ | .673 | MgCl$_2$ | .707 | $Cl^-$ | 1.0 | 0 |
| BaCl$_2$ | .630 | MnCl$_2$ | .670 | $Br^-$ | 1.259 | −.121 |
| BeCl$_2$ | .710 | NaCl | .668 | $I^-$ | 1.648 | −.313 |
| CaCl$_2$ | .676 | NaClO$_3$ | .639 | $NO_3^-$ | 2.248 | −.918 |
| CeCl$_3$ | .639 | NH$_4$Cl | .570 | $SO_4^{--}$ | .100 | .443 |
| CoCl$_2$ | .690 | PbCl$_2$ | .599 | $ClO_4^-$ | 1.223 | −.085 |
| CsCl | .496 | RbCl | .546 | $OH^-$ | −1.043 | 1.437 |
| CuCl$_2$ | .641 | SrCl$_2$ | .660 | | | |
| FeCl$_2$ | .679 | ThCl$_4$ | .717 | | | |
| HCl | 1.009 | YCl$_3$ | .646 | | | |
| KCl | .573 | ZnCl$_2$ | .612 | | | |

the method used to obtained $A_j$ and $B_j$. Let $\Gamma^\circ_{iCl}$ be for alkaline chlorides, and the values of $\Gamma^\circ_{iCl}$ are known, and $\Gamma^\circ_{ij}$ be for alkaline bromides ($j = Br$) for which some but not all values of $\Gamma^\circ_{ij}$ are known. A linear plot of $\Gamma^\circ_{ij}$ versus $\Gamma^\circ_{iCl}$ yields the values of $A_j$ and $B_j$. Since the values of $\Gamma^\circ_{iCl}$ for most chlorides are known, substitution of $\Gamma^\circ_{iCl}$ in this equation yields the value of $\Gamma^\circ_{ij}$. The values of $\Gamma^\circ$ for both types of ions refer to the same value of the ionic strength, $I$, and for convenience, $I = 2$ has been chosen. The results adopted are listed in Table IV.2. The application of (IV.15) is illustrated by the following example. Let $i$ and $j$ be Li$^+$ and Br$^-$, respectively, so that $\Gamma^\circ_{ij}$ be for LiBr whose reduced activity coefficient is to be estimated. Table IV.2 gives $A_j = 1.259$, $B_j = -0.121$, and $\Gamma^\circ_{LiCl} = 0.921$; therefore, for LiBr $\Gamma^\circ_{ij} = \Gamma^\circ_{LiBr}$ is given by $\Gamma^\circ_{ij} = 1.259 \times 0.921 - 0.121 = 1.039$, whereas the experimental value is 1.015. After $\Gamma^\circ_{ij}$ is estimated, the unknown value of $q$ can be computed from (IV.6) with $I = 2$, by successive approximation.

APPENDIX V

# STABILITY DIAGRAMS

A number of diagrams representing the regions of stability of phases have been devised and are used for convenience. The most important such diagrams are obviously the phase diagrams discussed in Chapter XV. Other stability diagrams of interest are (1) the Pourbaix diagrams,* and (2) thermodynamic stability diagrams. Pourbaix diagrams are particularly important since they are useful in electrolysis and corrosion science and technology. The usefulness of the second type of diagrams is limited to very few species and phases, and further, descriptions of multispecies thermodynamic equilibria can be obtained much more conveniently by using the computer programs cited in Chapter XII.

### Pourbaix Diagrams

Pourbaix diagrams are the maps of the regions of stability of substances in the presence of water having various degrees of acidity. Each diagram is represented on emf [i.e., $E$] versus pH coordinates as shown in Fig. V.1 for Mg. For simplicity, the ionic activities are set equal to the ionic molarities, and denoted as $[A^{z+}]$ and $[B^{z-}]$, and therefore pH $= -\log[\text{molarity of H}^+] = -\log[H^+]$, the molarities being very close to the molalities. The standard pressure is 101325 Pa $= 1$ atm in nearly all the existing Pourbaix diagrams. The most important equation in constructing these diagrams is (14.16) at 298.15 K with ln converted into log; i.e.,

$$E = E° - (0.0592/z)\log J; \qquad \text{(all } E \text{ in volts)} \tag{V.1}$$

The broken straight line a in Fig. V.1 is based on the electrode reaction $H_2(g) = 2H^+ + 2e^-$ in Table 14.1. For this reaction, $E°$ is zero, $z = 2$, and $\log J = \log\{[H^+]^2/P(H_2)\} = -2\text{pH} - \log P(H_2)$; hence,

$$E = 0.0592 \times \text{pH} + 0.0296 \log P(H_2) \tag{V.2}$$

Along ⓐ, $P(H_2) = 1$ atm, and below it, $H_2$ is generated at a pressure in excess of 1 atm, above it, water is stable, i.e., $H_2$ is consumed to form water. Line ⓑ, is based on the electrode reaction $2H_2O = O_2(g) + 4H^+ + 4e^-$ in Table 14.1, for which

$$E = -1.229 + 0.0592 \times \text{pH} + 0.25 \times 0.0592 \times \log P(O_2) \tag{V.3}$$

---

*M. Pourbaix, "Atlas of Electrochemical Equilibria in Aqueous Solutions," National Assoc. Corrosion Engineers (1974).

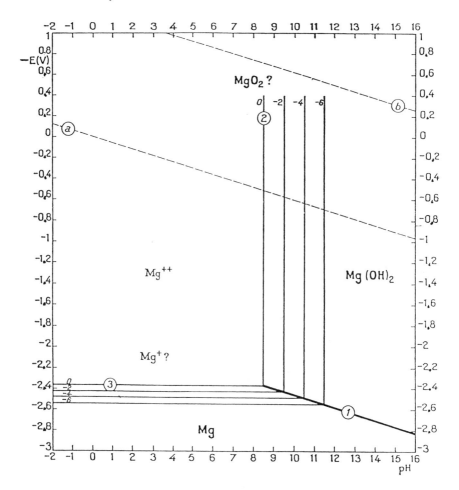

**Fig. V.1** Pourbaix diagram for magnesium. The sign of $E(V)$ in the original publication has been changed here to $-E(V)$ to agree with the IUPAC convention used in this book; see the footnote to Table 1.1. Symbols with question marks were not considered in constructing this diagram due to the lack of data. Reproduced from Pourbaix, loc. cit.

Along ⓑ, $P(O_2) = 1$ atm, and above it, $O_2$ is generated at a pressure in excess of 1 atm $O_2$ pressure, and below it, water is stable, i.e., $O_2$ is consumed to form water. Therefore, between ⓐ and ⓑ, water is stable and outside of them, water is unstable.

Line ① represents the limits of domains of relative stability of Mg and $Mg(OH)_2$, based on

$$Mg + 2H_2O = Mg(OH)_2 + 2H^+ + 2e^-;$$
$$(0) \quad (-474{,}282) \quad (-833{,}644) \quad (0)$$
$$\Delta G° = -359{,}362 = -2FE°; \qquad E° = 1.862 \text{ volts} \tag{V.4}$$

where $\Delta_f G°$ in Joules under each species is from Gen. Refs. (1 and 3), and that under $H^+$ is zero by convention. The expression for log $J$ is simply $-2pH$; hence, (V.1) becomes

$$E = 1.862 + 0.0592 \times pH; \quad \text{(all } E \text{ in volts)} \tag{V.5}$$

Line ② is for the solubility of $Mg(OH)_2$ in water, obtained with the sum of the following reactions:

$$2e^- + Mg^{++} = Mg(s);$$
$$(E° = -2.372 \text{ from Table 14.1}, -2.363 \text{ in Pourbaix}) \tag{V.6}$$

$$Mg(s) + 2H_2O = Mg(OH)_2 + 2H^+ + 2e^-; \quad E° = 1.862$$
$$\text{(see above)} \tag{V.4}$$

$$\overline{Mg^{++} + 2H_2O = Mg(OH)_2(s) + 2H^+;} \quad E° = -0.510;$$
$$(-0.501 \text{ in Pourbaix}) \tag{V.7}$$

Since the solubility refers to equilibrium, $E = 0$, and (V.1) for the preceding reaction is $0 = -0.510 - 0.0296 \log\{[H^+]^2/[Mg^{++}]\}$, which can be rearranged after the substitution of pH to obtain the following equation for Line ②:

$$\log[Mg^{++}] = 17.23 - 2pH \tag{V.8}$$

(This equation in Pourbaix is $\log[Mg^{++}] = 16.95 - 2pH$, but other equations are nearly the same and they have been recalculated here to be consistent with the latest data). The line representing this equation is vertical because it is independent of $E$. The numbers $0, -2, -4$, and $-6$ beside the vertical lines refer to the values of pH with which $\log[Mg^{++}]$ have been calculated.

Line ③ represents the boundary of $Mg(s)$ and $Mg^{++}$, based on $Mg(s) = Mg^{++} + 2e^-$ in Table 14.1, for which the corresponding equation is

$$E = 2.372 - 0.0296 \log[Mg^{++}] \tag{V.9}$$

This equation is independent of pH and therefore it is parallel to the horizontal axis. The numbers above each line is the value of $\log[Mg^{++}]$ with which $E$ has been calculated.

Magnesium is stable in the lower region of the diagram where water is not stable, i.e., there is no common region of stability for $H_2O$ and Mg. Therefore, it is impossible to prepare metallic Mg by electrolysis of aqueous solutions containing $Mg^{++}$ ions, because such an attempt would lead to the evolution of $H_2$ at the cathode without the formation of Mg.

The pourbaix diagrams are more complex for other materials but the basic principle of their calculation and interpretation is the same. In some publications, Pourbaix diagrams are called $E_H$-pH diagrams, signifying that $E$ refers to the hydrogen electrode. Computerized Pourbaix diagrams are available in F*A*C*T*, cited in Chapter XII.

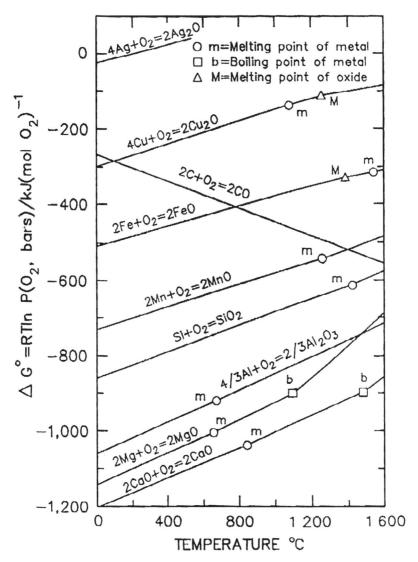

**Fig. V.2** Thermodynamic stability diagram for oxides.

### Thermodynamic Stability Diagrams

A form of thermodynamic stability diagram, based on the data in Gen. Refs. (1–3), is shown in Fig. V.2. Each reaction is written for 1 mol of reacting $O_2$ to show the relative stabilities of oxides. Similar diagrams can be constructed for other gases, e.g., $S_2$, $N_2$, and halogens. The diagram depicts that the stability increases from $Ag_2O$ to $CaO$, provided that the reactions proceed as written. However for

Appendix V     385

**Table I**  The calculated equilibrium constants ($K$) of reactions determining the predominant zones for Cu-S-O system including $Cu_2SO_4$. ($K$ here $= K_p$).

| Reaction | log K | | |
|---|---|---|---|
| | 300° | 350° | 400° |
| 1. $CuS + 2O_2 = CuSO_4$ | 46.57 | 41.39 | 36.98 |
| 2. $2CuS + O_2 = Cu_2S + SO_2$ | 27.76 | 25.77 | 24.07 |
| 3. $Cu_2S + SO_2 + 3O_2 = 2CuSO_4$ | 65.38 | 57.01 | 49.89 |
| 4. $Cu_2S + 3/2O_2 = Cu_2O + SO_2$ | 30.14 | 27.28 | 24.83 |
| 5. $Cu_2S + O_2 = 2Cu + SO_2$ | 18.77 | 17.12 | 15.70 |
| 6. $2Cu_2O = 4Cu + O_2$ | −22.74 | −20.32 | −18.27 |
| 7. $2CuSO_4 = Cu_2O + 2SO_2 + 3/2O_2$ | −35.23 | −29.73 | −25.05 |
| 8. $CuO \cdot CuSO_4 = Cu_2O + SO_2 + O_2$ | −21.66 | −18.39 | −15.62 |
| 9. $2CuSO_4 = CuO \cdot CuSO_4 + SO_2 + 1/2O_2$ | −13.58 | −11.34 | −9.44 |
| 10. $4CuO = 2Cu_2O + O_2$ | −14.66 | −12.62 | −10.88 |
| 11. $CuO \cdot CuSO_4 = 2CuO + SO_2 + 1/2O_2$ | −14.33 | −12.08 | −10.18 |
| 12. $Cu_2S + 2O_2 = Cu_2SO_4$ | 42.47 | 37.54 | 33.34 |
| 13. $Cu_2O + SO_2 + 1/2O_2 = Cu_2SO_4$ | 12.33 | 10.26 | 8.50 |
| 14. $Cu_2SO_4 + SO_2 + O_2 = 2CuSO_4$ | 22.91 | 19.47 | 16.55 |

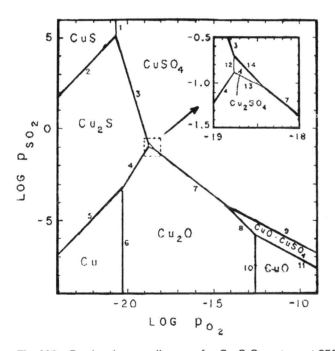

**Fig. V.3**  Predominance diagram for Cu-S-O system at 350°C. Numbers indicate the reactions listed in Table I. Reproduced from Nagamori and Habashi.*

*M. Nagamori and F. Habashi, Met. Trans., 5, 523 (1974); for a similar diagram see also N. Jacinto, S. N. Sinha, M. Nagamori, and H. Y. Sohn, Met. Trans., 14B, 136 (1983); and A. K. Biswas and W. G. Davenport, "Extractive Metallurgy of Copper," Pergamon Press (1994).

# APPENDIX VI

# LIST OF SYMBOLS

Purely mathematical symbols, mostly used in Chapters I and II are not included in the list. Roman numerals designate the chapters, arabic numbers in parentheses, the equations where the symbols are used or defined. Based mostly on IUPAC 1993 recommendations.*

## Capital Letters

| | |
|---|---|
| $A$ | Helmholtz energy, $A = U - TS$ |
| $A$ | Area, (3.5) |
| $A$ and $A^{z-}$ | A nonmetal and its cation in (14.7) |
| $A, B, C$ | Constants in Antoine equation (8.19) |
| $A_i, B_i$ | Parameters used for $\ln \gamma_i$ in (10.33) and (10.34) |
| $A_{ij}, B_{ij}, C_{ij}$ | Parameters of Margules Equations (11.21) |
| $A_1, A_2, \ldots$ | Reactants; $A_\mathrm{I}, A_\mathrm{II} \ldots$ reaction products in chemical reactions |
| $\dot{A}, \dot{B}$ | Constants in (11.55) |
| $B$ | Second virial coefficient of a pure gas; gas mixture after (9.10) |
| $B_o$ | Second virial coefficient term in (9.9) |
| $B_1, B_2, B_3, \ldots$ | First, second, third, ... virial coefficients, (3.49) and (9.5); $B_1 = RT$ |
| $C$ | Coulombs (electricity) |
| $C_{el}$ | Electronic heat capacity, (3.63) |
| $C_p$ | Heat capacity at constant pressure, (3.18) |
| $C_V$ | Heat capacity at constant volume, (3.20) |
| °C | Degrees Celsius |
| $D$ | Dielectric constant, dimensionless, (13.17) |
| $E$ | Electromotive force, emf, always in volts |
| $E_\mathrm{tr}, E_\mathrm{rot}, E_\mathrm{vib}$ | Translational, rotational, and vibrational energy of molecules |
| $F$ | Force in III only; (3.5) |
| $\mathcal{G}$ | Extensive Gibbs Energy, for $n$ moles |
| $G$ | Molar Gibbs Energy, $G = H - TS = \mathcal{G}/n$ |
| $G_i^\circ(T)$ | Standard Gibbs energy of pure idealized gas $i$; $G_i^\circ$ for a pure ideal gas $i$ at 1 bar (1 atm in older publications) |

---

*I. Mills et al., see footnote to Table I.1; see also the symbols therein.

388  *Thermodynamics*

| | |
|---|---|
| $G_i^\circ(P, T)$ | Standard Gibbs energy of pure condensed phase; (9.33) |
| $\mathcal{G}^I, \mathcal{G}^{II}, \ldots \mathcal{G}^\Phi$ | Extensive Gibbs energy of I, II, … Φ phases, (8.33) – (8.40) |
| $Gef$ | Gibbs energy function; (12.32) |
| $\overline{G}_i$ | Partial Gibbs Energy of $i$ |
| $\overline{G}_i^{el}$ | $\overline{G}_i$ due to electrostatic interaction of ions; (13.16), (13.17) |
| $H$ | Enthalpy, (3.57) |
| $\mathcal{H}$ | Magnetic field in (7.9) and in XVI |
| $I$ | Integration constant in (3.73); ionic strength in XIII |
| $J$ | Defined by (12.6); $J = K_p$ only at equilibrium; see $K_p$ |
| $K_p$ | Equilibrium constant; (12.8) and (12.10) |
| $K_p'$ | Defined by (12.15) |
| $K_i$ | Ionization constant based on molality; (13.46) |
| $K_{i,w}$ | Ionization constant of water (14.26) and (14.26a) |
| $L, L^{z+}$ | Metal and its ion |
| $L_1, L_2$ | Liquid 1 and 2 in XV |
| $M$ | Molar mass |
| $\mathcal{M}$ | Magnetic moment per mole in (7.9) and in XVI |
| $N_A$ | Avogadro number, $6.022137 \times 10^{23}$ mol$^{-1}$ |
| $P$ | Pressure always in bars unless specified otherwise |
| $P^\circ$ | Standard pressure, 1 bar = $10^5$ Pa. |
| $P_\circ$ | A very low pressure |
| $Q, Q_1, Q_2$ | Heat exchange between system and surroundings |
| $Q_i^A, Q_i^B$ | Heat exchange with Carnot engines A and B ($i = 1$ or 2) |
| $Q_P, Q_T, Q_V$ | Heat exchange at constant $P$, $T$ and $V$ |
| $S$ | Entropy; (4.24) |
| $S_0$ | Entropy at 0 K |
| $\Delta S_\theta^\circ$ | Constant in (5.4), (8.16) and (8.17) |
| $T$ | Thermodynamic temperature; always in K |
| $T_o, T', T'', T_{\alpha\beta}$ | Phase transition temperatures in III only |
| $T_\lambda$ | Lambda point in Fig. 7.2 |
| $T_\varepsilon$ | Eutectic temperature; XV |
| $U$ | Energy of a system, (3.1)–(3.4) |
| $V$ | Volume |
| $V_\circ$ | Volume constant in (3.53) |
| $\overline{V}_i$ | Partial volume of $i$; IX |
| $W$ | Work of expansion or compression; $dW = -P\,dV$ |
| $\mathbf{W}$ | Weight loss, g s$^{-1}$ from Knudsen cell; (12.46) |
| $W'$ | Work other than work of expansion or compression |
| $W''$ | Electric work; gravitational work in XV |
| $W_2$ | Weight of solute in (10.26) and related equations only |
| $W_{12}$ | Molecular exchange energy; (11.51) |
| $X_i$ | Mole fraction of $i$ only in II, later always $x_i$ |

Appendix VI 389

| | |
|---|---|
| $Z$ | Compressibility factor, $PV/nRT$, dimensionless; (3.36) |
| $\overline{Z}$ | Average value, only in II; (2.70) |
| $Z'$ | Most probable value only in II; (2.74a) |

### Lowercase Letters

| | |
|---|---|
| **a** | Orifice area in cm$^2$; (12.46) |
| $a, b$ | van der Waals constants |
| $a_i$ | Activity defined by $f_i/f_i^\circ$, standard state is pure $i$; (9.20) |
| $a_i^\infty$ | Activity, reference state is $x_i \to 0$ |
| $a_+, a_-$ | Activity of a cation (+) and activity of an anion (−); XIII |
| $a_\pm$ | Mean ionic activity if an electrolyte defined by (13.4) |
| $\mathring{a}$ | Distance of closest approach between two ions; (13.17) |
| $\overline{a}$ | Average deviation; (2.71) |
| (aq) | Aqueous solution |
| $b_2, b_3 \ldots$ | Coefficients in (3.49), $b_3 = (B_3 - B_2^2)/R^2T^2$ |
| (c) | Crystal; (s) is also used |
| **d** | Distance between two ions or molecules |
| $d$ | Total differential, exact differential |
| $đ$ | Inexact differential |
| $e$ | Base of napierian logarithm, (natural logarithm) |
| $e_{ij}$ | Energy of a bond between $i$ and $j$ particles (molecules) |
| $e_\circ$ | Permittivity of perfect vacuum, $8.85419 \times 10^{-12}$ C m$^{-1}$ volt |
| $f$ | Fugacity as defined by (9.2), always in bars |
| $f_i^*$ | Fugacity of pure $i$; often also denoted by $f^\circ$ |
| $f_i(P)$ | Fugacity of $i$ at $P$ of mixture, not at $P_i$; (9.10) |
| fin | Subscript for final state |
| g | Abbreviation for grams |
| $g_i$ | Partial Excess Gibbs Energy of $i$ |
| **g** | Acceleration; XVI |
| $g$ | Rational osmotic coefficient; (13.40) |
| (g) | Gas phase |
| $h_i$ | Partial Excess Enthalpy of $i$ |
| $h$ | Height from a reference point; in XVI only. |
| $i$ | Subscript for component $i$, or species $i$ |
| in | Subscript for initial state or initial property |
| irr | Subscript for irreversible |
| $j$ | Subscript for component $j$, or species $j$ |
| k | Abbreviation for kilo |
| $k$ | Boltzmann constant |
| $k_{H,i}$ | Henry's law coefficient for component $i$; (9.30) |
| (l) | Liquid phase |

| | |
|---|---|
| ln | Napierian (natural) logarithm, base $e = 2.71828$ |
| log | Common logarithm, base 10 |
| m | Abbreviation for molal after a number or quantity |
| $m_i$ | Molal concentration of $i$ |
| $m_+, m_-$ | Molal concentration of a cation $(+)$, and anion $(-)$ |
| $m_\pm$ | Mean ionic molality; (13.12) |
| $n$ | Total moles of all constituents of a system |
| $n_i$ | Moles of $i$ in a system |
| $q$ | Number of Coulombs in XV |
| $r$ | Radius; in XVI only |
| rev | Subscript for reversible |
| $s_i$ | Partial Excess Entropy of $i$ |
| (s) | Solid phase |
| (sln) | Solution |
| surr | Subscript for surroundings |
| sys | Subscript for system |
| $t$ | Temperature in °C |
| v | Subscript for a constant volume condition |
| $x$ | A mathematical variable in II only |
| $x_i$ | Mole fraction of component $i$; $X_i$ in II only. |
| $x_i(s), x_i(l)$ | Mole fraction of component $i$ in solid and liquid solutions in XV |
| $y$ | Mathematical variable |
| $y_i$ | Mole fraction of component $i$ in a gas or solid phase, not in liquid |
| $z$ | Mole of electrons; (14.2) |
| $z_+, z_-$ | Charge number of cations $(+)$, and anions $(-)$; $z_- < 0 < z_+$ |
| $z_i$ | Effective volume fraction of component $i$; (11.37); in XI only |
| $\bar{z}$ | Coordination number, i.e., number of nearest neighbors of a molecule in a condensed phase; (11.51) |

**Miscellaneous**

| | |
|---|---|
| 1, 2 | Subscripts for initial property or state, 1, and final property or state, 2; subscripts for components 1 and 2 |
| I, II, ... | Various crystal forms |
| I, II, ... | Superscripts for phases, or separated systems |
| I, II, ... | Subscripts for products of a chemical reaction |
| $\infty$ | Superscript for properties based on Henry's law |
| $\int$ | Integral |
| $\oint$ | Contour integral (circular integral) |

Appendix VI 391

## Greek Letters

| | |
|---|---|
| $\Gamma$ | Product of ratios of fugacity coefficients and activity coefficients;(12.16) |
| $\Delta$ | Change; final property minus initial property |
| $\Lambda, \Lambda_\infty$ | Conductance; $\Lambda_\infty$ at infinite dilution |
| $\Pi$ | Pressure over solution minus 1, in bars; (13.38) |
| $\sum$ | Summation |
| $\Sigma$ | Sigma function defined by (8.18a) |
| $\Upsilon$ | Degrees of freedom; variance (8.43) |
| $\alpha, \beta, \lambda, \epsilon$ | Coefficients in power series for $C_p^\circ$; (3.66), (8.15a) |
| $\alpha$ | Defined by $\alpha = A_{12}/RT$ in (11.43a) |
| $\alpha, \beta, \lambda, \eta$ | Stoichiometric coefficients in (8.52) |
| $\alpha, \beta$, etc. | Designations for various phases in XV |
| $\dot{\alpha}$ | Cubical coefficient of thermal expansion in $K^{-1}$; (5.31) |
| $\dot{\beta}$ | Compressibility coefficient in bar$^{-1}$; (5.32) |
| $\gamma_i$ | Activity coefficient of $i$ in solution, dimensionless; standard state pure $i$ at 1 bar; (9.23) |
| $\gamma_i^\infty$ | Activity coefficient of solute $i$; reference state $x_i \to 0$; (9.31) |
| $\gamma_{el}$ | Constant for electronic heat capacity in (3.63) |
| $\gamma_+, \gamma_-$ | Ionic activity coefficients |
| $\gamma_\pm$ | Mean ionic activity coefficient; (13.5) |
| $\beta$ | Parameter in (11.52) to (11.54) |
| $\delta$ | Degree of ionization in moles dissociated per mole of initially added electrolyte in XIII. |
| $\partial$ | Partial differential; (2.2) |
| $\epsilon_{ij}$ | Wagner interaction coefficient after (11.71) |
| $\eta$ | Total number of measurements; summation index in (2.70) |
| $\theta$ | Debye characteristic temperature in III only |
| $\theta_E$ | Einstein characteristic temperature; (3.65); $\theta_E = h\nu/k$ |
| $\theta$ | Temperature on an arbitrary scale in IV; (4.6) |
| $\kappa$ | Defined by (13.18) |
| $\lambda_i$ | Stoichiometric coefficients in III; (3.96) |
| $\mu_{JK}$ | Joule-Kelvin coefficient, K bar$^{-1}$; (5.41) |
| $\nu$ | Characteristic frequency, s$^{-1}$, in a crystal; $\nu_m$ cutoff frequency |
| $\nu_i$ | Stoichiometric coefficients; (3.96) |
| $\nu_1, \nu_2, \ldots, \nu_I, \nu_{II}, \ldots,$ | Stoichiometric coefficients in (12.1); $\nu_+, \nu_-$ Stoichiometric coefficients in ionization reactions; (13.6) |

| | |
|---|---|
| $\nu$ | Sum of $\nu_+$ and $\nu_-$; (13.10) |
| $\pi$ | Reduced pressure, i.e., actual pressure divided by critical pressure, (3.46); in all other cases, 3.141593 |
| $\rho_i$ | Density of $i$ in kg m$^{-3}$, g cm$^{-3}$ |
| $\rho_1$ | Error in II only; (2.69) |
| $\rho_\varepsilon$ | Probable error; (2.74) |
| $\sigma$ | Surface tension, force per length; (7.9) |
| $\sigma$ | Root mean square deviation (2.72) |
| $\sigma°$ | Standard deviation, (2.73) |
| $\tau$ | Reduced temperature, i.e., actual temperature divided by critical temperature in III only; (3.46) |
| $\tau$ | Thermodynamic temperature scale in (4.12) to (4.17a), thereafter $\tau = T$ |
| $\phi$ | Number of phases in VIII |
| $\phi$ | Osmotic coefficient defined by (13.39) |
| $\phi_i$ | Fugacity coefficient of $i$ |
| $\varphi$ | Efficiency in VI only; (4.6); $\varphi(\theta_1, \theta_2)$ function of temperature $\theta_1$ and $\theta_2$ on an arbitrary scale |
| $\varphi$ | Reduced volume, i.e. actual volume divided by critical volume; (3.46) |
| $\psi$ | Probability defined by (5.19); in V only |
| $\omega$ | Angular velocity in radians s$^{-1}$; XVI |

**Subscripts to Δ**

| | |
|---|---|
| vap | Vaporization |
| sub | Sublimation |
| fus | Fusion (melting) |
| trs | Transition (between two solids) |
| mix | Mixing (usually forming solutions) |
| ads | Adsorption |
| r | Reaction |
| f | Formation reaction |

**Superscripts**

| | |
|---|---|
| ° | Standard |
| * | Pure component; ° is also used. |
| ∞ | Infinite dilution |
| id | Ideal |
| E | Excess property |

# INDEX

Acetic acid, ionization of, 262
Acetone, 230
Acetone-benzene, 148
Acetone-carbon disulfide, 164, 168–170
Acetone-chloroform, 167, 168, 171
Acetone, isopropyl alcohol, hydrogen
    equilibrium, 236–237
Activity, definition, 143–144
    variation with composition, 170
Activity coefficient, definition, 144
    definition for electrolytes, 248–250
    determination:
        from Debye-Hückel theory, 250–252
        from depression of freezing point, 257
        from emf measurements, 269–276
        from osmotic pressure, 258–260
        from solubility, 261
        from vapor pressure, 258
    reference state, 146
    standard state, 144, 146
    variation with composition, 170–175
    variation with P and T, 172
Adiabatic, compression, 44, 77–82
    flame temperature, 66
    irreversible, 41
    process, 43–45, 77–82
    walls, 5, 6
Additive pressure law, 53
Additive volume law, 53
Adsorption, 325
    Gibbs equation for, 325
Allotropes, entropy of, 110
Aluminum, 217–225
    fluoride, 225–226
    nitride, 235–236
    oxide, 341
Aluminum-copper, 175, 180

Amagat's law, 53
Ammonia, entropy, 113
    synthesis, 209
Antoine equation, 124
Atmosphere, pressure, 3
Atomic weights, (see periodic chart)
Available heat for a process, 69
Average deviation, 28
Avogadro's number, 3
Azeotropic point, 164–169

Bar of pressure, 3
Benzene-carbon dioxide, 165
Benzene-carbon tetrachloride, 174
Benzene, vapor pressure, 136
Berthelot equation of state, 51, 91, 95
Boiling point elevation, 160
Boltzmann constant, 3
Boyle's law, 8
    temperature, 48
Bubble point, 153, 155–156
Butanone-methanol, 169

Calcium, thermodynamic properties, 59
Calorie, definition, 3
Calorimeter, 60
Carathéodory, method, 85
Carbon, thermodynamic properties, 342
Carnot, cycle, 72
    engine, 72, 77
    theorem, 74
    temperature entropy diagram, 84
Carbon dioxide-ethanol, 149
Carbon monoxide and dioxide,
    thermodynamic properties, 346–348
Carbon tetrachloride-benzene, 174, 175
Catatectic, 296

## 394  Thermodynamics

Cell, concentration, 270
  Daniell, 274
  galvanic, 267
  half cells, 272
  hydrogen-calomel, 272, 275
  Knudsen, 232
  reversible, 267, 268
  solid electrolyte, 280
Centrifugal force, 329
Change of independent variables, 27
Characteristic temperature, 55
  Debye, Einstein, 55
Charles law, 8
Chemical equilibrium, 204–244
Chemical potential, 130
Chlorine, thermodynamic properties, 427
Clapeyron equation, 120–121
Clausius-Clapeyron equation, 120
Coefficient of thermal expansion, 96
Coherent (Curie-type) transition, 331
Complex equilibria, 238
Components, definition, 2, 126
Compressibility coefficient, 96
Compressibility factor, 47
Concentration, bulk, 323
  surface, 324
Concentration cell, 270
Condition of, exactness, 86
  integrability, 86
Conditions of phase equilibrium, 131
Conductance, 262
Conservation, energy, 37
  mass (mass balance), 238
Constant pressure diagrams, 156, 159
  binary, 156
  ternary, 184–187, 302–313
Constants, table, 3
Conversion units, 3
Copper and its oxides, thermodynamic
    properties, 349–351
Corresponding states, law of, 51
Coulomb's law, 250
Criterion, Euler, 22
Criteria, for equilibrium, 136, 324
  reversibility and irreversibility, 105
Critical point, 49–50
  data, 50
  temperature, 335
Critical (mixing) point, 190–192, 287–288
Cross differentials, 25
Cross differentiation, 22, 25
Curie point, 114

Cyclic processes, reversible, 79
  irreversible, 83

Dalton's law, 42, 52
Daniell cell, 274
Debye equation for solids, 55
Debye-Hückel, equation, 252
  limiting equation, 252
  theory, 250
Degree of dissociation, 262
Degrees of freedom (variance), 130
Depression of the freezing point, 159, 257
Derivatives, 11
  second, 14
  useful, 14
Determinants, 31
Determination of $\Delta G°$ for reactions, 212
Determination of molecular weights, 160
Deviation, average, root mean square and
    standard, 28, 29
Dew point, 153, 155–156
Diamond-graphite equilibrium, 120, 121
Diathermic walls, 5
Dielectric constant, 251
  for water, 253
Differentials, 11, cross, 25
  exact, 21
  inexact, 22, 24, 85–87
Dilute solutions, 172
Discrepancy, 28
Dissolution of gases in liquids and
    solids, 241–242
Distribution of a solute between two
    solvents, 242
Droplets, vapor pressure of, 326
Dulong and Petit, law of, 54

Efficiency, 72, 78
Einstein equation, 55
Electric field, 327, 329–330
Electrochemical reactions, 268–282
Electrolytes, 247–263
  activity, definition, 248–249
  activity coefficient, 249
  Debye-Hückel equation, 252
  osmotic coefficients, 260
  solid, 280–282
  strong, 247–262, 266–277
  weak, 262–263, 277–280
Electromotive force, emf, 214, 267–282
  electrode potential, 272
  half cells, 272
  oxidation potential, 274

sign convention, 269, 272
standard emf, 272
variation with P and T, 276
Electron volt, 3
Electronic charge, 3
Electronic heat capacity, 56
Elevation of boiling point, 160
Energy, 8
bond, 65
bound, 102
conservation, 37
Gibbs, 101
gravitational, 327
Helmholtz, 101
ideal gases, 40
kinetic, 45
mechanical, 8
molecular exchange, 193
rotational, 47
translational, 47
vibrational, 47
Enthalpy, 53
definition, 54
excess, 174–175
of formation, standard, 62
estimation, 65
function, 222
mixing molar, 173–175
partial (molar), 136
excess, 174–175
phase transformation (transition), 58
reaction, variation with $T$, 63
real gases, correction for, 95
relative, 222
second law method of calculation, 229–231
third law method of calculation, 231
variation with $P$, 95
variation with $T$, 63
Entropy, 82, 109–118
adiabatic processes, 84, 87
allotropes, 110
ammonia, 113
change, 83, 89
chemical reactions, 92
excess molar, 173–175
partial (molar), 173–175
gases, 113–114
glasses, 114–115
ideal gases, 91, 113–114
irreversible processes, 83, 106
melting (fusion), 93

mixing ideal, 91
nonideal, 162–175
monatomic gases, 113–114
oxygen, 114, 117
partial (molar), 130
phase change, 92
probability, 93
randomness, 93
reversible processes, 83
Richards rule, 93
Sackur-Tetrode equation, 113
second law method of calculation, 113, 229
statistical mechanics, 113
temperature diagram, 84
sulfur, 112
supercooled liquids, 114
thermal evaluation, 116
third law convention, 113
third law method of calculation, 117
third law of thermodynamics, 111
Trouton's rule, 93
vaporization, 93
water, 114
Equation of state, 7, 8, 48–52
Berthelot, 51
reduced, 51
thermodynamic, 94–95
virial, 52, 140
van der Waals, 48
Equilibrium, among phases, 131
chemical, 203–244
complex, 238
criteria for, 136, 324
definition, 9, 131, 204
in gas mixtures, 206–207
involving condensed phases, 211
Equilibrium constant, 204
determination, 212
generalized, 240
Erroneous phase diagrams, 288–290
Error, 28
probable, 29
Ethyl alcohol-methyl cyclopentane-$n$-hexane, 196
Ethyne, thermodynamic properties, 345
Ethylpyrydine-water, 287
Euler criterion, 22
theorem on homogeneous functions, 17
Eutectic point, 159, 286
reaction, 286
Eutectoid, 286

396  *Thermodynamics*

Exact differentials, 21
Excess thermodynamic properties, 173, 182–201
Expansion, 38–39
   Joule-Kelvin, 97
   work, 38–39
Extensive and intensive thermodynamic properties, 18

Faraday constant, 3
Feasibility of reactions, 204
First law of thermodynamics, 37
Freezing point depression, 159
Fugacity, 139–143
   coefficient, 139, 142
   definition, 139, 140
   equation for, 139–143
   gas mixtures, 141
   Gibbs energy, 139
   Lewis and Randall rule, 141
   pure gases, 140
   reference state, 139
   variation with P and T, 140, 142
Functions, 11
   homogeneous, 17, 18
   inflection points, 15, 16
   maximum, minimum, 15, 16

Galvanic cells, see Cell
Gas, constant, 3
   ideal, 8, 40–41
   equation of state, 7, 8
   energy, 40
   entropy, 90–91, 113–114
   processes with, 42
   liquefaction, 49–99
   real, 47–53
GBW approximation, 334
Generalized equilibrium constants, 240
General references, 339–340
Gibbs adsorption equation, 411
Gibbs-Duhem equation (relation), 19, 165, 170
Gibbs energy, 101
   chemical potential, 130
   definition, 102–103
   diagrams, 291–302
   emf, 214, 267–282
   effect of T, 101–105, 195
   excess, molar, 173, 179
      partial (molar), 173, 179–201
      ternary systems, 184
      with Henrian reference state, 199–200
   formation, 222
   function, 222, 227
   isothermal changes, 104
   mixing, molar, 173
   mixing, diagrams, 291–302
   partial (molar), 130, 135
      and criterion for equilibrium, 136
   vaporization, 122
Gibbs energy function, 222, 227
Gibbs-Konovalow theorems, 166, 289
Gibbs-Helmholtz equation, 103, 268
Gibbs-nickel system, 302, 303
Graphite-diamond transition, 120, 121
Gravitational field, 327–328
Gravity, standard, 3
Guldberg-Waage rule, 50

Haber process, 113, 209
Heat (see also enthalpy), 39
Heat capacity, 42, 53, 54–58
   Debye equation, 55
   definition, 42
   difference $C_p - C_v$, 54, 96, 97
   Dulong and Petit, law of, 54
   Einstein equation, 55
   electronic, 56
   empirical equations, 57
   ideal gases, 42–45
   Kopp, or Kopp-Neumann rule, 56
   reaction, 64
   solids, 54
   Sommerfeld equation, 56
   variation with P and V, 97
   with T, 55, 57
Heat engines, 72
Heat theorem, Nernst's, 110
Helium, 115
Heterogeneous chemical equilibria, 211
Helmholtz energy, 101
   isothermal changes, 104
Henry's law, 145
Homogeneous functions, 17
   thermodynamic, 18
Hugoniot equation, 53
Hydrazine-UDMH system, 163
Hydrogen, 356
   chloride, 253
   compressibility, 47–48, 140–141
   electrode, 272–275
   process of dissolution, 147

reaction with isopropyl alcohol and acetone, 236
sulfide, 358
thermodynamic properties, 356

Ideal gas, 8, 40–41
Ideal solutions, 151
Ideal solubilities of gases, 161
Inexact differentials, 22
Inflection points, 15
Infinite series (see power series)
Integrability, condition of, 86
Integrals, 20, 21
    line, 23
Intensive thermodynamic properties, 18
Integrating denominator (or factor), 22, 23, 86
Interaction parameter, Wagner, 200
Iodine, distribution between $CS_2$ and $H_2O$, 242
Ionic strength, 251
Ionization, constant, 262–263, 277–279
Ions, activity, 248–250
    activity coefficient, 248–255
    estimation, 373–379
Iridium sulfide, 232
Iron, alum, 114–115
    oxides, 353–355
    thermodynamic properties, 352
Iron-nickel, activities, 150
Irreversibility, criteria for, 105
Irreversible processes, 83, 105
Isenthalpic expansion, 98
Isentropic (adiabatic) process, 84
Isobaric process, 45
Isopropyl alcohol, 230, 236
Isothermal processes, 40, 43–45
Isotopes and entropy, 112

Joule, 3, 40
Joule-Kelvin (Joule-Thomson) coefficient and experiment, 97–99, 143

Kelvin (scale of temperature), 3, 75
Kinetic energy, 45
    theory, 45
Knudsen effusion cell, 232
Kopp (or Kopp-Neumann) rule, 56

Lagrange's method of undetermined multipliers, 26, 133, 325
Lambda point, 114

Lanthanum chloride, 253, 254
Laplace equation, 327
Law of corresponding states, 51
Least squares method, 29
Lead-antimony system, 159
Le Chatelier, principle of, 211
Legendre transformation, 28, 103
Lever rule, 290
Lewis and Randall rule, 141
L'Hôpital's theorem, 16
Line integrals, 23
Liquefaction of gases, 99
Liquid-junction, and potential, 270–271
Liquids, 53
Liquidus, 157, 159
    positions relative to solidus, 158
London forces, 250
Long-range order, 330–337

Maclaurin series, 32
Magnetic field, 327, 329–330
Margules equation, 183–186
Maxima, 15, 16
Maxwell relations, 90
Mean ionic activity and activity coefficient, 249
Mechanics of molecules, 113, 215
Mercury, vapor pressure, 123
Metatectic, 286
Methane, thermodynamic properties, 344
Method of Carathéodory, 85
Method of intercepts for $\overline{G}_i$, 181–182
Method of least squares, 29
Methyl ethyl ketone-methanol system, 167, 169
Minima, 15, 16
Molar volume, perfect (ideal) gas, 3
Molality, 2
Molarity, 2
Mole fraction, 2
Molecular exchange energy, 193
Molecular weight from
    depression of freezing point, 160
    elevation of boiling point, 161
Monotectic, 286

Nernst's heat theorem, 110
Nicotine-water system, 287
Nitrogen, compressibility factor, 140
    fugacity, 140
    in iron, 241

Nitrogen (*continued*)
    thermodynamic properties, 359

Onsager-Fuoss method, 261
Open systems, 1
Ordering and clustering, 331
Order parameter, 333
Osmotic coefficients, 260
Osmotic pressure, 260
Oxygen, compressibility factor, 141
    entropy, 117
    in metals, 242
    thermodynamic properties, 361–362
Oxidation potential, 272–274

Partial (molar) properties, 130, 135–136, 180
    enthalpy, 136
    entropy, 136
    Gibbs energy, 135–136
    heat capacity, 195
Periodic chart, on inside back cover,
Peritectic, 286
Peritectoid, 286
Permittivity of vacuum, 3
Perpetual motion, 37
Pfaff (differential) equation, 85
Phase, definition, 2, 4
    diagrams, 125, 127, 157–160, 285–321
        erroneous, 288
        from thermodynamic data, 299–302
    equilibria, 119–135
        conditions of, 131
        representation, 125
    transformation, 58
pH, 280
Phase rule, 133–135
Planck constant, 3
Point function (single-valued function), 23, 37
Potassium chloride, activity and activity coefficient, 253
    hydroxide, 253
Power series, 32
    Maclaurin, 32
    ratio test for, 32
    Taylor, 32
Poynting equation, 124
Practical osmotic coefficient, 260
Pressure, 1, 3
    critical, 49–51
    osmotic, 260
    reduced, 50–51
Probability, 93

Probable error, 29
Process, 4
Pyridine-toluene, 197

Randomness, 93
Rankine cycle, 74
Raoult's law, 144
Rational Osmotic coefficient, 260
Ratio test for power series, 32
Reactions, feasibility, 204
    generalized, 240
    standard Gibbs energy change of, 203, 240
Reaction isotherm, 205
    and emf, 271
Real gases, 47
Real solutions, 162–171
    equilibrium with its vapor at constant $T$ and $P$, 162, 166
Redox reactions, 270
Reduced equation of state, 51
Reduced $P$, $T$ and $V$, 51
Reference state, (fugacity), 139
    activity, 146
Reference temperature, 62
Refrigeration engine, 79
Regular solutions, 171, 189–194
    first approximation to, 193–194
    theoretical derivation, 192–195
Representation of data, 28
    of phase equilibria, 125, 285–321
Reversibility criteria, 105
Reversible, cyclic processes, 79
    electrochemical reactions, 268–271
    galvanic cells, 267–282
    processes, 40, 83
    processes in closed systems, 40
Richards rule, 93
Root mean square deviation, 28

Sackur-Tetrode equation, 113
Second derivatives, 14
Second law of thermodynamics, 71, 87
    method of Carathéodory, 85–87
Second law method, 229
Segregation coefficient, 243
Semiconductors, 280
Sigma plot, 123, 237–238
Silver chloride, 275–276
    electrode, 275–276
Sodium chloride and sulfate, activity and activity coefficient, 253
Solid electrolytes, 280–282

Solids, enthalpy, 57
  heat capacity, 54
Solidus, 157, 159
  positions relative to liquids, 158–159
Solubility of gases, 161, 241–243
  carbon dioxide in ethanol, 149
  ideal, 161
  in various solvents, 161, 241–243
  oxygen in water, 245
  nitrogen in iron, 241–243
    in metals and alloys, 241–243
    in water, 241
  Sieverts law, 147, 241
Solute, definition, 2
Solution, dilute, 172
  effect of gravitational field, 328
  electrolyte, 247–264
  ideal, 144, 151–158
    equilibrium with its vapor, 152
    liquid-solid equilibrium, 156–159
  nonelectrolyte, 151–201
  regular, 171, 189–194
Solvent, definition, 2
Solvus, 159
Sommerfeld equation, 56
Spectroscopic data, 113, 215
Spinodal points, 192
Spontaneous processes, 79
Standard deviation, 29
Stability diagrams, 381–386
Standard emf, 272–276
  of half cells, 272
Standard enthalpy of formation, 62
Standard Gibbs energy change of reactions, 204
  determination of, 212–215
Standard state, 62
State of system, 4
States, corresponding, 51
  reference, 146, 216
Strong electrolytes, 247–262, 266–277
Sulfur, entropy of, 112
  thermodynamic properties of, 363–364
Surface, concentration, 409
  phase, 323, 325
  tension, 324
Surfactants, 326
Symbols, 1
  list of, 387–392
Syntectic, 286
System, binary, 151–176, 179–184, 285–321
  boundaries, 1, 5

  closed, 5
  heterogeneous, 119–135
  homogeneous, 2
  multicomponent (see ternary systems)
  one component, 119–127
  open, 1, 5
  ternary, 184–187, 302–313
  thermodynamic state of, 4

Tables of thermodynamic data, 341–369
Taylor series, 32
-tectic, -tectoid, 286
Temkin Rule, 264
Temperature, 8–10
  Celsius (Centigrade) scale, 7
  critical, 49–50
  Debye characteristic, 55
  definition, 7
  determination of, 3, 75–77
  Einstein characteristic, 55
  ideal gas, 8
  Kelvin scale (thermodynamic temperature scale), 3, 75–77
  reduced, 51
Terms, 1
Ternary systems, 184–187, 302–313
Thermochemical calorie, 3
Thermochemistry, 60
Thermodynamic calculations (computer), 371
  tables, 341–369
Thermodynamic equation of state, 94
Thermodynamic state of system, 4
Thermodynamic simulator for calculation, 371
Thermometer, 7
Third law method of calculation, 117, 231
Third law convention, 112
Third law of thermodynamics, 109, 111
  consequences of, 115
Tie lines, 49, 309
Total differentials, 11
Translational energy, 45–46
Triple point, 126
  of water, 7, 126–127
Trouton rule, 93
Tutton salt, 114–115

Unavailable heat, 84–85
Units and conversion factors, 3
Useful differentials, 14
Useful series, 32

van der Waals equation, 48, 142
van Laar equation, 188
van't Hoff equation, 122, 205
Vaporization equilibria, 121
Vapor-liquid equilibrium, 121–127
Vapor pressure, 121, 136
  activity, 143
  benzene, 136
  Clapeyron equation, 120–121
  Clausius-Clapeyron equation, 120
  of droplets, 326
  of mercury, 123
  Poynting equation, 124
  variation with $P$ and $T$, 124
  zinc, 137
Vapor-solid equilibrium, 125–126
Vaporization equilibria, 121, 240
Variables of state, 4, 130
Variance (degrees of freedom), 134
Velocity of light, 3
Virial equation of state, 52, 140
Volume, critical, 49–50
  fraction, 188
  molar, 152
  partial molar, 152, 179–180
  reduced, 51

Wagner interaction parameters, 200
Water, density and dielectric constant, 253
  entropy, 114
  ionization constant, 277–279
  phase diagram, 127
  triple points, 3, 7, 127
  thermodynamic properties, 357–358
  vaporization, 121
Weak electrolytes, 262–263, 277–280
Wohl equation, 188
Work, 8
  adiabatic, 43
  compression, 38–39
  constant pressure, (isobaric), 40, 45
  electric, 329–330
  expansion, 38–39
  gravitational, 327
  isothermal, 44–45
  magnetic, 116, 329–330
  mechanical, 8
  reversible, 38–39
Working substance, 72

Zeroth law of thermodynamics, 6
Zinc, vapor pressure, 137
Zone refining, 243